W9-CJX-860

OPERATIONAL AMPLIFIERS

OPERATIONAL AMPLIFIERS
Design and Applications

JERALD G. GRAEME
Editor, Part 1
Manager, Monolithic Engineering
Burr-Brown Research Corporation

GENE E. TOBEY
Editor, Part 2
Amplifier Product Marketing Engineer
Burr-Brown Research Corporation

LAWRENCE P. HUELSMAN, Ph.D.
Consulting Editor
Professor of Electrical Engineering
The University of Arizona

McGRAW-HILL BOOK COMPANY

New York St. Louis San Francisco Düsseldorf Johannesburg
Kuala Lumpur London Mexico Montreal New Delhi
Panama Rio de Janeiro Singapore Sydney Toronto

OPERATIONAL AMPLIFIERS

 Library of Congress Catalog Card Number 74-163297

ISBN 07-064917-0

34567890 MAMM 754321

This book was set in 8A by The Maple Press Company, and printed on permanent paper and bound by the Maple Press Company. The editors were Tyler G. Hicks and Karen Kesti. The designer was Naomi Auerbach. Stephen J. Boldish supervised production.

CONTENTS

Part 2. APPLICATION

CONTRIBUTORS

THOMAS M. CATE *Product Marketing Engineer*
BRIAN K. CONANT, Ph.D. *Product Manager*
JERALD G. GRAEME *Manager, Monolithic Engineering*
HOWARD HANDLER, Ph.D. *Manager, Function Module Design*
LOUIS F. LAMPE *Manager, Amplifier Design*
JIMMY R. NAYLOR *Design Engineer*
LARRY L. SCHICK *Western Area Sales Manager*
JOHN D. SKIPPER *Design Engineer*
GENE E. TOBEY *Product Marketing Engineer*

PREFACE

The operational amplifier has become a basic analog building block common to a multitude of electronic functions performed in instrumentation, computation, and control. From the availability of these economical and versatile amplifiers has come a transition in the development of such electronics which has made the operational amplifier a basic component. As is generally the case, however, this rapidly growing new area of electronics has not been accompanied by thorough treatment of the associated technology. Beginning with the *Handbook of Operational Amplifier Applications* published by Burr-Brown in 1964, several segments of the operational amplifier technology have been given abbreviated coverage. Within this book, a complete treatment of the design and application as well as the theory and testing, of operational amplifiers has been developed by the staff of Burr-Brown. From this treatment results a practical engineering reference related both to accepted electronics theory and actual electronics practice. It is hoped that this reference will aid the development of the operational amplifier technology by making available much previously unpublished information.

Due to the influence of integrated circuits, operational amplifier design and application techniques are becoming essential tools of the linear circuit designer. The integrated circuit designer will find the differential and direct-coupled stages treated in Part 1 to be fundamental building blocks. Similarly, the instrumentation designer will find the application techniques of Part 2 to be the basis of future designs.

To treat the various major aspects of operational amplifier technology, the material of this book is presented in two principal parts and two appendixes. Part 1 considers the design of operational amplifiers to provide insight into the factors which determine amplifier performance characteristics and to outline the techniques available for their control. Part 2 presents an extensive selection of practical operational amplifier applications with sufficient descriptions of operation to permit design adaptation from the specific circuits described. In Appendix A the basic theory of operational amplifiers is reviewed to provide an accompanying reference. Following, in Appendix B, concise definitions of the performance parameters used to characterize operational amplifiers are given, and associated test circuits are presented and described. For those acquainted with operational amplifiers, a study of Parts 1 and 2 using the Appendixes as references should familiarize the reader with the material covered. Those desiring to acquire a more thorough understanding of the subject would benefit from a prior study of the Appendixes. Following this familiarization, the book should serve its most valuable function as a reference to engineers on the nature of operational amplifiers and the array of electronic functions which they can perform.

Part 1 develops the elements of operational amplifier design from the characteristics of bipolar transistors and FETs to the characteristics of individual stages and then to the complete multistage operational amplifier. In Chapter 1 the signal characteristics of differential stages are resolved in terms of commonly available semiconductor device parameters. By relating the differential stage to the familiar common-emitter and common-source transistors, the analysis and understanding of differential stages is greatly simplified. Then, in Chapter 2, the DC errors and noise of differential stages are analyzed to define their respective sources. From these results the techniques of compensating input offset voltage drift are summarized in readily applied equations and graphs. Next a survey of practical input, intermediate, and output stages of operational amplifiers is made in Chapter 3 including descriptions of individual stage characteristics. Upon combining these various stages to form a complete amplifier, the overall operating characteristics are found from the interaction of individual stages as described in Chapter 4. The interaction of the signal and error parameters of various stages are discussed there including a simple technique of predicting the frequency response of cascaded stages. From the high gain of the cascaded stages the resulting operational amplifier provides control of electronic functions through negative feedback, and the amplifier response characteristics determined by this feedback are covered in Chapter 5. Feedback stability as provided by phase compensation is related to steady-state and transient amplifier response characteristics including a straightforward approach for predicting peaking.

Throughout the applications section of the book (Part 2), an attempt has been made to show the basic principles of operation so that the reader will be able to extend the results and modify the circuitry to meet his own particular needs. Chapter 6 represents the most common linear circuit applications including the various types of feedback amplifiers. Also discussed here are integrators, regulators, and reference circuits. Chapter 7 is devoted to nonlinear functions realized with operational amplifier circuitry. Included are limiters, function generators, log amplifiers, and a collection of analog multiplier techniques. Chapter 8 is a rather unique treatment of the realities and practicalities of active filter design through the use of operational amplifiers. Chapter 9 is a survey of the various switching and sampling circuits which use operational amplifiers. Included are multiplexers, A/D and D/A converters, sample-hold circuits and various types of peak detectors and comparators. An extensive collection of signal generators—sine wave, triangle wave, square wave, sawtooth, etc.—is discussed in Chapter 10. And, finally, Chapter 11 illustrates the use of operational amplifiers in circuits which perform modulation and demodulation.

The writers are grateful to H. Koerner, Dr. L. P. Huelsman, and D. R. McGraw for their assistance in maintaining consistency and accuracy in the manuscript. We also wish to thank Carole Williams, Joan Burgess, and Maryon Hartman for their exceptional accuracy in typing, and the Burr-Brown Graphics section for the preparation of highly detailed artwork.

Special thanks are due to Tom Fern for his support of the project and to Thomas R. Brown, Jr. for providing a conducive environment.

Jerald G. Graeme
Gene E. Tobey

HISTORICAL NOTE

The term "operational amplifier" was apparently coined by John R. Ragazzini, and colleagues, in a paper[1] published by the IRE in May of 1947. The paper described the basic properties of such amplifiers when used with linear and nonlinear feedback and was based on work performed in 1943 and 1944 for the National Defense Research Council. This, and most other early work with operational amplifiers, concentrated heavily on their use in analog simulations and in the solution of integro-differential equations. Credit for much of the initial development of the operational concept must go to George A. Philbrick who worked as a technical aide on the NDRC work described above, and who later was instrumental in the development of the first commercial "plug-in" operational amplifiers—using vacuum tubes.

It was not, however, until the introduction of modular solid-state operational amplifiers in 1962, by Burr-Brown Research Corporation and G. A. Philbrick Researches, Inc., that the full value of the concept began to be apparent. Since that time the operational amplifier, in modular and integrated circuit form has, to an ever-increasing degree, dominated the design of nondigital systems. Although predicting future developments is always risky, it seems safe to say that the operational amplifier will continue to be an extremely important tool in system and circuit design during the remainder of this decade.

[1] J. R. Ragazzini, R. H. Randall, and F. A. Russell, Analysis of Problems in Dynamics by Electronic Circuits, *Proc. IRE*, May, 1947.

Part 1
DESIGN

1

DIFFERENTIAL AMPLIFIER STAGE SIGNAL CHARACTERISTICS

A differential amplifier stage as represented in Fig. 1.1 provides high voltage gain to differential signals applied between its inputs while responding with much lower gain to voltages common to the two inputs. As a result, the desired differential signals are amplified with little effect from extraneous common-mode signals. Such extraneous signals frequently result from signal current flow in long lines or from noise pickup, but they are essentially rejected by a differential stage, as will be described. The differential stage also provides isolation of input and output quiescent voltage levels by means of its common-mode signal characteristics. Because of its low common-mode gain, the stage has only small variations in the quiescent or average level of the two output signals for large variations of this type at the inputs.

Developed in this chapter are the signal characteristics of bipolar transistor and field-effect transistor (FET) differential stages. The differential gains, the common-mode gains, and the associated frequency responses of these stages are derived in a simplified manner by drawing on the similarity of the stages to common-emitter and common-source amplifiers. Then differential circuit unbalances giving rise to common-

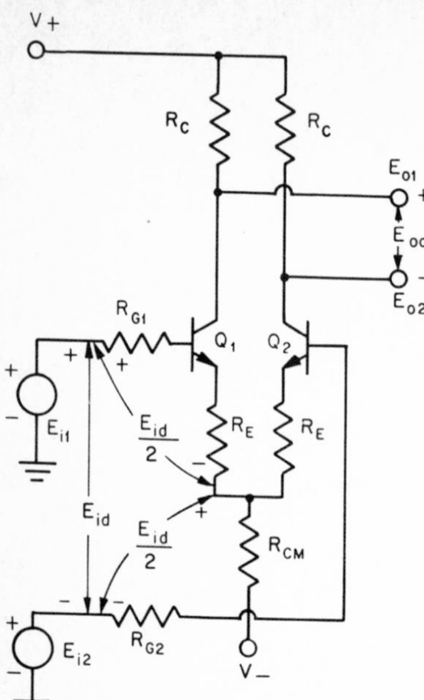

Fig. 1.1 Basic bipolar transistor differential stage.

mode signal sensitivity are analyzed, providing a common-mode rejection figure of merit. These and other considerations involved in the design of a differential amplifier stage are outlined and related to the corresponding detailed analyses in this and other chapters. Concluding the chapter is a description of several differential-stage designs which improve specific characteristics.

1.1 Low-frequency Differential Signal Characteristics

Since the differential stage is composed of two common-emitter amplifiers, Q_1 and Q_2 in Fig. 1.1, the well-known common-emitter analysis can be applied by considering the manner in which the signal is amplified. Differential signal E_{id} is impressed upon the source resistances, emitter-base junctions, and emitter resistors of the two transistors. For matched resistances and transistors under small signals, one-half of E_{id} will drop on each side of the stage as indicated. Effects of mismatched components are considered in Sec. 1.4. Equal division of the differential input signal produces equal and opposite current changes in the two transistor emitters, resulting in no change in the total stage current supplied by common-mode biasing resistor R_{CM}. Associated collector signal currents produce equal and opposite output signal voltages, E_{o1} and E_{o2}, to produce the differential output signal E_{od}.

Since the current in R_{CM} is not affected by differential signals, R_{CM} may be omitted for differential analysis, leaving the simplified common-emitter circuit of Fig. 1.2a. One-half of the input signal is applied to each common-emitter transistor input circuit, giving

$$\frac{E_{id}}{2} A_1 - \frac{-E_{id}}{2} A_2 = E_{o1} - E_{o2} = E_{od}$$

where A_1 and A_2 are the voltage gains of transistors Q_1 and Q_2 as common-emitter amplifiers.

For a balanced circuit, $A_1 = A_2$, and the differential gain becomes

$$A = A_1 = \frac{E_{od}}{E_{id}}$$

The amplification of a differential signal is, then, equal to the gain of one side of the differential circuit. Application of the signal to a common-emitter amplifier identical to one side of the differential circuit will result in a gain equal to that of the stage. As a result, differential gain is found by using common-emitter analysis[1] applied to the equivalent circuit of Fig. 1.2b.

The transistor model of Fig. 1.3 includes the primary characteristics important for low-frequency common-emitter amplifiers. Base resistance r_b' is not considered in this model as it is too small to influence

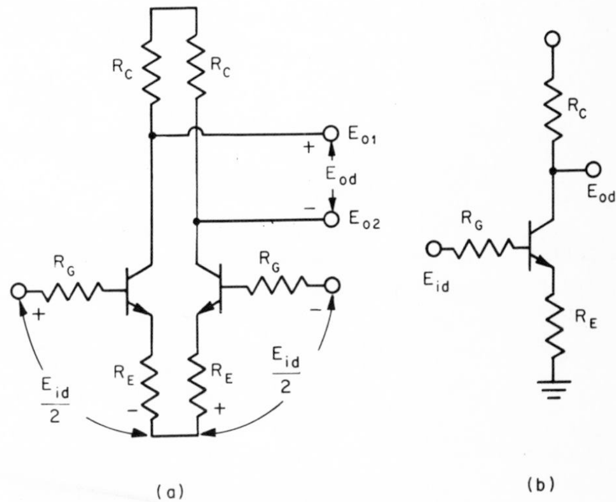

Fig. 1.2 (a) Balanced differential mode signal circuit and (b) equivalent single transistor circuit.

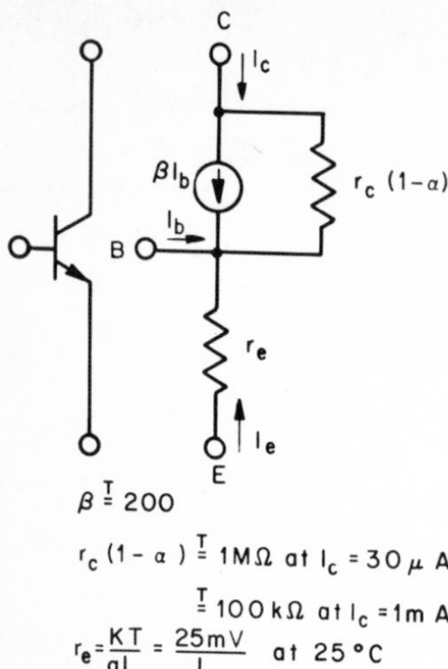

Fig. 1.3 Bipolar transistor model.

significantly the common-emitter characteristics to be described. To indicate common small-signal transistor characteristics, typical values are defined in the diagram, using the symbol $\overset{T}{=}$. Resistance of the reverse-biased collector-base junction results in the $r_c(1 - \alpha)$ resistance which conducts feedback current from collector to emitter and which loads the output current generator. Application of this model to the equivalent common-emitter circuit results in the gain analysis model of Fig. 1.4 for $R_e \ll r_c(1 - \alpha)$. From the currents identified, the low-frequency differential voltage gain of a differential stage is

$$A_O \doteq \frac{-\alpha R_C r_c}{R_e(R_C + r_c) + R_G[R_C + r_c(1 - \alpha)]} \tag{1-1}$$

where

$$R_e = R_E + r_e \qquad R_e \ll r_c(1 - \alpha)$$

Generally the collector resistor is much less than $r_c(1 - \alpha)$, and the gain expression simplifies to

$$A_O \doteq \frac{-R_C}{R_e + R_G/\beta} \overset{T}{=} 10 \text{ to } 100 \tag{1-2}$$

where $(1 - \alpha) \doteq 1/\beta$. While $r_c(1 - \alpha)$ drops from its megohm level at collector currents higher than 30 μA, the collector resistor also must decrease to maintain a bias voltage drop within the limit placed by the power supply level. As a result, the approximate gain expression is accurate in most cases. Gain will be less than predicted by the simplified expression where the approximation becomes less precise.

By the same comparison with a common-emitter transistor, differential input and output resistances of the differential stage may be found. Note that the series-connected transistors of the stage yield a differential input resistance which is twice that of the common-emitter amplifier represented by one side of the stage, as in Fig. 1.4. From the common-emitter input resistance of this circuit, differential input resistance is

$$R_I \doteq 2\beta R_e \frac{R_C + r_c}{\beta R_C + r_c} \qquad \text{for } (1 - \alpha) \doteq \frac{1}{\beta} \tag{1-3}$$

$$R_I \doteq 2\beta R_e \qquad \text{for } R_C \ll r_c(1 - \alpha) \doteq \frac{r_c}{\beta} \tag{1-4}$$

When the collector resistor approaches $r_c(1 - \alpha)$, input resistance falls, demonstrating the feedback effect of the reverse collector resistance r_c. Two output resistances of interest with a differential stage are that presented to the collector resistors and the resulting resistance appearing at the output terminals. The ability of the stage to drive currents into the collector resistors is indicated by the high resistance presented by the transistor outputs. Output resistances of the two transistors appear in series between the output terminals, resulting again in a resistance which is twice that of the common-emitter case represented in Fig. 1.5. Doubling the output resistance found for the circuit shown

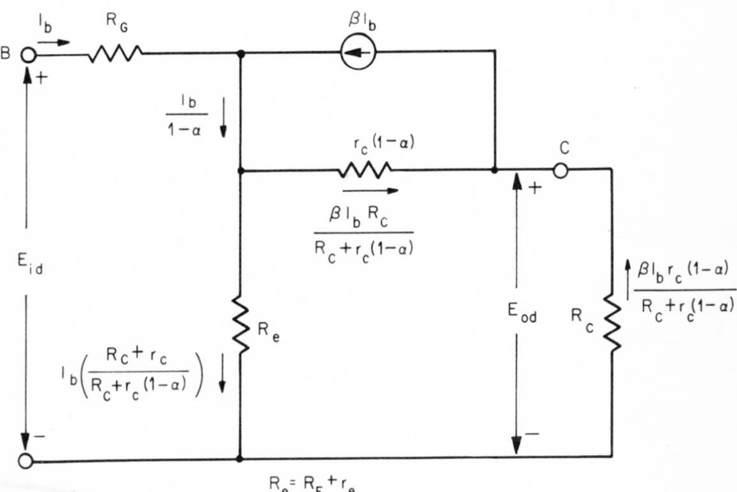

Fig. 1.4 Gain analysis circuit.

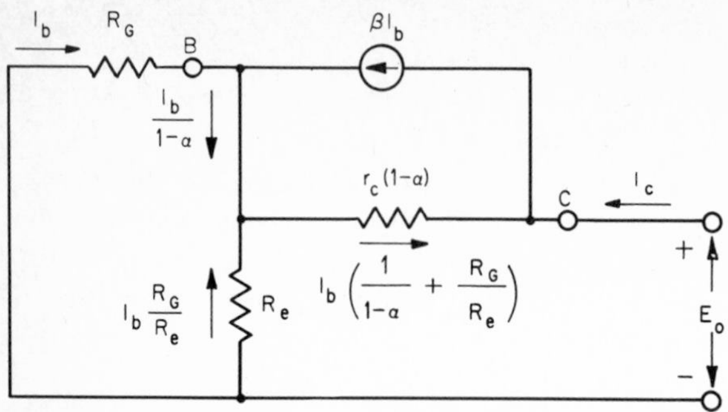

Fig. 1.5 Common-emitter output resistance analysis circuit.

gives the differential output resistance presented to the collector resistors.

$$R'_O = 2\frac{R_eR_G + R_er_c + R_Gr_c(1-\alpha)}{R_e + R_G}$$

$$R'_O \doteq 2r_c\frac{R_e + R_G/\beta}{R_e + R_G} \qquad \text{for } R_G \ll r_c \tag{1-5}$$

The ability of the stage to develop a voltage across a load is described by the net output resistance at the output terminals. This resistance is the parallel combination of the two collector resistors and the resistance of Eq. (1-5).

$$R_O \doteq \frac{2r_eR_C(R_e + R_G/\beta)}{r_c(R_e + R_G/\beta) + R_GR_C} \qquad \text{for } R_C \ll r_c \tag{1-6}$$

$$R_O \doteq 2R_C \qquad \text{if } R_C \ll r_c(1-\alpha) \tag{1-7}$$

By drawing a similar parallel between a junction-FET (JFET) differential stage and a common-source FET, characteristics of this stage are obtained. Defined in Fig. 1.6 are the stage and its differential gain equivalent common-source circuit. For analysis, the JFET dc model[2] of Fig. 1.7a is applied to the preceding gain equivalent circuit. The complete circuit model is analyzed in Fig. 1.7b, resulting in the currents indicated. Differential voltage gain of the FET stage is then

$$A_O = \frac{E_{od}}{E_{id}} = -\frac{g_{fs}R_Dr_{ds}}{R_D + r_{ds}} \cdot \frac{1}{1 + \dfrac{R_G + R_S}{r_{gs}} + g_{fs}\dfrac{R_Sr_{ds}}{r_{ds} + R_D}} \tag{1-8}$$

Being the resistance of a reverse-biased junction, r_{gs} is far greater than typical driving or signal source resistance, and the drain resistor is often less than 300 kΩ level of r_{ds}, resulting in

$$A_O \doteq -\frac{g_{fs}R_D}{1 + g_{fs}R_S} \qquad \text{for } r_{gs} \gg R_G + R_S \qquad r_{ds} \gg R_D \qquad (1\text{-}9)$$

As before, differential input and output resistances seen in series across the two FETs are twice those found for a common-source FET. Input resistance can be written from the currents recorded in Fig. 1.7b.

$$R_I = 2r_{gs}\left(1 + g_{fs}\frac{R_S r_{ds}}{r_{ds} + R_D}\right) + R_S \qquad (1\text{-}10)$$

$$R_I \doteq 2r_{gs}(1 + g_{fs}R_S) \qquad \text{for } r_{gs} \gg R_S \qquad r_{ds} \gg R_D \qquad (1\text{-}11)$$

$$R_I \overset{T}{=} 10^{11}\ \Omega$$

The determination of the output resistance presented to the drain resistors follows from the currents defined in Fig. 1.8 for the conditions shown.

$$R'_O = 2r_{ds} + 2R_s\frac{r_{gs}(1 + g_{fs}r_{ds}) + R_G}{r_{gs} + R_G + R_S} \qquad (1\text{-}12)$$

$$R'_O \doteq 2r_{ds}(1 + g_{fs}R_S) \qquad \text{for } r_{gs} \gg R_G + R_S \qquad r_{ds} \gg R_S \qquad (1\text{-}13)$$

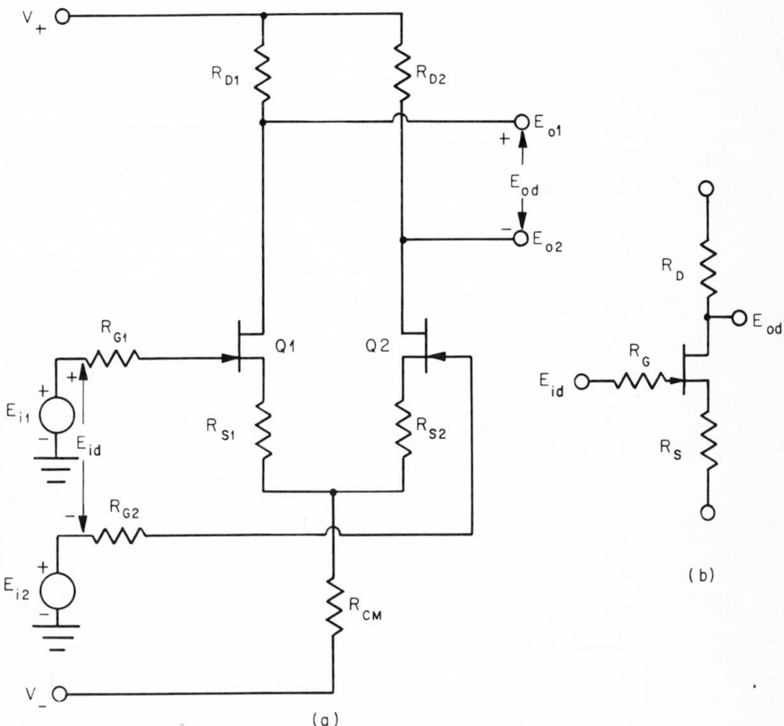

Fig. 1.6 (a) Basic FET differential stage; (b) single FET differential equivalent circuit.

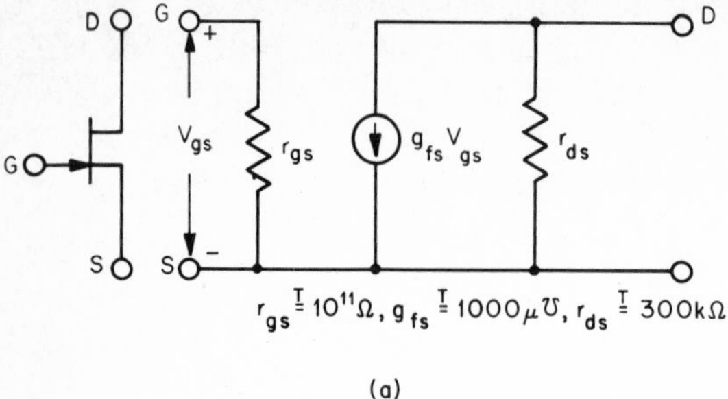

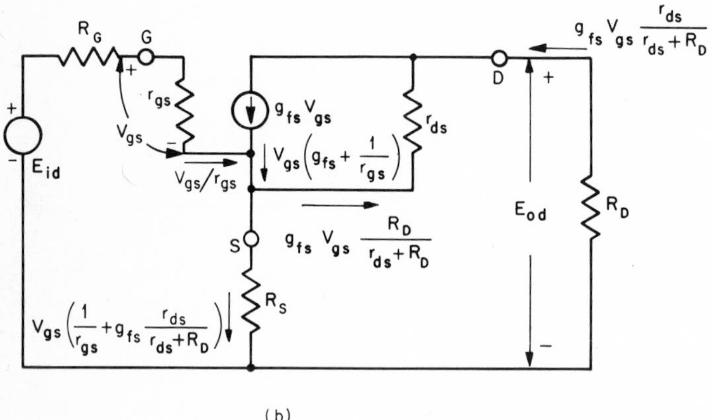

Fig. 1.7 (a) Junction-FET dc model; (b) circuit model of gain equivalent circuit.

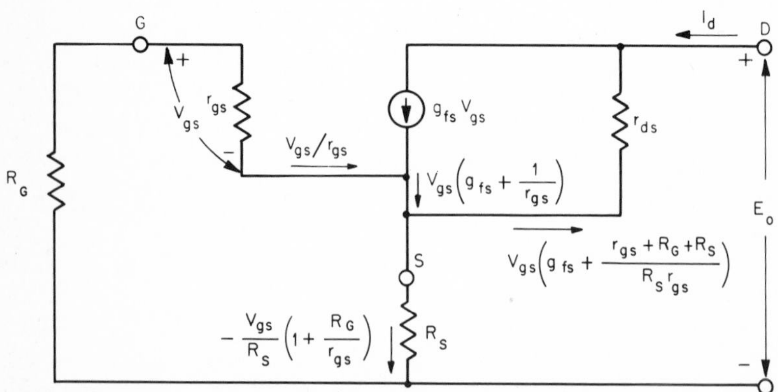

Fig. 1.8 Common-source output resistance analysis.

Comparing the approximate output resistance expression with the gain relationship of Eq. (1-9) reveals that addition of a source resistor increases output impedance by the same factor by which it reduces gain. Combining the resistance appearing at the FET outputs with the drain resistors gives the net resistance between the output terminals:

$$R_O \doteq 2r_{ds}(1 + g_{fs}R_S)\|2R_D \qquad (1\text{-}14)$$

$$R_O \doteq 2R_D \qquad \text{if } r_{ds} \gg R_D \qquad (1\text{-}15)$$

1.2 High-frequency Differential Signal Characteristics

Frequency response of a differential stage is conveniently analyzed by reflecting collector-base capacitance C_c or gate-drain capacitance C_{gd} to the input as Miller-effect equivalent capacitance. For high-gain differential amplifiers, the Miller effect creates the dominant input capacitance, and the emitter-base capacitance C_{eb} or gate-source capacitance C_{gs} will be negligible. The frequency of the stage pole may then be found by considering the shunting effect of the Miller capacitance on the signal source resistance and the equivalent input resistance. When the signal source resistance is small, this shunting effect is less significant, and the effect of collector-base or gate-drain capacitance on the load resistor must be considered. To include shunting effects on the collector load resistor with Miller-effect representation, the hybrid-pi model of a common-emitter transistor may be used to develop an equivalent circuit which includes both effects.

From the hybrid-pi model of Fig. 1.9a a unilateral two-port model of a differential stage can be defined to simplify analysis of high-frequency differential-stage characteristics. Feedback and output shunting effects of r_c were included in the input resistance and the output resistance found in the preceding section. For a single common-emitter transistor the resistances are one-half those defined for a differential stage by Eqs. (1-3) and (1-5). Replacing r_c with these equivalent resistances and neglecting C_{eb} provides the common-emitter transistor model of Fig. 1.9b. Again, the small r_b' is neglected and α is assumed near unity. To complete the transformation, the feedback and shunting effects of C_c will be represented by shunt capacitors across the input and output as shown in Fig. 1.9c. Capacitors C_1 and C_2 will have the same effects as C_c if the currents drawn are the same.

$$I = j\omega C_c(E_i' - E_o) = j\omega C_1 E_i' = -j\omega C_2 E_o$$

$$C_1 = \left(1 - \frac{E_o}{E_i'}\right) C_c$$

$$C_2 = \left(1 - \frac{E_i'}{E_o}\right) C_c \doteq C_c \qquad \text{for high gain}$$

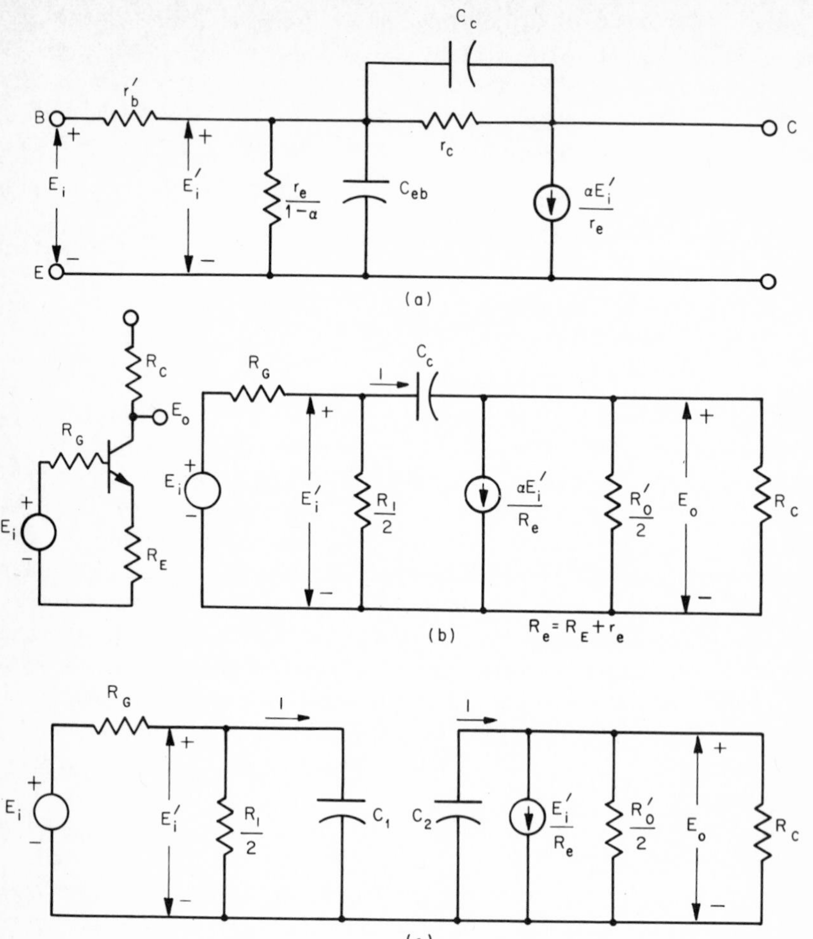

Fig. 1.9 (a) Hybrid-pi model of a bipolar transistor. (b) Common-emitter amplifier and simplified hybrid-pi representation. (c) Miller-effect equivalent circuit.

When C_2 is known, C_1 is resolved from the diagram as

$$C_1 \doteq \left(1 + \frac{Z_c}{R_e}\right) C_c$$

where

$$Z_c = \frac{R_O}{2 + j\omega R_O C_c} \qquad R_O = R'_O \| R_C$$

To confirm the accuracy of neglecting C_{eb}, its contribution to input capacitance can be compared with that of the above Miller-effect capacitance C_1. The effect of the emitter-base capacitance may be compared by considering its relationship to the transistor gain bandwidth product[3] ω_t.

$$C_{eb} = \frac{1}{\omega_t r_e} - C_c$$

For a typical silicon small-signal transistor at a collector current of 50 μA, f_t is greater than 50 MHz, resulting in

$$C_{eb} \leq 3\ \text{pF} \simeq C_c$$

Considering the effect of this capacitor on input capacitance,

$$C_I \doteq \frac{r_e}{R_e} C_{eb} + \left(1 + \frac{Z_c}{R_e}\right) C_c$$

For high stage gain, C_{eb} will be negligible in comparison with Miller-effect input capacitance.

To apply the above results to a differential stage, the stage may again be considered as two series-connected common-emitter transistors as shown in Fig. 1.2a. Differential input capacitance is the series combination of the input capacitances of the two common-emitter amplifiers.

$$C_I \doteq \left(1 + \frac{Z_c}{R_e}\right) \frac{C_c}{2} \qquad Z_c = \frac{R_O}{2 + j\omega R_O C_c} \tag{1-16}$$

A typical differential stage with a gain of 30 will have a low-frequency differential input capacitance of around 50 pF. However, this input capacitance is not a constant but decreases with increasing frequency as the stage gain falls.

$$C_I \doteq \left(1 + \frac{R_O/R_e}{1 + j\omega R_O C_c}\right) \frac{C_c}{2}$$

Operational amplifier input capacitance is significantly affected in this manner as the high-gain input stage generally has heavy capacitive loading produced by phase compensation between the stage outputs. Considering the loading on an input stage as C_L, the pole frequency of the input capacitance is reduced by the load as described in

$$C'_I \doteq \left(1 + \frac{R_O/R_e}{1 + j\omega R_O (C_c + C_L/2)}\right) \frac{C_c}{2}$$

Combining the input capacitance and input resistance expressions and neglecting the first term above, the differential input impedance will be

$$Z_I = \frac{R_I}{1 + j\omega R_I C_I} \doteq R_I \left(\frac{1 + j\omega R_O C_c}{1 + j\omega R_O C_c R_I / 2R_e} \right) \qquad (1\text{-}17)$$

if $R_I/2R_e \gg 1$, as is true for $R_C \ll r_c$.

Capacitance loading typically increases high-frequency input impedance since the Miller capacitance is decreased. Reduced Miller capacitance lowers the frequency of the input impedance zero without an accompanying decrease in pole frequency. Addition of C_L changes input impedance to

$$Z_I' \doteq R_I \frac{1 + j\omega R_O (C_c + C_L)}{1 + j\omega R_O C_c R_I / 2R_e}$$

As a result, differential input capacitance is usually only 10 pF at the frequency for which the input resistance is shunted 50 percent and input impedance is halved. With this frequency dependence, the significance of the input capacitance under operating conditions should be evaluated. A 50-pF differential input capacitance at 10 Hz is a negligible shunt to input impedance. Rather, the more meaningful input capacitance for consideration is that at a frequency for which input impedance is significantly lowered.

Combination of the Miller-effect equivalent circuits of the two common-emitter transistors provides the differential signal model of the stage in Fig. 1.10 for $\beta \gg 1$. Input and output resistances shown are as described in Sec. 1.1, and the total source and load resistances are the

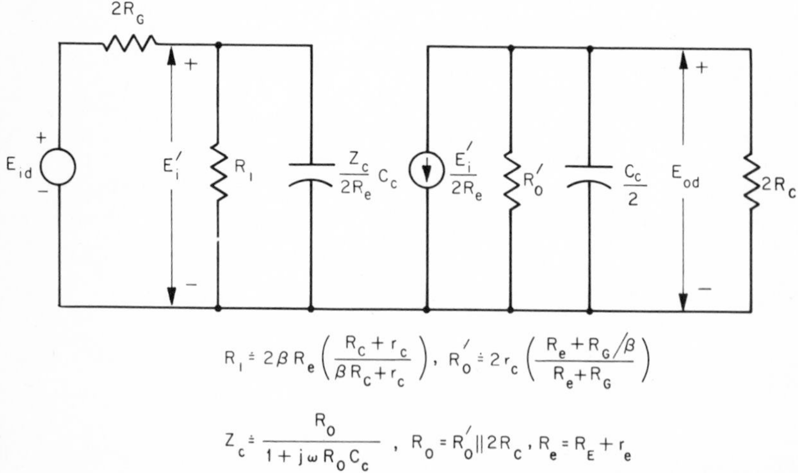

Fig. 1.10 Differential signal model of a bipolar transistor differential stage.

sum of those of the common-emitter transistors. Differential signal frequency response may be described by solving for the gain of this model. In the typical case $R_C \ll r_c(1 - \alpha)$, and the model simplifies with $R_I \doteq 2\beta R_e$ and $R_O \doteq 2R_C$. For high-gain stages the dominant portion of C_I shown is the second term, resulting in an output of

$$E_{od} = \frac{-R_C E_i'}{R_e(1 + j\omega R_C C_c)}$$

where

$$E_i' = \frac{E_{id} Z_I}{Z_I + 2R_G}$$

Then the differential gain is

$$A(j\omega) = \frac{E_{od}}{E_{id}} \doteq \frac{-R_C}{R_e + R_G/\beta} \frac{1}{1 + j\omega R_C C_c \left(\dfrac{R_e + R_G}{R_e + R_G/\beta}\right)}$$

$$\text{for } R_C \ll r_c(1 - \alpha) \qquad (1 - \alpha) \doteq \frac{1}{\beta} \qquad Z_c \gg R_e \qquad (1\text{-}18)$$

Note the heavy dependence upon source impedance of both the dc gain

$$A_O \doteq \frac{-R_C}{R_e + R_G/\beta}$$

and of the pole frequency

$$f_p \doteq \frac{1}{2\pi R_C C_c} \frac{R_e + R_G/\beta}{R_e + R_G} \qquad (1\text{-}19)$$

Increasing the source resistance decreases the gain and bandwidth, forcing the frequency response curve toward the origin, as illustrated in Fig. 1.11 for the two cases of negligible and predominant R_G.

By relating C_c to commonly measured transistor capacitance the response pole frequency of a differential stage can be predicted. Collector-base capacitance is generally measured as C_{ob} which is the common-base output capacitance. Comparison of the operating collector-base bias voltage V_{CB} with the C_{ob} test voltage V_{CB} defines C_c in terms of C_{ob}. In common planar-diffused transistors, junction capacitance is inversely proportional to the cube root of the junction voltage, and

$$C_c \doteq C_{ob} \sqrt[3]{\frac{V_{CBT}}{V_{CB}}} + C_p$$

where

$$C_p \triangleq \text{package capacitance}$$
$$C_p \stackrel{T}{=} 0.5 \text{ pF}$$

In the case of a phase-compensated differential stage, it is advantageous to reflect capacitance effects to the stage output for response considera-

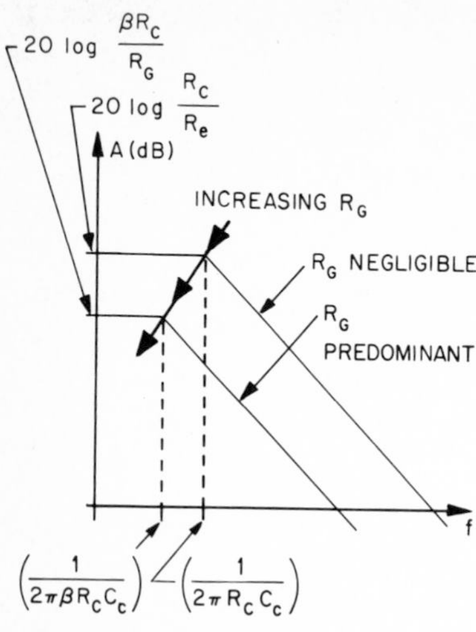

Fig. 1.11 Differential-stage Bode plot showing the effect of source resistance on response.

tion, rather than to use the previous input capacitance representation. In this way the frequency-determining elements of the stage may be combined directly with the phase compensation elements typically connected to the stage outputs and also with the loading impedance of any following stage. Once again the common-emitter amplifier analysis may be applied. The differential output impedance of the stage is that of two common-emitter amplifiers connected in series as in Fig. 1.2a. If the transistor collector capacitance C_c is considered to be the major frequency limitation, as before, the transistor output characteristics are modified by the shunting of r_c by C_c. Output impedance presented to the collector resistors follows from the output resistance of Eq. (1-5) by replacing r_c with the impedance of r_c and C_c in parallel.

$$R'_O \doteq 2r_c \frac{R_e + R_G/\beta}{R_e + R_G}$$

$$Z'_O \doteq \frac{2r_c}{1 + j\omega r_c C_c} \frac{R_e + R_G/\beta}{R_e + R_G} \tag{1-20}$$

Since output impedance is formed by the output resistance in parallel with the output capacitance,

$$Z'_O = \frac{R'_O}{1 + j\omega R'_O C'_O}$$

Combining the last three expressions, the equivalent output capacitance of a differential stage is

$$C'_O = \frac{C_c}{2} \frac{R_e + R_G}{R_e + R_G/\beta} \tag{1-21}$$

The resulting ac equivalent circuit of the differential stage is shown in Fig. 1.12.

The above expressions for output resistance, output capacitance, and the response pole frequency demonstrate the reverse transfer effects of the bipolar transistor stage by their high dependence upon the driving source impedance. As a result of these reverse transfer characteristics, cascading one bipolar differential stage with another changes the responses of both stages. Note that as the source impedance approaches zero the output resistance increases and simultaneously the output capacitance decreases, causing the pole frequency to increase.

$$\left.\begin{aligned} R'_O &\doteq 2r_c \overset{T}{=} 100\ M\Omega\ @\ I_C = 30\ \mu A \\ C'_O &\doteq \frac{C_c}{2} \overset{T}{=} 1.5\ pF \\ f_p &\doteq \frac{1}{2\pi R_C C_c} \end{aligned}\right\} \quad \text{if } R_G \ll \beta R_e$$

As the source impedance becomes large, the stage output impedance drops and the increasing equivalent output capacitance reduces bandwidth.

$$\left.\begin{aligned} R'_O &\doteq 2r_c(1-\alpha) \doteq \frac{2r_c}{\beta} \overset{T}{=} 1\ M\Omega\ @\ I_C = 30\ \mu A \\ C'_O &\doteq \frac{C_c}{2(1-\alpha)} \doteq \frac{\beta C_c}{2} \overset{T}{=} 150\ pF \\ f_p &\doteq \frac{1-\alpha}{2\pi R_C C_c} \doteq \frac{1}{2\pi R_C C_c \beta} \end{aligned}\right\} \quad \text{if } R_G \gg \beta R_e$$

Comparison of the above two cases indicates that a source resistance which is large compared with the emitter resistance reduces output resistance, increases output capacitance, and decreases bandwidth by a factor approximately equal to the beta of the transistors.

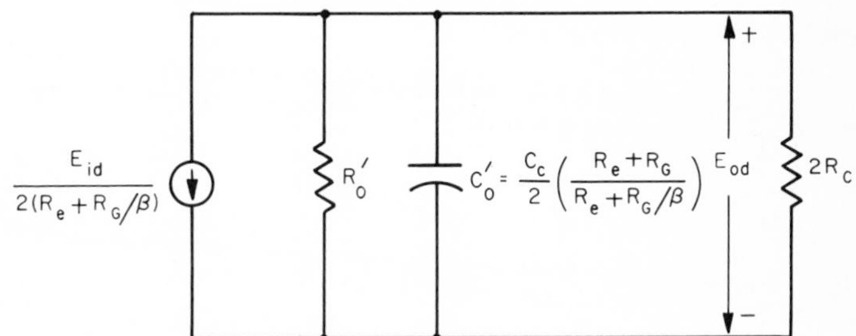

Fig. 1.12 Differential signal equivalent output circuit of a bipolar transistor differential stage.

For an FET differential stage, differential ac signal behavior is determined by a similar analysis. The capacitances of an FET are indicated in the device model of Fig. 1.13a. Common small-signal FETs have a gate-drain capacitance C_{gd} of 3 pF and a 6-pF gate-source capacitance C_{gs}. Once again, the Miller multiplication of the reverse transfer capacitance creates the dominant input capacitance effect and the gate-source capacitance is negligible. By connecting the FET as a common-source amplifier, the input and output resistances are increased by the source resistance, as developed in Sec. 1.1. From that analysis, the resistances are one-half that resulting with a differential stage, as indicated in Fig. 1.13b. Since this equivalent circuit is of the same form as that of the

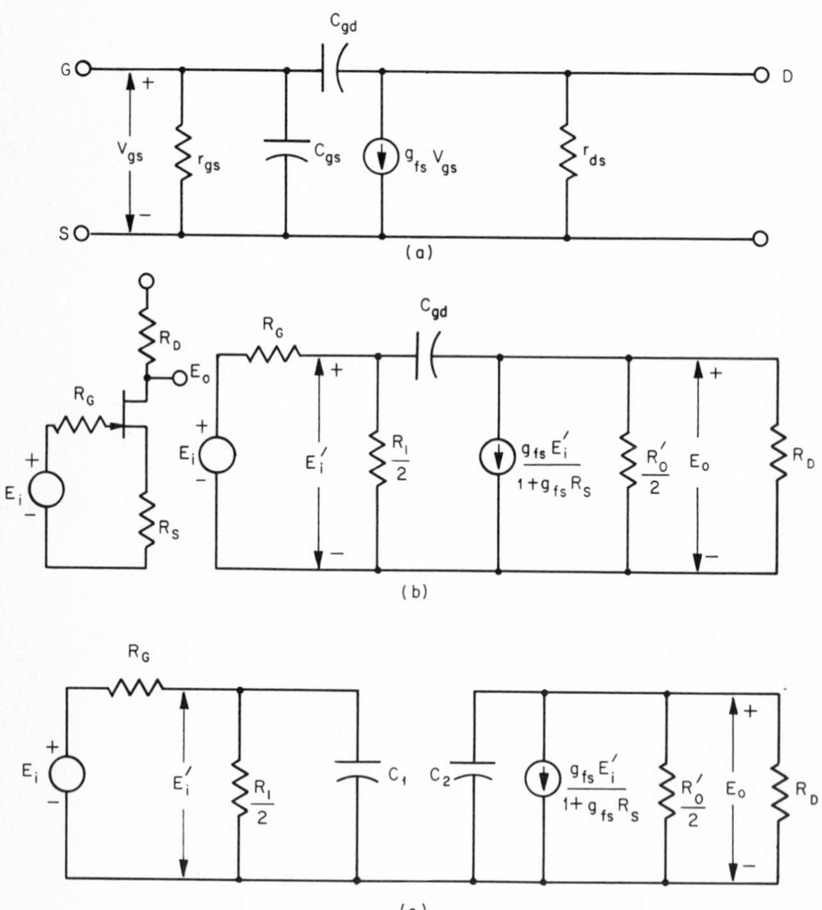

Fig. 1.13 (a) Junction-FET pi model. (b) Common-source representation. (c) Miller-effect equivalent circuit of (b).

bipolar transistor in Fig. 1.9b, the reverse transfer capacitance may be replaced by similar input and output capacitances C_1 and C_2 in Fig. 1.13c.

$$C_1 \doteq \left(1 + \frac{g_{fs}Z_d}{1 + g_{ds}R_S}\right) C_{gd}$$

$$C_2 \doteq C_{gd} \qquad \text{for high gain}$$

Differential input capacitance of the FET differential stage equals the series combination of those of the two common-source FETs forming the circuit.

$$C_I \doteq \left(1 + \frac{g_{fs}Z_d}{1 + g_{fs}R_S}\right) \frac{C_{gd}}{2}$$

where

$$Z_d = \frac{R_O}{2 + j\omega R_O C_{gd}} \tag{1-22}$$

Being comparable in magnitude to bipolar stage input capacitance, the input capacitance of the FET stage presents a low-frequency shunt to its high input resistance. The resulting input impedance is

$$Z_I = \frac{R_I}{1 + j\omega R_I C_I}$$

and, neglecting the first term of C_I,

$$Z_I \doteq R_I \frac{1 + j\omega R_O C_{gd}}{1 + j\omega r_{gs} g_{fs} R_O C_{gd}} \qquad \text{for } g_{fs}r_{gs} \gg 1 \tag{1-23}$$

As a result of Miller-effect input capacitance, the high input resistance of the FET differential stage does not provide as dramatic an improvement in impedance isolation for ac signals. Cascode biasing of the FETs, however, helps to extend the frequency range for high impedance isolation, as will be discussed in Sec. 1.5.

Combining the input resistance and dc gain results of the previous section with equivalent capacitors derived as in the bipolar transistor case, a differential signal model for the FET stage is shown in Fig. 1.14. For almost all practical levels of signal source resistance, the input resistance presents only a very small shunt and can be neglected. Using this model, with the first term of C_I omitted for the high-gain case, the differential response of an FET stage is

$$A(j\omega) \doteq \frac{-g_{fs}R_O}{1 + g_{fs}R_S} \frac{1}{1 + j\omega R_O C_{gd}[(1 + g_{fs}R_G)/(1 + g_{fs}R_S)]}$$

$$\text{for } R_I \gg R_G \qquad Z_d \gg R_S \tag{1-24}$$

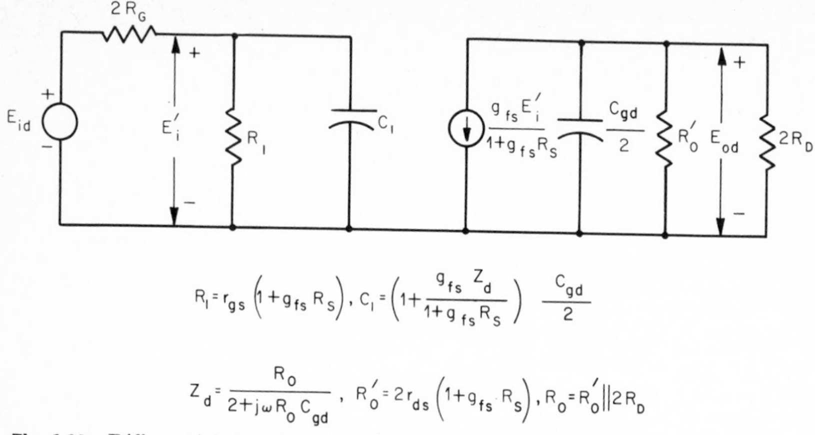

Fig. 1.14 Differential signal model of an FET differential stage.

Although the dc gain of this equation is independent of signal source resistance R_G, reverse transfer capacitance again makes the pole frequency highly sensitive to this driving resistance. From the above response expression the pole frequency is

$$f_p = \frac{1}{2\pi R_O C_{gd}} \frac{1 + g_{fs}R_S}{1 + g_{fs}R_G} \tag{1-25}$$

To relate the FET differential-stage response to FET characteristics normally measured, C_{gd} can be defined in terms of C_{rss}. The latter capacitance is measured as the reverse transfer capacitance from drain to gate with the source shorted to the gate. Again the gate-drain capacitance will be inversely proportional to the cube root of the reverse junction voltage as described for the bipolar transistor case. Then C_{gd} will be related to C_{rss} by the cube root of the ratio of V_{GD} to the C_{rss} test voltage V_{GDT}.

$$C_{gd} \doteq C_{rss} \sqrt[3]{\frac{V_{GDT}}{V_{GD}}} + C_p$$

where

$$C_p \triangleq \text{package capacitance}$$
$$C_p \stackrel{T}{=} 0.5\ \text{pF}$$

An equivalent circuit of the stage output is similarly useful in resolving the effects of phase compensation networks on the FET stage. By considering the pole frequency above, an equivalent output capacitance providing the same response in conjunction with a load R_O is

$$C'_O = \frac{1 + g_{fs}R_G}{1 + g_{fs}R_S}\frac{C_{gd}}{2} \tag{1-26}$$

Representing the FET-stage differential response behavior, Fig. 1.15 is the resulting equivalent output circuit.

In this and the preceding section low-frequency and high-frequency differential signal characteristics have been defined and modeled. Common-mode characteristics of a differential stage are similarly treated in the following section.

1.3 Common-mode Behavior of a Differential Stage

As analyzed in the preceding sections, the desired output from a differential stage is that produced by a differential signal applied between the stage inputs. However, the common-mode voltage present at both inputs also creates an output voltage. At the output both a common-mode and a differential-mode error voltage result, as will be discussed in this and the following section. The relative importance of these two output error signals depends upon whether a differential or a single-ended output is taken from the stage. With a perfectly balanced stage, the common-mode input voltage will result in only a common-mode output voltage. For an analysis of common-mode signal effects in the balanced case, consider the circuit of Fig. 1.16 in which the inputs have a common connection and the outputs have a common connection. With no differential input or output voltages, this circuit represents a balanced differential stage under common-mode signals. Simplifying the circuit results in the single common-emitter amplifier having two parallel transistors as shown. From this representation differential-stage common-mode

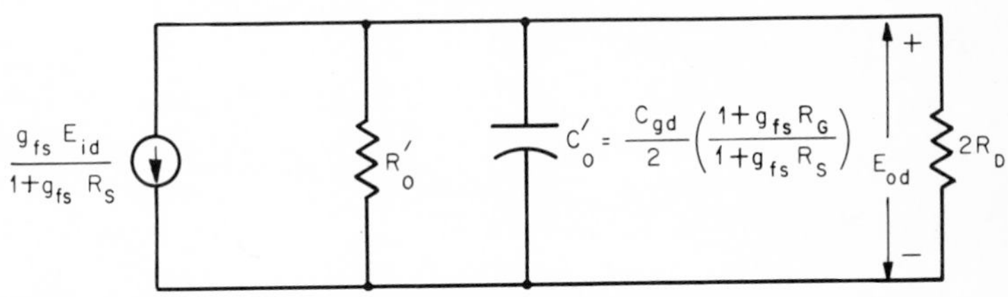

Fig. 1.15 Differential signal equivalent output circuit of an FET differential stage.

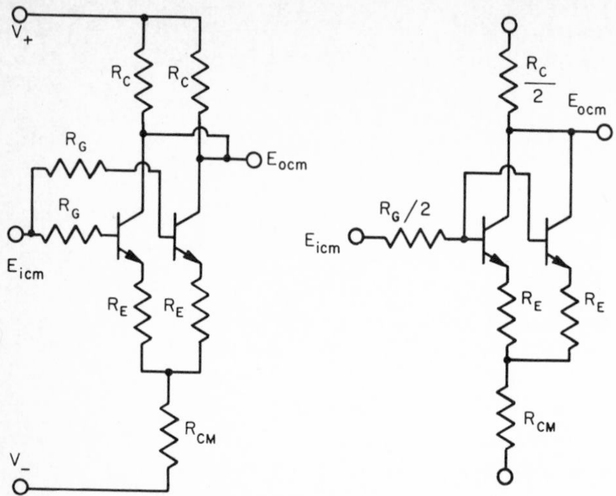

Fig. 1.16 Balanced differential-stage common-mode equivalent circuits.

characteristics can be found by using the common-emitter relationships employed in Sec. 1.1.

To provide a stable common-mode current bias to the stage, the biasing resistor R_{CM} is made large to keep current changes from input voltages small. In general, R_{CM} is far greater than R_E or r_e. Dominating the resistance of the input circuit, R_{CM} absorbs essentially all the common-mode input signal unless a very large signal source resistance is involved. Common-mode input resistance provided by a large R_{CM} is much greater than the differential input resistance developed by R_e. From the common-emitter expression, common-mode input resistance is

$$R_{Icm} \doteq \beta R_{CM} \frac{r_c}{r_c + 2\beta R_{CM}} \qquad \text{for } R_C \ll r_c(1 - \alpha) \tag{1-27}$$

Common-mode input resistance will then be between βR_{CM} and $r_c/2$, which can reach the 100-MΩ level.

It is this very high common-mode input resistance which makes the noninverting configuration desirable for impedance isolation with bipolar transistor operational amplifiers. Output resistance is also improved by the large R_{CM} and, for the common-mode case, is the parallel combination of the output resistances of the two common-emitter transistors. Resistance presented to the collector resistors is

$$R'_{Ocm} = r_c \frac{R_{CM}}{2R_{CM} + R_G} \qquad \text{for } \frac{R_G}{2\beta} \ll R_{CM} \ll \frac{r_c}{2} \tag{1-28}$$

Combining this resistance with the collector resistors, the output resistance presented to any load on the stage is found to be

$$R_{Ocm} \doteq \frac{R_C}{2} \frac{R_{CM}r_c}{R_{CM}r_c + R_C R_G/2} \qquad \text{for } R_C \ll r_c \tag{1-29}$$

Except in the case of exceptionally large collector or source resistances, the output resistance is, as might be anticipated,

$$R_{Ocm} \doteq \frac{R_C}{2} \qquad \text{for } R_C \ll \frac{2R_{CM}r_c}{R_G} \tag{1-30}$$

Because of the large common-mode biasing resistor R_{CM}, relatively small changes in the stage currents result from common-mode input signals. These changes produce correspondingly small signal voltages on the collector resistors, and the amplification of a common-mode signal is far less than that provided for differential signals which are impressed upon R_e. Common-emitter analysis applied to the circuit of Fig. 1.16, neglecting R_e in comparison with R_{CM}, results in a common-mode gain expression of

$$A_{Ocm} = \frac{-R_C(\alpha r_c - 2R_{CM})}{2R_{CM}(R_C + r_c) + R_G[r_c(1 - \alpha) + R_C + 2R_{CM}]} \qquad \text{for } R_{CM} \gg R_e \tag{1-31}$$

For typical resistance levels the common-mode gain is approximated by

$$A_{Ocm} \doteq \frac{-R_C}{2R_{CM}} \qquad \text{for } \frac{R_G}{2\beta} \ll R_{CM} \ll \frac{r_c}{2} \qquad R_C \ll r_c(1 - \alpha) \tag{1-32}$$

The frequency dependence of the common-mode gain follows from the common-emitter representation of the reaction of input impedance with source impedance, along with a feedthrough effect of the collector-base capacitance. Common-mode input capacitance is composed of a Miller-effect equivalent of the two collector capacitances. Because of the low gain of the common-mode circuit, the Miller multiplication is correspondingly small. However, the effect of the emitter-base capacitance C_{eb} is again negligible since it bypasses only r_e whereas the Miller capacitance bypasses the much larger R_{CM} as well. Since the two transistors of the stage are essentially in parallel, as described above, their input capacitances add. Paralleling the result of Sec. 1.2, common-mode input capacitance is

$$C_{Icm} \doteq 2\left(1 - \frac{Z_c}{2R_{CM}}\right)$$

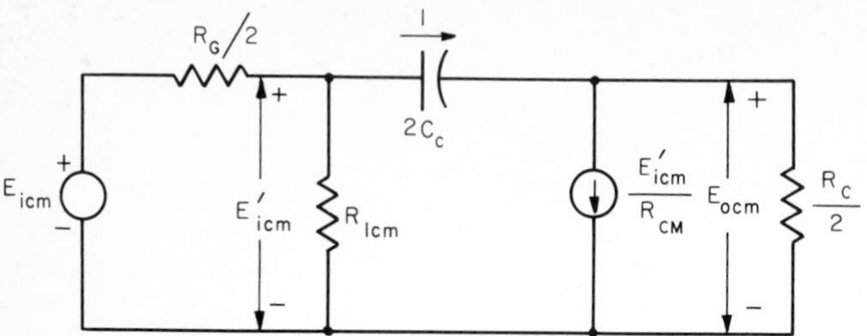

Fig. 1.17 A pi model of the common-mode signal equivalent circuit for a bipolar transistor differential stage.

where

$$Z_c = \frac{R_O}{2 + j\omega R_O C_c}$$

Typically $Z_c \ll 2R_{CM}$ and

$$C_{Icm} \doteq 2C_c \overset{T}{=} 6 \text{ pF} \tag{1-33}$$

The feedback effect of the collector-base capacitance above is small for the low common-mode gain, but the feedthrough effect from input to output becomes appreciable, as the output swing is far less than the input swing. Referring to the common-mode-circuit pi model of Fig. 1.17, an input signal which is large compared with the output signal will result in a large feedthrough current from input to output through the capacitance $2C_c$. This effect is the reverse of the feedback current which flows in C_c when the output signal is much greater than that at the input. From the Sec. 1.2 derivation of the Miller-effect equivalent circuit, the current in $2C_c$ will be

$$I = 2(E'_{icm} - E_{ocm})j\omega C_c$$

An equivalent representation of this feedthrough current in the two parallel collector capacitances will be a current generator $2E'_{iCM}j\omega C_c$ and a capacitor $2C_c$ across the output. The resulting differential-stage model for common-mode signals is shown in Fig. 1.18. Common-mode gain for the typical case is

$$A_{cm}(j\omega) = \doteq -\frac{R_C}{2R_{CM}} \frac{1 + 2j\omega R_{CM}C_c}{(1 + j\omega R_G C_c)(1 + j\omega R_C C_c)}$$

$$\text{for } R_{Icm} \gg \frac{R_G}{2} \qquad R_C \ll r_c(1 - \alpha) \tag{1-34}$$

Examination of the above gain expression reveals that common-mode gain increases with increasing frequency, because of C_c feedthrough until shunting effects become significant, as represented in Fig. 1.19.

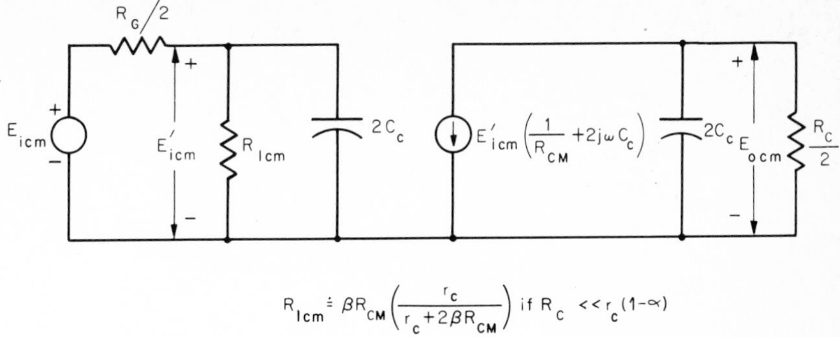

$$R_{Icm} \doteq \beta R_{CM}\left(\frac{r_c}{r_c+2\beta R_{CM}}\right) \text{ if } R_C << r_c(1-\alpha)$$

Fig. 1.18 Common-mode equivalent circuit for a balanced bipolar transistor differential stage.

The complete model for a differential stage is formed by combining the differential signal and common-mode signal models of Figs. 1.10 and 1.18, respectively. With the two sides of the stage separated, the complete model is given in Fig. 1.20. For general analysis, the separate differential and common-mode models are easier to use, but the complete model represents the interaction of the common-mode signal with the differential circuit unbalances.

Common-mode behavior of an FET differential stage is represented in a manner similar to that used with the bipolar stage by considering the equivalent circuit of Fig. 1.21. The equivalent circuit results from considering the two common-source FETs of the stage connected in

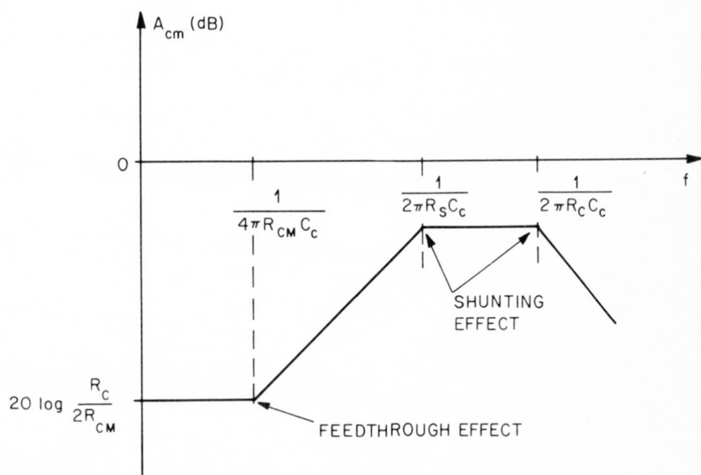

Fig. 1.19 Common-mode gain Bode plot showing the feedthrough and shunting effects of C_c.

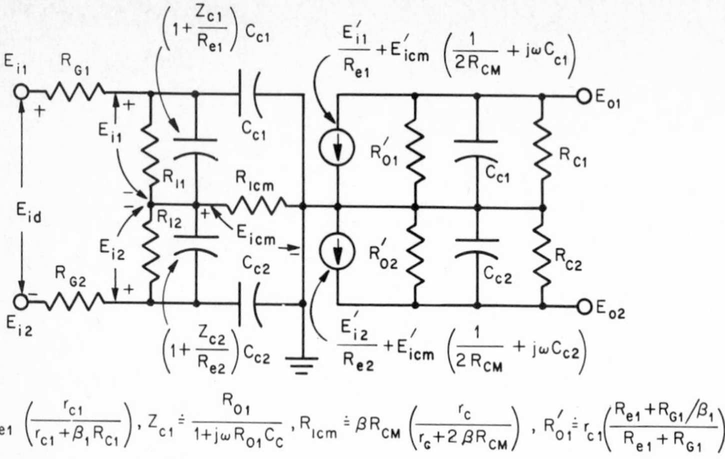

Fig. 1.20 Bipolar transistor differential-stage model including differential and common-mode signal equivalent circuitry.

parallel with a common input and a common output. The common-source expressions used in the preceding sections apply directly to this circuit. In this common-mode case, input signals fall primarily upon the large common-mode biasing resistor R_{CM}, resulting in only small stage current changes. Input resistance is, then, boosted by the gain degeneration of R_{CM}, modifying the result of Eq. (1-10) to

$$R_{Icm} \doteq \frac{r_{gs}}{2}(1 + g_{fs}R_{CM}) \qquad \text{for } r_{gs} \gg R_{CM} \qquad r_{ds} \gg R_D \quad (1\text{-}35)$$

In practice, input resistance is limited by dc leakages of the FET surface and the package to around 10^{12} Ω, and the extreme levels predicted by the last expression are not achieved. Output resistance for the

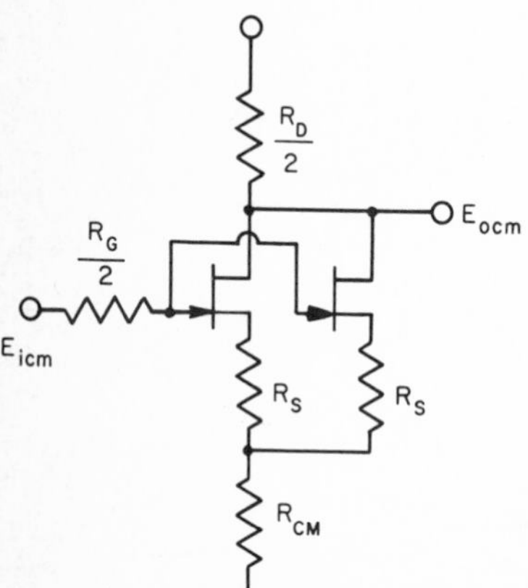

Fig. 1.21 Balanced FET-stage common-mode equivalent circuit.

parallel-connected common-mode equivalent is similarly improved by R_{CM}. In this case

$$R_{Ocm} \doteq \frac{r_{ds}}{2}(1 + g_{fs}R_{CM}) \qquad \text{for } r_{gs} \gg R_G + R_{CM} \qquad g_{fs}r_{ds} \gg 1 \qquad (1\text{-}36)$$

Since the common-mode signal drops across R_{CM}, the gain is far less than that resulting from differential signals impressed upon the gate-source junctions and the source resistors.

$$A_{Ocm} \doteq -\frac{R_D}{2R_{CM}} \qquad \text{if } r_{ds} \gg R_D \qquad r_{gs} \gg R_G + R_S + 2R_{CM} \qquad 2g_{fs}R_{CM} \gg 1 \qquad (1\text{-}37)$$

The common-mode frequency response of an FET differential stage resembles that of the bipolar transistor stage, being primarily limited by the gate-drain capacitance. With the very low common-mode gain, Miller multiplication of C_{gd} is small. However, the entire input voltage signal appears upon C_{gd}, whereas only that small portion of the signal appearing from gate to source falls across C_{gs}. The relative shunting effects of the two capacitors are apparent from the resulting input currents.

$$I_1 = E_{icm}\, j\omega C_{gd} \qquad \text{due to } C_{gd}$$

$$I_2 = \frac{E_{icm}}{g_{fs}R_{CM}}\, j\omega C_{gs} \qquad \text{due to } C_{gs}$$

For $C_{gs} \simeq 2C_{gd}$, $I_2 \ll I_1$.

Common-mode input capacitance is, therefore, essentially due to C_{gd} and, for low common-mode gain, is approximated by the two FET gate-drain capacitances in parallel.

$$C_{Icm} \doteq 2C_{gd} \stackrel{T}{=} 6\text{ pF} \qquad (1\text{-}38)$$

Note that common-mode input capacitance is much lower than encountered previously for differential signals. Comparing the two,

$$C_{Icm} = \frac{4R_S}{Z_d} C_I$$

As a result, shunting of the high FET impedance isolation by input capacitance is greatly reduced when an FET operational amplifier is operated in the noninverting mode.

The gate-drain capacitance shapes the FET-stage common-mode frequency response by its input-to-output feedthrough and by its shunting of signal source and load resistances. As in the bipolar transistor case, these effects can be represented by a feedthrough output current generator

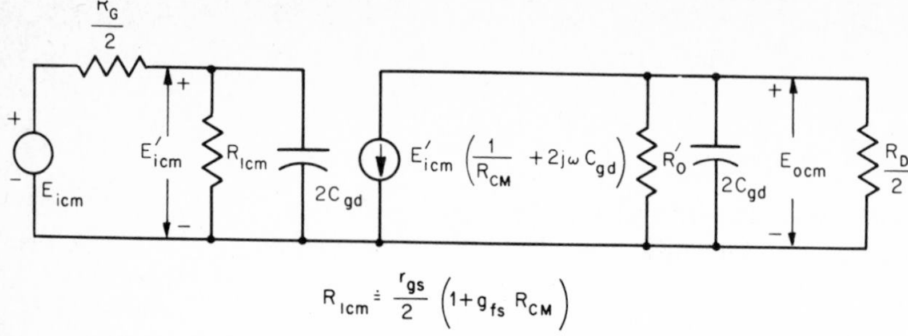

Fig. 1.22 Common-mode equivalent circuit for a balanced FET differential stage.

and by shunt capacitors across the input and output circuits. With a feedthrough signal voltage equal to E_{icm} the current generator will be $2E'_{icm}j\omega C_{gd}$. Combining this current, the input capacitance already discussed, and the load shunting equivalent capacitance with the low-frequency common-mode characteristics provides the model of Fig. 1.22. From this model the common-mode response is described by

$$A_{cm}(j\omega) \doteq -\frac{R_D}{2R_{CM}}\frac{1+2j\omega R_{CM}C_{gd}}{(1+j\omega R_G C_{gd})(1+j\omega R_D C_{gd})} \qquad \text{for } R_{Icm} \gg \frac{R_G}{2}$$

$$r_{ds} \gg R_D \qquad r_{gs} \gg R_G + R_S + 2R_{CM} \qquad 2g_{fs}R_{CM} \gg 1 \tag{1-39}$$

Again the common-mode gain initially increases with increasing frequency because of feedthrough on the gate-drain capacitance. The resulting

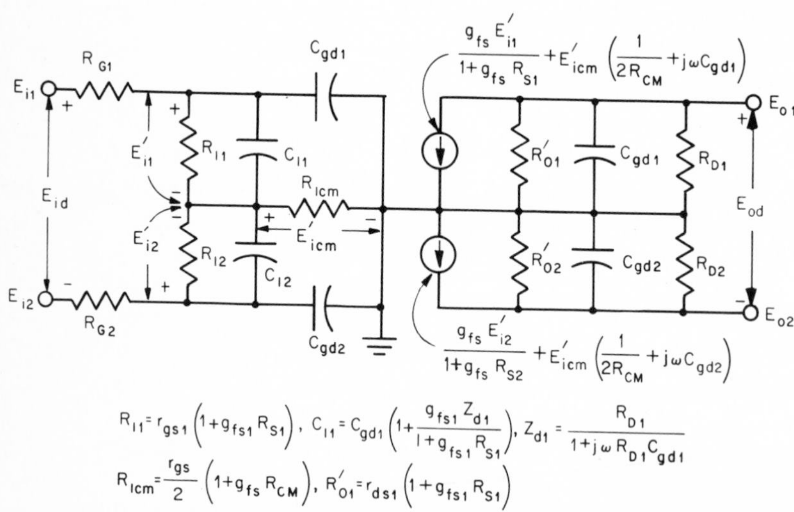

Fig. 1.23 FET differential-stage model including differential and common-mode equivalent circuits.

response curve is similar to that of Fig. 1.19. Combining the characteristics of the differential signal and common-mode signal models of Figs. 1.14 and 1.22, the complete model of an FET differential stage is shown in Fig. 1.23.

1.4 Differential-stage Unbalances and Common-mode Rejection

In the preceding section the common-mode behavior of a differential stage was considered under the assumption that the stage was perfectly balanced. For that case common-mode input signals produced only a common-mode signal at the outputs.

The unbalances in a differential stage are the sources of differential output error voltages resulting with common-mode input voltage swing. Mismatches in emitter-base junctions, emitter or collector resistors, betas, signal source resistances, output resistances, and collector capacitances all create these differential error voltages in a bipolar transistor stage. Rather than consider all these unbalances together, a far more manageable solution results by considering the typical case of small mismatches for which the interaction of the different unbalances is a second-order effect. The different unbalances may then be considered separately. Considering first the differences in emitter-base junctions and in emitter resistors, these unbalances cause unequal division of the common-mode current between the two halves of the differential stage. In Fig. 1.24 common-mode current I_{cm} is divided into emitter currents I_{e1} and I_{e2}. Note that the emitter-base junction mismatch is represented in terms of the difference in junction forward dynamic resistances Δr_e. The output error voltage E_{od} is

$$E_{od} = E_{o1} - E_{o2} = -\alpha(I_{e1} - I_{e2})R_C \doteq -(I_{e1} - I_{e2})R_C$$

with

$$I_{e2} = \left(1 + \frac{\Delta R_e}{R_e}\right) I_{e1} \qquad I_{e1} = -\frac{R_e}{2R_e + \Delta R_e} I_{cm} \qquad R_e = R_E + r_e$$

and

$$I_{cm} \doteq \frac{E_{icm}}{R_{CM}} \qquad \text{for } R_{CM} \gg R_e + \frac{R_G}{1 - \alpha}$$

Combining the above four relationships gives

$$\frac{E_{o1} - E_{o2}}{E_{icm}} \doteq \frac{\Delta R_e}{R_e} A_{Ocm} \qquad \text{due to } \Delta R_E \text{ and } \Delta r_e \text{ for } R_e \gg \Delta R_e$$

Considering next unequal source resistances or transistor betas, differential input error signals are developed by base current changes occurring

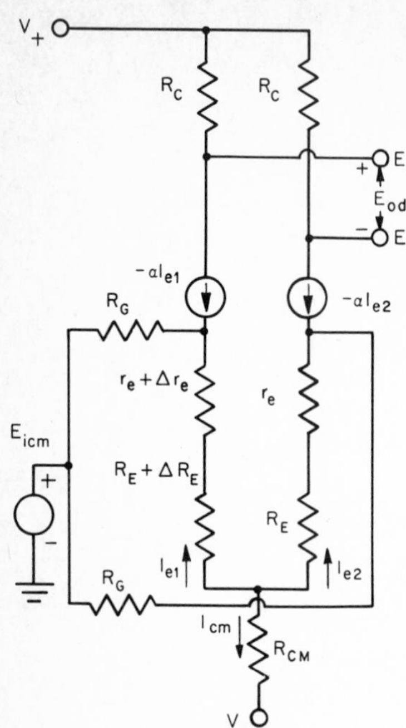

Fig. 1.24 Equivalent circuit of a bipolar transistor differential stage with unbalanced emitter resistances.

with common-mode voltage signals. Represented in Fig. 1.25, the error voltage E_{ie} is developed by either the source resistance or base current mismatch indicated. Appearing directly at the stage inputs with no additional intervening source resistance, the differential error is amplified by a gain of

$$A'_O \doteq -\frac{R_C}{R_e}$$

For unbalanced source resistances alone, the common-mode generated output error is

$$E_{o1} - E_{o2} \doteq A'_O E_{ie} = \frac{R_C}{R_e} I_b \, \Delta R_G$$

Since the common-mode input signal is essentially across R_{CM}, the base current signal which results is

$$I_b = \frac{E_{icm}}{2\beta R_{CM}}$$

and the output error is related by

$$\frac{E_{o1} - E_{o2}}{E_{icm}} \doteq \frac{\Delta R_G}{\beta R_e} A_{Ocm} \qquad \text{due to } \Delta R_G$$

If the transistor betas are mismatched, the base current differential ΔI_B shown in the figure results. It may be approximated by

$$\Delta I_b = I_{b1} - I_{b2} \doteq \frac{E_{icm}}{2\beta R_{CM}} \frac{\Delta\beta}{\beta} \qquad \text{if } \Delta\beta \ll \beta$$

The flow of this differential current in equal source resistances creates the following output error:

$$E_{o1} - E_{o2} \doteq A'_O E_{ie} = \frac{R_C}{R_e} \Delta I_b R_G$$

$$\frac{E_{o1} - E_{o2}}{E_{icm}} \doteq \frac{R_G}{\beta R_e} \frac{\Delta\beta}{\beta} A_{Ocm} \qquad \text{due to } \Delta\beta$$

Unequal output resistances cause additional common-mode error. For bipolar transistor differential stages, the transistor output resistance is typically far greater than the load resistance and does not greatly lower stage gain. However, when considering common-mode error terms, a high degree of match is important between the gains the two transistors present to common-mode input voltages. In the case of

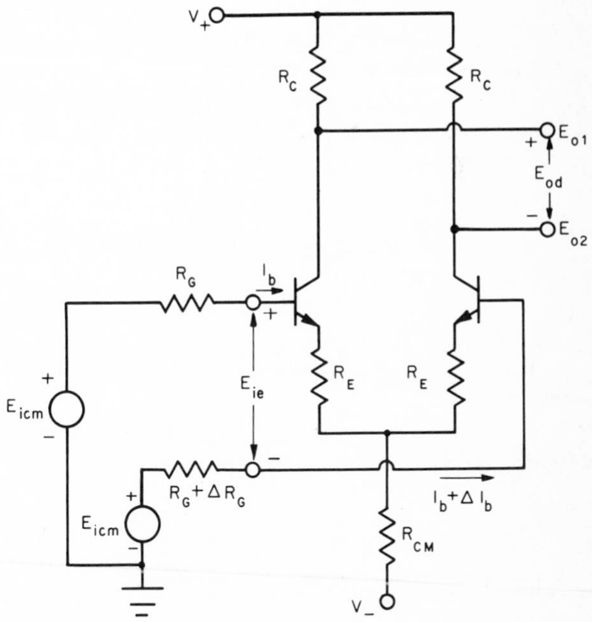

Fig. 1.25 Differential stage showing input error voltage resulting from unbalanced source resistances or unequal base currents.

the balanced stage discussed previously, the gains are identical, resulting in equal outputs and a common-mode output voltage only. Thus,

$$\frac{E_{ocm}}{E_{icm}} \doteq A_{Ocm} = -\frac{R_C}{2R_{CM}} = \frac{E_{o1}}{E_{icm}} = \frac{E_{o2}}{E_{icm}}$$

From this relationship, the effect of unbalanced collector loads can be seen:

$$\frac{E_{o1} - E_{o2}}{E_{icm}} = -\frac{\Delta R_C}{2R_{CM}} \qquad \text{due to } \Delta R_C$$

The effect of finite output impedance appears in loading on the equivalent output current generator, as represented in Fig. 1.26a. This model is derived from Fig. 1.20 for $R_G + \beta R_E \ll \beta R_{CM}$. The resulting unequal gains again create an error voltage as expressed by

$$E_{o1} - E_{o2} = -\frac{E_{icm}R_C}{2R_{CM}}\left(\frac{R_O + \Delta R_O}{R_O + \Delta R_O + R_C} - \frac{R_O}{R_O + R_C}\right)$$

For $R_O \gg \Delta R_O + R_C$,

$$\frac{E_{o1} - E_{o2}}{E_{icm}} \doteq \frac{\Delta R_O}{R_O} A_{Ocm} \qquad \text{due to } \Delta R_O$$

A final source of error to be examined is that of mismatched collector-base capacitances in the differential-stage transistors. These create unequal feedthrough currents from the common-mode input signal to the output. The result is a differential output error signal which may be considered by using the feedthrough current representation of Fig. 1.20, as repeated in Fig. 1.26b for $R_G + \beta R_e \ll R_{CM}$.

$$\frac{E_{o1} - E_{o2}}{E_{icm}} = j\omega R_C \, \Delta C_c \qquad \text{due to } \Delta C_c$$

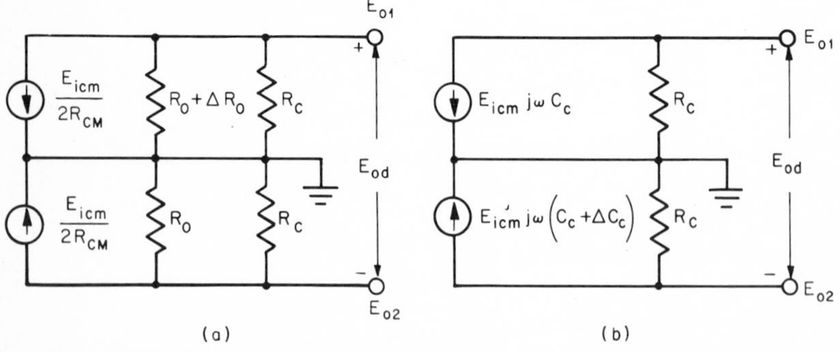

Fig. 1.26 Equivalent circuit for common-mode effects of unbalanced (a) output resistances and (b) collector-base capacitances.

In general, this capacitance unbalance is the major source of high-frequency output error from common-mode signals.

Summarizing the output voltages resulting from input common-mode voltage, the output common-mode and differential relationships are expressed below:

$$A_{cm}(j\omega) = \frac{E_{ocm}}{E_{icm}} \doteq A_{Ocm} \frac{1 + j\omega 2R_{CM}C_c}{(1 + j\omega R_C C_c)(1 + j\omega R_G C_c)}$$

$$A_{Ocm} \doteq -\frac{R_C}{2R_{CM}} \qquad (1\text{-}34)$$

Since the unbalances may each be either of two directions, each effect may add to or subtract from the others.

$$\frac{E_{o1} - E_{o2}}{E_{icm}} \doteq \pm A_{Ocm}\left(\frac{\Delta R_e}{R_e} \pm \frac{\Delta\beta R_G}{\beta^2 R_e} \pm \frac{\Delta R_G}{\beta R_e} \pm \frac{\Delta R_C}{R_C} \pm \frac{\Delta R_O}{R_O}\right) \pm j\omega R_C\, \Delta C_c \qquad (1\text{-}40)$$

Each term in this expression identifies the source of its common-mode error component by the differential of its numerator. Comparison of the above two expressions indicates that the low-frequency differential and common-mode output error terms are related by the degree of unbalance.

$$\frac{E_{od}}{E_{icm}} = \pm \frac{E_{ocm}}{E_{icm}}\left(\frac{\Delta R_e}{R_e} \pm \frac{\Delta\beta R_G}{\beta^2 R_e} \pm \frac{\Delta R_G}{\beta R_e} \pm \frac{\Delta R_C}{R_C} \pm \frac{\Delta R_O}{R_O}\right) \qquad \text{at dc}$$

For the unbalances generally encountered above, the resulting differential output error terms are a small fraction of the common-mode output error. If a differential output is taken from the stage, only this smaller error term represents a direct error to the amplified differential input signal. In this case the error voltage common to the output terminals is a common-mode signal presented to the load or following stage, and its associated signal error is determined by the common-mode signal sensitivity of such a load or stage. When a single-ended output referenced to common is taken from the stage, the output error voltage added to the amplified input signal consists of both the differential and common-mode output error components referred to above. A common-mode gain for each case is then defined in terms of the above error voltages included in the output signal. For a differential output,

$$A_{cmd} = \frac{E_{od}}{E_{icm}} = \pm A_{Ocm}\left(\frac{\Delta R_e}{R_e} \pm \frac{\Delta\beta R_G}{\beta^2 R_e} \pm \frac{\Delta R_G}{\beta R_e} \pm \frac{\Delta R_C}{R_C} \pm \frac{\Delta R_O}{R_O}\right) \pm j\omega R_C\, \Delta C_c \qquad (1\text{-}41)$$

For a single-ended output,

$$A_{cms} = \frac{E_{o1}}{E_{icm}} = \frac{E_{ocm}}{E_{icm}} + \frac{A_{cmd}}{2} \doteq \frac{E_{ocm}}{E_{icm}}$$

$$A_{cms} \doteq A_{Ocm} \frac{1 + j\omega 2R_{CM}C_c}{(1 + j\omega R_C C_c)(1 + j\omega R_G C_c)} \qquad (1\text{-}42)$$

Comparison of the differential and common-mode gain expressions reveals a key feature of the differential stage. This feature is expressed by the common-mode rejection ratio (CMRR) in terms of the ratio of differential gain to common-mode gain. For a single-ended output,

$$\text{CMRR}_s = \frac{A_O}{A_{Ocms}} \doteq \frac{2R_{CM}}{R_e + R_G/\beta} \quad \text{at low frequency} \qquad (1\text{-}43)$$

In general, the common-mode biasing resistor R_{CM} is much larger than the emitter resistance R_e, or R_G/β. The differential gain is, then, much greater than the common-mode gain, permitting an output signal due primarily to the differential input signal, as expressed by the common-mode rejection ratio. Voltages common to the two inputs, such as those resulting from noise, or ground loop currents are thereby rejected. A high degree of improvement in common-mode rejection is achieved with a differential output as expressed for low frequency by

$$\text{CMRR}_d = \frac{A_O}{A_{Ocmd}} \doteq \frac{2R_{CM}}{R_e + R_G/\beta} \cdot \frac{1}{\Delta R_e/R_e \pm \Delta\beta R_G/\beta^2 R_e \pm \Delta R_G/\beta R_e \pm \Delta R_C/R_C \pm \Delta R_O/R_O} \qquad (1\text{-}44)$$

Comparing the signal-ended and differential output cases,

$$\frac{\text{CMRR}_s}{\text{CMRR}_d} = \frac{\Delta R_e}{R_e} \pm \frac{\Delta\beta R_G}{\beta^2 R_e} \pm \frac{\Delta R_G}{\beta R_e} \pm \frac{\Delta R_C}{R_C} \pm \frac{\Delta R_O}{R_O} \quad \text{at low frequency}$$

As this expression indicates, common-mode rejection is far greater for a differential output than that achieved with a single-ended output.

Circuit unbalances in the basic FET differential stage of Fig. 1.27 give rise to similar differential output sensitivities to common-mode input voltage. Differential output errors are caused by mismatched source or drain resistors, forward transconductances, signal source resistances, output resistances, gate leakage currents, and gate-drain capacitances. Having a close parallel to the bipolar transistor case, the effects of FET-stage unbalances are drawn from the preceding analysis by considering the similar stage models of Figs. 1.20 and 1.23. Unbalanced

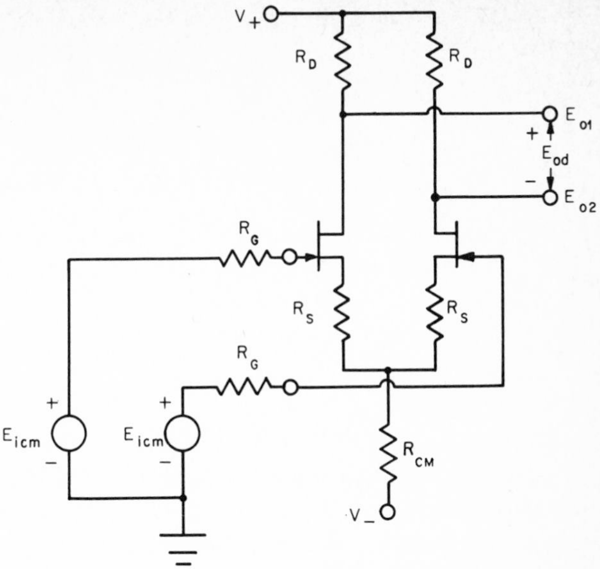

Fig. 1.27 Basic FET differential stage with a common-mode input signal.

source resistors and forward transconductances divide the common-mode current unequally between the two sides of the stage, as did the mismatched emitter resistors and emitter-base junctions in the bipolar case. The resulting output is expressed by

$$\frac{E_{o1} - E_{o2}}{E_{icm}} \doteq \frac{\Delta R_S + 1/\Delta g_{fs}}{R_S + 1/g_{fs}} A_{Ocm} \qquad \text{due to } \Delta R_S \text{ and } \Delta g_{fs}$$

As before, unequal load or output resistances mismatch the gains received by the common-mode signal on the two sides of the stage.

$$\frac{E_{o1} - E_{o2}}{E_{icm}} \doteq -\frac{\Delta R_D}{2R_{CM}} \qquad \text{due to } \Delta R_D$$

Output resistance unbalance and its voltage sensitivity are commonly the major sources of FET-stage sensitivity to common-mode voltage. Being relatively low, 300 kΩ, the output resistance of a typical FET stage presents a noticeable shunt to gain. Output resistance mismatches, then, have greater effect on the gain balance than that encountered with bipolar transistors. In addition, FET output resistance is fairly sensitive to the drain-source voltage, becoming significantly smaller at low voltages. As common-mode voltage swing lowers V_{GS}, the gain shunting by output resistance and the accompanying effect on gain balance increase. The output error is described by

$$\frac{E_{o1} - E_{o2}}{E_{icm}} \doteq \frac{\Delta R_O}{R_O} A_{Ocm} \qquad \text{due to } \Delta R_O$$

Unbalanced gate-drain capacitances result in different feedthrough currents in the same manner as did mismatched collector-base capacitances.

$$\frac{E_{o1} - E_{o2}}{E_{icm}} = j\omega R_D \, \Delta C_{gd} \qquad \text{due to } \Delta C_{gd}$$

Common-mode signal sensitivities resulting from differences in signal source resistances or gate leakage currents are due to the voltage sensitivity of the gate leakage current. A common-mode input voltage creates a direct change in gate-drain voltage, causing gate leakage current variation. For silicon FETs the reverse-biased junction leakage comprising the gate leakage current is very poorly described by the well-known junction equation. Formed essentially by thermal generation of carriers in the junction space charge layer, the leakage current is proportional to the volume of that layer. Depending upon the junction doping gradient, the space charge layer volume is proportional to the square root or cube root of the reverse-biasing voltage.[4] Since the gate-drain signal voltage is commonly far greater than the gate-source signal voltage, the former will primarily determine the signal-dependent gate leakage current. As an approximation,

$$I_G \sim \sqrt{V_{GD}} \tag{1-45}$$

and the leakage current signal resulting from common-mode voltage will be

$$I_g = I_B \sqrt{E_{icm}}$$

where I_B is the input bias current or static gate leakage current of the FET. Similarly, a differential between the gate leakage currents, the input offset current, develops a differential input signal current.

$$\Delta I_g = I_{OS} \sqrt{E_{icm}}$$

As analyzed for the bipolar transistor case, flow of the input current in unequal signal source resistances produces an error voltage between the stage inputs. This error is amplified by the differential gain of the stage as indicated in

$$E_{o1} - E_{o2} = I_g \, \Delta R_G A_O = I_B \, \Delta R_G A_O \sqrt{E_{icm}}$$

$$\frac{E_{o1} - E_{o2}}{E_{icm}} = \frac{I_B \, \Delta R_G A_O}{\sqrt{E_{icm}}} \qquad \text{due to } \Delta R_G$$

A similar error results from the input voltage produced by unequal gate currents in the signal source resistances as given by

$$\frac{E_{o1} - E_{o2}}{E_{icm}} = \frac{I_{OS}R_G A_O}{\sqrt{E_{icm}}} \qquad \text{due to } I_{OS}$$

Combining the preceding terms results in the complete expression describing the common-mode signal sensitivity of an FET differential stage due to circuit unbalances.

$$\frac{E_{o1} - E_{o2}}{E_{icm}} = \pm A_{Ocm}\left(\frac{\Delta R_S + 1/\Delta g_{fs}}{R_S + 1/g_{fs}} \pm \frac{\Delta R_D}{R_D} \pm \frac{\Delta R_O}{R_O}\right) \pm \frac{A_O}{\sqrt{E_{icm}}}(I_B\,\Delta R_G \pm I_{OS}R_G) \pm j\omega R_D\,\Delta C_{gd} \qquad (1\text{-}46)$$

where

$$A_{Ocm} \doteq -\frac{R_D}{2R_{CM}}$$

Individual unbalances are identified in the above by the differentials, except for I_{OS} which represents a differential current. The first series of terms indicates sensitivities proportional to the fractional unbalances. To consider separately common-mode rejection in terms of error sources significant for differential and single-ended loading of the stage, a common-mode gain is expressed for each case. With differential loading, only the differential output error signal directly adds to the amplified signal and, from Eq. (1-46),

$$A_{cmd} = \pm A_{Ocm}\left(\frac{\Delta R_S + 1/\Delta g_{fs}}{R_S + 1/g_{fs}} \pm \frac{\Delta R_D}{R_D} \pm \frac{\Delta R_O}{R_O}\right) \pm \frac{A_O}{\sqrt{E_{icm}}}(I_B\,\Delta R_G \pm I_{OS}R_G) \pm j\omega R_D\,\Delta C_{gd} \qquad (1\text{-}47)$$

For a single-ended output, both the differential and common-mode output error voltages add to the output signal. Common-mode gain of the balanced stage of Sec. 1.3 identifies the common-mode output term.

$$A_{CM}(j\omega) \doteq A_{Ocm}\frac{1 + 2j\omega R_{CM}C_{gd}}{(1 + j\omega R_G C_{gd})(1 + j\omega R_D C_{gd})} \qquad (1\text{-}39)$$

Under conditions of small circuit unbalances, the common-mode output established by the above gain will be much greater than the differential output defined in Eq. (1-46).

$$A_{cms} = \frac{E_{o1} - E_{o2}}{2E_{icm}} + \frac{E_{ocm}}{E_{icm}} \doteq \frac{E_{ocm}}{E_{icm}}$$

$$A_{cms} \doteq A_{Ocm} \frac{1 + 2j\omega R_{CM}C_{gd}}{(1 + j\omega R_G C_{gd})(1 + j\omega R_D C_{gd})} \tag{1-48}$$

Low-frequency common-mode rejection is defined by dividing the dc portions of the above common-mode gains into the low-frequency differential gain from Sec. 1.1:

$$A_O \doteq \frac{-g_{fs}R_D}{1 + g_{fs}R_S} \tag{1-9}$$

$$CMRR_s = \frac{2g_{fs}R_{CM}}{1 + g_{fs}R_S} \qquad \text{at low frequency} \tag{1-49}$$

As mentioned, the differential output case is much less sensitive to common-mode signals having a common-mode rejection of

$$CMRR_d = \frac{1}{\dfrac{1 + g_{fs}R_S}{2g_{fs}R_{CM}}\left(\dfrac{\Delta R_S + 1/\Delta g_{fs}}{R_S + 1/g_{fs}} \pm \dfrac{\Delta R_D}{R_D} \pm \dfrac{\Delta R_O}{R_O}\right) \pm \dfrac{I_B\,\Delta R_G \pm I_{OS}R_G}{\sqrt{E_{icm}}}} \tag{1-50}$$

at low frequency.

Because of the low and variable output resistance of an FET differential stage, its differentially loaded common-mode rejection ratio is typically an order of magnitude lower than that attained with a bipolar transistor stage. When a resistor is used for common-mode bias as considered, the rejection ratios are of the order of 10:1 and 100:1. Use of a transistor current source for biasing greatly improves common-mode rejection as covered in the next section. The FET-stage common-mode rejection is far less sensitive to signal source resistance and will be superior for resistances above about 50 kΩ.

1.5 Differential-stage Design and Specialized Differential Stages

In the design of a differential stage the interrelationships or various performance characteristics are considered. As a guide in the use of this book for differential-stage design, those basic characteristics affected by each design decision are identified in this section, with references to individual detailed discussions in other sections of the book. In the course of selecting the elements of the stage, compromises are made between gain and bandwidth, input bias current and slewing rate, common-mode rejection and component matching tolerances, and so forth. Additional circuitry which may be used to improve certain operating characteristics without significantly disturbing others is considered to avoid some of the design compromises. Characteristics considered in design include gain, bandwidth, input impedance, common-mode signal range and rejection, input currents and offset voltage, thermal drifts, slewing rate, and noise.

For discussion purposes, the elementary stages of Fig. 1.28 are used. In the bipolar transistor stage, the current level I_C is chosen to limit the input bias currents I_B while also providing sufficient output current and slewing rate. Being the base currents of the transistors, input bias currents are directly proportional to the collector current level. Flow of these base currents in unequal source resistances creates an input error voltage, as does the difference current, or input offset current, in matched signal source resistances. These currents have strong temperature dependences which are described more fully in Chapter 2. When drawn from a preceding differential-stage output, the input currents and their temperature sensitivities cause drift in the loaded stage, as will be covered in Chapter 4. For an FET stage, the drain current I_D is most commonly set at its zero temperature coefficient level. Minimum input voltage drift is achieved when the FETs are biased at this current level, as will be discussed in the next chapter.

Directly related to the stage current level are sensitivity to output current and slewing rate under capacitive load. The small unbalances between output currents considered in Chapter 4 produce sizable voltage

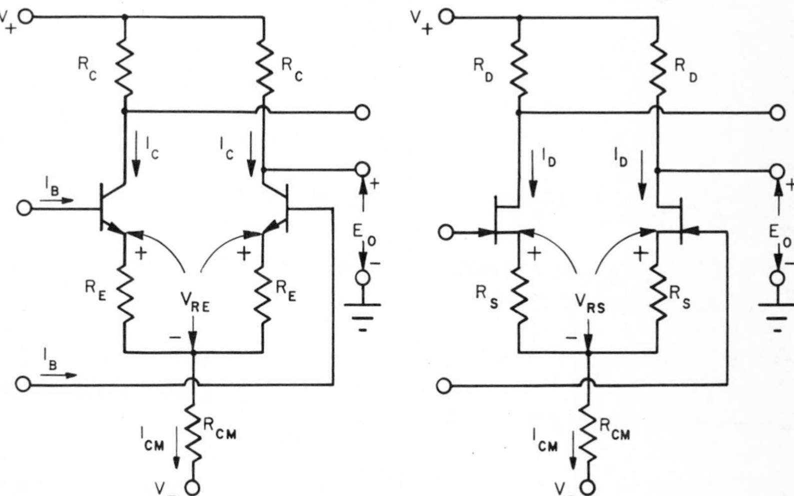

Fig. 1.28 Elementary differential stages.

offsets and drifts between the stage inputs. When driving a capacitive load, as in the case of phase-compensated stages in Chapter 5, the rate of change of output voltage is limited by the current available to charge the capacitor. The current available is commonly the sum of the first-stage collector currents since the input signal can shift the total stage current to one side for charging the capacitor. Thus, the slewing rate of the bipolar transistor stage output is limited to

$$\frac{de_o}{dt} = \frac{2I_C}{C}$$

where C is the capacitance load of the stage. By using the above expression, with consideration for sensitivity to output current, the collector current level can be chosen to minimize the input bias current, input offset current, and their drifts within the limits imposed by slewing rate requirements. When considering large ranges of collector current level, the effect of this current upon small-signal response may become a factor in the above compromise, since the required value of frequency compensation capacitor may vary with the current level. With the inputs at 0 V, the stage current level will be

$$I_C \doteq -\frac{1}{2}\frac{V_{BE} + V_{RE} + V_-}{R_{CM}}$$

or

$$I_D \doteq -\frac{1}{2}\frac{V_{GS} + V_{RS} + V_-}{R_{CM}}$$

Once current levels are chosen, resistances define voltage biases. Collector resistors R_C or drain resistors R_D are chosen to provide the desired gain and output dc level within the bias limits needed for common-mode voltage swing. As defined by Eqs. (1-2) and (1-9), stage gain is proportional to the values of the collector or drain resistors. Emitter or source degeneration resistors, R_E or R_S, stabilize the referenced gains by decreasing sensitivities to variations in dynamic emitter resistance r_e or transconductance g_{fs}. Both r_e and g_{fs} change with temperature, and a wide range of transconductances is found among FETs. In addition, the emitter resistor is significant in increasing differential input resistance defined in Eq. (1-4). With the establishment of the stage current in choosing R_{CM} above, selection of a load resistor, R_C or R_D, fixes the output dc level E_O. For the stages shown, positive common-mode voltage swings decrease the collector-base or gate-drain voltages, and saturation limits the input common-mode voltage range. Unsaturated operation is ensured for collector-base voltages above zero or for a minimum gate-drain voltage defined as V_S in the equation that follows. The positive common-mode voltage limit will be

$$(E_{ICM+})_{max} = \frac{2R_{CM}V_+ + R_C(V_{BE} + V_{RE} + V_-)}{2R_{CM} + R_C}$$

for the bipolar input stage or

$$(E_{ICM+})_{max} = \frac{2R_{CM}(V_+ - V_S) + R_D(V_{GS} + V_{RS} + V_-)}{2R_{CM} + R_C}$$

for the FET stage. Negative common-mode input swing on the illustrated stages is limited by the accompanying decrease in the stage current. By using a transistor current source to set current level, as will be described, the variation of collector or drain current with common-mode swing is greatly reduced and larger common-mode ranges are attained.

Matching the characteristics of the components of one side of the stage to those of corresponding elements on the opposite side improves common-mode rejection and decreases thermal drifts. In the preceding section common-mode errors due to unbalances in resistors and device characteristics were detailed. Lower input offset voltages and related thermal drifts to be discussed in the next chapter are achieved by matching bipolar transistor emitter-base voltages or by using FETs with equal dc parameters. For reduced input offset currents and drifts, equal bipolar transistor betas or FET gate leakage currents are chosen, as also discussed in Chapter 2. Noise performance considerations given in that chapter also dictate careful choice of transistors.

As discussed in Sec. 1.4, signal sensitivity to common-mode input voltages is decreased by the high resistance of the common-mode biasing resistor R_{CM}. Common-mode rejection (CMR) ratios derived there were shown to be proportional to this resistance. A significant improvement in CMR can be made by replacing the common-mode biasing resistor with a transistor current source as shown in Fig. 1.29. With the voltage divider base bias, as shown, a fixed voltage is established on the current source emitter resistor R_E to create a constant output current defined by

$$I_{CM} = \frac{\alpha}{R_E}\left(\frac{R_2(V_+ - V_-)}{R_1 + R_2} - V_{BE}\right)$$

The dynamic output resistance of a bipolar transistor provides common-mode resistance of the 10-MΩ level for two orders of magnitude increase in CMRR with the typical differential stage. To maximize the current source output resistance, the resistance at the transistor base, R_B, should

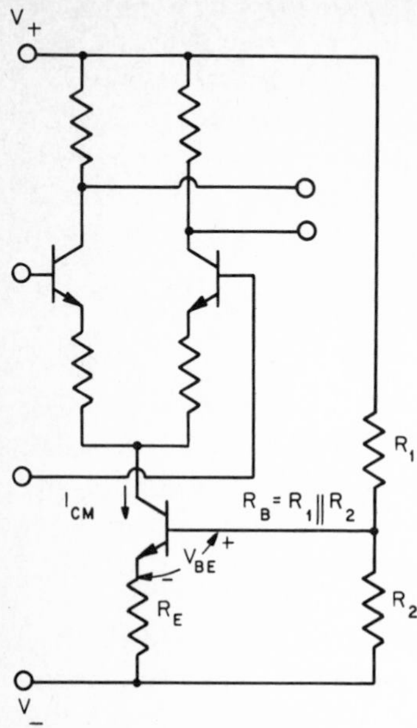

Fig. 1.29 Current-source-biased differential stage.

be low, as indicated in the common-emitter output resistance expression from Sec. 1.1.

$$R_{CM} = R_O' \doteq r_c \frac{R_e + R_B/\beta}{R_e + R_B} \overset{T}{=} 10\ M\Omega \text{ for } \beta \gg 1 \qquad \text{and} \qquad R_e \ll r_c$$

At higher frequencies the output resistance of the current source is bypassed by its collector-base capacitance, as was the differential-stage output resistance considered in Sec. 1.2. From this previous analysis, the equivalent output capacitance of the single current source transistor is twice that of the differential stage composed of two series-connected transistors. From Eq. (1-21)

$$C_O' \doteq C_c \frac{R_e + R_B}{R_e + R_B/\beta}$$

The output impedance will have a pole at

$$f_p \doteq \frac{1}{2\pi R_O' C_O'} = \frac{1}{2\pi r_c C_c} \overset{T}{=} 5\ \text{kHz}$$

To avoid extremely low common-mode impedances at high frequencies, a resistor R_{CM}' is added in series with the current source output, providing

$$Z_{CM} = R_{CM}' + r_c \frac{R_e + R_B/\beta}{R_e + R_B} \frac{1}{1 + j\omega r_c C_c} \tag{1-51}$$

With a transistor current source, common-mode rejection ratios of 50,000:1 and 2,000:1 are commonly achieved for differential and single-ended output bipolar stages, respectively. Analogous ratios for FET stages are of the order of 1,000:1 and 500:1. Common-mode input resistance is also improved by the high common-mode resistance provided by a transistor current source. For this case input resistance is limited only by the input transistor collector-base feedback presented by r_c. From Eq. (1-27),

$$R_{Icm} \doteq \frac{r_c}{2} \qquad \text{for } R_{CM} \gg \frac{r_c}{\beta}$$

at low current levels, R_{Icm} above can reach $10^8\ \Omega$. In general, the input differential stage of an operational amplifier includes a transistor current source.

Even further improvement in common-mode rejection of a differential stage is attained with common-mode feedback. Being the ratio of differential gain to common-mode gain, CMR is increased by feedback which reduces common-mode gain. Each of the common-mode gain expressions for unbalanced stages defined in the preceding section is proportional to the balanced stage common-mode gain.

$$A_{CM} \doteq -\frac{R_C}{2R_{CM}} \qquad \text{or } A_{CM} \doteq -\frac{R_D}{2R_{CM}}$$

Considering the bipolar transistor case, Fig. 1.30 represents a differential stage with common-mode feedback to its biasing current source. At the junction of the second-stage emitters the signal is essentially the common-mode output of the first stage for $R_1 + R_2 \gg r_{e2}$. A resulting signal in the current source base bias which creates a current change is indicated. This feedback signal current opposes that developed by the input common-mode voltage across the current source output resistance R_{CM}. For $R_3 \gg r_{e3}$ and $R_{CM} \gg R_E + r_{e1}$ the resulting current is as shown in the diagram, and one-half of this signal will flow in each collector load. Since the collector resistors are shunted by the second-stage input resistance, the load resistance seen by each collector is

$$R_L = R_C \| R_{Icm2} \doteq R_C \| \beta(R_1 + R_2) \qquad \text{for } R_2 \ll \beta R_3$$

and the common-mode gain is

$$\frac{E'_{ocm}}{E_{icm}} = -\frac{R_L}{2R_{CM}} \frac{(R_1 + R_2)R_3}{R_1R_3 + R_2R_3 + R_2R_L}$$

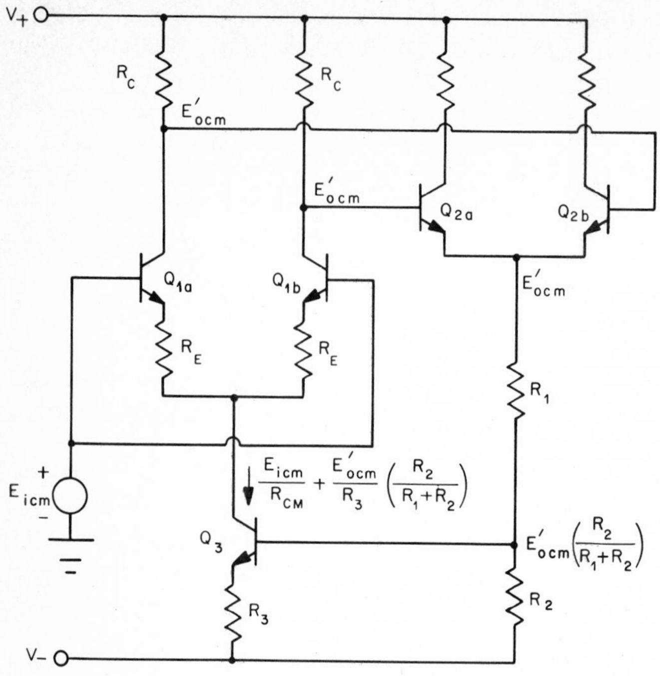

Fig. 1.30 Differential stage with common-mode feedback to its current source.

Expressing the above in terms of the gains resulting with and without feedback displays the degree of improvement. For A'_{CM} the gain with feedback is

$$A'_{cm} = \frac{A_{cm}}{1 + \beta R_2/R_3} \tag{1-52}$$

where

$$A_{cm} = -\frac{R_L}{2R_{CM}}$$

A factor of 3 reduction in common-mode gain is commonly achieved with the described feedback.

Cascode biasing applied to a differential stage provides improved common-mode rejection and lower input leakage current along with decreased input capacitance and greater bandwidth. Bipolar transistor cascode biasing of an FET differential stage is shown in Fig. 1.31. By means of a constant current source and a small resistor R_B, the cascode transistor bases are referenced to the FET sources. The emitter-follower cascode transistors then set the drain voltages at a fixed level above the voltage at the sources. Signal currents in the FETs produce very small gate-drain voltage changes, as the load resistance presented to the drain is the emitter follower output resistance $r_e + R_B/\beta$.

$$\frac{V_{gd}}{E_i} \doteq -\frac{g_{fs}(r_e + R_B/\beta)}{1 + g_{fs}R_S} \overset{T}{\ll} 1$$

Instead, the output signal swing occurs on the collector-base junctions of the cascode transistors which transmit the signal current to the load resistor R_C. For a transistor alpha near unity the signal current is essentially all transmitted and the dc voltage gain will be the same as that achieved with the conventional stage. To prevent input common-mode voltage swing from affecting the cascode circuit, it is biased from a common-mode point, which is the junction of the FET source resistors in this case. When both inputs are shifted by a common-mode input signal, the sources follow the inputs forcing the cascode bias and FET drains to track the common-mode signal at the gates. For $R_{CM} \gg R_S + 1/g_{fs}$ and $R_B \ll \beta r_{ds}$ the drain voltage follows the gate common-mode voltage exactly.

Differential input capacitance of the cascoded circuit is greatly reduced by elimination of Miller-effect multiplication and greater bandwidth

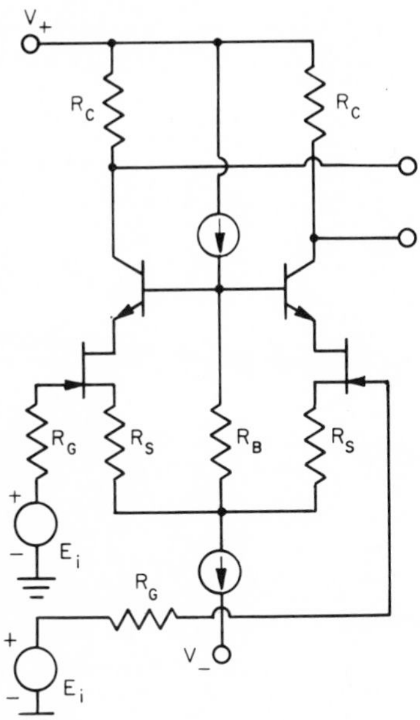

Fig. 1.31 Cascode-biased differential stage.

results. With only the reduced signal across the gate-drain capacitance it shunts only a negligible current from the input as given by

$$I \doteq \left(1 - \frac{g_{fs}(r_e + R_B/\beta)}{1 + g_{fs}R_S}\right) j\omega C_{gd}E_i/2$$

Gate-source capacitance is now one major component of differential input capacitance as represented in the analysis model of Fig. 1.32. For $r_{ds} \gg r_e + R_B/\beta$, the model yields

$$Z_I = \left(2R_S + 2\frac{1 + g_{fs}R_S}{1/r_{gs} + j\omega C_{gs}}\right)\left(\frac{2}{j\omega C_{gd}}\right)$$

Since r_{gs} rarely presents a shunt to typical signal source resistance levels, it may be neglected in the above to simplify the differential input capacitance expression.

$$C_I \doteq \frac{C_{gd}}{2} + \frac{C_{gs}}{2(1 + g_{fs}R_S)} \overset{T}{=} 3\ \text{pF} \tag{1-53}$$

In Sec. 1.2 the differential input capacitance was identified as a major bandwidth limitation for nonzero signal source resistance. Input capacitance shunting of the signal source by a cascoded stage is characterized by a higher pole frequency.

$$f_p = \frac{1}{2\pi(2R_G)C_I}$$

A second pole is added to the stage response by the collector-base capacitances of the cascode transistors. In biasing these transistors, the base bias resistance level is chosen to be small so that C_c bypasses the load

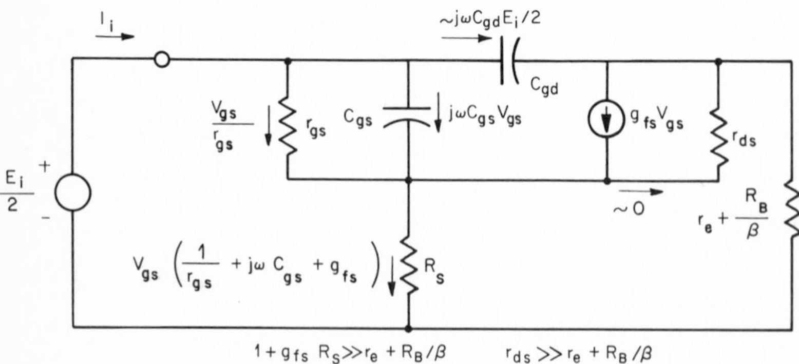

Fig. 1.32 Differential input capacitance analysis model of Fig. 1.31.

resistor to a low impedance point. A resulting time constant is, then, $R_C C_c$, and the pole frequency is

$$f_p \doteq \frac{1}{2\pi R_C C_c}$$

Bypass of the cascode emitter by its own capacitance and C_{gs} creates a negligible pole at much higher frequency than the above because of the low resistance at the emitter. Combining the input capacitance and collector capacitance poles with the low-frequency gain from Eq. (1-9) yields the response of the cascode stage.

$$A(j\omega) = \frac{A_O}{(1 + 2j\omega R_G C_I)(1 + j\omega R_C C_c)} \tag{1-54}$$

$$A_O \doteq \frac{g_{fs}R_C}{1 + g_{fs}R_S} \qquad \text{for } \alpha \doteq 1 \qquad r_{gs} \gg R_G + R_S$$

Common-mode rejection and gate leakage current are also improved by cascode bias. Lower gate leakage current results from the lower gate-drain bias voltage permitted by the cascode circuit. Since signal swing is absorbed by the cascode transistors, the large gate-drain voltage bias normally required to permit common-mode swing is unnecessary. In addition, the elimination of common-mode swing across the FET output resistance and gate-drain capacitance greatly improves common-mode rejection. As discussed in the preceding section, the low and voltage-sensitive output resistance of junction FETS is the major source of common-mode error in FET differential stages. A factor of about 20 higher output resistance is presented to the common-mode swing by the bipolar cascode transistors, and from Eq. (1-7) for a low resistance base bias the output resistance is

$$R'_O \doteq 2r_c \overset{T}{=} 20\ \text{M}\Omega$$

With no gate-drain swing the previous common-mode errors resulting from gate leakage currents, signal source resistances, and gate-drain capacitances do not occur. Each of these errors results from input or output currents generated by gate-drain voltage swing.

By using the high dynamic output resistance of a transistor current source as a load, much higher voltage gain is achieved in a differential stage. Such a dynamic load is shown as Q_3 in Fig. 1.33a. Dynamic load resistances up to 10 MΩ are provided by the transistor without the large load bias voltage drop which would result from a resistor of this size. Being the junction of two collectors, the output bias voltage is not rigidly

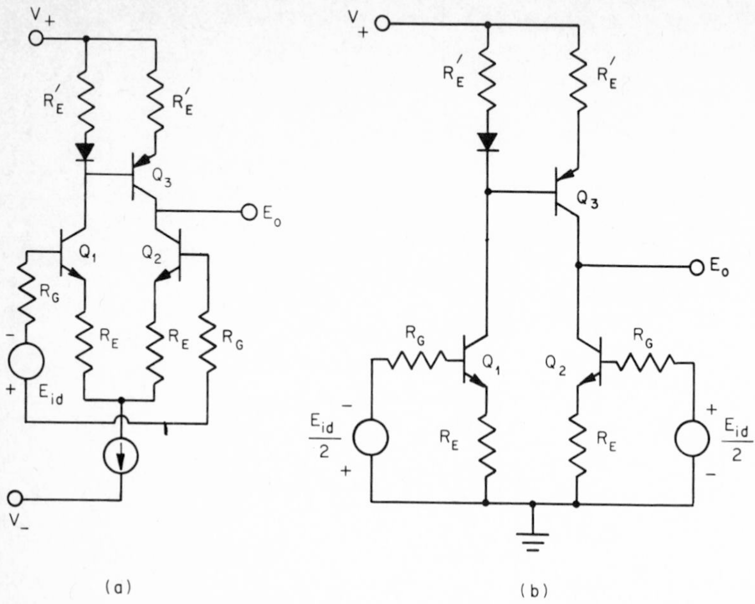

Fig. 1.33 (a) High-gain differential stage using a dynamic load and (b) differential gain analysis circuit.

fixed and will be established by the dc level of the load. Gain is further boosted in the circuit by the signal drive applied to the load transistor base; thus the voltage gains of all three transistors contribute to the stage gain. Neglecting common-mode bias for the differential gain analysis and considering one-half the input signal applied to each input, the analysis circuit of Fig. 1.33b results. From this circuit

$$E_o = A_2 \frac{E_{id}}{2} + A_1A_3 \frac{-E_{id}}{2}$$

$$A = \frac{E_o}{E_{id}} = \frac{1}{2}(A_2 - A_1A_3)$$

Solving for the voltage gain of Q_2 first, the exact common-emitter gain expression used in Eq. (1-1) is needed since the load resistance is comparable to the transistor output resistance. In this case

$$A_{O2} = \frac{-\alpha R_{C2} r_c}{R_e(R_{C2} + r_c) + R_G[R_{C2} + r_c(1 - \alpha)]}$$

To simplify the overall gain result, the same β, α, and r_c will be used for the three transistors. Load resistance R_{C2} seen by Q_2 is the output resistance of Q_3 as described by the common-emitter expression used for Eq. (1-5).

$$R_{C2} = R'_{O3} \doteq r_c \frac{R_{e3} + R_{G3}(1 - \alpha)}{R_{e3} + R_{G3}} \qquad \text{for } r_c \gg R_{G3}$$

Since the exact gain equation is sensitive to the magnitude of α, this last output resistance expression does not approximate $1 - \alpha$ as $1/\beta$, as done in Eq. (1-5). For Q_3, $R_{e3} = R'_E + r_{e3} \triangleq R'_e$ and $R_{G3} = R'_E + r_f = R'_e$ since the diode forward resistance equals the dynamic emitter resistance for equal current levels. Then,

$$R_{C2} \doteq r_c \frac{2 - \alpha}{2} \gg r_c(1 - \alpha)$$

and

$$A_{O2} \doteq \frac{-\alpha r_c}{R_G + R_e[(\alpha - 4)/(3\alpha - 4)]}$$

$$A_{O2} \doteq \frac{-r_c}{R_G + 3R_e} \qquad \text{for } \alpha \doteq 1$$

Voltage gain provided by driving the load transistor base, A_1A_3, is defined by using the simplified gain expression for A_1 and the exact equation for A_3. Since the load resistance presented to Q_1, R'_e for $\beta \gg 1$, is small compared with $r_c(1 - \alpha)$, the simplified common-emitter gain result used in Eq. (1-2) applies.

$$A_{O1} \doteq \frac{-R'_e}{R_e + R_G/\beta} \qquad \text{for } R'_e \ll r_c(1 - \alpha)$$

To solve for the gain of Q_3,

$$A_{O3} = \frac{-\alpha R_{C3} r_c}{R'_e(R_{C3} + r_c) + R_{G3}[R_{C3} + r_c(1 - \alpha)]}$$

For Q_3, $R_{G3} = R'_e$ and the load resistance R_{C3} is the output resistance of Q_2.

$$R_{C3} = R'_{O2} = r_c \frac{R_e + R_G(1 - \alpha)}{R_e + R_G} \qquad \text{for } r_c \gg R_G$$

Combining the last two expressions, the gain reduces to

$$A_{O3} = -\frac{r_c}{R'_e} \frac{\alpha R_e + \alpha^2 R_G/\beta}{(4 - \alpha)R_e + (4 - 3\alpha)R_G}$$

For the terms above containing α directly it is reasonable to assume $\alpha \doteq 1$, for which

$$A_{O3} \doteq -\frac{r_c}{R'_e} \frac{R_e + R_G/\beta}{3R_e + R_G}$$

Substituting the gain results above for the three transistors in the overall gain equation

$$A = \tfrac{1}{2}(A_2 - A_1A_3)$$

provides the final expression:

$$A_O \doteq \frac{-r_c}{3R_e + R_G} \tag{1-55}$$

Voltage gains as high as 10,000 are possible with this stage as evidenced by the gain approximation for low source resistance.

$$A_O \doteq -\frac{r_c}{3R_e} \qquad \text{for } R_G \ll 3R_e$$

Maintaining this high gain will require isolation of the high resistance stage output from lower impedance loads. Resistance at the stage output is the parallel-combination of those found above for Q_2 and Q_3.

$$R'_O \doteq r_e \frac{R_e + R_G/\beta}{3R_e + R_G} \tag{1-56}$$

REFERENCES

1. J. M. Pettit and M. M. McWhorter, *Electronic Amplifier Circuits*, pp. 55, 78–79, McGraw-Hill Book Company, New York, 1961.
2. B. L. Cochrun, *Transistor Circuit Engineering*, p. 385, The Macmillan Company, New York, 1967.
3. C. L. Searle, A. R. Boothroyd, E. J. Angelo, P. E. Gray, and D. O. Pederson, *Elementary Circuit Properties of Transistors*, p. 106, John Wiley & Sons, Inc., New York, 1964.
4. P. E. Gray, D. DeWitt, A. R. Boothroyd, and J. F. Gibbons, *Physical Electronics and Circuit Models of Transistors*, p. 61, John Wiley & Sons, Inc., New York, 1964.

2

INPUT ERROR SIGNALS AND THERMAL DRIFTS OF A DIFFERENTIAL STAGE

Error signals limiting the signal sensitivity of a differential stage result from dc bias and noise. Representing these signals as equivalent input error signals are the input offset voltage, input bias currents, input offset current, input noise voltage, and input noise currents of a differential stage. Thermal dependence of bias characteristics makes the dc error signals drift with temperature so that error compensation is difficult. For direct-coupled (DC) amplifiers, however, the differential stage offers significant reduction in input bias voltage and associated drift over the common-emitter or common-source stage. This is due to the fact that, although input voltages to common-emitter or common-source transistors must include the emitter-base or gate-source voltage bias, these biases are balanced by those of a second transistor in differential stages. Because of this balancing action only the differential bias voltage and bias voltage drift must be supplied as a DC input to establish the desired output quiescent level. The quiescent output of a differential stage is defined for zero voltage between the two output terminals, as is established by applying the input offset voltage between the inputs to supply the differential bias voltage.

Since it is required for biasing, the input offset voltage and its thermal drift present input error voltages to dc signals. These errors are reduced by matching emitter-base or gate-source voltages on the two sides of the stage and also through compensating circuit adjustments as described in the following sections. Additional input error voltage results from the flow of DC input bias currents in signal source resistances. In this chapter these currents are discussed in terms of the bipolar transistor beta or FET gate leakage current governing them and their temperature sensitivities. Just as was the case for input bias voltage, the differential stage provides a balancing bias to reduce error, as the two input bias currents will produce similar error voltages at both inputs whenever equal resistances are presented to the inputs. The error will then be due to the differential input current as represented by the input offset current. Adding to the dc errors discussed above will be ac errors from noise generated in the various components of the stage. Each source of noise is reflected to the stage input, providing an equivalent input noise voltage and input noise current representing noise characteristics of the stage.

2.1 Input Offset Voltage and Drift of Bipolar Transistor Stages

Mismatch of transistor emitter-base forward bias voltages is the source of input offset voltage in the elementary bipolar transistor differential stage. Defined as the input voltage required to provide zero output voltage, the input offset voltage applied as an input signal makes the two collector currents equal, as represented in Fig. 2.1. From the diagram,

$$E_{od} = 0 = -I_{C1}R_C + I_{C2}R_C$$

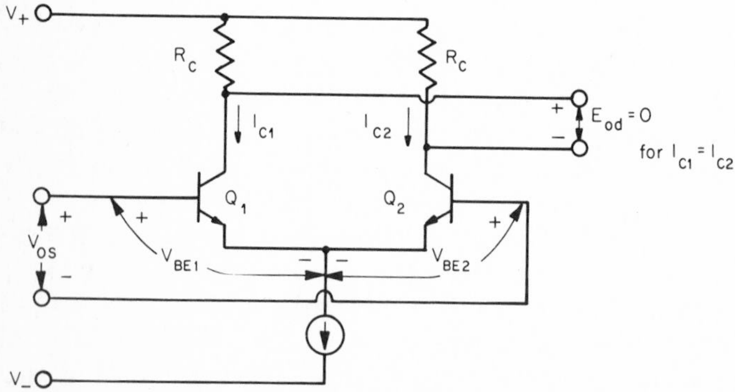

Fig. 2.1 Defining conditions for the input offset voltage of the elementary bipolar transistor differential stage.

where $I_{C1} = I_{C2}$ for balanced collector resistors. The input offset voltage is

$$V_{OS} = V_{BE1} - V_{BE2} \tag{2-1}$$

where V_{BE1} and V_{BE2} result for equal current levels. Selection of transistors for low emitter-base voltage difference readily yields input offset voltages around 1 mV. Similar offset is achieved in monolithic integrated-circuit differential stages through matching provided by simultaneous adjacent fabrication of the two transistors. Unbalanced output loading or mismatched resistors producing additional offset are considered in Chapter 4. Representing a small fraction of the 600-mV level emitter-base voltage, the 1-mV differential is made possible by the excellent consistency of emitter-base voltages among transistors of the same type.

Forward voltage drop of the emitter-base junction is described beginning with the junction equation.[1]

$$I_E = I_S(e^{qV_{BE}/KT} - 1) \tag{2-2}$$

where $I_S \triangleq$ thermal junction leakage current
$q \triangleq$ electron charge $= 1.6 \times 10^{-19}$ coulombs
$K \triangleq$ Boltzmann's constant $= 1.38 \times 10^{-23}$ joules/°K
$T \triangleq$ temperature, °K $=$ °C $+ 273$

Under forward bias the -1 term above is negligible and V_{BE} is found to be

$$V_{BE} = \frac{KT}{q} \ln \frac{I_E}{I_S} \tag{2-3}$$

Considering this result, it is seen that emitter-base voltage is determined by emitter current, thermal leakage current, and physical constants K, T, and q. At a given emitter current level the variations in V_{BE} between transistors of the same type is represented in Eq. (2-3) by the variability of I_S. Differences in thermal leakage currents reflect variations among junction depletion regions and doping levels which are the basic causes of emitter-base voltage differences. Generally the emitter-base voltages of a random group of transistors of the same type at the same current level are within 20 mV of each other. As a result, selection of transistors matched for V_{BE} to within 1 mV of each other is relatively easy. Although Eq. (2-3) accurately describes emitter-base voltage, the thermal leakage current included is masked by much larger surface leakage currents in silicon transistors at room temperature, preventing I_S from being a useful matching parameter.

Input offset voltage drift can be directly related to the input offset voltage of the balanced stage considered. Excellent uniformity of emitter-base voltage temperature coefficients and a high correlation of thermal matching to room temperature V_{BE} matching make low drift possible. To examine this drift the emitter-base voltage thermal characteristics are first resolved. From Eq. (2-3) emitter-base voltage temperature sensitivity is described by

$$\frac{dV_{BE}}{dT} = \frac{V_{BE}}{T} - \frac{KT}{qI_S}\frac{dI_S}{dT}$$

The temperature coefficient of I_S is essentially that of the square of the intrinsic carrier concentration[1] N_i^2, which is

$$N_i^2 = KT^3e^{-E_{go}/KT}$$

where E_{go} is the semiconductor band gap potential. Then,

$$\frac{1}{I_S}\frac{dI_S}{dT} = \frac{1}{N_i^2}\frac{d(N_i^2)}{dT}$$

from which the leakage current temperature coefficient is found to be

$$\frac{1}{I_S}\frac{dI_S}{dT} = \frac{3}{T} + \frac{E_{go}}{KT^2} \tag{2-4}$$

Thus,

$$\frac{dV_{BE}}{dT} = \frac{V_{BE} - E_{go}/q}{T} - \frac{3K}{q} \tag{2-5}$$

Much of this temperature dependence is determined by the physical constants K, E_{go}, and q and will be identical for all bipolar transistors of the same semiconductor type. For silicon, the band gap potential is 1.1 eV, giving

$$\frac{dV_{BE}}{dT} = \frac{V_{BE} - 1.1}{T} - 0.26\ \text{mV/°C} \simeq -2.2\ \text{mV/°C}$$

For a differential pair of transistors the components of V_{BE} thermal drift defined by the physical constants above cancel to give

$$\frac{dV_{OS}}{dT} = \frac{dV_{BE1}}{dT} - \frac{dV_{BE2}}{dT} = \frac{V_{BE1} - V_{BE2}}{T} \tag{2-6}$$

As can be seen in this expression, matching the emitter-base voltages of transistors further reduces input offset voltage drift. In practice, this matching tends to locate two transistors having similar junction geometry and doping characteristics, which have random variables not included in the analysis. Note also from the above expression and the offset voltage expression of Eq. (2-1) that the input voltage drift is predicted by the input offset voltage

$$\frac{dV_{OS}}{dT} = \frac{V_{OS}}{T} \qquad \text{T in degrees Kelvin} \tag{2-7}$$

Input offset voltage drift of 3.3 μV/°C, then, results for each millivolt of offset voltage at room temperature, 298°K. Solution of this differential equation for $V_{OS}(T)$ indicates that the offset voltage due to emitter-base voltage mismatch is a linear function of temperature expressed by

$$V_{OS}(T) = CT$$

where C is a constant. As a result, the offset voltage drift will be a constant over any temperature range.

$$\frac{dV_{OS}}{dT} = C = \frac{V_{OS}}{T}$$

Thus, by knowing the offset voltage at one temperature, the voltage drift curve of a differential stage may be drawn. Poor thermal tracking of other stage elements disturbs this relationship, but generally this is a second-order effect. For the multiple-stage operational amplifier, this correlation between the input offset voltage and its thermal drift is commonly disturbed by the offset and drift effects of following stages. The drift interaction of cascaded stages is considered in Chapter 4.

Typically, the voltage drift of a differential pair of bipolar transistors is reduced about a factor of 700 from the -2.2 mV/°C of a common-emitter silicon transistor. This dramatic accuracy in matching thermal emitter-base voltage drifts is possible because the individual temperature coefficients are partly controlled by physical constants of the semiconductor material and because matching of emitter-base voltages also matches their temperature coefficients. The mechanism of voltage drift reduction due to V_{BE} matching is demonstrated by substitution of the junction equation for V_{BE} in the drift expression of Eq. (2-6).

$$\frac{dV_{OS}}{dT} = \frac{K}{q}\ln\frac{I_{S2}}{I_{S1}} \qquad \text{for } I_{E1} = I_{E2}$$

Note in the above that the residual input voltage drift of a differential pair of transistors is related to a mismatch in the thermally generated leakage currents of the emitter-base junctions. This relationship indicates that differences in emitter-base junction temperature coefficients are largely due to differences in junction geometries and doping profiles.

Such differences result from variations in transistor fabrication masking and diffusion.

Of course, it is not possible to achieve exact emitter-base voltage matching, and additional second-order effects make zero input offset voltage drift unlikely. To compensate for these limitations, the input voltage drift may be further reduced by unbalancing emitter currents to force the emitter-base voltages of a given transistor pair to be equal.[2] The effect of current mismatch on the offset voltage drift is found by combining the drift expression and the junction equation considering unequal currents.

$$\frac{dV_{OS}}{dT} = \frac{K}{q} \ln \left(\frac{I_{E1}}{I_{E2}} \frac{I_{S2}}{I_{S1}} \right)$$

To achieve zero drift, the emitter current balance must then compensate for junction differences expressed in I_{S2}/I_{S1}.

$$\frac{dV_{OS}}{dT} = 0 \qquad \text{for} \qquad \frac{I_{E1}}{I_{E2}} = \frac{I_{S1}}{I_{S2}}$$

Since the thermal leakage I_S is masked by a much greater surface leakage current for silicon transistors at room temperature, leakage current matching except at elevated temperature cannot improve input offset voltage drift. However, much of the remaining drift can be experimentally nulled by creating an appropriate current unbalance. The required unbalance is found by considering the compensating input offset voltage drift resulting from emitter current mismatch separate from that related to I_S. Letting $I_{S1} = I_{S2}$ gives the compensation expression

$$\left(\frac{dV_{OS}}{dT} \right)_C = \frac{K}{q} \ln \frac{I_{E1}}{I_{E2}} \doteq (200\ \mu\text{V}/°\text{C}) \log \frac{I_{E1}}{I_{E2}} \tag{2-8}$$

From the plot of this expression in Fig. 2.2, the current unbalance needed to null a given offset voltage drift is found. As shown, a 10 percent unbalance will cancel an $-8\ \mu$V/°C drift. The resulting drifts are constant with temperature and thus provide a straight-line correction to match the drift caused by V_{BE} mismatch. In addition to predicting drift corrections, these results describe input offset voltage drift which will be caused by unequal current loading on the stage outputs.

Control of current balance for compensation of input offset voltage drift may be achieved by variation of the stage resistor balance, as illustrated in Fig. 2.3. Consistent with the definition of input offset voltage the case shown is for zero differential output voltage. For a multiple-stage DC amplifier this quiescent condition is established for the stage by dc feedback. The dc feedback is typically necessary in such high-gain amplifiers

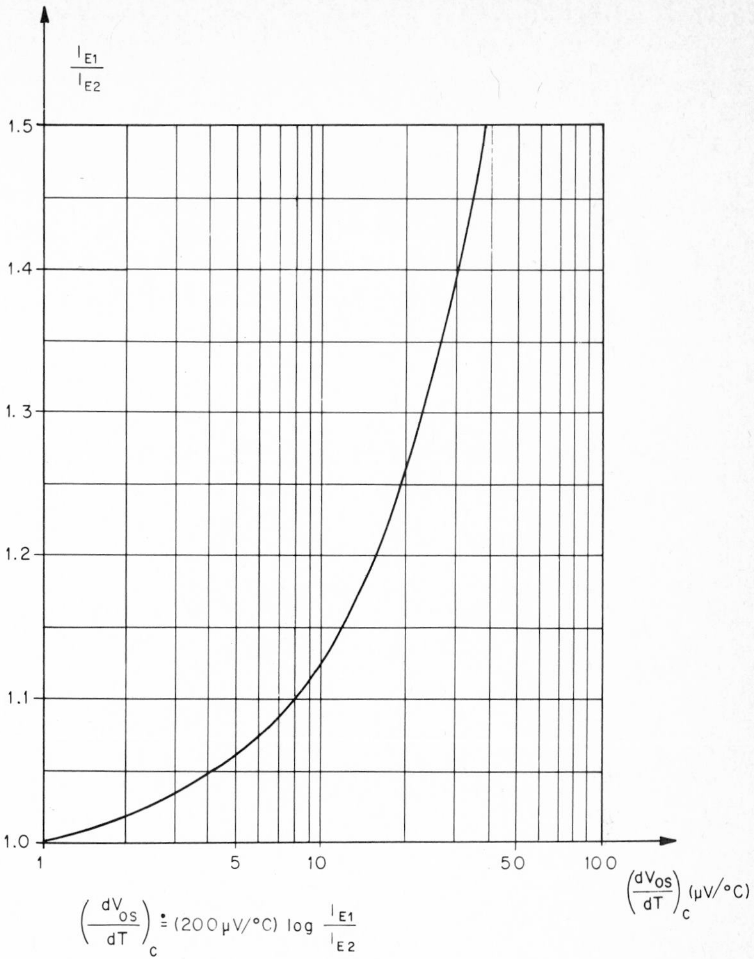

Fig. 2.2 Input offset voltage drift resulting from current unbalance in a bipolar transistor differential stage.

to prevent output saturation due to amplification of offset voltage by the high gain. When feedback is applied to the input, the input stage current is unbalanced to force the voltage between the stage outputs to zero for zero amplifier output. With 0 V between the stage outputs the collector load voltages are equal, and the stage will have a current division determined by the load resistance balance. Potentiometer R_{C1} will, then, vary the ratio of the two transistor currents to adjust input offset voltage drift as expressed by

$$E_{od} = 0 = -I_{C1}R_{C1} + I_{C2}R_{C2}$$

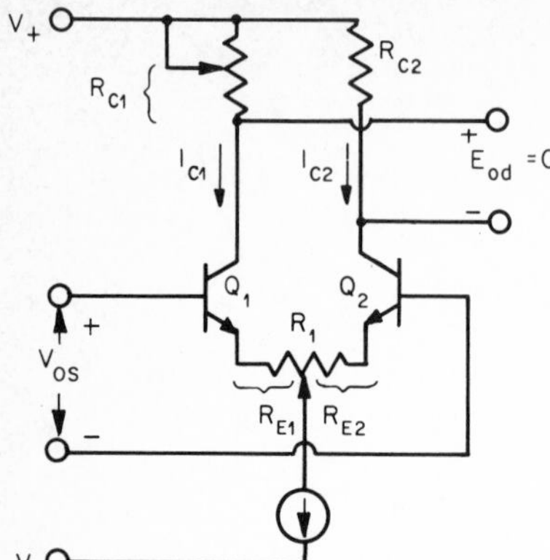

Fig. 2.3 Differential stage with input offset voltage control R_1 and drift control R_{C1}.

Expressing this result in terms of the current balance,

$$\frac{I_{C1}}{I_{C2}} = \frac{R_{C2}}{R_{C1}} \doteq \frac{I_{E1}}{I_{E2}} \qquad \text{for } \alpha \doteq 1$$

Applying the earlier drift compensation equation,

$$\frac{dV_{OS}}{dT} = (200\ \mu V/°C) \log \frac{R_{C2}}{R_{C1}}$$

The resulting total input offset voltage will be nulled by using R_1 to adjust the emitter resistance through which each current flows. In addition to offset from V_{BE} mismatch, a current unbalance results in unbalanced voltages on the transistor dynamic emitter resistances r_{e1} and r_{e2} as well as on R_{E1} and R_{E2}. The total input offset will be

$$V_{OS} = V_{BE1} - V_{BE2} + I_{E1}R_{e1} - I_{E2}R_{e2}$$

where $R_e = R_E + r_e$ and $V_{BE1} - V_{BE2}$ is the offset for equal currents. From these considerations the disadvantage of the often used collector circuit offset voltage balance control can be seen. Unbalance of the first-stage collector load resistances does provide offset voltage balance; however, an additional 3.3 μV/°C drift results for each millivolt of the offset reduced on the emitter-base junctions. Since the emitter resistor balance does not affect the current balance, input offset voltage can be nulled by emitter resistor trim without disturbing drift.

2.2 Input Offset Voltage and Drift of FET Stages

Input offset voltage and drift of an FET differential stage are typically far greater than those resulting with the bipolar transistor stage. To

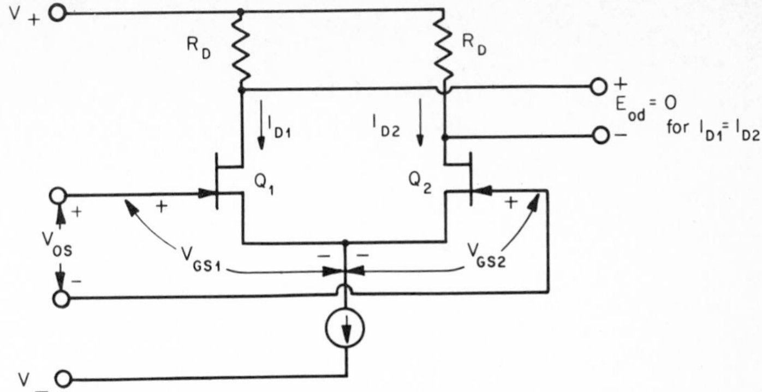

Fig. 2.4 Defining conditions for the input offset voltage of the elementary FET differential stage.

make use of the high FET input resistance and the low input gate current described in the next section, a compromise must be made with dc errors due to input offset voltage and drift. Less uniform dc characteristics and thermal drifts make FET matching more difficult and less accurate. Compensation of these error signals does, however, greatly improve dc performance. Input offset voltage of the basic FET differential stage results primarily from mismatch of gate-source voltages. Additional offset due to loading and resistor unbalance will be considered in Chapter 4. For zero output voltage the input to the stage is the input offset voltage, and the two drain currents are balanced for equal drain resistors, as described earlier for the bipolar transistor stage of Fig. 2.1. These conditions are applied to the basic function FET stage of Fig. 2.4 to define the input offset voltage:

$$V_{OS} = V_{GS1} - V_{GS2} \tag{2-9}$$

where V_{GS1} and V_{GS2} result for equal currents. Matching FET gate-source voltages for low input offset voltage is more difficult than is matching bipolar transistor emitter-base voltages. The bipolar transistors commonly have a 20-mV spread of V_{BE} drops, but gate-source voltages of FETs of the same type measured at the same current may vary by several volts. By considering characteristics specified for FETs, the potential range of gate-source voltages can be predicted. An expression for the gate-source voltage may be found, starting with the defining relationship for drain current[3] given below with typical small-signal n-channel FET parameters.

$$I_D = I_{DSS}\left(1 - \frac{V_{GS}}{V_P}\right)^2 \tag{2-10}$$

where

$$I_{DSS} \triangleq I_D \qquad \text{for } V_{GS} = 0$$
$$\stackrel{T}{=} 2 \text{ mA to } 20 \text{ mA}$$
$$V_P \triangleq \text{pinchoff voltage} = V_{GS} \qquad \text{for } I_D = 0$$
$$\stackrel{T}{=} -2 \text{ to } -4 \text{ V}$$

From the above

$$V_{GS} = V_P\left(1 - \sqrt{\frac{I_D}{I_{DSS}}}\right)$$

Thus, specified ranges of pinchoff voltage and I_{DSS} may be used to define the possible range of V_{GS} values at a given drain current. Because of the large range normally encountered, a V_{GS} match to within 20 mV is common, as compared with V_{BE} matching to within 1 mV. Nonlinear FET output characteristics further complicate the matching of FETs. This nonlinearity results from the voltage-sensitive output resistance discussed in Sec. 1.4 and causes the gate-source voltage established by the biasing drain-current to vary with drain-source voltage. When common-mode signals vary V_{DS}, the offset voltage will change unless the nonlinear output characteristics are also matched. Additional input offset voltage difficulties are encountered with MOSFETs (metal-oxide semiconductor FETs) because of time-sensitive gate-source voltages. As a result of these surface-related instabilities, random input offset shifts of several millivolts occur, and MOSFETs are seldom used successfully for differential stages. Only junction FETs will be considered here.

Considering the derivative of the offset of Eq. (2-9) with respect to temperature, input offset voltage drift is seen to be the result of inaccurate thermal tracking of gate-source voltages as expressed by

$$\frac{dV_{OS}}{dT} = \frac{dV_{GS1}}{dT} - \frac{dV_{GS2}}{dT}$$

For an FET biased at a fixed drain current, as in a differential stage, V_{GS} changes with temperature because of two temperature-sensitive characteristics.[4] The first is the width of the thermally generated depletion layer at the junction of the gate and channel. Thermal variation of the built-in voltage causes a 2.2 mV/°C increase in the magnitude of V_{GS} for fixed drain current. Temperature sensitivity of the majority carrier mobility is a second thermal factor affecting V_{GS}. This factor would reduce drain current by 0.6 to 0.8 percent per degree centigrade except for the fixed drain current bias. Instead, V_{GS} decreases in magnitude to maintain constant current as related by the FET transconductance[3] g_{fs} defined by

$$g_{fs} \triangleq \frac{dI_D}{dV_{GS}} = \frac{2I_{DSS}}{V_P}\left(1 - \frac{V_{GS}}{V_P}\right) \qquad (2\text{-}11)$$

Considering an average 0.7%/°C effect for the mobility variation, the gate-source voltage drift of an n-channel FET, for which V_{GS} is negative, becomes

$$\frac{dV_{GS}}{dT} = -2.2 \text{ mV/°C} + \frac{7 \times 10^{-3} I_D}{g_{fs}} \tag{2-12}$$

As displayed by this result, the two components of V_{GS} thermal drift are opposing, and for some value of drain current the drift will be zero. Zero temperature coefficient biasing occurs for drain current I_{DZ} at which level the transconductance is g_{fsZ}, and the relationship between the two parameters is defined by setting the drift of Eq. (2-12) to zero to arrive at

$$\frac{I_{DZ}}{g_{fsZ}} = 0.315 \text{ V} \tag{2-13}$$

The gate-source voltage at its zero temperature coefficient bias point is found by substituting the drain current expression [Eq. (2-10)] and the transconductance expressed by Eq. (2-11) in the above condition. The result is

$$V_{GSZ} = V_P - 0.63 \text{ V} \overset{T}{=} -1.5 \text{ to } -3.5 \text{ V} \tag{2-14}$$

where V_P is the pinchoff voltage. Using the preceding result, the zero-drift drain current is found with the drain current expression of Eq. (2-10) to be

$$I_{DZ} \doteq \frac{0.4 I_{DSS}}{V_P^2} \overset{T}{=} 200 \text{ to } 600\ \mu\text{A} \tag{2-15}$$

Selection of an FET for zero drift at a desired drain current level is made by applying this relationship to specified pinchoff voltages and I_{DSS} levels. To evaluate the gain attainable at this bias point, the V_{GSZ} result of Eq. (2-14) is substituted into the transconductance defined in Eq. (2-11), yielding

$$g_{fsZ} = \frac{1.26 I_{DSS}}{V_P^2} \overset{T}{=} 600 \text{ to } 2{,}000\ \mu\text{mhos} \tag{2-16}$$

Typically, g_{fsZ} is much smaller than the maximum attainable transconductance which results for zero gate-source voltage given by

$$g_{fs} = \frac{2 I_{DSS}}{V_P} \qquad \text{for } V_{GS} = 0$$

Unless a pinchoff voltage of −0.63 V is available, for which $V_{GSZ} = 0$, g_{fsZ} will be less than the maximum transconductance.

By combining the general drain current expression and the defined I_{DZ} or V_{GSZ} with the gate-source voltage drift of Eq. (2-12), the drift may be expressed as a function of other FET characteristics. First, substitution of the drain current expression [Eq. (2-10)] and the transconductance defined by Eq. (2-11) into the drift result display the drift dependence upon pinchoff voltage and gate-source voltage. For this case

$$\frac{dV_{GS}}{dT} = -2.2 \text{ mV/°C} + 3.5 \times 10^{-3}(V_P - V_{GS}) \tag{2-17}$$

Rewriting the drain current equation to resemble the last term above provides

$$V_P - V_{GS} = V_P \sqrt{\frac{I_D}{I_{DSS}}}$$

Combining the last two expressions, the gate-source voltage drift is related to the ratio of drain current to I_{DSS} by the expression

$$\frac{dV_{GS}}{dT} = -2.2 \text{ mV/°C} + 3.5 \times 10^{-3} V_P \sqrt{\frac{I_D}{I_{DSS}}} \tag{2-18}$$

As a measure of gate-source voltage drift resulting from bias at other than the zero-drift point, the zero-drift drain current of Eq. (2-15) is combined with Eq. (2-18) to give

$$\frac{dV_{GS}}{dT} = -2.2 \text{ mV/°C} \left(1 - \sqrt{\frac{I_D}{I_{DZ}}}\right) \tag{2-19}$$

Deviation of biased drain current from the zero-drift level I_{DZ} produces drift as shown in Fig. 2.5. Note that for a 10 percent deviation in drain current from I_{DZ} the gate-source voltage drift exceeds 100 μV/°C. Similarly, gate-source voltage drift may be expressed as a function of the difference of V_{GS} from its zero-drift level. Using V_{GSZ} as defined in Eq. (2-14) with Eq. (2-17),

$$\frac{dV_{GS}}{dT} = 3.5 \times 10^{-3}(V_{GSZ} - V_{GS}) \tag{2-20}$$

Using the above four drift results, the input offset voltage drift of an FET differential stage is expressed as functions of basic FET characteristics and zero-drift parameters. Being the difference in gate-source voltage drifts, input offset voltage drift from Eq. (2-17) will be

$$\frac{dV_{OS}}{dT} = 3.5 \times 10^{-3}(\Delta V_P - V_{OS}) \tag{2-21}$$

where

$$\Delta V_P \triangleq V_{P1} - V_{P2} \qquad V_{OS} \triangleq V_{GS1} - V_{GS2}$$

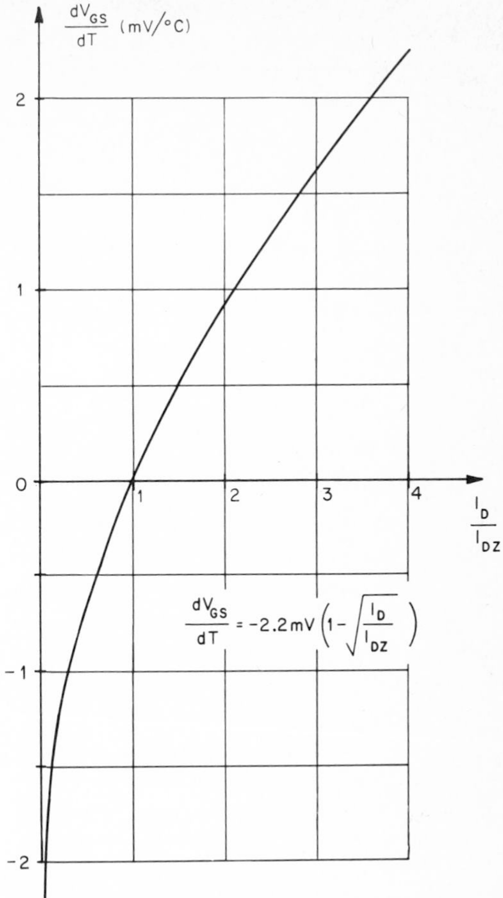

Fig. 2.5 Gate-source voltage thermal drift for drain currents about the zero-drift point.

As indicated, the drift is directly related to mismatch in pinchoff voltages and gate-source voltages. Each millivolt of V_{GS} or V_P difference at a given current level results in 3.5 μV/°C of input offset voltage drift, which is comparable to the 3.3 μV/°C experienced under similar conditions with bipolar transistors. Differential drift dependence upon I_{DSS} mismatch and drain current unbalance is displayed, using Eq. (2-18) to derive

$$\frac{dV_{OS}}{dT} = 3.5 \times 10^{-3}\left(V_{P1}\sqrt{\frac{I_{D1}}{I_{DSS1}}} - V_{P2}\sqrt{\frac{I_{D2}}{I_{DSS2}}}\right) \tag{2-22}$$

Whereas input offset voltage drift is directly related to V_P and V_{GS} mismatches, the drift varies only as the differences of the square roots of I_{DSS} levels and drain currents. Considering Eq. (2-19), dependence upon I_{DZ} match is expressed by

$$\frac{dV_{OS}}{dT} = -2.2\ \text{mV/°C}\left(\sqrt{\frac{I_{D1}}{I_{DZ1}}} - \sqrt{\frac{I_{D2}}{I_{DZ2}}}\right) \tag{2-23}$$

A 1 percent mismatch in I_{DZ} of the drain currents develops approximately 11 μV/°C input offset voltage drift. Alternatively, the drift can be written in terms of the gate-source voltages at the zero-drift bias points. From Eq. (2-20),

$$\frac{dV_{OS}}{dT} = 3.5 \times 10^{-3}(\Delta V_{GSZ} - V_{OS})$$
$$\Delta V_{GSZ} \triangleq V_{GSZ1} - V_{GSZ2} \tag{2-24}$$

To achieve minimum input offset voltage drift with an FET differential stage, the FETs are biased at the zero-drift current I_{DZ} and they are selected for matched characteristics. By choosing a drain current at the average zero-drift level for the type of FET used, the individual gate-source voltage drifts will be minimized. The resulting low V_{GS} drifts are more nearly equal than would occur for larger individual drifts at bias points away from the zero temperature coefficient point. Lower differences in gate-source voltage drifts or lower input voltage drift result for stages biased as close as possible to I_{DZ}. Examination of drift equation (2-22) indicates that input offset voltage drift in a stage having balanced drain currents is nulled by matching I_{DSS} levels and pinchoff voltages. Under these conditions equal gate-source voltages are predicted by Eq. (2-10), and the I_{DZ} levels described by Eq. (2-15) would be equal. However, variables not accounted for in these basic describing relationships also contribute to differences in gate-source voltage drift. As a result, increasing the temperature-matching accuracy does not continue to improve input offset voltage drift at lower levels, and matching at several temperatures is frequently performed. Typical input offset voltage drifts achieved are around 30 μV/°C, almost an order of magnitude higher than commonly attained with bipolar transistors.

To achieve lower input offset voltage drift with FET differential stages, drain currents are unbalanced to create additional drift which compensates that of the FETs and any other circuit unbalance. Drift equation (2-23) is rewritten to consider only the drain current unbalance, giving the drift compensation expression

$$\left(\frac{dV_{OS}}{dT}\right)_C = -2.2\ \text{mV/°C}\,\frac{\sqrt{I_{D1}} - \sqrt{I_{D2}}}{\sqrt{I_{DZ}}} \tag{2-25}$$

Using this expression, the input offset voltage drift corresponding to a range of drain current ratios is plotted in Fig. 2.6. Comparison of this curve with the analogous curve for bipolar transistor stages in Fig. 2.2 reveals that input offset voltage drift of the FET differential stage is more than an order of magnitude more sensitive to current unbalance. A 1 percent drain current unbalance creates an 11 μV/°C drift, whereas a collector current unbalance of 14 percent is required to develop the same drift. As a result, unbalanced current loading at the stage outputs

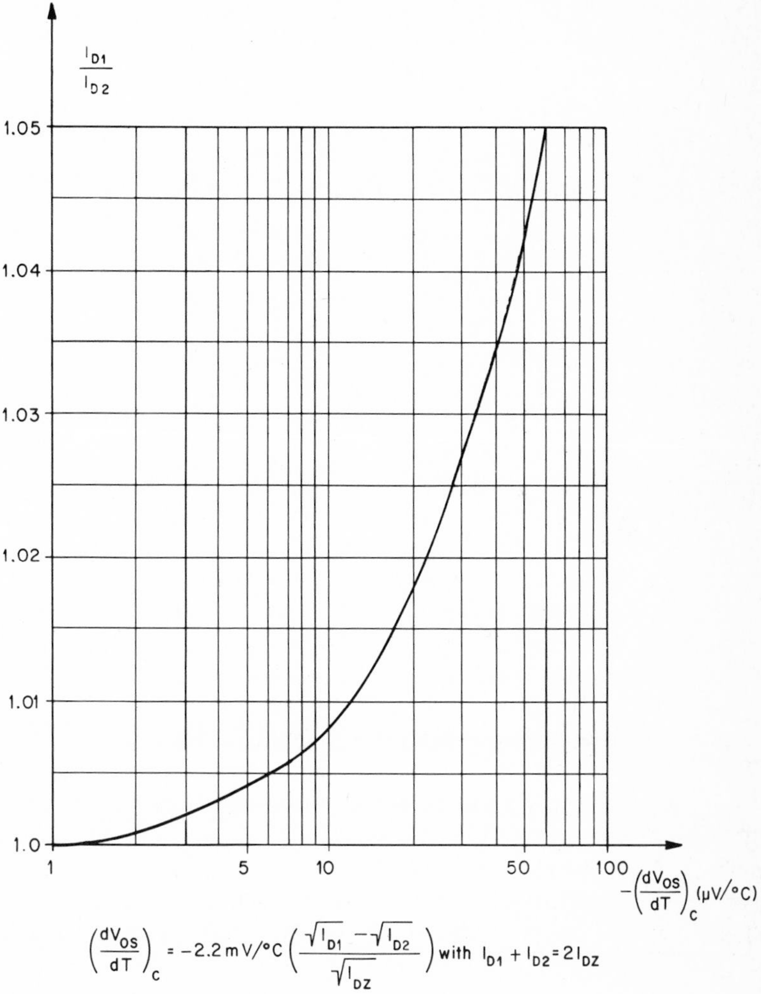

Fig. 2.6 Input offset voltage drift resulting from current unbalance in an FET differential stage.

will be far more serious to the FET stage. Since drain current unbalances of less than 5 percent are adequate for compensating typical drifts, an approximation to the drift compensation equation can be made. For FETs biased at their zero-drift point

$$I_{D1} + I_{D2} = 2I_{DZ}$$

and for

$$\Delta I_D = I_{D1} - I_{D2}$$

$$I_{D1} = I_{DZ} + \frac{\Delta I_D}{2} \qquad I_{D2} = I_{DZ} - \frac{\Delta I_D}{2}$$

Combined with compensation equation (2-25), these terms give

$$\left(\frac{dV_{OS}}{dT}\right)_C = -2.2\ \text{mV/°C}\left(\sqrt{1 + \frac{\Delta I_D}{2I_{DZ}}} - \sqrt{1 - \frac{\Delta I_D}{2I_{DZ}}}\right)$$

Small unbalances defined for $\Delta I_D \ll I_{D1} \simeq I_{DZ}$ permit use of the approximation

$$\sqrt{1 - x} \doteq 1 - \frac{x}{2}$$

Then

$$\left(\frac{dV_{OS}}{dT}\right)_C \doteq -1.1\ \text{mV/°C}\,\frac{\Delta I_D}{I_{DZ}} \qquad \text{for } \Delta I_D \ll I_{DZ} \qquad I_{D1} \simeq I_{D2} \simeq I_{DZ} \tag{2-26}$$

Also, since $\Delta I_D \ll I_{D2}$ for small unbalances, $I_{DZ} \doteq I_{D2}$, transforming the last result to

$$\left(\frac{dV_{OS}}{dT}\right)_C \doteq -1.1\ \text{mV/°C}\left(\frac{I_{D1}}{I_{D2}} - 1\right) \qquad \text{for } \Delta I_D \ll I_{DZ} \qquad I_{D1} \simeq I_{D2} \simeq I_{DZ} \tag{2-27}$$

In this expression the drift as a function of the drain current ratio approximates the curve of Fig. 2.6.

To vary the drain current balance for input offset voltage drift compensation a variable drain resistor such as R_{D1} in Fig. 2.7 may be used. Analogous to the bipolar transistor stage controls, the changing drain resistance affects both input offset voltage and its drift, whereas the source resistance balance control R_1 affects only the offset. With the input offset voltage between the inputs as shown, the dc differential output is zero, resulting in equal voltages on the drain resistors, or

$$I_{D1}R_{D1} = I_{D2}R_{D2}$$

$$\frac{I_{D1}}{I_{D2}} = \frac{R_{D2}}{R_{D1}}$$

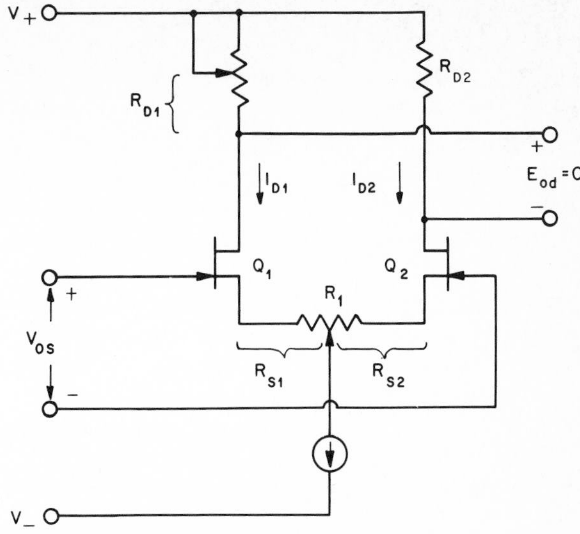

Fig. 2.7 FET differential stage with input offset voltage control R_1 and drift control R_{D1}.

A drain current ratio controlled by the above drain resistor ratio provides a new compensation equation from Eq. (2-27) given by

$$\left(\frac{dV_{OS}}{dT}\right)_C = -1.1\ \text{mV/°C}\left(\frac{R_{D2}}{R_{D1}} - 1\right) \qquad \text{for } \Delta I_D \ll I_{DZ}$$

$$I_{D1} \simeq I_{D2} \simeq I_{DZ}$$

Unbalanced currents in the FETs increase the input offset voltage, which is nulled by unbalancing the source resistances. The resulting total input offset voltage is

$$V_{OS} = V_{GS1} - V_{GS2} + I_{S1}\left(R_{S1} + \frac{1}{g_{fs1}}\right) - I_{S2}\left(R_{S2} + \frac{1}{g_{fs2}}\right)$$

where V_{GS1} and V_{GS2} result for equal currents. Because of the high sensitivity of FET differential-stage input offset voltage drift to drain current unbalance, drain resistor adjustment is a poor choice for an input offset voltage trim.

2.3 Input Bias Currents, Offset Current, and Drifts

Currents in the inputs of a differential stage result from the base bias currents of bipolar transistors or from the gate leakage currents of FETs. For dc-coupled amplifiers these input currents must be supplied by the signal source except for that portion which might be supplied by additional biasing circuits. Such current supply circuits can provide approximately the input current required; however, the base or gate current is independently set by device characteristics and other stage biasing. As a result, dc-coupled differential stages will always draw some input bias currents from the signal sources, thus developing voltage drops on the signal source resistances. If the signal source resistances presented to the two inputs are not equal, a differential error voltage is developed between the inputs. Similarly, unequal bias currents creating an input offset current produce a differential input error voltage even when source resistances are identical. Each of the input error currents has a thermal drift giving rise to temperature-sensitive input error voltages and making error compensation difficult. Because of the very low gate leakage current of FETs, FET differential stages typically provide a factor of 1,000 lower error voltages on signal source resistances than do comparable bipolar transistor stages.

Since the input bias currents are the base currents in a bipolar transistor differential stage, the currents and their temperature sensitivities result from transistor beta characteristics. The input bias current is related by

$$I_B = \frac{I_C}{\beta} \tag{2-28}$$

and its drift may be expressed in terms of this bias current and the fractional beta drift.

$$\frac{dI_B}{dT} = -\left(\frac{1}{\beta}\frac{d\beta}{dT}\right) I_B \tag{2-29}$$

The temperature dependence of beta is due to changing minority carrier lifetime in the base region,[5] and the temperature coefficient averages about 0.5%/°C for typical silicon transistors above 25°C and 1.5%/°C below 25°C. Input bias current drift is then approximated by

$$\frac{dI_B}{dT} \doteq CI_B(25°\mathrm{C}) \qquad \begin{cases} C = -0.005/°\mathrm{C}, & T > 25°\mathrm{C} \\ C = -0.015/°\mathrm{C}, & T < 25°\mathrm{C} \end{cases} \tag{2-30}$$

Comparing these approximations with a typical bias current temperature dependence, Fig. 2.8 shows the actual and approximate curves of input bias current versus temperature. Deviations from the typical curve shown result from variations in doping characteristics and from leakage currents. In the case of low-current stages the collector-base leakage current produces noticeable additional reduction of input bias current at high temperatures. Typical small-signal silicon transistors under 10 V of collector-base bias have an additional decrease in input bias current reaching about 30 nA at 125°C because of this leakage.

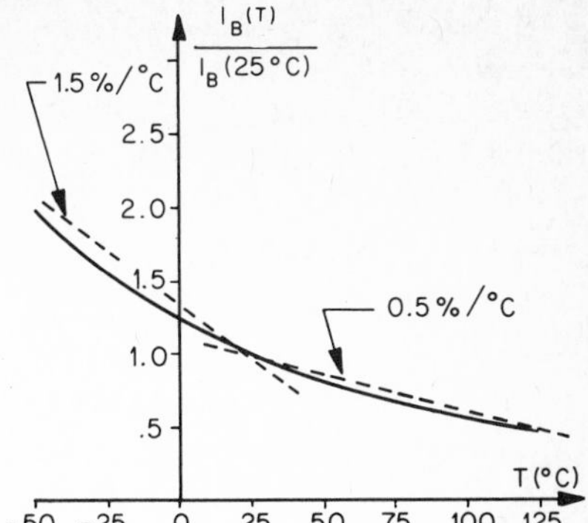

Fig. 2.8 Typical temperature dependence of input bias current for a bipolar transistor differential stage.

Input offset current of a bipolar transistor differential stage having balanced collector currents results from mismatched transistor betas, and its drift results from inexact thermal tracking of the resulting unequal input bias currents. As defined,

$$I_{OS} = I_{B1} - I_{B2} = \Delta I_B \tag{2-31}$$

where I_{B1} and I_{B2} are the bias currents at the two inputs. Expressing input offset current in terms of beta, the bias current of Eq. (2-28) is combined with the preceding relationship to obtain

$$I_{OS} = \frac{(\beta_2 - \beta_1)I_C}{\beta_1\beta_2} \qquad \text{for } I_{C1} = I_{C2} = I_C$$

This expression may be simplified using a beta average to get

$$\beta_1 = \beta - \Delta\beta/2$$

and $\beta_2 = \beta + \Delta\beta/2$, where $\Delta\beta = \beta_2 - \beta_1$. Using these relationships,

$$I_{OS} = \frac{\Delta\beta}{\beta^2 - \Delta\beta^2/4} I_C$$

Considering a small mismatch and relating I_{OS} to I_B,

$$I_{OS} \doteq \frac{\Delta\beta}{\beta} I_B \qquad \text{for } \Delta\beta \ll \beta \tag{2-32}$$

From this result the beta match required to ensure a given offset current level can be seen. Drift of the input offset current is described by using the derivative of the offset current definition in Eq. (2-31) with the bias current drift expressed in Eq. (2-29) to get

$$\frac{dI_{OS}}{dT} = \left(\frac{1}{\beta_1}\frac{d\beta_1}{dT}\right) I_{B1} - \left(\frac{1}{\beta_2}\frac{d\beta_2}{dT}\right) I_{B2}$$

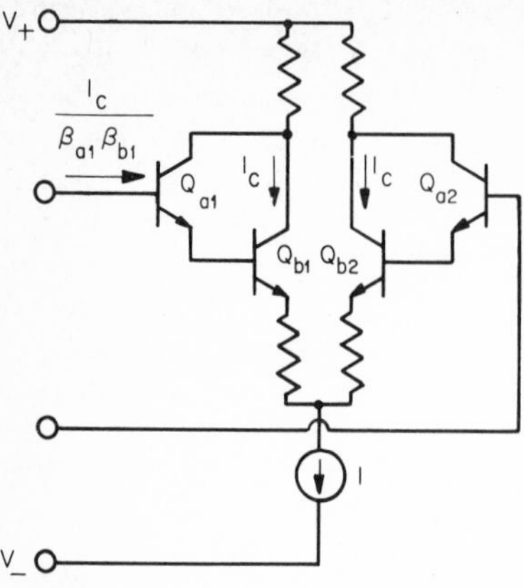

Fig. 2.9 Darlington differential stage.

Two factors creating the above drift are unequal input bias currents and unequal beta temperature coefficients. For transistors of the same type with only small beta differences, the beta temperature coefficients will be very nearly equal as expressed by

$$\frac{1}{\beta_1}\frac{d\beta_1}{dT} \doteq \frac{1}{\beta_2}\frac{d\beta_2}{dT} = \frac{1}{\beta}\frac{d\beta}{dT}$$

Essentially all the input offset current drift will then be due to input bias current mismatch, and the drift expression simplifies to

$$\frac{dI_{OS}}{dT} \doteq \left(\frac{1}{\beta}\frac{d\beta}{dT}\right) I_{OS} \tag{2-33}$$

From the previous discussion of the temperature coefficient of beta as used for Eq. (2-30), this drift expression can be approximated by

$$\frac{dI_{OS}}{dT} \doteq CI_{OS}\ (25°C) \qquad \begin{cases} C = -0.005/°C, & T > 25°C \\ C = -0.015/°C, & T < 25°C \end{cases} \tag{2-34}$$

As indicated, low input offset current drift results for low initial offset current. A beta match to within 5 percent provides input offset current and drift which are one-twentieth of the input bias current and drift, as indicated by comparing the last result and the offset current expression of Eq. (2-32) with the corresponding input bias current relationship of Eq. (2-30).

Reduction of the input currents and their drifts is achieved with the Darlington differential stage of Fig. 2.9. Input bias currents are reduced by the current gain of the added transistors to

$$I_B \doteq \frac{I_C}{\beta_a\beta_b} \overset{T}{=} 0.01\,\frac{I_C}{\beta_b}$$

Since input offset current is related to the product of the beta mismatches, it is not reduced as greatly as the input bias currents. Applying the combined beta of a Darlington pair to the previous offset current expression, Eq. (2-32) results in an offset current expression given by

$$I_{OS} = \frac{\Delta(\beta_a\beta_b)}{\beta_a\beta_b} I_B$$

Similarly, thermal drift of the added current gain limits input bias current drift reduction as expressed by the derivative of the preceding expression for I_B as

$$\frac{dI_B}{dT} = -\left(\frac{1}{\beta_a}\frac{d\beta_a}{dT} + \frac{1}{\beta_b}\frac{d\beta_b}{dT}\right) I_B$$

The two betas will be unequal because of highly different collector current levels, but the fractional changes or temperature coefficients of the current gains will be nearly equal as expressed by

$$\frac{1}{\beta_a}\frac{d\beta_a}{dT} \doteq \frac{1}{\beta_b}\frac{d\beta_b}{dT} = \frac{1}{\beta}\frac{d\beta}{dT}$$

The input bias current drift is, then,

$$\frac{dI_B}{dT} \doteq -2\left(\frac{1}{\beta}\frac{d\beta}{dT}\right) I_B \qquad \text{for the Darlington stage}$$

As expressed above, the temperature coefficient of the input bias current for a Darlington differential stage is twice that of the conventional stage described by Eq. (2-29). Applying this result to the drift approximation of Eq. (2-30) results in

$$\frac{dI_B}{dT} \doteq 2CI_B \qquad \begin{cases} C = -0.005/°C, \ T > 25°C \\ C = -0.015/°C, \ T < 25°C \end{cases}$$

The input offset current drift of the conventional stage given in Eq. (2-33) is the product of the beta temperature coefficient and the input offset current. The corresponding drift of the Darlington stage will be the product of the compound beta temperature coefficient and the new offset current. From the above bias current drift result, the compound beta temperature coefficient is twice that of the single beta, giving

$$\frac{dI_{OS}}{dT} \doteq -2\left(\frac{1}{\beta}\frac{d\beta}{dT}\right) I_{OS} \doteq 2CI_{OS}$$

Frequently the input offset current drift of the Darlington stage is worse than predicted above. Greater drift results from the difficulty of matching betas of the added transistor which carry only the base currents of the original transistors. As the base currents of the original transistors vary with temperature, a beta match in the added transistors is needed over a wide range of collector currents. Such a match is more difficult to achieve. In addition, V_{BE} thermal tracking under these conditions is less accurate. With a reduction in input bias current $\beta_a:1$ and a 2:1 increase in the effective beta temperature coefficient, the Darlington stage typically reduces input bias current drift by a factor of $0.5\beta_a:1$. Compound beta matching errors limit reduction of input offset current to about a factor of $0.9\beta_a:1$ whereas the temperature-dependent collector current and added beta temperature coefficient reduce improvement in the related current drift to around $0.3\beta_a:1$.

Accompanying these improved input current characteristics is about a factor of 3 increase in input offset voltage and drift from the added pair of emitter-base junctions and their varying currents. Input offset voltage is greater because of the V_{BE} mismatch of the added transistors and their unequal currents as created by the beta mismatch of the high-current transistors. Sensitivity of V_{BE} to current level is expressed in Eq. (2-3) which can be applied to predict the offset voltage resulting from the beta difference. In this case

$$\Delta V_{BEa} = \frac{KT}{q} \ln \frac{I_{Ea1}}{I_{Ea2}} \qquad \text{for } I_{Sa1} = I_{Sa2}$$

Since the low-current transistors are biased by the high-current transistor base currents, $I_{Ea} \doteq I_C/\beta_b$, and

$$\Delta V_{BEa} \doteq \frac{KT}{q} \ln \frac{\beta_{b2}}{\beta_{b1}}$$

A 5 percent beta unbalance creates an added 1.5-mV input offset voltage at room temperature, where $KT/q \doteq 25$ mV. Input offset voltage drift resulting from the associated current unbalance would be 5 μV/°C, from the curve of Fig. 2.2. Similar increases in offset and drift result just from the normal mismatch and drift of the added transistors.

Input bias and offset currents of bipolar transistor stages are frequently reduced by compensating circuits which supply canceling input currents. Compensation resistors R_1 and R_2 of Fig. 2.10a supply currents opposing the transistor base currents. By selecting these resistors for a given stage, a factor of 10 reduction in input currents is readily achieved. However, the compensating currents are sensitive to power supply variations and input signals and provide compensation at only one temperature. Added input bias current sensitivity to signals reflects decreased input resistance, which is typically only significant for the high common-mode input resistance. Both the input currents and their thermal drifts are decreased by using the compensation circuitry shown in Fig. 2.10b, for which the base currents of compensation transistors Q_3 and Q_4 are supplied to the inputs as indicated. Since the compensation transistors are the opposite of the amplifier transistors in conductivity type, the base currents of the amplifier and compensation transistors are of opposite polarity and tend to cancel. Compensation is provided by matching the collector currents and betas of the compensation transistors to those of the amplifier

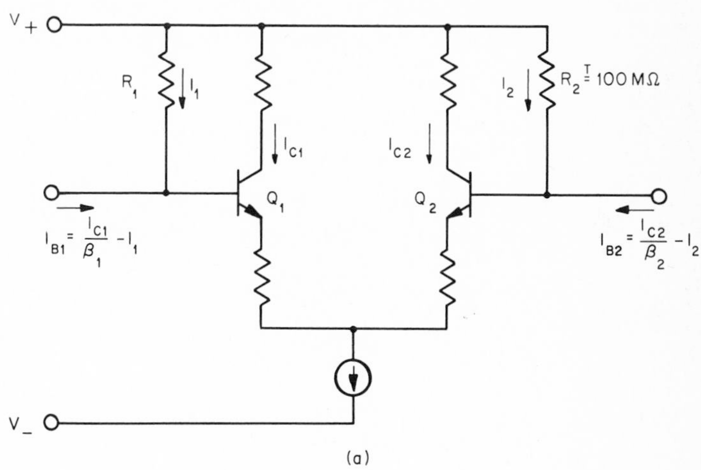

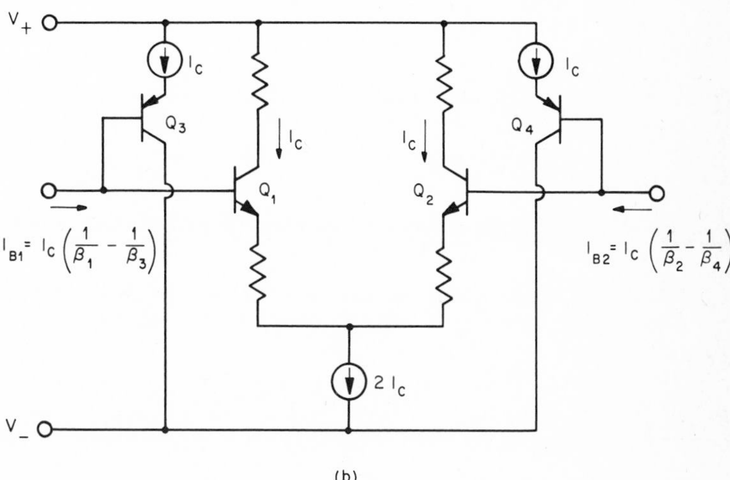

Fig. 2.10 Input bias current compensation using (a) resistors and (b) transistors of opposite conductivity type.

transistors. However, matching of opposite-type transistors for a range of temperatures is difficult. Beta matching within 10 percent and temperature tracking within 20 percent provide a 10:1 reduction in input bias current and a factor of 5 decrease in thermal drift.

Even greater reductions in the input bias currents and their drifts have been achieved by deriving a compensating current from the actual characteristics which generate the bias currents. Such compensating currents are supplied by the two techniques of Fig. 2.11. In Fig. 2.11a[6]

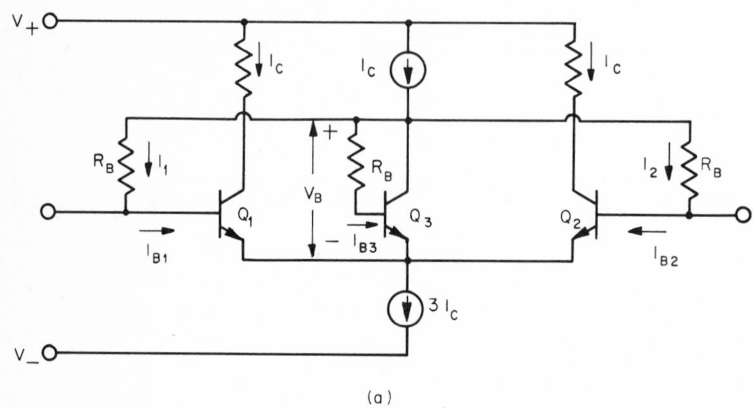

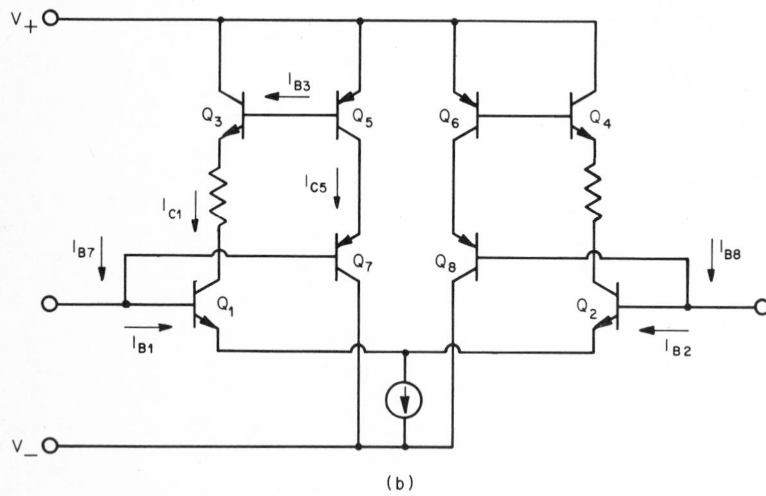

Fig. 2.11 Input bias current compensation using (a) a derived voltage bias (*Pearlman, U.S. Patent* 3,230,468) and (b) a derived compensation current (*Graeme, U.S. Patent* 3,551,832).

the resistor compensation approach is refined by deriving the resistor bias voltage V_B from a base current matched to the input bias current. For a base current in Q_3 which matches those of Q_1 and Q_2, the three transistors are biased with equal collector currents and their betas are matched. Then the resistor bias voltage will be

$$V_B = I_{B3}R_B + V_{BE3}$$

where

$$I_{B3} = I_{B1} = I_{B2}$$

This voltage is supplied to the two compensation resistors as expressed by

$$\begin{aligned} V_B &= I_1R_B + V_{BE1} \\ &= I_2R_B + V_{BE2} \end{aligned}$$

If the emitter-base voltages of the three transistors are then matched, the last three expressions define the compensation currents as

$$I_1 = I_2 = I_{B3}$$

Since I_{B3} matches the input bias currents I_{B1} and I_{B2}, they will be matched by the compensation currents. In this way an order of magnitude reduction in input bias currents and drifts is readily achieved. However, differential input resistance is somewhat shunted by the compensation resistors.

The compensation technique of Fig. 2.11b[7] provides feedback control to the compensation circuit of Fig. 2.10b and eliminates its difficult matching requirement. As will be seen, the feedback control avoids the need for beta matching between transistors of opposite conductivity type. The use of transistors rather than high value resistors for bias current compensation makes this approach more suitable for monolithic fabrication. On the left side of the stage the feedback begins with Q_3 which conducts the same current as does Q_1. If the betas of these two n-p-n transistors are matched, their base currents will be nearly equal as expressed by

$$I_{B3} = \frac{\beta_1}{\beta_3 + 1} I_{B1}$$

As shown, I_{B3} biases Q_5, establishing a collector current which is the emitter current of Q_7. Since Q_5 and Q_7 carry the same current, beta matching of these p-n-p's will also establish approximately equal base currents which are related by

$$I_{B7} = \frac{\beta_5}{\beta_7 + 1} I_{B5}$$

where $I_{B5} = I_{B3}$. Then, from the expressions above for I_{B3} and I_{B7}, the input bias current and the compensation current are related by

$$I_{B7} = \frac{\beta_1\beta_5}{(\beta_3 + 1)(\beta_7 + 1)} I_{B1}$$

$$I_{B7} \doteq I_{B1} \qquad \text{for } \beta_3 \gg 1 \qquad \beta_7 \gg 1 \qquad \beta_1 = \beta_3 \qquad \text{and } \beta_5 = \beta_7$$

The latter beta matching and magnitude requirements are readily achieved to provide an order of magnitude reduction in the input bias currents and their drifts.

Three to five orders of magnitude reduction in input bias and offset currents is achieved with junction FETs as compared with bipolar transistor stages biased at similar current levels. Input current drifts are also reduced; however, less improvement is attained because of the greater temperature sensitivity of gate leakage current as compared with that of beta. Nevertheless, a factor of at least 100 reduction in drift magnitude is typical over the widest operating temperature range. For high source resistance applications, the FET stage will produce less total dc input error than the bipolar transistor stage in spite of the higher input offset voltage and drift associated with FETs. By greatly lowering bipolar transistor collector currents, input currents comparable to those of FETs at high temperature can be provided, but the ability of the stage to supply current and thereby its slewing rate decrease proportionally with the collector current.

Superior input current characteristics of the junction FET result from gate currents composed of the low reverse leakage current of a junction. Input bias current of the FET stage is this leakage current, which may be related to the commonly specified gate leakage current I_{GSS}, tested with source shorted to drain under a reverse bias V_{GSS}. As described in Sec. 1.4, the gate leakage of silicon FETs does not correspond to the junction equation expressing

$$I = I_S(e^{qV/KT} - 1)$$

$$I = I_S \qquad \text{under reverse bias}$$

Reverse leakage current included in this equation as I_S is that due to thermal generation of carriers in the charge neutral regions and does not include that generated in the space charge layer.[1] For silicon junctions the reverse-biasing voltage makes the latter leakage component far greater than I_S. Since it is thermally generated leakage current from the space charge layer of the reversed-biased junction, the gate leakage current is proportional to the volume of the layer. This was considered proportional to the square root of the reverse voltage in Eq. (1-45). Under normal operation the entire gate-channel junction is not under as great a reverse bias as the I_{GSS} test voltage since V_{GS}, and generally V_{GD}, will be less than the test voltage V_{GST}. From drain to source the reverse-biasing voltage across the junction will vary from V_{GD} to V_{GS} in a linear manner for uniform channel doping. Although common planar-diffused FETs have uniform doping in directions parallel to the junction, the doping profile into the channel is graded, making the linear variation of reverse bias only a first-order approximation. A linear variation permits the effective reverse bias to be approximated by the average of V_{GD} and V_{GS} for space-charge-layer volume considerations. Gate leakage current as related to I_{GSS} will be proportional to the square root of the ratio of this average voltage to V_{GST}. When the average reaches the test voltage V_{GST}, the leakage current will become I_{GSS}. From these conditions, the relationship of gate leakage current or input bias current to I_{GSS} is approximated by

$$I_B = I_G \doteq I_{GSS}\sqrt{\frac{V_{GD} + V_{GS}}{2V_{GST}}}$$

$$I_{GSS} \overset{T}{=} 10\text{ pA} \qquad @\ V_{GST} = 30\text{ V and T} = 25°\text{C} \tag{2-35}$$

$$I_B \overset{T}{=} 5\text{ pA} \qquad @\ 25°\text{C}$$

For only small differences in voltages on the two FETs of a stage the input offset current will be essentially due to I_{GSS} mismatch expressed by

$$I_{OS} = \Delta I_G \doteq \Delta I_{GSS}\sqrt{\frac{V_{GD} + V_{GS}}{2V_{GST}}} \tag{2-36}$$

The temperature sensitivity of FET differential-stage bias currents is far lower than that predicted from leakage current characteristics governing I_S. Equation (2-4) defined the temperature coefficient of I_S as

$$\frac{1}{I_S}\frac{dI_S}{dT} = \frac{3}{T} + \frac{E_{go}}{KT^2}$$

which makes the familiar prediction that junction reverse leakage current will double for every 10° temperature increase with germanium or for every 6°C increase with silicon. However, the rate of generation of carriers in the space charge layer is much less temperature-sensitive than that in the neutral regions,[1] making the dominant gate leakage current in silicon more stable. In practice, input bias currents of differential stages using silicon FETs very nearly double every 10°C rather than every 6°C. This characteristic is expressed by

$$I_B(T) \doteq I_B(T_1)2^{(T-T_1)/10} \tag{2-37}$$

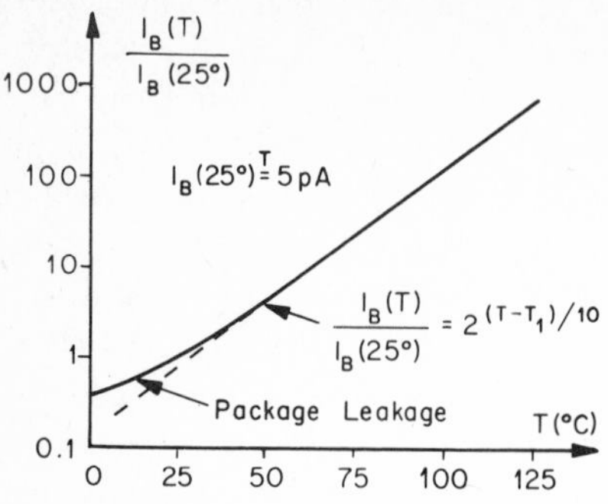

Fig. 2.12 Typical input bias current temperature dependence for an FET differential stage. (*Empirical result from Burr-Brown Model* 3307/12C.)

Leakage currents associated with the FET packages add to the thermally generated leakage, preventing significant low temperature reduction of input bias currents. At room temperature roughly one-third of the input currents are package leakage currents, resulting in a temperature dependence as displayed in Fig. 2.12. Note that the 1,000:1 increase in gate leakage current from room temperature to 125°C develops about an 800:1 increase in input bias current.

2.4 Equivalent Input Noise Voltage and Currents

Input offset voltage, input currents, and their drifts constitute dc input error signals, whereas equivalent input noise voltage and currents represent ac input error signals in a differential stage. The noise generated within a differential stage places a limit on the signal sensitivity of the circuit since any amplified signal must be of sufficient magnitude to be detectable over the inherent noise level of the amplifier. To some extent differential stages can be designed with improved noise performance by selection of components and attention to biasing current levels. Source resistances will also significantly affect overall noise performance as noise currents from the inputs flow in these resistances. Ideally, in a signal amplifying device, the signal enters the device undisturbed by noise, and the limit placed upon the sensitivity of the amplifier is set by the noise it introduces itself. The noise associated with a differential amplifier stage, as is the case with all amplifying devices, is the result of three noise phenomena: Schottky or shot noise; Johnson or thermal noise; and flicker or 1/f noise. Shot noise is due to the discrete particle nature of current carriers in semiconductors. Although a current of average value I_{DC} may be flowing, the random arrival times of the charges generates a nondeterministic time-varying noise current[8] i_n. Associated with this current is a spectral density $S_i(f)$ defining the mean-square noise current as

$$\overline{i_n^2} = \int_{f_1}^{f_1+\Delta f} S_i(f)\, df$$

The spectral density of shot noise is constant from zero frequency to frequencies of the order of the inverse of the charge transit time, which may extend into the infrared region. For semiconductors the density is

$$S_i(f) = 2qI_{DC}$$

where q is the charge of an electron, 1.6×10^{-19} coulombs. Applying this constant spectral density to the preceding expression defines the mean-square shot noise current in semiconductors as

$$\overline{i_n^2} = 2qI_{DC}\, \Delta f \tag{2-38}$$

where Δf is the system bandwidth. In addition to shot noise, a semiconductor generally exhibits additional noise at low frequencies. The origin of this excess noise in general is not well understood; however, the power spectral density of the noise exhibits an 1/f behavior and is traditionally called flicker noise. Although its noise generating mechanisms are in question, the power spectral density may be measured and the device characterized experimentally.

Thermal or resistance noise is caused by the random motion of charges which is independent of their mean or average motion. For an ohmic resistance at thermal equilibrium the thermal noise has an associated constant spectral density which is expressed for mean-square voltage noise by

$$S_v(f) = 4KTR$$

where K is Boltzmann's constant, 1.38×10^{-23} joules/°K, and T is temperature in degrees Kelvin, °C + 273. Integration of the spectral density over the system bandwidth as before results in a mean-square thermal noise voltage of a resistor of

$$\overline{e_n^2} = 4KTR\, \Delta f \tag{2-39}$$

where R is the ohmic resistance. As illustrated in Fig. 2.13a, the thermal noise source can be represented schematically as a voltage generator in series with a noise-free resistance or the equivalent noise current generator in parallel with the resistance. Forming a Norton equivalent circuit of the noise voltage representation, the mean-square thermal noise current of the resistor will be

$$\overline{i_n^2} = \frac{\overline{e_n^2}}{R^2} = \frac{4KT\, \Delta f}{R} \tag{2-40}$$

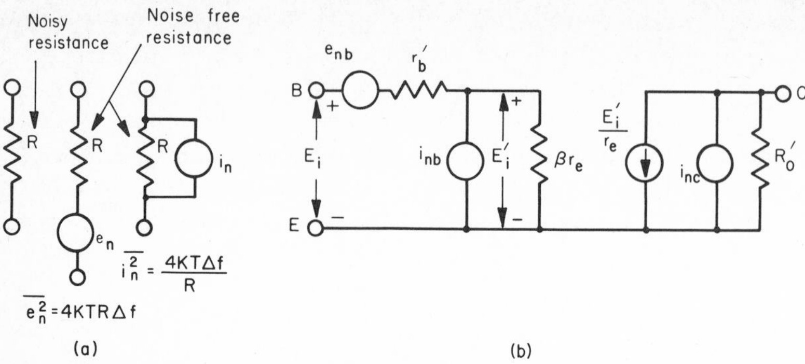

Fig. 2.13 Equivalent noise representations of (a) a resistor and (b) a bipolar transistor.

In considering the overall noise behavior of an amplifier, many noise sources, both internal and external to the amplifier, must be taken into account. An important result from noise theory states that a set of mean-square noise voltages or currents may be added to obtain the cumulative mean-square value, provided the noise sources are uncorrelated. In characterizing an amplifier it is therefore convenient to separate the noise sources as statistically independent and therefore uncorrelated noise generators.

In the preceding analyses a system passing only frequencies from f_1 to $f_1 + \Delta f$ has been considered. Since practical systems do not have infinite response selectivity, other frequencies are transmitted at varying gains described by the transfer function of the system. For an input noise voltage having spectral density $S_{vi}(f)$ applied to a network having a transfer function $H(j\omega)$ the output mean-square noise voltage will be

$$\overline{e_{no}^2} = \int_0^\infty S_{vi}|H(j\omega)|^2 \, df$$

and for S_{vi} the constant spectral density associated with shot or thermal noise

$$\overline{e_{no}^2} = S_{vi} \int_0^\infty |H(j\omega)|^2 \, df$$

For a single resistor-capacitor low-pass filter the output noise would be as described below.

$$H(j\omega) = \frac{1}{1 + j\omega RC}$$

$$\overline{e_{no}^2} = \frac{S_{vi}}{4RC}$$

This specific case is particularly useful in amplifier noise considerations since the response of an amplifier stage is commonly a single-pole response.

Associated with the average currents in a bipolar transistor are shot noise sources described by the defining relationship, Eq. (2-38). Represented in the pi transistor model in Fig. 2.13b, the noise currents are modeled by uncorrelated noise sources related to the collector and base currents by[8]

$$\overline{i_{nb}^2} = 2qI_B\,\Delta f \tag{2-41}$$

$$\overline{i_{nc}^2} = 2qI_C\,\Delta f \tag{2-42}$$

In these expressions leakage currents are neglected as they are commonly small in comparison with the biasing currents. Also represented in the circuit model is a noise voltage source accounting for the thermal noise of the base spreading resistance r_b'. From the thermal noise relationship of Eq. (2-39), the mean-square thermal noise voltage generated in the base is

$$\overline{e_{nb}^2} = 4KTr_b'\,\Delta f \tag{2-43}$$

Applying this model to the basic differential stage leads to the result in Fig. 2.14 which can be used to develop equivalent noise sources at the stage inputs. Identical characteristics are assumed for the two transistors as shown since typical unbalances will result in only minor noise differences. Output resistance R_O' is omitted in the complete diagram, as is valid for the general case with $R_C \ll r_c(1 - \alpha)$ from Chapter 1. Noise generators i_{nrc} and e_{ne} represent the thermal noise characteristics of R_C and R_E, respectively.

By reflecting each noise source to the stage input as an equivalent noise generator it is possible to simplify the noise representation to one noise voltage and one noise current at each input. Being directly at the inputs, these net equivalent noise generators can be compared with the input signals in describing errors due to noise. Since the collector current shot noise and the collector resistor thermal noise current both flow through the collector resistor, it is of interest to compare their mean-square magnitudes to observe which noise component is predominant. The ratio is

$$\frac{\overline{i_{nc}^2}}{\overline{i_{nrc}^2}} = \frac{2qI_C\,\Delta f}{4KT\,\Delta f/R_C} = \frac{I_C R_C}{2KT/q}$$

At room temperature KT/q is near 25 mV; the collector resistor drop $I_C R_C$ is commonly 2.5 V, and

$$\frac{\overline{i_{nc}^2}}{\overline{i_{nrc}^2}} \doteq 500$$

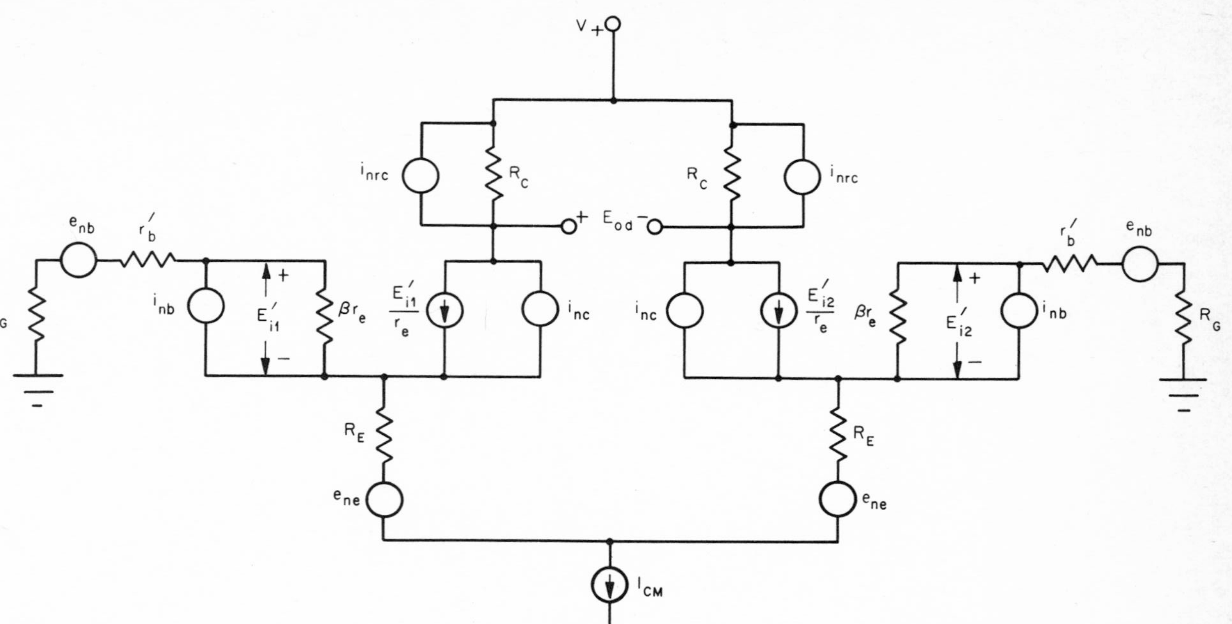

Fig. 2.14 Noise model for a bipolar transistor differential stage.

Being much smaller than the shot noise current i_{nc}, the thermal noise current of the collector resistance results in negligible noise on the resistor and can be omitted. For a balanced differential stage the noise current associated with the common-mode biasing current I_{CM} divides equally between the two sides of the stage to develop equal noise signals on the collector resistors. No differential output noise then results from this noise source, unless the small resulting base noise currents flow in very high source resistances. For the general case, noise generated with I_{CM} is neglected.

To replace the collector noise current generators with equivalent input noise generators the circuit is separated at the junction of the emitter resistors. If the high resistance of the common-mode current source I_{CM} is neglected, the equivalent resistance presented by the differential stage to the emitter of the left side of the stage is the output resistance of an emitter follower formed by the right side of the stage. The equivalent resistance is

$$R_e + \frac{R_G + r_b'}{\beta}$$

where

$$R_e = R_E + r_e$$

Replacing the remainder of the stage by this resistance results in the equivalent circuit of Fig. 2.15a for which the circuit has been redrawn to combine the resistances in series with the emitter of the left side. Only the collector current shot noise generator is considered in the circuit to resolve its equivalent input noise voltage generator e_{nc} as represented in Fig. 2.15b. The equivalent input noise voltage is calculated by solving for the output noise current it creates.

$$E_{i1}' = \frac{e_{nc}(2\beta R_e + R_G + r_b')}{2(\beta R_e + r_b' + R_G)}$$

$$i_{nc} = \frac{E_{i1}'}{2R_e + (R_G + r_b')/\beta}$$

From these expressions the input noise voltage equivalent to the collector noise current is found to be

$$e_{nc} = 2i_{nc}\left(R_e + \frac{r_b' + R_G}{\beta}\right)$$

where

$$\overline{i_{nc}^2} = 2qI_C\,\Delta f$$

Total equivalent input noise generators are found by combining the various equivalent input components. The voltage noise representation at the input includes the above e_{nc}, the thermal noise e_{nb} of the base

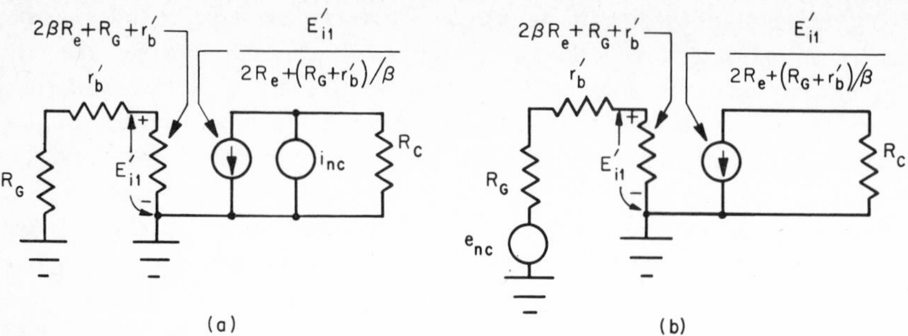

Fig. 2.15 The equivalent circuit of a differential stage referred to one input with (a) the collector current noise generator and (b) the equivalent input noise voltage generator.

spreading resistance, and the reflected thermal noise of the emitter resistor. The emitter resistance noise generator can be transferred directly to the input as in either position it will produce the same emitter current. Adding the mean-square noise voltages from the three sources results in the equivalent input noise voltage on one side of the circuit:

$$\overline{e_{ni1}^2} = \overline{e_{nc}^2} + \overline{e_{nb}^2} + \overline{e_{ne}^2}$$

where e_{nc} is defined from the preceding equation, and the thermal noises of r'_b and R_E follow from the basic thermal noise equation (2-39) to give

$$\overline{e_{ni1}^2} = 4qI_C\,\Delta f\left(R_e + \frac{r'_b + R_G}{\beta}\right)^2 + 4kT(R_E + r'_b)\,\Delta f$$

Similarly, the right side of the circuit has an input noise representation, and it is equal to the one given above for the balanced stage considered. Addition of the two input noise generators results in the total equivalent differential input noise of the stage, e_{ni}, which is

$$\overline{e_{ni}^2} = \overline{e_{ni1}^2} + \overline{e_{ni2}^2}$$

$$e_{ni} = 2\sqrt{2qI_C\,\Delta f\left(R_e + \frac{r'_b + R_G}{\beta}\right)^2 + 2KT(R_E + r'_b)\,\Delta f} \qquad (2\text{-}44)$$

where

$$R_e = R_E + r_e$$

Noise currents in the differential-stage inputs are the shot noise currents associated with the base currents as represented in the stage model of Fig. 2.14. From the base current noise expressed by Eq. (2-41), the input noise currents are

$$i_{ni1} = i_{nb1} = \sqrt{2qI_{B1}\,\Delta f} \qquad (2\text{-}45a)$$

$$i_{ni2} = i_{nb2} = \sqrt{2qI_{B2}\,\Delta f} \qquad (2\text{-}45b)$$

In addition to flowing in the source resistances, the input noise currents flow through the emitter resistors of Fig. 2.14. The effect of the base current noise in the emitter resistors is small compared with that produced by the collector noise current since the two currents as defined in Eqs. (2-41) and (2-42) are related by beta. As a result, the base current generators may be returned to common instead of to the emitters. Also, the noise voltage resulting from the base current noise in the small r_b' is negligible in comparison with the preceding equivalent input noise voltage. Because of this, the base current noise generators can be returned to the opposite side of r_b' which is the base terminal. Having reflected all noise sources to the inputs of a differential stage, it may be represented by a noise-free stage preceded by the equivalent input noise generators, as is done in Fig. 2.16.

From the preceding expressions for equivalent input noise voltage and currents, several observations can be made concerning the noise characteristics of a bipolar transistor differential stage. First it is noted that both the input noise voltage and the input noise currents are proportional to the square root of the bandwidth considered. A second conclusion is that input noise voltage is minimized by lowering collector current. Also, low collector current and high beta result in small input shot noise

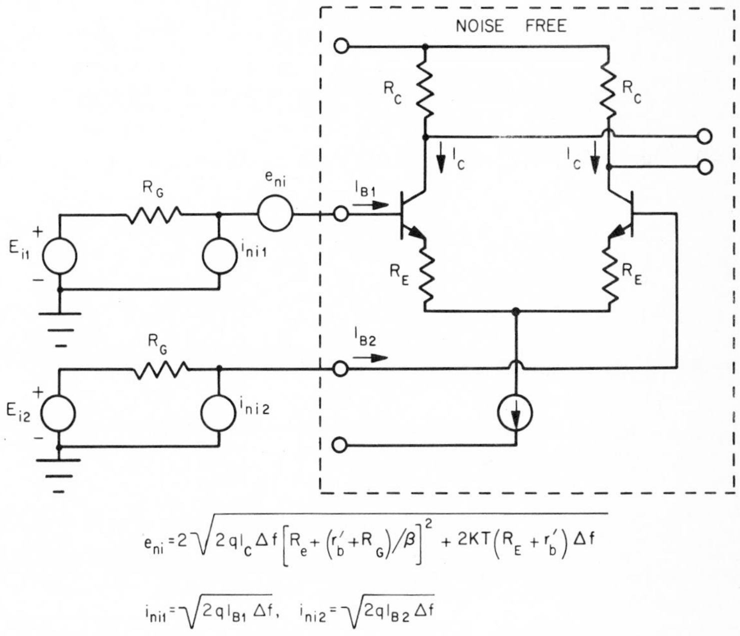

$$e_{ni} = 2\sqrt{2qI_C\Delta f\left[R_e + (r_b' + R_G)/\beta\right]^2 + 2KT(R_E + r_b')\Delta f}$$

$$i_{ni1} = \sqrt{2qI_{B1}\Delta f}, \quad i_{ni2} = \sqrt{2qI_{B2}\Delta f}$$

Fig. 2.16 Equivalent input noise representation of a bipolar transistor differential stage.

currents. The latter conditions are consistent with the requirements of the previous section for maintaining low dc input currents and drifts. Finally, excluding the thermal noise of the source resistances, the equivalent input noise voltage has a slight dependence upon source resistances.

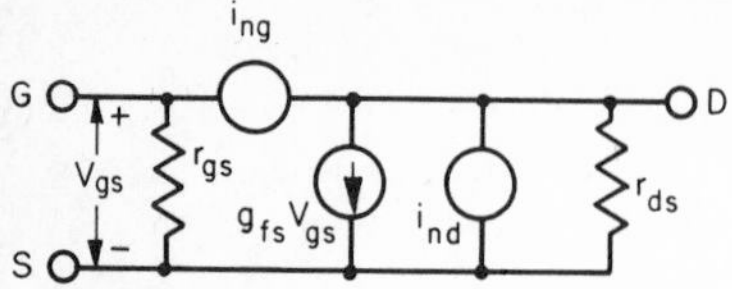

Fig. 2.17 Pi noise model for an FET.

In general, the latter portion of the total input noise is masked by that developed by the source resistances with the input noise current. Representative levels of input noise voltage and currents can be obtained by considering a typical case. With

$$I_C = 25\ \mu A \qquad \beta = 200 \qquad R_G = 1\ k\Omega$$
$$R_E = 500\ \Omega \qquad r_b' = 50\ \Omega \qquad \Delta f = 1\ kHz$$

the equivalent input noise voltage will be 0.3 μV rms and the input shot noise currents will be 6.3 pA rms. Input noise voltage encountered in practice with careful shielding is close to that predicted above. However, the noise currents can be several times the predicted shot noise because of flicker noise, particularly in the frequency range of dc to 1 kHz.

The representation of FET differential-stage noise sources by equivalent input noise generators results from an analysis paralleling that for the bipolar transistor case. Noise sources within an FET are represented in Fig. 2.17 by the gate and drain noise current generators.[10] Shot noise associated with the gate leakage current results in the gate noise current for which

$$\overline{i_{ng}^2} = 2qI_G\ \Delta f \tag{2-46}$$

Since the gate-drain voltage is typically many times the gate-source voltage, the leakage current is primarily gate to drain and the associated noise generator is connected between the latter terminals. Within the FET, channel noise is generated that is approximated by the thermal noise of a resistor equal to $1/g_{fs}$. In the model this noise is represented by i_{nc} which will be

$$\overline{i_{nc}^2} = 4KTg_{fs}\ \Delta f \tag{2-47}$$

Since this output noise current model is of the same form as that in Fig. 2.13 used with the bipolar transistor noise analysis, the equivalent input noise voltage of an FET stage can be drawn from that of the bipolar transistor stage.

By reflecting the drain noise current to the input as a voltage noise, the equivalent input noise voltage is defined. Replacing one side of the stage by its source-follower output resistance, $R_S + 1/g_{fs}$, and combining this with the model of the remaining side result in the equivalent circuit of Fig. 2.18 for $r_{ds} \gg R_D$ and $r_{gs} \gg R_G$. The total noise current in the drain is composed of the channel noise current i_{nc} and the thermal noise current i_{nd} of the drain resistor defined by

$$\overline{i_{nd}^2} = \frac{4KT\ \Delta f}{R_D}$$

Solving for the input noise voltage which produces an equivalent total drain noise current gives

$$\overline{e_{ni1}^2} = 16KT\ \Delta f\left(g_{fs} + \frac{1}{R_D}\right)\left(\frac{1}{g_{fs}} + R_S\right)^2$$

Addition of this noise voltage to an equal component from the other side of the stage results in the total equivalent input voltage noise in the form

$$e_{ni} = 4\left(\frac{1}{g_{fs}} + R_S\right)\sqrt{2KT\ \Delta f\left(g_{fs} + \frac{1}{R_D}\right)} \qquad (2\text{-}48)$$

Input noise currents are the shot noise currents of the gate leakage currents as described by

$$i_{ni1} = \sqrt{2qI_{G1}\ \Delta f} \qquad (2\text{-}49a)$$

$$i_{ni2} = \sqrt{2qI_{G2}\ \Delta f} \qquad (2\text{-}49b)$$

Since the effects of these small noise currents in the drain resistor are small compared with the thermal noise generated there, the input current noise generators may be returned to common instead of to the drains.

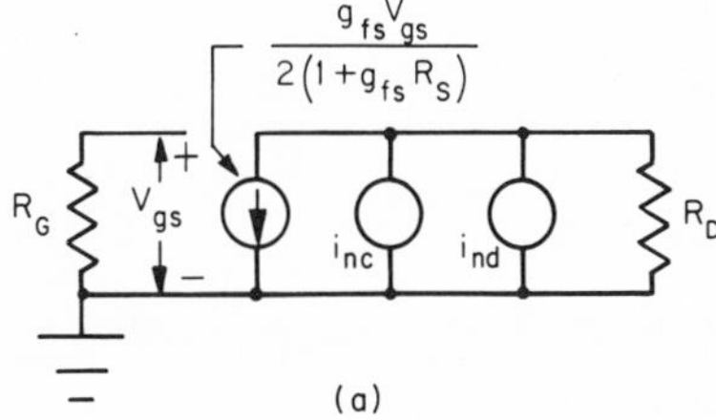

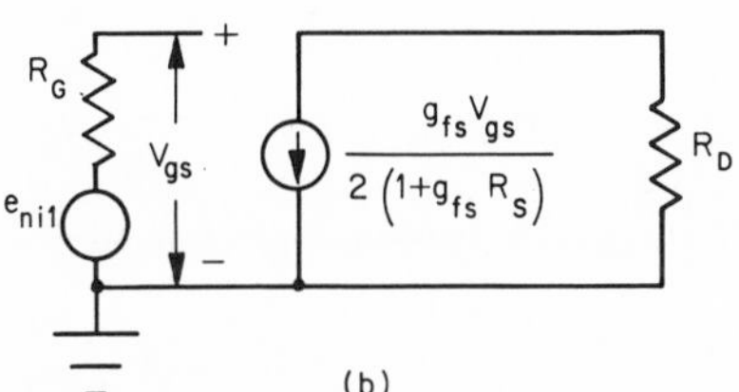

Fig. 2.18 The equivalent circuit of an FET differential stage referred to one input with (a) channel and drain resistor noise current generators and (b) the equivalent input noise voltage generator.

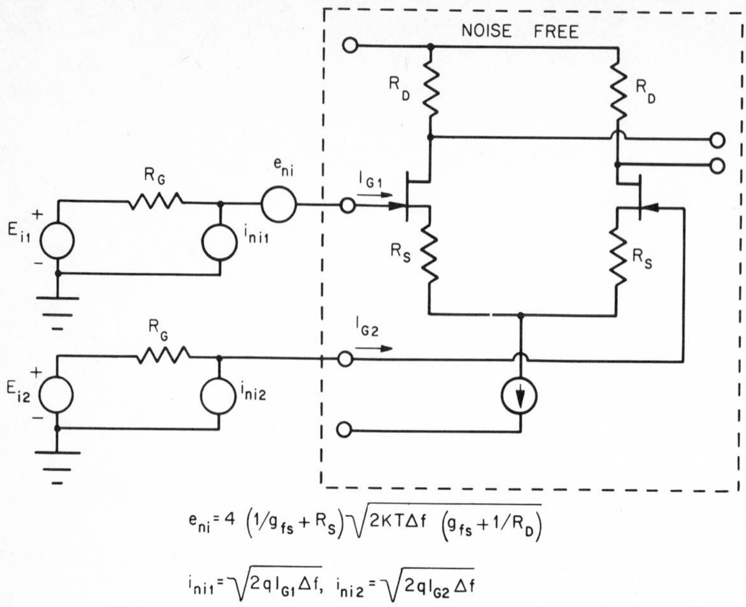

Fig. 2.19 Equivalent input noise representation of an FET differential stage.

With all noise sources reflected to the inputs, the noise characteristics of the FET differential stage are represented in Fig. 2.19.

From the expressions for input noise voltage and currents the conditions for low-noise FET differential stages can be interpreted. Once again noise component is proportional to the square root of the bandwidth considered. This is the case for any noise whose mean-square value has constant spectral density, as do shot noise and thermal noise. It is also noted that low input leakage currents in the FETs corresponds to low input noise currents. Finally it can be seen that high FET transconductance tends to lower equivalent input noise voltage. To calculate some representative noise levels a typical stage is considered having

$$g_{fs} = 1000\ \mu\text{mhos} \qquad R_D = 10\ \text{k}\Omega \qquad R_S = 100\ \Omega$$
$$\Delta f = 1\ \text{kHz} \qquad I_G = 10\ \text{pA}$$

The equivalent input noise voltage resulting is 0.42 μV rms, and the input noise currents are 0.056 pA rms. In general, the input noise currents produce negligible noise voltages on the source resistances as compared with the thermal noise of the resistances. However, the noise voltage is greatly increased by 1/f noise and will typically be 1.0 μV rms under the above conditions for the bandwidth from dc to 1 kHz.

REFERENCES

1. P. E. Gray, D. DeWitt, A. R. Boothroyd, and J. F. Gibbons, *Physical Electronics and Circuit Models of Transistors*, pp. 43, 48–50, 61–62, John Wiley & Sons, Inc., New York, 1964.
2. A. H. Hoffait and R. D. Thornton, Limitations of Transistor DC Amplifiers, *Proc. IEEE*, February, 1964.
3. B. L. Cochrun, *Transistor Circuit Engineering*, pp. 371–372, The Macmillan Company, New York, 1967.
4. L. L. Evans, Biasing FET's for Zero DC Drift, *Electro-Technol.*, August, 1964.
5. C. L. Searle, A. R. Boothroyd, E. J. Angelo, Jr., P. E. Gray, and D. O. Pederson, *Elementary Circuit Properties of Transistors*, pp. 142–143, John Wiley & Sons, Inc., New York, 1964.
6. A. R. Pearlman, Apparatus For Compensating a Transistor for Thermal Variations in Its Operating Point, U.S. Patent 3,230,468, 1962.
7. J. G. Graeme, Transistor Base Current Compensation System, U.S. Patent 3,551,832, 1971.
8. R. D. Thornton, D. DeWitt, E. R. Chanette, and P. E. Gray, *Characteristics and Limitations of Transistors*, pp. 138–142, John Wiley & Sons, Inc., New York, 1966.
9. L. J. Sevin, *Field Effect Transistors*, pp. 46–50, McGraw-Hill Book Company, New York, 1965.

3

THE STAGES OF AN OPERATIONAL AMPLIFIER

The individual stages used in operational amplifiers are separately chosen to develop different amplifier characteristics. Those amplifier characteristics which are determined by a given stage depend on whether it functions as an input stage, intermediate stage, or output stage. For example, within the input stage many of the operational amplifier accuracy limitations are set, and the design of this stage is primarily directed toward control of dc and ac errors. In contrast to this, the intermediate stages provide additional voltage and current gain in the amplifier with less effect upon accuracy. From the results of the preceding chapters the performance characteristics of the differential stages used as input intermediate stages can be described. For the single-ended intermediate stages similar results will be developed in this chapter. Also to be described are the characteristics of output stages which serve to isolate the amplifier from loading effects and to supply output current. Output current limiting techniques which limit transistor dissipation are then discussed for the typical output stage. The ways in which the various stages interact to determine overall amplifier performance will be covered in Chapter 4.

3.1 Input Stages

Characteristics of the differential input stage of an operational amplifier are the most critical factors which affect the accuracy of an operational amplifier in providing voltage gain. Errors effects of following stages are reduced in significance by the gain isolation provided by the first stage, as will be analyzed in the next chapter. Design of the input stage is directed by the requirements placed on the operational amplifier characteristics by accuracy needed within the intended applications of the amplifier. Since it is a differential stage, the characteristics of an input stage are described by the results of the preceding chapters. Design decisions for the stage are made by relating these previous results to amplifier performance requirements, using the design procedure in Sec. 1.5. Both dc and ac performances are considered in the design process. To achieve accurate dc performance at high voltage gain, a large input resistance, and a high common-mode rejection are needed in an input stage which has low dc input error signals. The input error voltages and related drifts from the input offset voltage, input bias currents, and input offset current of the first stage are directly added to the input signal. Any high resistance or capacitance connected to the amplifier inputs in high gain or integrating applications makes dc precision highly sensitive to the input currents of this stage. Common-mode rejection must be preserved in the input stage since common-mode error voltages once added to the amplified signal output cannot be separated. Overall common-mode rejection of an operational amplifier will be no greater than that of the input stage. In addition to providing gain accuracy under feedback, high stage gain provides a large output signal which is less sensitive to the dc errors and common-mode errors of the following stages. Input resistance of the first stage determines the loading error resulting from the source or feedback resistances connected to the amplifier.

Accurate ac performance of an operational amplifier depends largely upon the noise, bandwidth, and slewing rate of the input stage, in addition to the gain, input impedance, and common-mode rejection as discussed above. Since it adds error to the signal prior to any amplification, input noise of the first stage creates the greatest percentage error of the amplifier noise sources. The bandwidth of the input stage is frequently a principal limiting factor to overall amplifier bandwidth when high gain and low current levels are established in this stage. For low input bias currents in bipolar transistor stages, as desired for dc accuracy, low first-stage collector currents are chosen. Associated amplifier bandwidth restrictions limit high frequency gain and the accuracy provided by feedback. As will be covered in Chapter 5, the phase compensation capacitance providing amplifier frequency stability is commonly connected as a load to the narrowband input stage. With this choice the response pole frequency reduced by phase compensation will be about the lowest, and the greatest overall amplifier bandwidth will result. This capacitance load slows the rate of change of the output voltage or slewing rate to that limit imposed by the current available to charge the capacitance. In Sec. 1.5 the available current was identified as the total stage current, $2I_C$ or $2I_D$, since the stage will unbalance to supply the load. The slewing rate of the phase-compensated stage will be

$$\frac{de_C}{dt} = \frac{2I_C}{C} \text{ or } = \frac{2I_D}{C}$$

where e_C is the voltage on capacitance C. Large signal response is commonly restricted in frequency by the rate limiting of the phase-compensated first stage. Approaching this frequency limit, large signals are distorted, developing a second type of ac response error.

Bipolar transistors or FETs are used in the input stage, as represented by the typical circuits of Fig. 3.1, with the choice between the two transistor types being primarily determined by external circuit impedance levels. As shown and as analyzed in Sec. 1.5, a transistor current source is generally used to bias input stages for common-mode rejection improvement over that resulting with resistor bias. Input offset voltage balance controls at the first-stage emitter or source resistors provides offset control without the bias disturbance resulting from offset adjustment at other points in the amplifier. When compensating for amplified input offset voltage at a later stage, large bias unbalances may be required. Any such bias change results in additional input offset voltage and drift. Selection of the indicated resistors and bias levels is made as outlined in Sec. 1.5. The common-mode voltage swing limits of the input stage are essentially the only restrictions to the common-mode voltage range of the amplifier since the very low common-mode gain of the first stage results in a small common-mode signal applied to the following stage. Bipolar transistors are commonly used in the input when signal source or feedback resistance of 50 kΩ or less are to be used with the operational amplifier. Because of higher tranconductance in bipolar transistors, $1/r_e$ as compared with g_{fs} of FETs, higher gain is commonly achieved with bipolar input stages. Above 50 kΩ resistance levels, bipolar transistor stage input bias currents, input offset current, and their drifts typically result in total dc input errors which are greater than those of the higher input offset voltage and drift of an FET input differential stage. Input currents of the FET stage, varying from about 5 pA to 4 nA over extreme temperature ranges, result in small dc input errors until a signal source resistance level much greater than 50 kΩ

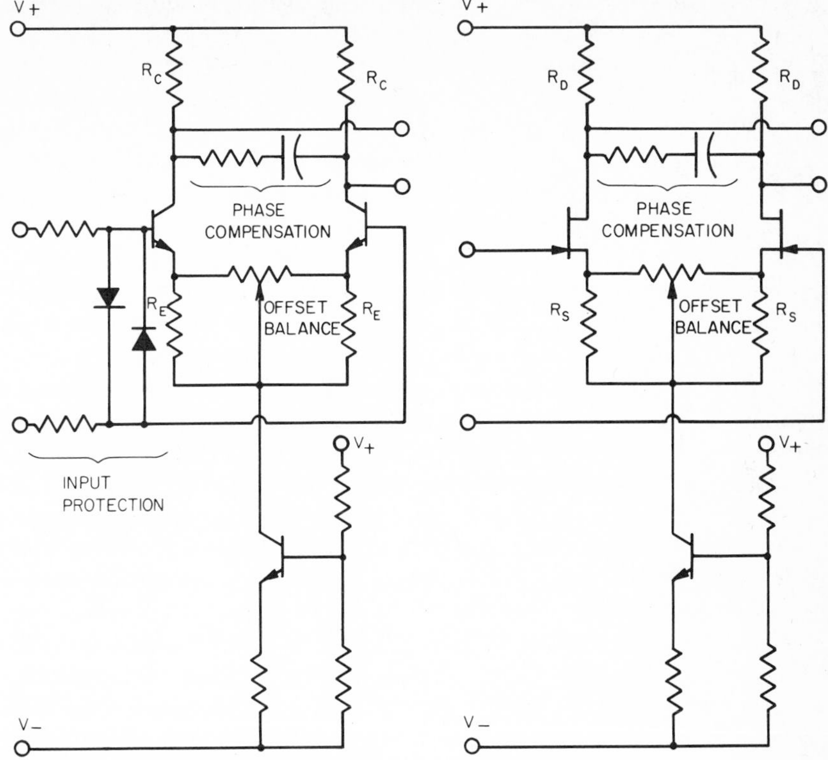

Fig. 3.1 Typical input-stage circuits.

is reached. Generally, an FET input stage biased at its zero-drift current level provides a slewing rate under phase compensation loading that is many times that developed by bipolar stages at their lower current levels.

In addition to the basic differential stages considered here, specialized stages described in the preceding chapters are frequently used as input stages. Cascode biasing, described in Sec. 1.5, is used to achieve improvements in many of the outlined operational amplifier characteristics affecting precision, including input capacitance, bandwidth, common-mode rejection, and input leakage currents. Voltage gain required to ensure accurate dc and ac operation, as discussed, is made large in the input stage, using the dynamic load circuit analyzed in the same section. Input-stage voltage gain of the order of 1,000 is achieved with this stage, providing high isolation from errors in following stages. Reduced input bias currents to operational amplifiers having bipolar transistor inputs is commonly attained by using resistors to supply current to the first-stage bases as demonstrated in Sec. 2.3. Occasionally the

Darlington-connected differential stage described in that section is also employed to lower input currents, although input offset voltage and drift are greatly increased.

Input protection circuitry, such as that shown with the bipolar transistor differential stage of Fig. 3.1, is needed to prevent damage from large differential input voltages. Without the protective clamp circuit shown, such input signals would reverse-bias one of the transistor emitter-base junctions and would reach an avalanche breakdown state for about 6-V reverse bias on common planar-diffused silicon transistors. In this state high currents may be conducted through the input terminals, dissipating significant power in the reverse-biased junction, unless the current is limited. Large series base resistors will limit current to prevent destruction of the transistor, but they also create a response pole with the stage input capacitance and dc errors with the input currents. Not only must current be limited to avoid destroying the transistor but the effects of this reverse current upon beta and V_{BE} levels must be carefully noted. Both characteristics undergo permanent shifts when stressed in this condition even well below the rated power dissipation of the transistor. Matched transistor betas and emitter-base voltages are unbalanced in this way, disturbing the input dc error and drift characteristics. To avoid any such parameter shifts the protection circuit shown limits the differential voltage impressed upon the transistors to the forward voltage drop of one of the protection diodes. Series resistors needed to limit the diode current are small compared with those discussed above since much larger current can be handled with the 0.6-V diode drop than with the 6-V reverse breakdown. Small-signal response of the input stage is not greatly affected by the protection circuit in its OFF state until frequencies above a megahertz are reached. Low reverse voltage on the protection diodes maximizes the diode capacitances to about 3 pF and roughly triples differential input capacitance at such high frequencies. Although input capacitance above 1 MHz is of little importance in general-purpose operational amplifiers, response of wideband types will be appreciably affected by the protection circuit.

In some applications it is undesirable to have large input currents drawn by protection circuits under input overdrive such as those resulting with the above circuit. When a differential stage is used as a switch, overdriven inputs are normal and serious loading of the signal source may result from the above protection circuit action. As an alternative protection method, diodes having high reverse breakdown voltage are placed in series with the emitters of the stage in Fig. 3.2. Large input voltages to this develop reverse emitter-base current that is limited to the small leakage current of a protection diode. Significant current will not flow unless the input signal exceeds the higher breakdown

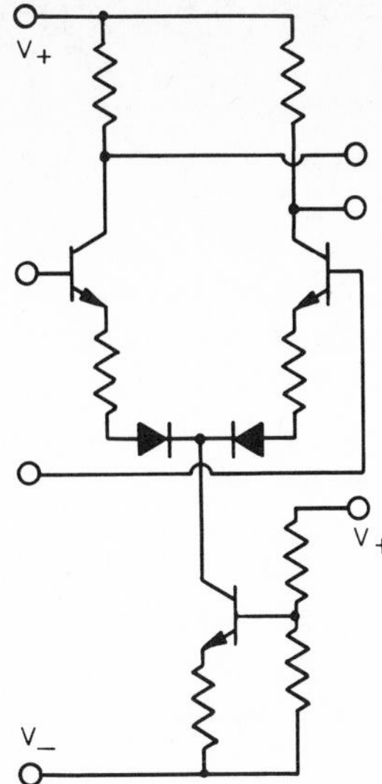

Fig. 3.2 Input protection relying upon breakdown voltages of diodes.

voltage of the diode. However, if input protection is provided with this approach, normal operation results in two additional forward junction voltage drops added in series with the input circuit. Mismatch of the added diode drops adds input offset voltage and drift. Since the added dc errors are uncorrelated with those of the transistors, offset and drift are not typically increased by as much as a factor of 2 but rather by the square root of 2.

3.2 Intermediate Stages

The input stage of an operational amplifier, as described in the preceding section, is designed to minimize ac and dc errors of the overall amplifier by controlling the error sources of this stage and by developing high gain which reduces the effect of following-stage errors. As a result, the intermediate-stage characteristics are less critical. The ways in which intermediate-stage errors combine with those of the input stage will be considered in the next chapter, and only the characteristics of the separate intermediate stages will be developed here. In the design of intermediate stages the emphasis is placed on developing additional voltage gain, providing current gain from the first stage to the output stage, and shifting the quiescent voltage level back to zero at the output. Added voltage gain is of course needed from these stages to provide high overall amplifier gain, and frequently the major portion of the total voltage gain is developed in the intermediate stages. Current gain in this part of the amplifier provides high current to the output stage without heavy loading of the input stage by impedance transformation. To provide a 0-V quiescent output level the dc bias level must be shifted from that at the outputs of the first stage by means of intermediate-stage biasing. In general, the outputs of the first stage are biased well away from common to permit large common-mode input voltage swings as described in Sec. 1.5.

Both differential and single-ended forms are used for intermediate stages. Unless a differential output is desired from the completed amplifier, it is common to employ single-ended stages for at least part of the intermediate section. The choice between the two types of stages is based primarily upon common-mode rejection requirements and the dc loading of the stage. Common-mode rejection is developed by the high differential signal gain and the low common-mode gain of a differential stage. When the transition is made to single-ended stages, common-mode rejection is limited to that developed by the preceding differential stages. However, continued addition of differential rather than single-ended stages provides diminishing returns since other errors become dominant. These latter errors are differential signals developed from the common-mode signal by stage unbalances, as considered in Sec. 1.4, and cannot be reduced by the common-mode rejection of added stages. In addition to this consideration, the dc loading effects of the two forms must be evaluated. The dc bias and drifts of a single-ended stage are compensated by a balancing transistor when a second transistor is added to form a differential stage, as will be analyzed in Sec. 4.2. Once again the overall improvement afforded by each successive differential stage is diminishing since the dc errors of each stage are reduced in effect by the preceding gain. Typically only two moderate-gain differential stages or one high-gain differential stage is used in an operational amplifier.

Widely used intermediate stages include bipolar or FET differential stages, common-emitter or common-source amplifiers, emitter followers or source followers, and a wide range of variations on these basic forms. The characteristics of the differential stages are described in the previous chapters, and the single-ended stage characteristics will be outlined or developed in the remainder of this section. Any of these forms can be used to provide dc level shifting as indicated by the representative cases in Fig. 3.3. Each of the loading stages in the figure translates the dc bias level from the first-stage output level E_{o1} to a 0-V quiescent output. For the differential stage the level shifting is provided by using transistors which are opposite in conductivity type to those of the input stage. In this way the voltage level difference is dropped on the collector-base junctions of the intermediate stage. An analogous situation results for the common-source stage. In the emitter-follower case, the dc level is shifted by a voltage divider. Since the voltage divider also reduces gain, modifications of this approach are generally made, as will be considered.

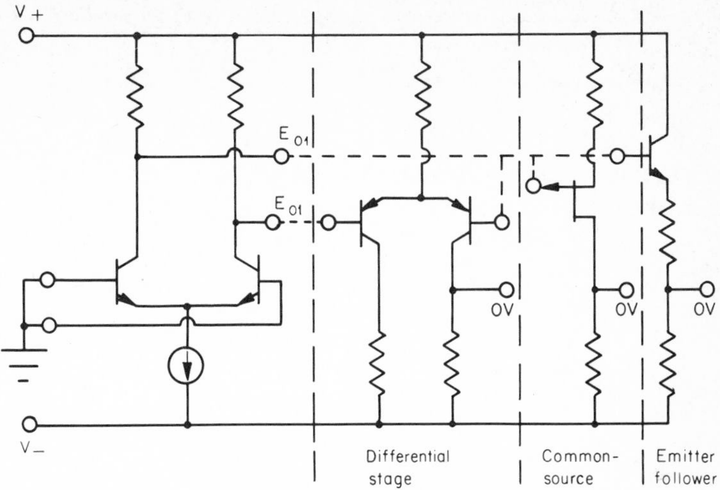

Fig. 3.3 Basic level shifting techniques.

The basic follower circuits used for intermediate stages are represented in Fig. 3.4, using bipolar transistors. Analogous circuits apply with FETs. When the followers are used for impedance isolation without level shifting, the same circuits are used with $R_E = 0$. As mentioned earlier, level shifting with the voltage divider form reduces gain, and the other two circuits shown are used to improve the divider impedance ratio in limiting the gain loss. The impedance of the zener diode shown will be small in comparison with the resistance levels normally used for R_L, making the gain loss small. However, the noise introduced by zener diodes in this application is often significant for operational amplifiers. Alternatively the divider is modified by replacing R_L with

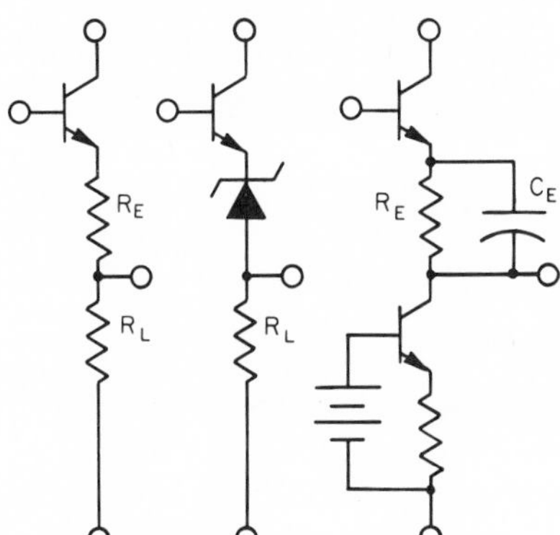

Fig. 3.4 Intermediate-stage follower circuits.

the high impedance of a transistor current source, as shown for the third circuit. To compensate for the bypassing of the current source output resistance by its C_c the capacitor C_E is added, bypassing R_E. The effect of C_c is analyzed in Sec. 1.5.

Except for different divider impedances the three follower circuits introduced above can be described by the same expressions. For analysis the circuit model of Fig. 3.5 is used. The analysis results are simple extensions of the basic emitter-follower characteristics and are given here for reference. Low-frequency voltage gain is expressed by

$$A_O \doteq \frac{R_L}{R_L + R_e + R_G/\beta} \qquad \text{for } R_G \ll r_c \tag{3-1}$$

Input resistance is found to be

$$R_I = \frac{r_c(R_e + R_L)}{R_e + R_L + r_c(1 - \alpha)} \tag{3-2}$$

where

$$R_e = R_E + r_e$$

$$R_I \doteq \beta(R_e + R_L) \qquad \text{for } R_e + R_L \ll r_c(1 - \alpha) \tag{3-3}$$

Output resistance is

$$R_O \doteq \left(R_e + \frac{R_G}{\beta}\right) \| R_L \qquad \text{for } R_G \ll r_c \tag{3-4}$$

Assuming r_c presents a negligible shunt to the output as modeled, the input capacitance from the model equals the sum of C_c and a fraction of C_{eb}. This fraction equals the fraction of the voltage swing bypassed by C_{eb}, and the input capacitance will be

$$C_I \doteq C_c + \frac{r_e}{R_e + R_L} C_{eb} \qquad \text{for } r_c \gg R_e + R_L \tag{3-5}$$

Associated with C_I is a voltage gain response pole at the break frequency

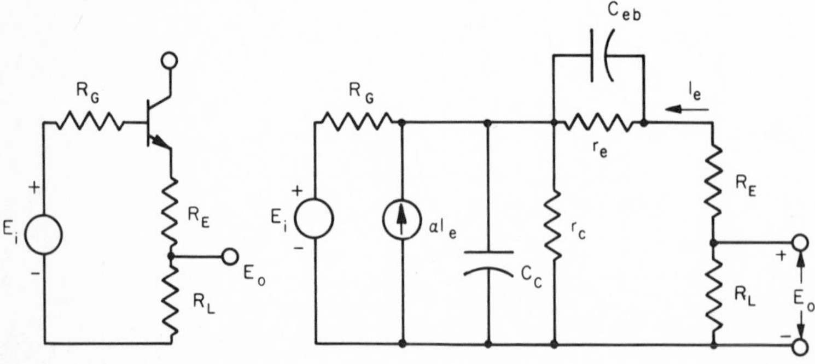

Fig. 3.5 Emitter-follower stage.

of C_I with the input and source resistances. Neglecting the second term of the above expression for C_I, the response pole will be

$$f_p = \frac{1}{2\pi(R_I \| R_G)C_I} \doteq \frac{\beta(R_e + R_L) + R_G}{2\pi\beta(R_e + R_L)R_G C_c} \qquad \text{for } C_I \doteq C_c \qquad (3\text{-}6)$$

Paralleling the above emitter-follower analysis, the source-follower stage of Fig. 3.6 provides analogous results. The low-frequency voltage gain is

$$A_O \doteq \frac{g_{fs}R_L}{1 + g_{fs}(R_S + R_L)} \qquad \text{for } r_{gs} \gg R_G \qquad r_{gs} \gg R_S + R_L \qquad r_{ds} \gg R_S + R_L \qquad (3\text{-}7)$$

Input resistance of the source follower is limited by packaging to around $10^{12}\ \Omega$, and output resistance is expressed by

$$R_O \doteq \frac{R_L}{1 + g_{fs}(R_S + R_L)} \qquad \text{for } r_{gs} \gg R_G \qquad r_{ds} \gg R_S + R_L \qquad (3\text{-}8)$$

Input capacitance is found to be

$$C_I \doteq C_{gd} + \frac{C_{gs}}{1 + g_{fs}(R_S + R_L)} \qquad \text{for } r_{gs} \gg R_G \qquad r_{gs} \gg R_S + R_L \qquad r_{ds} \gg R_S + R_L \qquad (3\text{-}9)$$

By neglecting the second term of the input capacitance expression and the input resistance, the voltage gain response pole becomes

$$f_p \doteq \frac{1}{2\pi R_G C_{gd}} \qquad \text{for } r_{gs} \gg R_G \qquad C_I \doteq C_{gd} \qquad (3\text{-}10)$$

Performance characteristics of the remaining types of single-ended intermediate stages, common-emitter and common-source, can be extrapolated from the differential-stage results of Chapter 1. As discussed in Sec. 1.1, the differential stage is composed of two common-emitter or common-source amplifiers which are effectively connected in series for differential signal analysis. The individual single-ended stage

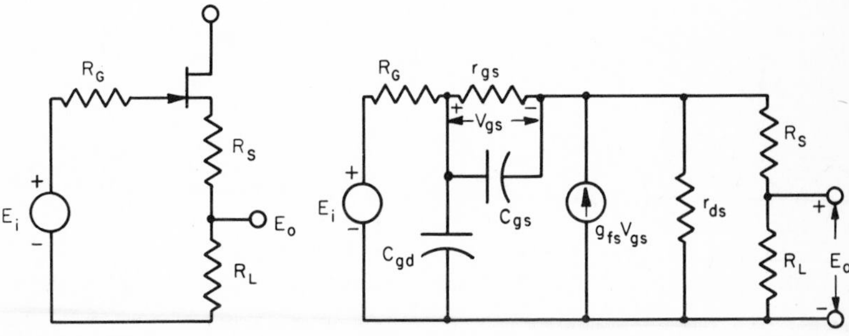

Fig. 3.6 Source-follower stage.

characteristics can be written from those of the differential stage in Secs. 1.1 and 1.2 by again considering the parallels drawn there. From this procedure the common-emitter and common-source describing expressions will be resolved in a form compatible with the differential-stage analysis results. This uniformity eases the analysis of combined differential and single-ended stages when a complete amplifier is considered, as in the next chapter. The common-emitter-stage analysis model is shown in Fig. 3.7. For this model the low-frequency voltage gain is

$$A_O \doteq \frac{-\alpha R_C r_c}{R_e(R_C + r_c) + R_G[R_C + r_c(1 - \alpha)]} \tag{3-11}$$

where

$$R_e = R_E + r_e \qquad R_e \ll r_c(1 - \alpha)$$

Under typical collector loads, this becomes

$$A_O \doteq \frac{-R_C}{R_e + R_G/\beta} \qquad \text{for } R_C \ll r_c(1 - \alpha) \tag{3-12}$$

Common-emitter input resistance when load resistance is very large will be

$$R_I \doteq \beta R_e \frac{R_C + r_c}{\beta R_C + r_c} \tag{3-13}$$

When load resistance is at normal levels, the input resistance simplifies to

$$R_I \doteq \beta R_e \qquad \text{for } R_C \ll r_c(1 - \alpha) \tag{3-14}$$

The output resistances without load and with load R_C are

$$R'_O = r_c \frac{R_e + R_G/\beta}{R_e + R_G} \qquad \text{for } R_G \ll r_c \tag{3-15}$$

$$R_O \doteq \frac{r_c R_C (R_e + R_G/\beta)}{r_c(R_e + R_G/\beta) + R_G R_C} \qquad \text{for } R_C \ll r_c \tag{3-16}$$

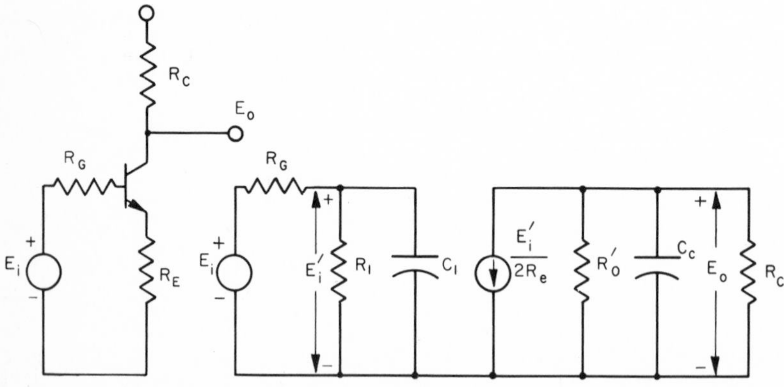

Fig. 3.7 Common-emitter-stage analysis model.

The input capacitance will be

$$C_I \doteq \left(1 + \frac{Z_c}{R_e}\right) C_c \qquad Z_c = \frac{R_O}{1 + j\omega R_O C_c} \tag{3-17}$$

Because of this capacitance the dominant response pole of the gain occurs at approximately

$$f_p \doteq \frac{1}{2\pi R_C C_c} \frac{R_e + R_G/\beta}{R_e + R_G} \tag{3-18}$$

In the same way the characteristics of the common-source stage can be written in a compatible form from the expressions for an FET differential stage. The resulting analysis model is that of Fig. 3.8. In this form the stage has a low-frequency gain expressed by

$$A_O \doteq \frac{-g_{fs} R_D r_{ds}}{R_D + r_{ds}(1 + g_{fs} R_S)} \qquad \text{for } r_{gs} \gg R_G + R_S \tag{3-19}$$

For typical load resistance levels the gain simplifies to

$$A_O \doteq \frac{-g_{fs} R_D}{1 + g_{fs} R_S} \qquad \text{for } R_D \ll r_{ds} \tag{3-20}$$

Input resistance is developed by the large reverse resistance of the gate junction and is packaged-limited to around $10^{12}\ \Omega$. The unloaded output resistance is described by

$$R_O' \doteq r_{ds}(1 + g_{fs} R_S) \qquad \text{for } r_{gs} \gg R_G + R_S \qquad r_{ds} \gg R_S \tag{3-21}$$

Under typical load resistances the output resistance approximately equals the load resistance.

$$R_O \doteq R_D \qquad \text{for } r_{ds} \gg R_D \qquad r_{ds} \gg R_S \tag{3-22}$$

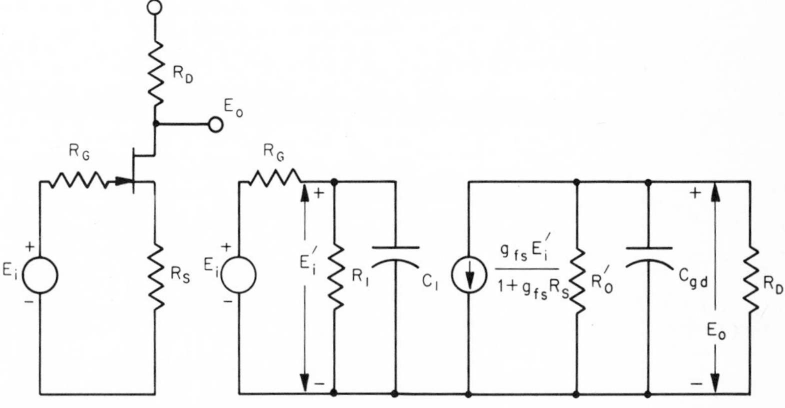

Fig. 3.8 Common-source-stage analysis model.

The input capacitance is

$$C_I \doteq \left(1 + \frac{g_{fs}Z_d}{1 + g_{fs}R_S}\right) C_{gd}$$

where

$$Z_d = \frac{R_O}{1 + j\omega R_O C_{gd}} \tag{3-23}$$

From this capacitance a gain response pole results at

$$f_p = \frac{1}{2\pi R_O C_{gd}} \frac{1 + g_{fs}R_S}{1 + g_{fs}R_G} \tag{3-24}$$

In order to achieve higher gain a current source is frequently used in place of a load resistor in intermediate stages. The high dynamic output resistance of the current source acts as a high resistance load without developing the large dc bias voltage that would result with an equivalent resistor. When such a load is used with a common-emitter stage, as represented in Fig. 3.9, the higher voltage gain is accompanied by reduced bandwidth and input resistance. As represented, identical biasing resistances are connected to both transistors, and the same parameters will be assumed for both transistors to simplify analysis. In this case the load shown as R_{I2} represents the input resistance of the following stage. The net load resistance presented to the common-emitter transistor is the current source output resistance in parallel with R_{I2}. Using Eq. (3-16), this load resistance can be expressed by substituting R_{I2} for R_C to get

$$R_L = \frac{r_c R_{I2}(R_e + R_G/\beta)}{r_c(R_e + R_G/\beta) + R_G R_{I2}}$$

As a lower limit this load resistance approaches R_{I2} which is commonly far greater than the normal collector resistor level would have been. The

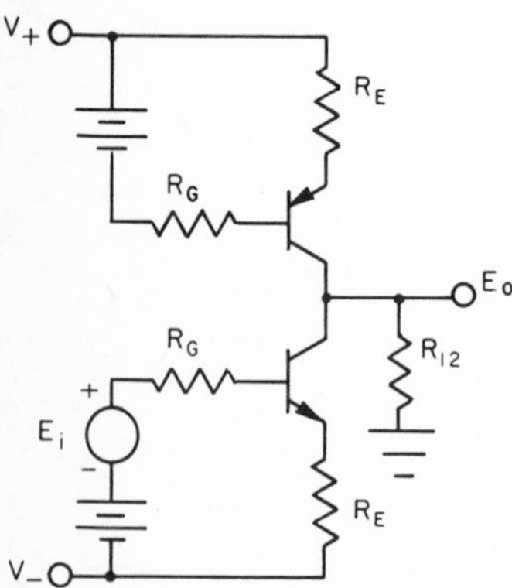

Fig. 3.9 A common-emitter amplifier using a current source in place of a collector resistor.

voltage gain resulting with this high load resistance must be found with the more detailed gain expression of Eq. (3-11). However, this expression can be somewhat simplified for this case to get

$$A_O \doteq \frac{-R_C r_c}{R_G R_C + r_c(R_e + R_G/\beta)} \qquad \text{for } R_L \ll r_c \qquad R_C \ll r_c \qquad 1 - \alpha \doteq \frac{1}{\beta}$$

By substituting the previous expression for R_L in this result in place of R_C, the low-frequency voltage gain of the common-emitter stage with a dynamic load is found to be

$$A_O \doteq \frac{-R_{I2}}{(R_e + R_G/\beta) + 3R_{I2}R_G/r_c} \tag{3-25}$$

The high load resistance also affects the circuit input resistance, output resistance, and bandwidth. Input resistance can be expressed from Eq. (3-13), using the approximation

$$R_I \doteq \beta R_e \frac{r_c}{\beta R_C + r_c} \qquad \text{for } R_C \ll r_c$$

Replacing R_C in the last equation with the expression previously developed for R_L, the input resistance becomes

$$R_I \doteq \frac{\beta R_e}{1 + \dfrac{\beta R_{I2}(R_e + R_G/\beta)}{r_c(R_e + R_G/\beta) + R_{I2}R_G}} \tag{3-26}$$

Note that R_I is significantly reduced from that expressed by Eq. (3-14) as βR_e for lower load resistances. The output resistance of the common-emitter stage is greatly increased by the current source load. As presented to R_{I2} of the following stage, the output resistance will be the parallel combination of the two transistor output resistances. For identical transistor parameters and bias resistances, the stage output resistance follows from Eq. (3-15) as

$$R_O \doteq \frac{r_c}{2}\,\frac{R_e + R_G/\beta}{R_e + R_G} \qquad \text{for } R_G \ll r_c \tag{3-27}$$

Also accompanying the higher gain is a reduced bandwidth which can be described from a model of the stage. This model is drawn from the common-emitter model of Fig. 3.7 and the output circuit representation of Fig. 1.12. Using the latter representation the response effects of the current source load are all represented in its output circuit where they may be combined directly with the common-emitter output circuit elements. For the single transistor of the current source, the output capacitance is twice that shown in Fig. 1.12 for a differential stage, and the resulting model for the common-emitter stage with a dynamic load appears in Fig. 3.10. For this model R_I and R_O are as expressed by Eqs. (3-26) and (3-27), and the circuit capacitances are

$$C_I = \left(1 + \frac{Z_L}{R_e}\right) C_c \qquad Z_L = \frac{R_O \| R_{I2}}{1 + j\omega R_O \| R_{I2}(C_c + C_o')}$$

$$C_o' = \frac{R_e + R_G}{R_e + R_G/\beta} C_c$$

Although the model can be used to analyze the complete circuit response, only the dominant pole is considered here. An approximate method of simply resolving the complete response of such circuits will be developed in Chapter 4. In general, the high resistance of the dynamic load makes the output circuit pole dominant, and it occurs at

$$f_p \doteq \frac{r_c(R_e + R_G/\beta) + 2R_{I2}R_G}{2\pi r_c R_{I2}(2R_e + R_G)C_c} \qquad \text{for } \beta \gg 1$$

3.3 Output Stages

Following the input and intermediate voltage gain stages of an operational amplifier, it is desirable to provide impedance isolation from loads. In this way the characteristics of the gain stages are preserved under load, and adequate signal current is made available to the load. As with any amplifier, the output stage provides isolation by presenting a high input impedance to the preceding stage and a low output impedance to the load. Accompanying this low output impedance must be the ability to supply the desired load current, and this capability is provided by the current gain of the stage. To provide isolation without degrading high-frequency performance the output stage is generally chosen with low input capacitance and wide bandwidth. The most common output stages are some form of emitter-follower stage which is adapted to the output-stage requirements. Basic characteristics of the emitter follower which also apply to output stages were developed in the preceding

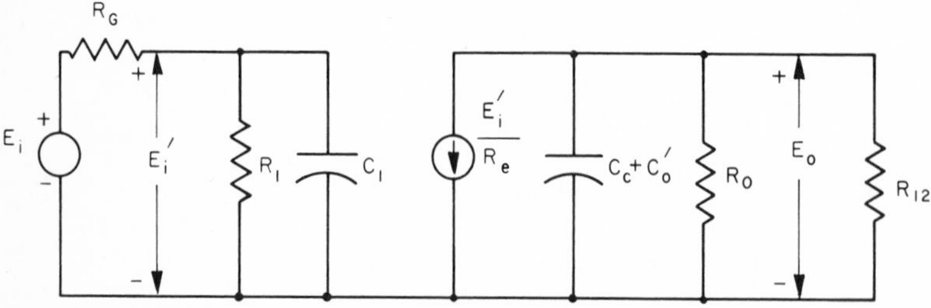

Fig. 3.10 Analysis model of the circuit of Fig. 3.9.

section. As described in that analysis, this type of stage provides the high input impedance, low output impedance, high current gain, and wide bandwidth needed for output stages. Improved capabilities for low quiescent current, high output voltage, and high output current requirements are achieved with modified emitter-follower stages. Another common output stage which will be discussed is the totem pole stage which provides both voltage gain and impedance isolation.

Although the basic emitter-follower stage provides many of the desired output-stage characteristics, it requires a relatively large quiescent biasing current when the stage must be capable of supplying output currents of either polarity. Considering the class A emitter-follower stage of Fig. 3.11, it can be seen that negative output current cannot be supplied by the transistor and that the current is limited to that which can be drawn through the emitter resistor by the supply voltage. To supply a given output current at a negative output voltage the no load bias current level in R_E must then be at least as great as the desired current. Then, output current is limited to

$$|I_o| \leq \left| \frac{V_- - E_o}{R_E} \right| \qquad \text{for negative swing}$$

The quiescent bias current level which must be established to ensure the availability of this current with negative output voltages is even larger than I_o. For output signals reaching $I_{o\,max}$ and $E_{o\,max}$ the required quiescent bias current I_{BQ} is

$$I_{BQ} \geq \frac{V_-}{V_- + |E_{o\,max}|} |I_{o\,max}| \tag{3-28}$$

In the event that only one polarity of output current is needed, high quiescent current is not required since output current above the quiescent level can be drawn through the transistor. However, for most operational amplifiers bidirectional output swing is desired, and quiescent current levels lower than the output current are achieved with class B output stages.

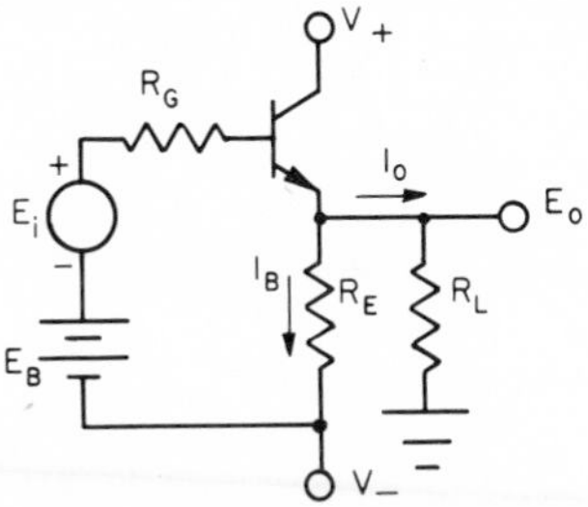

Fig. 3.11 Class A emitter-follower output stage.

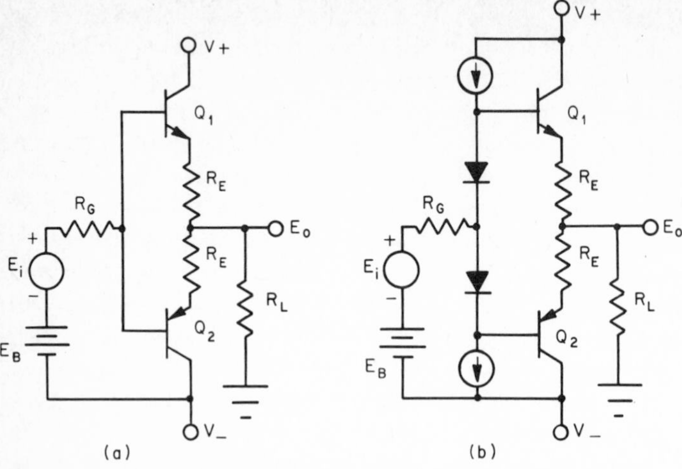

Fig. 3.12 Emitter-follower output stages biased as (a) a class B stage and as (b) a class A-B stage.

Class B biasing of an emitter-follower output stage provides separate transistors to supply the two polarities of output current. As represented in Fig. 3.12a, the class B stage can supply output currents of either polarity without being limited by the quiescent current level. In fact, class B biasing with bipolar transistors develops essentially no quiescent current since neither emitter-base junction is forward-biased with the input and output voltages at zero. Single swing forward-biases the appropriate transistor to develop output current. However, at the crossover from one transistor to the other there is a range of input signals, approximately +0.5 to −0.5 V, for which neither transistor is turned on. In this range no output current can be supplied, and the output voltage remains at zero, resulting in crossover distortion as depicted in Fig. 3.13 for a sine-wave signal. Decreased distortion results when feedback is applied from the output to preceding gain stages, as the feedback reduces the time taken by the input signal to traverse the voltage range for which the output stage is turned off. However, to

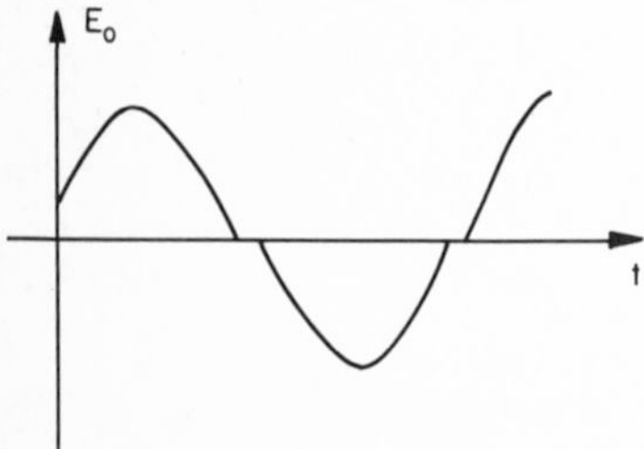

Fig. 3.13 Crossover distortion of a sine-wave signal characteristic of class B bipolar transistor output stages.

eliminate the distortion it is most common to modify the stage bias and avoid the turned-off state. This is achieved with the class A-B biasing of Fig. 3.12b for which the added diodes provide a forward bias to the output transistors when the input signal is zero. With this bias one transistor begins turning on as signal swing brings the other transistor close to its turnoff point. Quiescent current in the class A-B case is controlled by the current biasing the diodes, by the relative voltage drops of the diodes and the output transistor emitter-base junctions, and by the emitter resistors. Generally the biasing diode junctions are smaller than the transistor emitter-base junctions and the diode voltages are greater at the quiescent current levels employed.

Except for quiescent current levels, the class B or class A-B emitter-follower-stage characteristics are similar to those of the basic emitter-follower stage analyzed in the previous section. At any nonzero signal level only one of the two transistors of the class B circuit is supplying current, and this transistor is essentially the previously mentioned emitter follower loaded by a transistor which is turned off. For most of the typical output voltage range, one transistor of the class A-B stage is also turned off, and the circuit characteristics can be closely approximated, considering this condition. If the n-p-n transistor is considered to be conducting in the class B or class A-B stages of Fig. 3.12, the ac equivalent circuit of Fig. 3.14 results. In this equivalent circuit the reverse impedance of the p-n-p emitter-base junction is neglected since it is large compared with the output resistance of the conducting transistor. Also, the impedance of the biasing diode is small in comparison with the input resistance of Q_1 and is omitted. As a result, the only significant effect of the off transistor is that of its collector-base capacitance C_{c2}. The dc characteristics of the class B or class A-B stage, such as gain, input resistance, and output resistance, are then the same as those developed for the simple emitter follower in Sec. 3.2. Input capacitances of the class B and class A-B stage are greater than that of the single emitter follower expressed by Eq. (3-5). Adding C_{c2} to the emitter-follower input capacitance,

$$C_I \doteq 2C_c \qquad \text{for } C_{c1} \doteq C_{c2} \tag{3-29}$$

The response pole frequency of Eq. (3-6) is also modified by the increased input capacitance and becomes

$$f_p \doteq \frac{R_G + \beta R_L}{2\beta R_L R_G C_c} \qquad \text{for } C_{c1} \doteq C_{c2} \qquad R_L \gg R_E \tag{3-30}$$

A second basic type of output stage, the totem pole stage, is common in vacuum tube circuits and can be adapted to solid state circuits with a junction FET as shown in Fig. 3.15. As an output stage this circuit

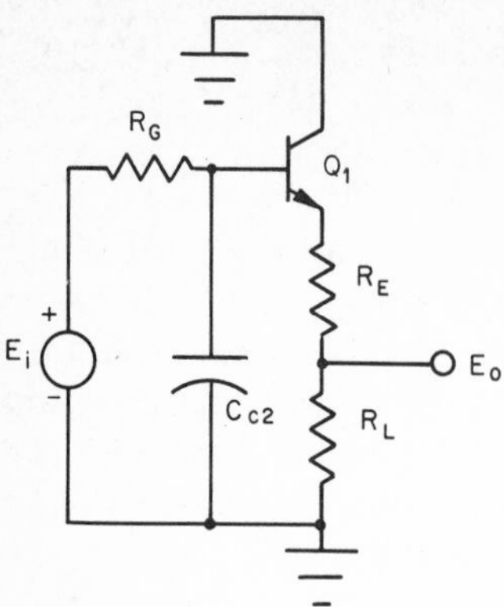

Fig. 3.14 Ac equivalent circuit of the class B or class A-B emitter-follower output stages when only Q_1 is conducting.

provides voltage gain along with impedance isolation. In the form considered the totem pole stage is basically a common-emitter amplifier with a dependent current source load. Input resistance is that of the common-emitter bipolar transistor as approximated earlier by βR_e. Both voltage gain and output impedance are improved by the dependence of the FET current source on the collector current of the common-emitter transistor. A signal increase in I_c results in an increase in the magnitude of the gate-source voltage V_{gs} and a decrease in the source current I_s. The two current changes add to produce a greater change in output current. In this way the FET boosts the output current resulting from a given input signal.

To develop expressions for the improved gain and output resistance the pi models of the two transistors, as presented in Figs. 3.7 and 3.8, are combined to result in the equivalent circuit of the totem pole stage in Fig. 3.16. For the common levels of R_S and R_L a low-frequency analysis can be made, neglecting R'_O and r_{ds}. Using this approximation, the low-

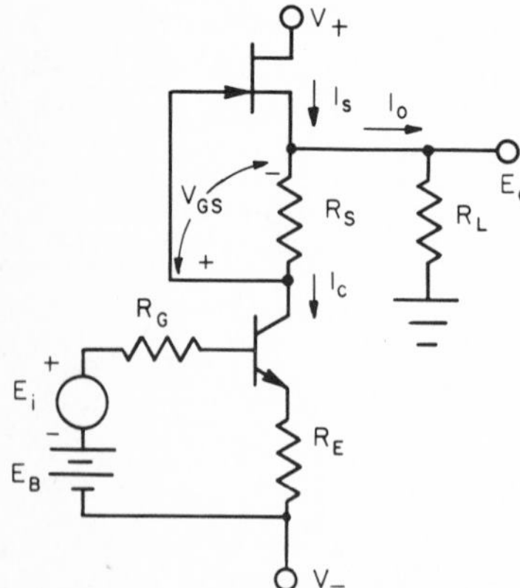

Fig. 3.15 Totem pole output stage.

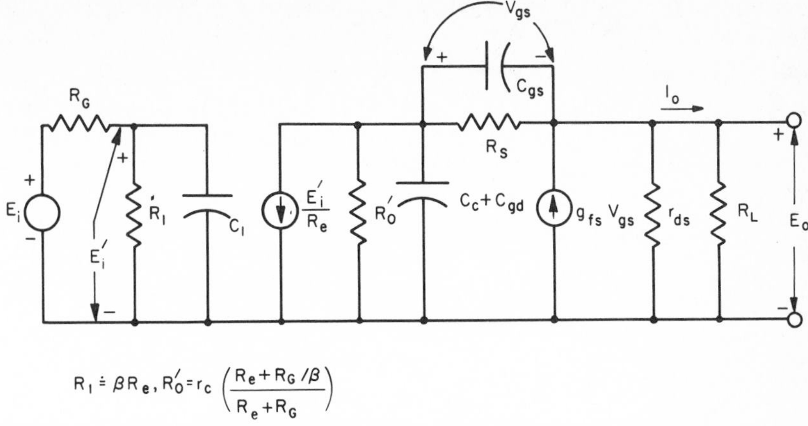

Fig. 3.16 Ac model of the totem pole stage of Fig. 3.15.

frequency voltage gain of the totem pole state is found to be

$$A_O \doteq -\frac{(1 + g_{fs}R_S)R_L}{R_e + R_G/\beta} \qquad \text{for } R_S + R_L \ll r_c(1 - \alpha) \qquad R_L \ll r_{ds} \quad (3\text{-}31)$$

Comparing this result with the gain of a common-emitter transistor, as given by Eq. (3-12), it is seen that the addition of the FET to form the totem pole stage increases gain by a factor of $1 + g_{fs}R_S$. The output resistance of the circuit is found from the open-circuit voltage and short-circuit current of the model. In this case R_L is not included, and R_O' and r_{ds} cannot be neglected. From this analysis

$$R_O = \frac{R_O' r_{ds}}{r_{ds} + R_O'(1 + g_{fs}R_S)} \quad (3\text{-}32)$$

Once again, the improvement provided by the dependence of I_s on I_c is represented by the $1 + g_{fs}R_S$ term in Eq. (3-32). This dependence is removed by setting R_S to zero, leaving a fixed current source load on the common-emitter transistor. The output resistance of the totem pole stage is quite high compared with that provided by the emitter-follower. Both r_{ds} and R_O' of the previous expression are of the order of 200 kΩ, resulting in an output resistance of about 25 kΩ. Lower output resistance is provided by negative feedback to preceding gain stages, which is common to most operational amplifier applications.

The frequency response of the totem pole stage is fairly complex, because of the several capacitances of the circuit. However, the dominant response pole frequency can be readily found by neglecting the influences of other poles on this frequency. For the resistance levels normally encountered, the dominant pole results from the high Miller-effect input capacitance of the common-emitter transistor. Neglecting the other circuit capacitances, and neglecting R'_O and r_{ds} in comparison with R_L, the input capacitance can be found, by using the procedure of Sec. 1.2, to be

$$C_I \doteq \left(1 + \frac{R_S + R_L}{R_e}\right) C_c$$

Then the response pole frequency is

$$f_p \doteq \frac{1}{2\pi(R_I \| R_G)C_I}$$

$$f_p \doteq \frac{\beta R_e + R_G}{2\pi\beta R_G(R_e + R_S + R_L)C_c} \tag{3-33}$$

For the typical case this pole frequency will be several megahertz whereas other response poles will result at least a decade higher in frequency.

Other common output stages are generally specialized forms of the class A-B emitter-follower stage. High-voltage output stages, for instance, are frequently more economically realized by using several low-voltage transistors in place of a high-voltage transistor. By connecting low-voltage transistors in series as in Fig. 3.17, the high voltage is divided between the transistors. As indicated on the diagram, the signal and supply

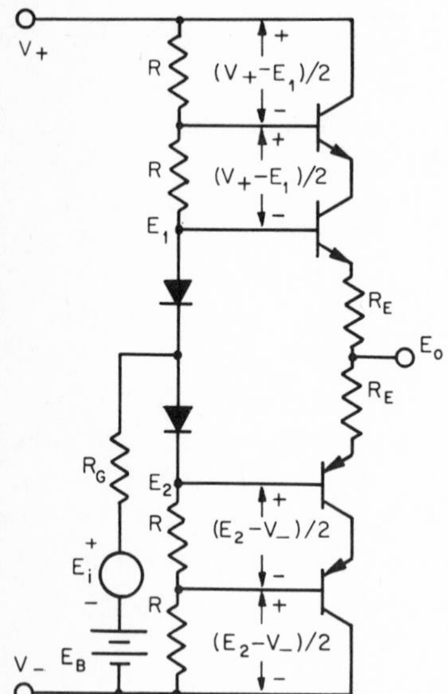

Fig. 3.17 Series-connected emitter followers forming a high-voltage output stage.

voltage on each emitter-follower output transistor is reduced by a factor of 2 when the second set of emitter followers is added in series. The added transistors serve simply as voltage biases to the collectors of the output transistors. As the signal changes, these voltage biases are also varied by the input signal to ensure that the bias levels will not limit the output voltage range. The highest voltage developed on any one transistor of this circuit will be one-half the total supply voltage when the output is at a voltage equal to one of the supply voltages. By adding even more emitter followers in series with the output transistors, the maximum voltage applied to any one transistor can be further reduced.

Added output transistors can also be used to increase the output current capability of an operational amplifier. Although higher-current transistors could also be used, it may again be more economical to use several low-current transistors. In this case transistors can be connected in parallel to divide the output current between two or more transistors, as in Fig. 3.18. Care must be taken with this technique to ensure that the output current divides equally between the paralleled transistors. The current division is largely controlled by the separate emitter resistors used with each transistor. If one emitter resistor were shared by two parallel-connected transistors, the current division would be determined only by the temperature-sensitive emitter-base voltages. In the latter case the transistor conducting the higher current would reach the higher temperature, resulting in an even further increase in its current. To avoid this condition when high output currents are being supplied, the emitter resistors are chosen to develop voltage drops under load which equal a large portion of the emitter-base voltage.

As higher voltages and currents are supplied by output transistors their power dissipations reach levels for which silicon p-n-p transistors are com-

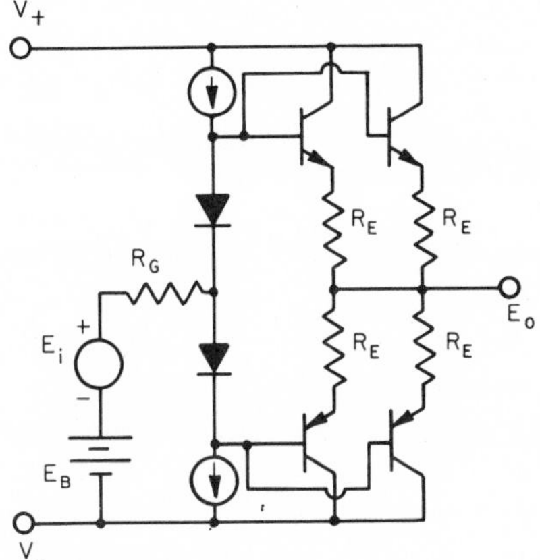

Fig. 3.18 Parallel-connected emitter followers forming a high-current output stage.

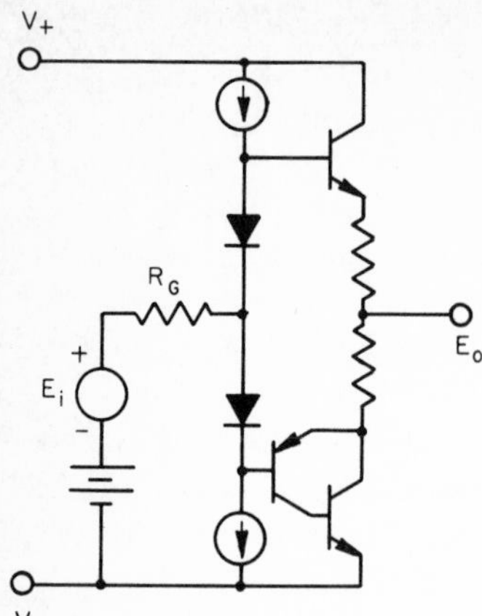

Fig. 3.19 Output stage using a p-n-p–n-p-n feedback pair for a high power p-n-p.

monly more expensive than the n-p-n counterpart or more difficult to achieve in monolithic form. In this case it may be desirable to replace the p-n-p transistor with the p-n-p–n-p-n feedback pair combination of Fig. 3.19. The p-n-p of the combination controls the output voltage at its emitter much as a single p-n-p power transistor would. However, the output current is conducted primarily by the n-p-n of the combination which is driven by the p-n-p collector. Only the n-p-n base current is supplied by the p-n-p and so it can be a low-power transistor, and the n-p-n will be the high-power transistor.

In many operational amplifier applications the load does not have to be grounded, as is the case for speakers and relays. For these applications it is possible to use a differential output which doubles the output voltage swing available to the load and isolates the load from ground. A differential output stage can be made by using two output stages which are driven from opposite sides of a differential stage, as represented in Fig. 3.20. Any of the output-stage forms discussed can be used. Since the two outputs have opposite phases, the differential output voltage is twice that developed from either output to common. The peak-to-peak output swing can then be greater than the total supply voltage and approaches twice the total supply voltage as a limit. When an operational amplifier with a differential output is used, an additional offset voltage may result. If feedback is applied to only one of the outputs, a differential output offset voltage results from the difference in offsets of the two signal paths through the amplifier. Unless additional circuit balancing is performed, this offset can reach hundreds of millivolts. When differential feedback is applied from both outputs, this offset term is avoided although the quiescent output levels will have a common-mode offset voltage from common. An additional common-mode feedback loop can be employed to reduce significantly the output common-mode offset voltage.

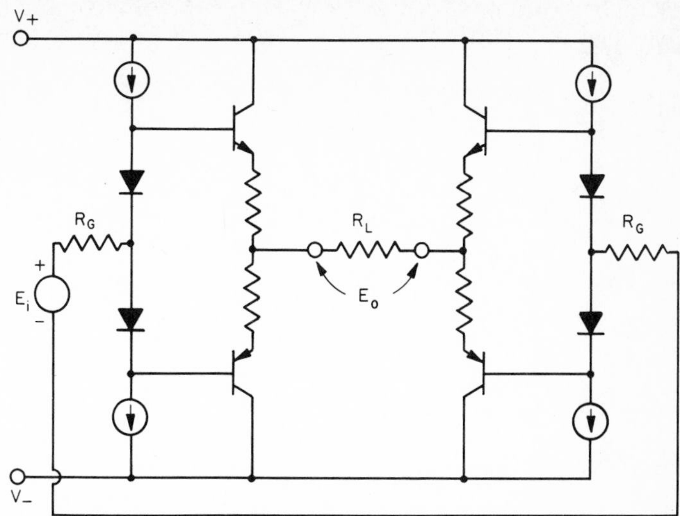

Fig. 3.20 Differential class A-B output stage.

3.4 Output Current Limiting

For operational amplifier output stages, such as those described in the preceding section, the ability to supply current to a load is limited by the safe operating temperature of the output transistors. In operation the actual temperature of the transistors is determined by the ambient temperature, the power dissipation within the transistors, and the ease with which heat is conducted away from the transistors as expressed by thermal resistance. For a given package configuration the thermal resistance is fixed and the permissible transistor power dissipation is then set by the maximum ambient temperature to be encountered. In general, the power dissipation in operational amplifier output stages is controlled by current limiting. The necessary current-limit level depends upon the operating load conditions for which protection is needed. For a given transistor power dissipation limit the maximum allowable output current is determined by the voltage across the transistor in its current-limited state. This voltage is determined by the load condition, as will be seen.

In addition to normal operating load conditions, fault conditions such as an output short circuit to common or to a power supply are commonly considered in selecting a current-limit level. For a positive signal and normal load conditions, the power dissipation in transistor Q_1 of the class A-B output stage of Fig. 3.21 will be

$$P_1 = I_o(V_+ - E_o) = I_o(V_+ - I_oR_L) \qquad \text{for } R_L \gg R_e$$

Differentiating this expression with respect to I_o, the output current resulting in maximum power dissipation is found. This current repre-

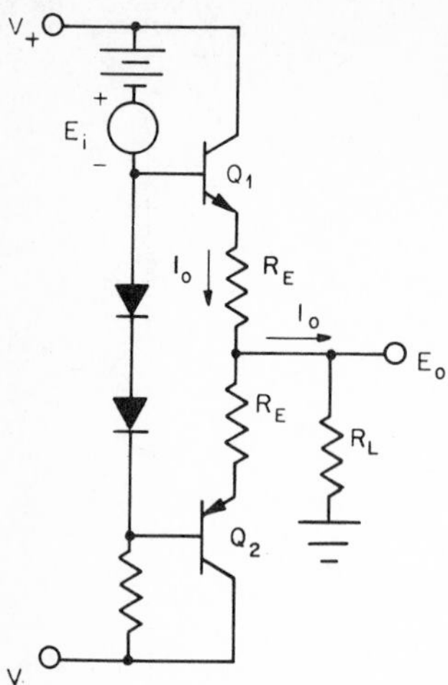

Fig. 3.21 Class A-B emitter-follower output stage under positive signal swing.

sents the output current limit I_{oL} needed for normal operating load conditions and is expressed by

$$I_{oL} = \frac{V_+}{2R_L} = \frac{2P_{1\,max}}{V_+} \tag{3-34}$$

Using this result, the minimum load resistance which may be connected without exceeding this current in the unlimited stage is

$$R_{L\,min} = \frac{V_+^2}{4P_{1\,max}}$$

Then, as long as the load meets this condition, no added protection is needed for the output stage. However, if capacitive loads are encountered, the high-frequency load impedance must be observed to avoid excessive output current drain. Similarly, accidental output short circuits can cause output current to exceed the above limit. To protect the output stage under such conditions the output current must be limited by additional circuitry. In general, the output current limit is set to provide protection under short circuit or capacitive loads to common. For this case, the voltage across Q_1 in Fig. 3.21 can reach essentially V_+, assuming R_E is small. Then the output current must be limited to

$$I_{oL} = \frac{P_{1\,max}}{V_+} \qquad \text{for short circuit to ground} \tag{3-35}$$

In some cases it may be desirable to provide output protection permitting capacitive loads returned to a power supply voltage or merely permitting short circuit to this voltage. In that event the voltage across the output transistor could reach $V_+ - V_-$ and the output current limit would be set at

$$I_{oL} = \frac{P_{1\,max}}{V_+ - V_-} \qquad \text{for shorts to supply} \qquad (3\text{-}36)$$

Several types of current-limit circuits are in common use for operational amplifiers, with the choice between them being determined primarily by the limiting precision needed and the limit-circuit complexity. Limiting precision relates the final limited current level to that current at which the limit circuit first begins to affect output current. As such, the ability of the current-limit circuit to control power dissipation without disturbing signals nearing the limit is determined by the precision of the limiting circuit. For efficient high current outputs it is desirable to use an accurate limit which permits signal currents that are close to the limiting level. When signal current levels result in transistor power dissipation which is well below its limit, a less efficient current limit can be used. Such an approximate current limit is provided by adding resistors in series with the collectors of the emitter-follower output stage as in Fig. 3.22. Resistive limits of this type simply limit the power dissipation in the transistor by linearly decreasing the voltage across the transistor, as the current increases, until the transistor saturates.

For an output short circuit to ground the current is limited by collector resistors to

$$I_{oL} \doteq \frac{V_+}{R_C} \text{ or } \frac{V_-}{R_C} \qquad \text{for } R_C \gg R_E$$

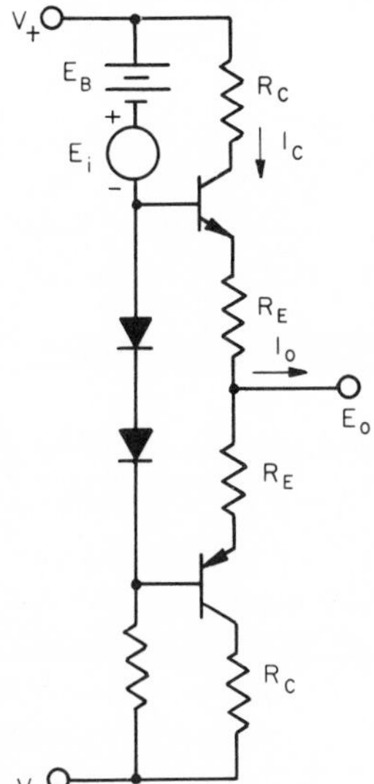

Fig. 3.22 Resistive current limit applied to an emitter-follower output stage.

Limiting of output current under signal voltage swing occurs at lower levels, expressed by

$$I_{oL} = \frac{V_+ - E_o}{R_C} \qquad \text{for } E_o > 0 \tag{3-37}$$

Although this current limit under signal swing is far less than that for a grounded output, the limiting efficiency is not as poor as this comparison might suggest. The usable output signal current is instead compared against the limiting current which occurs at the point of maximum transistor power dissipation, since this dissipation is the true limiting factor. Power dissipated in Q_1 is

$$P_1 \doteq I_o(V_+ - I_oR_C) \qquad \text{for } I_o \doteq I_C$$

and the output current at the point of $P_{1\,max}$ is limited to

$$I_{oL} = \frac{V_+}{2R_C}$$

Substitution of this result in the expression for P_1 defines the collector resistance needed for protection as

$$R_C = \frac{V_+^2}{4P_{1\,max}} \tag{3-38}$$

For typical output voltage ranges and supply voltages, the ratio of this last limiting current to the usable output signal current is about 2.5:1. Using the same power dissipation limit, an ideal current limit would permit signal currents 2.5 times as large as those provided by the resistive limit under peak output voltage swing.

More precise current limiting is provided by nonlinear circuits which clamp the output current, as do D_3 and D_4 in Fig. 3.23. Operation of the limit circuit is initiated by the voltage developed on R_E by the output current. As the current in Q_1 increases, this voltage drop will eventually forward-bias D_3 as expressed by

$$\begin{aligned} V_{F3} &= I_eR_E + V_{BE1} - V_{F1} \\ &\doteq I_eR_E \qquad \text{for } V_{BE1} \doteq V_{F1} \end{aligned}$$

Conduction of D_3 shunts the biasing current provided by I_1, limiting the base current available for Q_1 to that level needed by the transistor to keep D_3 forward-biased. Output current is then limited to

$$I_{oL} \doteq \frac{V_{F3}}{R_E} \tag{3-39}$$

where

$$I_o \doteq I_e$$

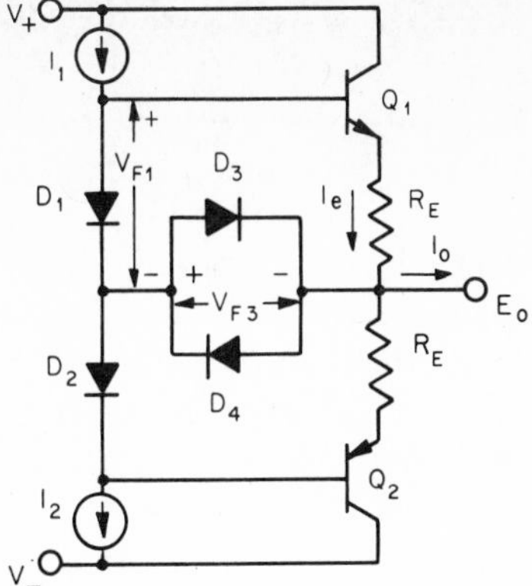

Fig. 3.23 Diode clamp output current limit.

In this case the ratio of the final limiting current to that at which the output signal is first affected is around 1.2:1. The precision of this limiting circuit is primarily determined by the forward resistance of D_3. Although much greater precision is provided by the diode clamp, its current limit is fairly temperature-sensitive because of the thermal dependence of V_{BE} in Eq. (3-39). Fortunately the temperature coefficient is negative, about $-0.3\%/^\circ C$, and helps to control the transistor junction temperature by lowering the limiting current as ambient temperature increases. Similar limiting characteristics are achieved with the transistor clamp of Fig. 3.24. Voltage developed on R_E by the output current turns on a clamp transistor which then shunts the bias current away from the base of the output transistor. The precision of the current limit and its temperature dependence are determined by the emitter-base junction of the clamp transistor in the same way as

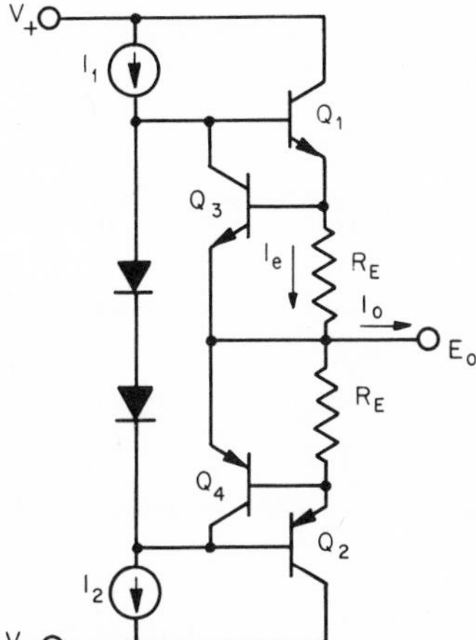

Fig. 3.24 Transistor clamp output current limit.

these characteristics were established by the clamping diode above.

Another popular type of output current limit, known as the foldback limit, forces the current below its initial limiting value once the limit has been reached. Although such operation reduces power dissipation under limit, it has characteristics which make it undesirable for operational amplifiers. First, the foldback limit can have two stable states such that removal of the overload will not necessarily turn off the limiting circuit, and power supply removal may be necessary to deactivate the limiter. Also, upon driving a capacitive load at high frequencies the output current will frequently reach the limit. In such a case a foldback limit reduces the capacitor charging current and slows the output response.

As mentioned, the effectiveness of a current-limit circuit is measured in terms of how well it limits current without disturbing output signals which are below the limit. The lack of limiting precision and capacitance loading or storage effects are the principal limiter characteristics which affect output signals below the actual limiting level. As discussed above, precise limiting is commonly hampered by resistances in the limiting circuit. Capacitances of diodes or transistors used for current limiting can alter small-signal response by capacitive loading, or they can result in time delay during overload recovery because of discharge times. As a result, the current limit must be tested at high frequency when making a choice between limiting techniques. For operational amplifiers the test should be performed without heavy feedback, since the loop gain causes the amplifier to correct for distortion introduced by the current-limiting circuit.

4

MULTISTAGE OPERATIONAL AMPLIFIERS

In the preceding chapters the differential and single-ended stages used in operational amplifiers have been described. By combining such stages in a wide variety of ways, multistage operational amplifiers can be formed to realize different combinations of performance characteristics. Characteristics of the complete amplifier are determined by those of all the separate stages which have been previously considered. In this chapter the ways in which individual-stage characteristics combine and interact to determine overall amplifier performance are analyzed.

Operational amplifier performance characteristics derived in the following sections from those of the composite stages include voltage gain, frequency response, input dc offsets and drifts, and noise. Also discussed are the improvements in dc error characteristics provided by the addition of an ac-coupled modulated carrier stage at the amplifier input, as commonly employed in chopper-stabilized and varactor diode amplifiers. In the case of voltage gain, the net amplifier gain follows directly from the separate stage when interstage loading effects are included. The overall frequency response of an operational amplifier, however, is not readily developed from the individual stage responses because of complex interaction between the stages. This interaction is complicated primarily by the variation in Miller-effect input capacitance with frequency. More direct interaction occurs between the dc bias characteristics which create dc input errors and drifts. The biasing input voltages and currents of a loading stage change the quiescent current level of the loaded stage, developing new dc errors. By combining the effects of the various dc errors in this way, the equivalent input offset voltages, the bias currents, and offset currents, as well as the thermal drift of each, can be predicted for operational amplifiers. In a similar manner the noise voltages and currents of the stages of an operational amplifier can be represented together in an equivalent input noise voltage and current. Continuing the equivalent input characteristic representation, the improved dc error characteristics of chopper-stabilized and varactor-type operational amplifiers are described. By ac-coupling the input, such amplifiers avoid the direct effects of input stage voltage and current biases. In these cases the dc component of the signal is transferred by a modulated carrier.

4.1 Gain and Frequency Response

Extensive interaction between the cascaded stages of an operational amplifier greatly alters the individual stage response characteristics. Therefore, a direct combination of the separate stage responses developed in Chapters 1 and 3 cannot be used to resolve the overall gain and frequency response of the amplifier. Such interaction is most pronounced for frequency response and less significant for other amplifier characteristics. Input and output resistances of an operational amplifier are essentially determined by the input and output stages, respectively. Somewhat more sensitive to stage interaction than R_O and R_I is the low-frequency voltage gain of the amplifier. Since the input resistance of a loading stage shunts the load resistance of the preceding stage, the gain of the first stage is decreased when stages are cascaded. The low-frequency gains of the individual stages when cascaded then follow from the results of previous chapters by including the shunting effect. This is achieved by replacing the collector to drain resistor terms of such expressions with the parallel combination of that resistor and the loading-stage input resistance R_{I2}. For the basic differential stages the low-frequency differential voltage gains of Eqs. (1-2) and (1-9) become

$$A_{O1} \doteq \frac{-R_{C1} \| R_{I2}}{R_{e1} + R_{G1}/\beta_1} \qquad \text{for } R_{C1} \| R_{I2} \ll r_{c1}(1 - \alpha_1) \tag{4-1}$$

or

$$A_{O1} \doteq \frac{-g_{fs1}(R_{D1} \| R_{I2})}{1 + g_{fs1} R_{S1}} \qquad \text{for } R_{D1} \| R_{I2} \ll r_{ds1} \tag{4-2}$$

where the subscripts 1 and 2 denote the two stages. The single-ended common-emitter or common-source stages have the same gain expressions as their differential counterparts, as indicated in Sec. 3.2. The input resistances lowering the gains above are as derived for the various stages in preceding chapters. In general, the input resistance of a loading FET stage can be neglected since this 10^{11}-Ω resistance level is far greater than practical collector or drain resistances. For a bipolar transistor differential loading stage the shunting resistance from Eq. (1-4) is

$$R_{I2} \doteq 2\beta_2 R_{e2} \qquad \text{for } R_{C2} \| R_{I3} \ll r_{c2}(1 - \alpha_2)$$

One-half of this resistance is presented as a load by the single-ended common-emitter stage as expressed in Eq. (3-14). In a similar manner, the low-frequency common-mode gains of cascaded differential stages are written from Eqs. (1-32) and (1-37) as

$$A_{Ocm1} \doteq \frac{-R_{C1} \| R_{Icm2}}{2R_{CM1}} \qquad \text{for } \frac{R_{G1}}{2\beta_1} \ll R_{CM1} \ll \frac{r_{c1}}{2} \qquad R_{C1} \| R_{Icm2} \ll r_{c1}(1 - \alpha_1) \qquad (4\text{-}3)$$

or

$$A_{Ocm1} \doteq \frac{-R_{D1} \| R_{Icm2}}{2R_{CM1}} \qquad \text{for } 2g_{fs1} R_{CM1} \gg 1 \qquad R_{D1} \| R_{Icm2} \ll r_{ds1} \qquad (4\text{-}4)$$

Once again, if the loading stage is an FET type, the input resistance loading is negligible. For a bipolar transistor differential loading stage, the loading resistance from Eq. (1-27) will be

$$R_{Icm2} \doteq \beta_2 R_{CM2} \qquad \text{for } R_{C2} \| R_{Icm3} \ll r_{c2}(1 - \alpha_2) \qquad 2\beta_2 R_{CM2} \ll r_{c2}(1 - \alpha_2)$$

When the loading stage is a single common-emitter transistor, the loading resistance is approximately $\beta_2 R_{e2}$ as before.

Since the loading interaction is included in the low-frequency gains of stages, the overall operational amplifier gain is simply the product of these gains:

$$A_O = A_{O1} A_{O2} \cdots A_{On}$$

However, it should be noted that the total gain of a differential stage is not combined with that of the amplifier when this stage drives a single-ended rather than a differential loading stage. In the typical operational amplifier of Fig. 4.1, only one-half of the second stage drives the following stage. As a result, the gain provided to the amplifier by the second stage is one-half the differential gain of the stage. Overall amplifier voltage gain from the four stages is then

$$A_O = \frac{A_{O1} A_{O2} A_{O3} A_{O4}}{2}$$

where the individual-stage gains include loading effects as given in Eqs. (4-1) and (4-2). Since most operational amplifiers have single-ended outputs, this conversion from differential to single-ended stages is common. The overall common-mode gain of an operational amplifier in the balanced case is composed of the common-mode gains of the differential stages and the ordinary voltage gains of the single-ended stages. No modification of this gain results from the transition to single-ended stages since both differential-stage outputs swing the entire common-mode output voltage. The resulting low-frequency common-mode gains of the circuits in Fig. 4.1 for the balanced case will be

$$A_{Ocm} = A_{Ocm1}A_{Ocm2}A_{O3}A_{O4}$$

However, the response of an operational amplifier to common-mode signals is significantly altered by circuit unbalances, as discussed for differential stages in Sec. 1.4. These unbalances largely determine the common-mode rejection of the amplifier. Representing the stage unbalances as a difference between the gains presented to a common-mode signal by the two sides of a stage, the outputs of the first stage of either circuit in Fig. 4.1 for a common-mode input signal will be

$$E_{ocm1} = A_{cm1}E_{icm}$$
$$E_{od1} = \Delta A_{cm1}E_{icm}$$

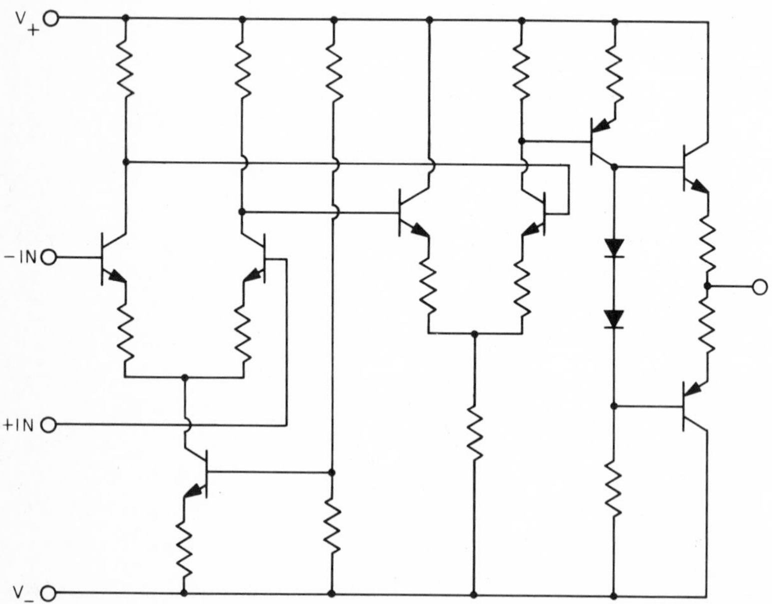

Fig. 4.1 Typical operational amplifier circuit.

The differential output E_{od1} resulting from circuit unbalances is then multiplied by the high differential gain of the remaining portion of the amplifier. Much less amplification is received by the common-mode output E_{ocm1} because of the common-mode rejection of the second stage. At the single-ended output of the second stage the combined signal is

$$\begin{aligned} E_{o2} &= A_{cm2}E_{ocm1} + A_2E_{od1} \\ &= (A_{cm1}A_{cm2} + \Delta A_{cm1}A_2)E_{icm} \end{aligned}$$

From this point, the combined signal is amplified by the remaining single-ended gain, giving an overall amplifier common-mode gain of

$$A_{cm} = \frac{E_o}{E_{icm}} = A_{cm1}A_{cm2}A_3A_4 + \Delta A_{cm1}A_2A_3A_4$$

With typical circuit unbalances of several percent the common-mode gain for the types of circuits considered is primarily determined by the unbalances represented in ΔA_{cm1} above. As a result, common-mode rejection is principally controlled by first-stage component matching in typical operational amplifiers.

High-frequency interaction of the stages of an operational amplifier is complicated by the frequency-dependent Miller-effect input capacitances of the stages. Cascading two bipolar transistor differential stages and using the stage model of Fig. 1.10 result in the equivalent circuit of Fig. 4.2 for differential signals. From the three separate segments of the circuit, it can be seen that three time constants are present in this model, predicting response poles at

$$f_{p1} = \frac{1}{2\pi(R_{I1}\|2R_G)C_{I1}}$$

$$f_{p2} = \frac{1}{2\pi(R_{O1}\|R_{I2})(C_{I2} + C_{c1}/2)}$$

$$f_{p3} = \frac{1}{\pi R_{O2}C_{c2}}$$

where

$$R_O = R'_O\|2R_C \qquad R'_O = 2r_e\frac{R_e + R_G/\beta}{R_e + R_G}$$

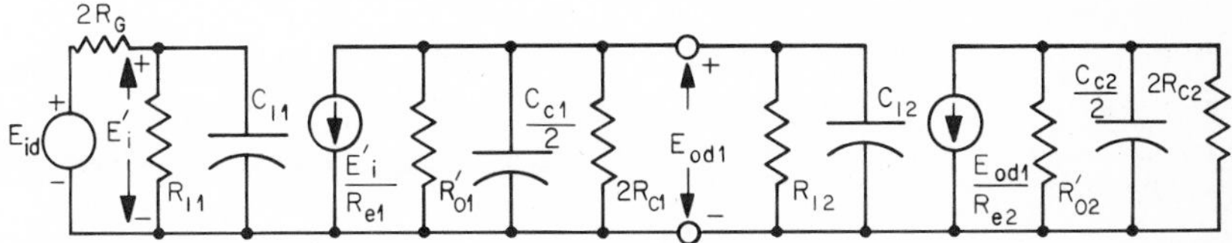

Fig. 4.2 Differential signal model of two cascaded bipolar transistor differential stages.

However, the actual response of the cascaded stages is complicated by the fact that C_{I1} and C_{I2} are not fixed capacitances but are frequency dependent Miller-effect capacitances. Input capacitance expressed by Eq. (1-16) is modified by the loading of the following stage input impedance to be of the form

$$C_{I1} = \left(1 + \frac{Z_{c1}\|Z_{I2}}{R_{e1}}\right)\frac{C_{c1}}{2} \tag{4-5}$$

where

$$Z_{c1} = \frac{R_{O1}}{2 + j\omega R_{O1}C_{c1}} \qquad Z_{I2} = \frac{R_{I2}}{1 + j\omega R_{I2}C_{I2}}$$

Note that the interaction does not necessarily end with the two directly connected stages. From the above expressions it is seen that C_{I1} is a function of C_{I2}, through its dependence on Z_{I2}, and C_{I2} will be similarly dependent upon any additional stages. As a result, the frequency response of each stage is influenced by every stage following it. Essentially the same interaction occurs for cascaded FET stages since their input capacitance representation in Sec. 1.2 is of the same form. Because of the complexity of the above interaction, a general solution for the frequency response of cascaded stages is not very useful. Instead, approximate solutions for specific cases can be found.

Some general observations about the effects of the frequency-dependent input capacitance upon frequency response permit simplifications when considering specific circuits. For a differential stage with a fixed load, R_L and C_L, the input capacitance from Eq. (1-16) will be

$$C_I = \left(1 + \frac{Z_c\|Z_L}{R_e}\right)\frac{C_c}{2} \tag{4-5}$$

where

$$Z_c\|Z_L = \frac{R_O\|R_L}{2 + j\omega(R_O\|R_L)(C_L + C_c)}$$

In this case C_I will vary as shown in the logarithmic plots of Fig. 4.3 over the range of

$$\frac{C_c}{2} \leq C_I \leq \left(1 + \frac{R_O\|R_L}{2R_e}\right)\frac{C_c}{2}$$

Since it is the shunting effect of C_I which creates a response pole, its reactance is also plotted. Input capacitance is decreased from its low-frequency value as shown by the falloff of stage gain. The gain reduction is created by capacitive shunting of the load resistance. From the above expression for C_I its response pole is seen to be that of $Z_c\|Z_L$ which is

$$f_1 = \frac{1}{\pi(R_O\|R_L)(C_L + C_c)}$$

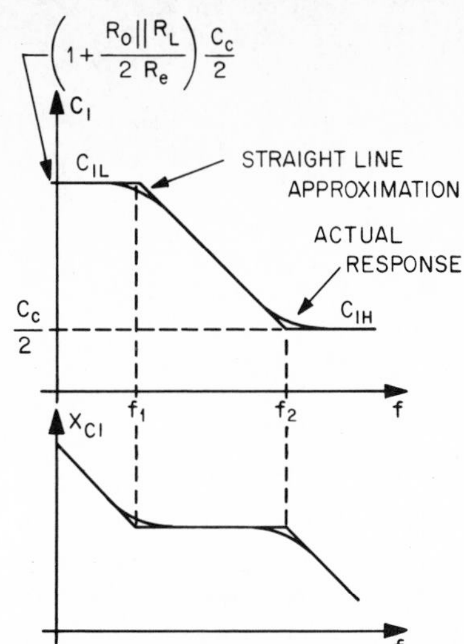

Fig. 4.3 Frequency dependence of differential-stage input capacitances and its reactance.

Above f_1 the input capacitance becomes inversely proportional to frequency, resulting in a constant reactance as shown. A reactance which is constant with frequency is resistive and will not create a frequency response pole. After C_I decreases to its lower limit of $C_c/2$, the reactance is again capacitive.

From these considerations it can be seen that C_I can create a response pole only outside the frequency range of f_1 to f_2, using the straight-line approximation shown. Frequency response poles created by C_I are then developed with essentially the end point limiting values of C_I indicated. If the pole of the input capacitance occurs before that of the output circuit, below f_1, then C_I creates this pole with essentially its low-frequency value of

$$C_{IL} = \left(1 + \frac{R_O||R_L}{2R_e}\right)\frac{C_c}{2} \qquad \text{for a bipolar differential stage} \qquad (4\text{-}6a)$$

Similarly the low-frequency values of input capacitances for other types of stages can be defined. From Eqs. (1-22), (3-17), and (3-23) they are

$$C_{IL} = \left[1 + \frac{g_{fs}(R_O||R_L)}{2(1 + g_{fs}R_S)}\right]\frac{C_{gd}}{2} \qquad \text{for an FET differential stage} \qquad (4\text{-}6b)$$

$$C_{IL} = \left(1 + \frac{R_O||R_L}{R_e}\right)C_c \qquad \text{for a common-emitter stage} \qquad (4\text{-}6c)$$

$$C_{IL} = \left[1 + \frac{g_{fs}(R_O||R_L)}{1 + g_{fs}R_S}\right]C_{gd} \qquad \text{for a common-source stage} \qquad (4\text{-}6d)$$

When frequency then reaches f_1 and the stage output circuit pole breaks, the shunting effect of C_I stops varying with frequency as represented by its reactance curve. In other words, a response zero is developed coinciding with the output circuit pole. If the pole due to C_I does not occur before that of the output circuit at f_1, the pole associated with C_I will occur above f_2 with the constant term of C_I representing its high-frequency limit of

$$C_{IH} = \frac{C_c}{2} \quad \text{for a bipolar differential stage} \tag{4-7a}$$

$$C_{IH} = \frac{C_{gd}}{2} \quad \text{for an FET differential stage} \tag{4-7b}$$

$$C_{IH} = C_c \quad \text{for a common-emitter stage} \tag{4-7c}$$

$$C_{IH} = C_{gd} \quad \text{for a common-source stage} \tag{4-7d}$$

No input circuit response zero is developed in this case since the reactance of C_I did not present a significant shunt prior to f_1. To determine which of the two approximations applies in a given case, the resistance shunted by C_I can be compared with a reactance plot such as that of Fig. 4.3. A response pole results at that frequency for which the reactance of C_I equals the resistance shunted by the capacitance.

Since the input capacitance of a differential stage can be approximated by a fixed input capacitance in a specific case, as described, this type of representation can be applied to resolve an approximate frequency response for a given operational amplifier. Beginning at the amplifier output, the pole of the last stage is found first. Then the pole resulting with the input capacitance of the preceding stage is approximated by comparing possible pole frequencies with the above output circuit pole. If the pole which would result from the low-frequency value of input capacitance C_{IL} of Eqs. (4-6) is at a lower frequency than that of the output circuit pole computed earlier, this input circuit pole is a good approximation to the actual response pole. In this case the previously computed output circuit pole is accompanied by a zero in the input circuit and creates no net response change. A second pole would then occur in the input circuit when C_I reaches its high-frequency limit, C_{IH} of Eqs. (4-7), and this is resolved by substituting this capacitance for C_I. In the opposite case from that just considered, the pole frequency computed above using C_{IL} is above that of the output circuit pole, and the input capacitance creates a pole only with its high-frequency limit C_{IH}. For response considerations in the latter case C_{IH} is substituted for C_I. In either case a net two-pole response results with the high-frequency pole determined by C_{IH} and the low-frequency pole set by either C_{IL} or the output circuit pole, whichever comes first.

Applying the above procedure to the amplifier of Fig. 4.4 such a response will be developed. The emitter-follower fourth stage, as analyzed in Chapter 3, presents a load to the third stage which is the parallel combination of

$$R_{I4} \doteq \beta_4 R_L \qquad \text{and} \qquad C_{I4} \doteq 2C_{c4}$$

This load combines with the output circuit elements of the third stage which are included in the model of Fig. 4.5a from the same chapter. From Eq. (3-27), the unloaded output resistance R'_{O3} shown is defined by

$$R'_{O3} = \frac{r_{c3}}{2} \frac{R_{e3} + R_{G3}/\beta_3}{R_{e3} + R_{G3}}$$

where $R_{G3} \doteq R_{O2}/2$ in this case from the model. One frequency response pole results from the fourth-stage loading above, and from the model the

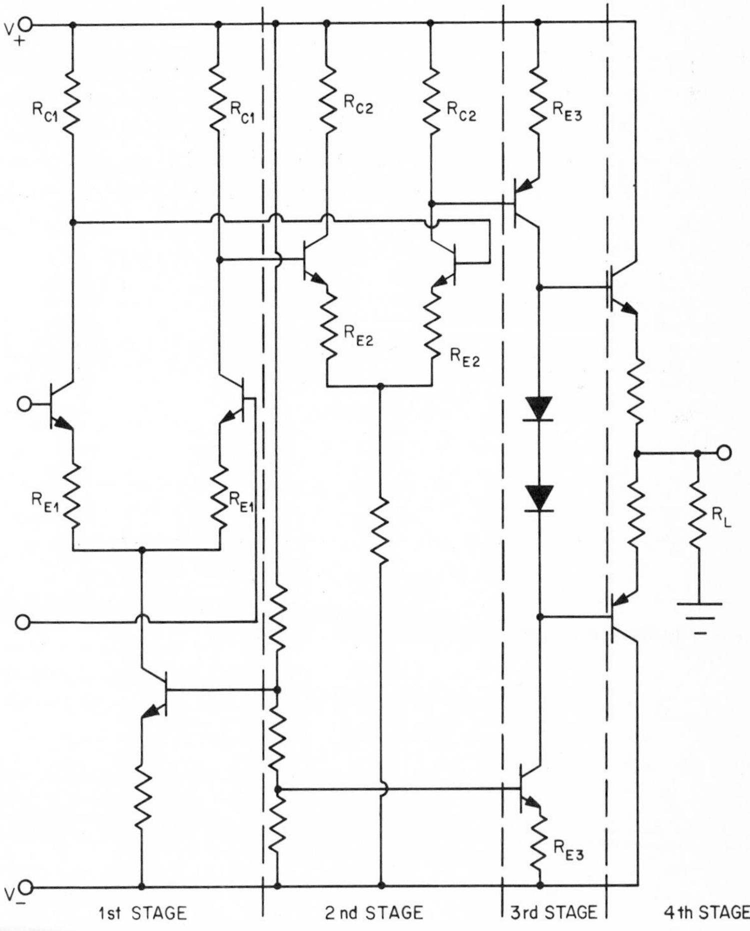

Fig. 4.4 Typical bipolar transistor operational amplifier.

pole frequency will be

$$f_{p4} \doteq \frac{1}{2\pi(R_{I4}\|R'_{O3})(2C_{c4} + 2C_{c3})}$$

The resistance determining f_{p4}, $R_{I4}\|R'_{O3}$, is quite high because of the current source load of the third stage and the emitter-follower isolation provided by the fourth stage. As a result, the output circuit pole of the third stage, f_{p4}, is commonly at a lower frequency than that due to its input capacitance and the relatively low driving resistance normally presented by the second stage shown. Then for response calculations the input capacitance of the third stage can be approximated by its high-frequency limit of

$$C_{I3} \doteq C_{IH} = C_{c3}$$

As represented in Fig. 4.5b, this capacitance and the input resistance R_{I3} combine with the output elements of the second-stage model to form the next pole. This equivalent circuit results from the differential signal model of Fig. 1.10 with the output elements adjusted to coincide with the single-ended loading. From this circuit model the pole asso-

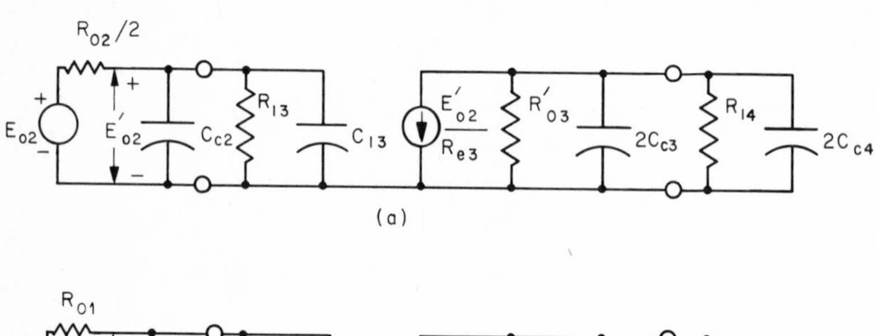

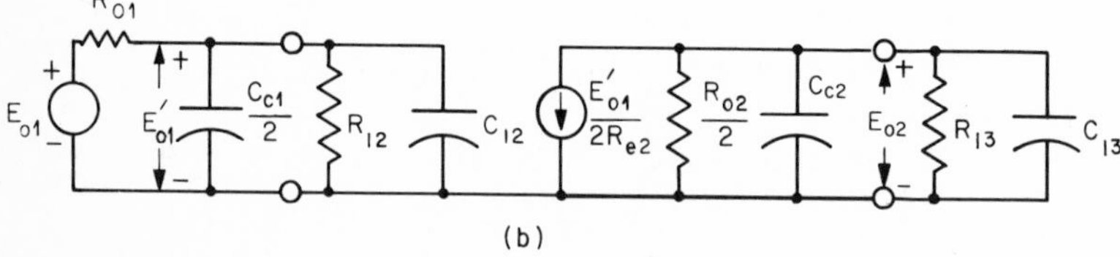

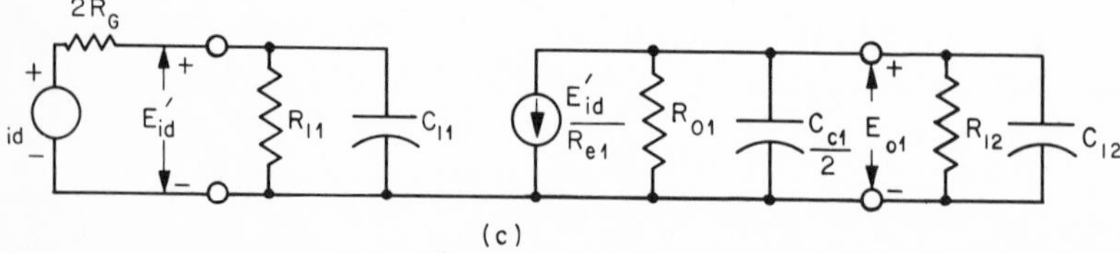

Fig. 4.5 Analysis models of the operational amplifier of Fig. 4.4 for (a) the third stage, (b) the second stage, and (c) the first stage.

ciated with third-stage input capacitance occurs at

$$f_{p3} \doteq \frac{1}{2\pi(R_{I3}\|R_{O2}/2)(C_{c2} + C_{c3})}$$

Because of the low capacitance and relatively low resistance creating the above pole in the second-stage output circuit, it is commonly higher in frequency than the pole due to the second-stage input circuit represented in Fig. 4.5c. Input capacitance of the second stage at the associated pole frequency would then be at its low-frequency limit C_{IL} since the pole at f_{p3} would not yet have reduced second-stage gain and Miller-effect capacitance. Thus, for calculation of this pole,

$$C_{I2} \doteq \left(1 + \frac{R_{O2}\|R_{I3}}{R_{e2}}\right)\frac{C_{c2}}{2}$$

Using the circuit model, the pole frequency is found to be

$$f_{p2} \doteq \frac{1}{\pi(R_{O1}\|R_{I2})[C_{c1} + (1 + R_{O2}\|R_{I3}/R_{e2})C_{c2}]}$$

However, when the load of this stage is bypassed at a higher frequency, as occurs at f_{p3} above, Miller-effect capacitance C_{I2} also decreases, resulting in a response zero defined by

$$f_{z1} = f_{p3}$$

The decrease in C_{I2} ends when its high-frequency limit of $C_{c2}/2$ is reached and a new input circuit pole occurs at

$$f'_{p2} = \frac{1}{\pi(R_{O1}\|R_{I2})(C_{c1} + C_{c2})}$$

The remaining response pole of the amplifier considered results from the reaction of the first-stage input capacitance with the stage input resistance and the net source resistance presented to the two amplifier inputs. As modeled in Fig. 4.5c, C_{I1} will cause a pole at

$$f_{p1} = \frac{1}{2\pi(R_{I1}\|2R_G)C_{I1}}$$

Depending on the source resistance level, C_{I1} can be approximated by its upper or lower limit, as done above for the other stages.

From this frequency response analysis and the preceding operational amplifier low-frequency gain analysis, the overall gain can be approximated by

$$A(j\omega) = \frac{A_{O1}A_{O2}A_{O3}A_{O4}}{2\left(1 + j\frac{f}{f_{p1}}\right)\left(1 + j\frac{f}{f_{p2}}\right)\left(1 + j\frac{f}{f'_{p2}}\right)\left(1 + j\frac{f}{f_{p4}}\right)}$$

where the canceling effects of f_{p3} and f_{z1} are omitted. The approximate frequency can be drawn using a straight-line approximation as in Fig. 4.6. Straight-line approximations of response curves are made, using the common Bode plot techniques. These techniques are reviewed in the next chapter with operational amplifier phase compensation discussions. As will be discussed there, the buildup of phase shift from multiple amplifier poles, such as those of the response shown, creates the need for phase compensation to ensure frequency stability when the operational amplifier is operated at lower closed-loop gain levels under feedback.

4.2 DC Input Errors and Thermal Drifts

Each stage of an operational amplifier has dc input errors including an input offset voltage, input bias currents, and an input offset current which have thermal drifts as discussed in Chapter 2. The dc error components and drifts of each stage contribute to the overall amplifier dc errors and drifts. By referring the contribution of each stage to the amplifier input, the total equivalent input dc errors and their drifts are found. This representation provides a measure of the overall amplifier dc error and drift which is independent of the amplifier application. To consider the reflection of dc errors to the input of an operational amplifier, the two-stage circuit of Fig. 4.7 is used. The results to be thus obtained for first-stage input errors due to the second stage will also hold for second-stage input errors caused by a third stage and so forth. In this way the results of the analysis may be extended to an amplifier with any number of stages. Note that the voltage between the first-stage collectors in the diagram is constrained to equal the second-stage input offset voltage, rather than to be a voltage equal to the first-stage offset voltage multiplied by its gain. This constraint

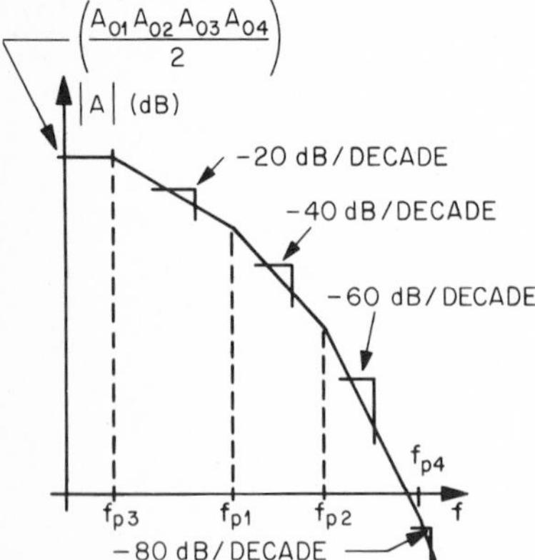

Fig. 4.6 Straight-line frequency response approximation for the amplifier of Fig. 4.4.

establishes a 0-V output as defined for an amplifier input which equals the equivalent input offset voltage V_{OS1}. In practice, dc feedback around an operational amplifier actually holds the differential dc input voltage essentially equal to the equivalent input offset voltage in order to prevent saturation of the output by very high-gain amplification of offsets. Even for output signals over the full output range, this dc input voltage remains quite close to the equivalent offset level. Only small changes in the dc differential input voltage are needed to develop full output swings since the typical dc open-loop gain is quite high.

It will be seen that, because of the gain loading effect of one bipolar transistor stage on another, the equivalent input offset voltage contribution of an intermediate stage is not simply its offset voltage divided by the voltage gain preceding it. This approximation typically holds if the loading is the high input impedance of an FET stage, but for a bipolar transistor loading stage the current unbalance created by the loading offset voltage in the preceding stage has to be considered to find the effect on equivalent input offset voltage. Adding to this current unbalance will be the input offset current of the loading stage. The equivalent input offset current resulting from loading stages can also be found for bipolar transistors stages by considering this current unbalance created by loading. When the equivalent input dc errors due to intermediate stages have been found as described, the results hold for either bipolar transistor or FET loading stages.

Unbalanced stage currents result from mismatches in load resistors and from the input offset voltage and current of the loading stage. For the small unbalances typically encountered, each effect may be separately considered to simplify the describing expressions. Since the

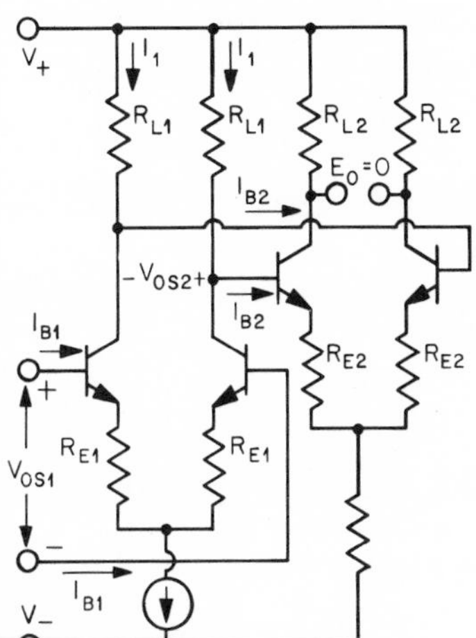

Fig. 4.7 Two-stage section of an operational amplifier.

first-stage differential output bias in Fig. 4.7 is virtually fixed by the second stage as mentioned earlier, a collector-resistor mismatch in the first stage will unbalance the stage currents by

$$\Delta I_1 = \frac{\Delta R_{L1}}{R_{L1}} I_1 \qquad \text{due to } \Delta R_{L1}$$

Considering next the second-stage offset voltage, it unbalances the first-stage load voltages, establishing a differential current of

$$\Delta I_1 = \frac{V_{OS2}}{R_{L1}} \qquad \text{due to } V_{OS2}$$

Finally, because of input offset current of the loading stage, there results an additional first-stage current unbalance of

$$\Delta I_1 = I_{OS2} \triangleq \Delta I_{B2} \qquad \text{due to } I_{OS2}$$

Depending on the direction of unbalance, each of the above factors adds to or subtracts from the others, resulting in a net stage current unbalance of

$$\Delta I_{1D} = \frac{\Delta R_{L1}}{R_{L1}} I_1 \pm \frac{V_{OS2}}{R_{L1}} \pm I_{OS2} \tag{4-8}$$

for differential loading. The thermal sensitivity of the unbalance, found by differentiating the last expression, results from the input offset voltage and current drifts of the loading stage as expressed by

$$\frac{d}{dT} \Delta I_{1D} = \frac{1}{R_{L1}} \frac{dV_{OS2}}{dT} \pm \frac{dI_{OS2}}{dT} \tag{4-9}$$

If the loading stage is single-ended, such as the case in Fig. 4.8, the dc feedback controls the current in the loaded side of the first stage to provide the 0-V output. The first-stage current unbalance is determined by this

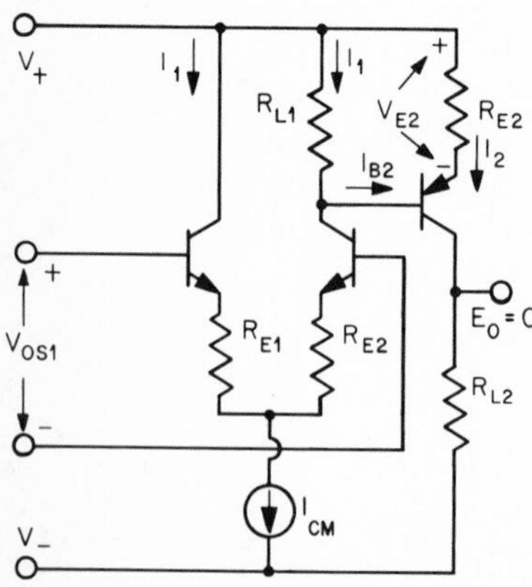

Fig. 4.8 Differential stage loaded by a single-ended stage.

feedback constraint and by deviations from nominal biasing values. Since the voltage on the first-stage collector resistor is again fixed by the loading stage, a deviation in the load resistor R_{L1} from nominal unbalances the first-stage currents, as did the resistor mismatch previously considered. Differences in second-stage input bias voltage and current and from their design nominals will unbalance the first-stage currents in the same manner as did the input offset voltage and current of a differential loading stage. An additional contributor to the current mismatch will be any deviation in the total stage current I_{CM}, from its design center. Considering all the above for a single-ended loading stage, the total current unbalance in the loaded stage will then be

$$\Delta I_{1S} \doteq \frac{\Delta R_{L1}}{R_{L1}} I_1 \pm \frac{\Delta V_{BE2} + \Delta V_{E2}}{R_{L1}} \pm \Delta I_{B2} \pm \Delta I_{CM} \quad \text{for single-ended loading} \qquad (4\text{-}10)$$

In general, the emitter-base voltage will be within 20 mV of nominal for no significant variation in its current. The other deviations are determined by component tolerances. The principal temperature sensitivities of this unbalance are due to drift in the input biasing voltage and current of the loading stage as found by differentiating the result of Eq. (4-10) to get

$$\frac{d}{dT} \Delta I_{1S} = \frac{1}{R_{L1}} \left(\frac{dV_{BE2}}{dT} + \frac{dV_{E2}}{dT} \right) \pm \frac{dI_{B2}}{dT} \qquad (4\text{-}11)$$

where

$$\frac{dV_{BE}}{dT} \simeq 2.2 \text{ mV/°C}$$

For both the differential and single-ended bipolar transistor loading stages considered above, the major sources of the current unbalances are the loading-stage input current terms.

In the case of FET loading stages, such as in Fig. 4.9, for the single-ended case, similar current unbalances result, causing offsets and drifts. The input current of a loading FET stage consists of only the gate leakage current, which is negligibly small, as compared with bipolar transistor loading for which offset current loading transmits the dominant loading effect. Otherwise, the current unbalances developed by FET loading stages parallel those derived previously. Considering Eq. (4-8), the current unbalance for the analogous FET loading will be

$$\Delta I_{1D} = \frac{\Delta R_{L1}}{R_{L1}} I_1 \pm \frac{V_{OS2}}{R_{L1}} \qquad (4\text{-}12)$$

for differential FET loading. The thermal drift of the unbalance is due only to the input offset voltage drift of the loading stage as indicated by

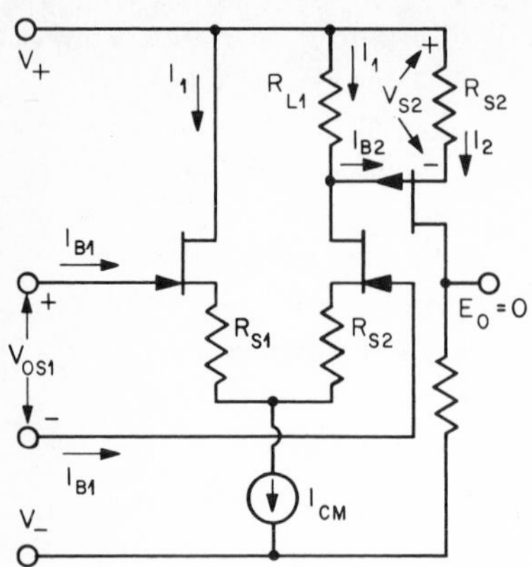

Fig. 4.9 Single-ended FET stage.

its derivative

$$\frac{d}{dT}\Delta I_{1D} = \frac{1}{R_{L1}}\frac{dV_{OS2}}{dT} \tag{4-13}$$

Similarly, when the loading is single-ended as represented in Fig. 4.9, the current unbalance follows from Eq. (4-10) and will be

$$\Delta I_{1S} = \frac{\Delta V_{GS2} + \Delta V_{S2}}{R_{L1}} \pm \Delta I_{CM} \tag{4-14}$$

for single-ended FET loading. Once again the stabilities of the loading stage input bias characteristics determine the drift of I_1 as shown by the derivative below:

$$\frac{d}{dT}\Delta I_{1S} = \frac{1}{R_{L1}}\frac{dV_{GS2}}{dT} \pm \frac{R_2}{R_{L1}}\frac{dI_2}{dT} \tag{4-15}$$

Using the preceding current unbalance representations, the dc bias loading effects of bipolar transistor and FET stages can be added to a loaded-stage current for cases of both differential and single-ended loading. By combining the effects of such loading with the input bias characteristics of bipolar transistor and FET differential stages developed in Chapter 2, the total equivalent input offset voltage, bias current, and offset current can be found for operational amplifiers formed by any combination of stages. Associated thermal drifts are similarly resolved. The input offset current of a bipolar transistor operational amplifier results from the collector current unbalance described above and the beta mismatch discussed in Sec. 2.3. The total input offset current is

$$I_{OS1} = \pm\frac{\Delta I_1}{\beta_1} \pm \frac{\Delta\beta_1}{{\beta_1}^2} I_1 \tag{4-16}$$

For small current unbalances, the input bias currents, on the other hand, are not appreciably affected by the loading stage, and the input bias currents of the operational amplifier are essentially those of the unloaded input stage, which are

$$I_{B1} \doteq \frac{I_1}{\beta_1} \tag{4-17}$$

Thermal drifts of the input currents result from the temperature sensitivities of beta and the current unbalance as described in Chapter 2. From Eq. (2-29) the input bias current drift of an operational amplifier having a bipolar transistor input stage is

$$\frac{dI_{B1}}{dT} = -\left(\frac{1}{\beta_1}\frac{d\beta_1}{dT}\right) I_{B1} \doteq CI_{B1}(25°C) \tag{4-18}$$

where

$$C = -0.005/°C \qquad T > 25°C$$
$$C = -0.015/°C \qquad T < 25°C$$

The offset current drift of a balanced stage in Chapter 2 was approximated as

$$\frac{dI_{OS}}{dT} = CI_{OS}(25°C)$$

As discussed in that earlier analysis, this relationship holds for the drift of the second term in the offset current expression of Eq. (4-16), and the drift of the total equivalent input offset current for bipolar transistors becomes

$$\frac{dI_{OS1}}{dT} = \left(\frac{1}{\beta_1}\frac{d}{dT}\,\Delta I_1 - \frac{\Delta I_1}{\beta_1^2}\frac{d\beta_1}{dT}\right) \pm C\left(\frac{\Delta\beta_1}{\beta_1^2}\,I_1\right)\Bigg|_{25°C}$$

The second term of this result can be simplified, using the beta temperature coefficient approximation used in Eq. (4-18). Then

$$\frac{dI_{OS1}}{dT} = \frac{1}{\beta_1}\left(\frac{d}{dT}\,\Delta I_1 - C\,\Delta I_1\right) \pm C\left(\frac{\Delta\beta_1}{\beta_1^2}\,I_1\right)\Bigg|_{25°C} \tag{4-19}$$

The equivalent input offset voltage is created by mismatches in emitter-base voltages and emitter degeneration resistors along with the stage current unbalance discussed above. The current unbalance results in an offset voltage as related by the stage transconductance in the form

$$V_{OS1} \doteq \Delta I_1 R_{e1} \qquad \text{due to } \Delta I_1 \qquad R_{e1} = R_{E1} + r_{e1}$$

Unequal emitter degeneration resistors produce a direct input offset of

$$V_{OS1} = I_1\,\Delta R_{E1} \qquad \text{due to } \Delta R_{E1}$$

Combining the last two expressions with emitter-base voltage difference gives

$$V_{OS1} = \Delta V_{BE1} \pm \Delta I_1 R_{e1} \pm I_1\,\Delta R_{E1} \overset{T}{=} 1.5 \text{ mV} \tag{4-20}$$

where ΔI_1 is given by Eqs. (4-8), (4-10), (4-12), or (4-14). Input offset voltage drift is then

$$\frac{dV_{OS1}}{dT} = \frac{d}{dT}\Delta V_{BE1} \pm \left(\Delta I_1 \frac{dr_{e1}}{dT} + R_{e1}\frac{d}{dT}\Delta I_1\right)$$

where $(d/dT)\,\Delta I$ is given by Eqs. (4-9), (4-11), (4-13), or (4-15). Using the drift result of Eq. (2-7) with the first term above, the input offset voltage drift resulting from emitter-base voltage mismatch is replaced by the mismatch divided by the temperature or

$$\frac{d}{dT}\Delta V_{BE1} = \frac{\Delta V_{BE1}}{T} \qquad T \text{ in } {}^\circ K = {}^\circ C + 273^\circ$$

The second term of the voltage drift expression may be written in terms of the fractional current unbalance.

$$\Delta I_1 \frac{dr_{e1}}{dT} = \Delta I_1 \frac{d}{dT}\frac{KT}{qI_1} = \frac{K\,\Delta I_1}{qI_1} \qquad \frac{K}{q} = 8.6 \times 10^{-5}\ V/{}^\circ K$$

The total equivalent input offset voltage thermal drift of a differential stage loaded by a following stage is then

$$\frac{dV_{OS1}}{dT} = \frac{\Delta V_{BE1}}{T} \pm \left(\frac{K}{q}\frac{\Delta I_1}{I_1} + R_{e1}\frac{d}{dT}\Delta I_1\right) \overset{T}{=} 5\ \mu V/{}^\circ C \qquad (4\text{-}21)$$

$$R_{e1} = R_{E1} + r_{e1} \qquad r_{e1} = \frac{KT}{qI_1}$$

where ΔI_1 and $(d/dT)\,\Delta I_1$ are as given earlier. As expressed above, input offset voltage drift is created by emitter-base voltage mismatch, the current unbalance, and temperature sensitivity of the unbalance.

Current unbalances in an FET stage affect only the input offset voltage and its drift. The input bias and offset currents and their related drifts for an amplifier having an FET input stage are due to the gate leakage currents as described in Sec. 2.3. Then, from Eqs. (2-35) and (2-36), the input bias and offset currents of an FET input operational amplifier will be approximated by

$$I_{B1} \doteq I_{GSS1}\sqrt{\frac{V_{GD1} + V_{GS1}}{2V_{GST}}} \qquad (4\text{-}22)$$

$$I_{OS1} \doteq \Delta I_{GSS1}\sqrt{\frac{V_{GD1} + V_{GS1}}{2V_{GST}}} \qquad (4\text{-}23)$$

where V_{GST} is the test voltage used in measuring I_{GSS}. Thermal drift of the amplifier input bias currents follows from Eq. (2-37).

$$I_B(T) = I_B(T_1)2^{(T-T_1)/10} \qquad (4\text{-}24)$$

Except above room temperature this input bias current is limited by package leakage, as discussed in Sec. 2.3.

The input offset voltage for an FET input is affected by loading in a manner similar to that of the bipolar transistor case. A current unbalance creates offset voltage as reflected by the transconductance, and unequal resistors in the source leads result in additional differential input error voltage. Combined with the gate-to-source voltage mismatch, these errors create the total FET amplifier equivalent input offset voltage of

$$V_{OS1} = \Delta V_{GS1} \pm \frac{\Delta I_1 R_{S1}}{1 + g_{fs1}R_{S1}} \pm I_1\,\Delta R_{S1} \overset{T}{=} 20\text{ mV} \qquad (4\text{-}25)$$

where ΔV_{GS1} is the mismatch under balanced currents. Again drawing from the earlier differential-stage drift results of Chapter 2, the overall amplifier input offset voltage drift is resolved. That portion of the drift related to the FET mismatch was found in Sec. 2.2 to be

$$\frac{dV_{OS}}{dT} = 3.5 \times 10^{-3}(\Delta V_{GS} - \Delta V_P) \qquad \text{due to FET mismatch}$$

The effect of unequal drain currents was resolved in terms of the ratio of the current unbalance to the zero-drift drain current in Eq. (2-26). From that result

$$\frac{dV_{OS}}{dT} = (-1.1\text{ mV/°C})\,\frac{\Delta I_1}{I_{DZ}} \qquad \text{due to } \Delta I_1$$

As a result, total amplifier drift for the FET input case is

$$\frac{dV_{OS1}}{dT} \doteq 3.5 \times 10^{-3}(\Delta V_{GS1} - \Delta V_{P1}) \pm (1.1\text{ mV/°C})\,\frac{\Delta I_1}{I_{DZ}} \overset{T}{=} 40\ \mu\text{V/°C} \qquad (4\text{-}26)$$

In summary, the effects of bipolar or field-effect transistor loading stages on the input offsets and drifts of a differential stage have been defined in terms of a current unbalance in the loaded stage. The result of this unbalance is combined with the errors of input stages of both transistor types in expressions defined in terms of the current unbalance. By choosing the unbalance relationship for the appropriate type of loading stage and by using the applicable input error expression, the preceding results will define input offsets and drifts for any of the combinations of bipolar or field-effect transistor differential or single-ended stages. Starting with the last two stages of a given operational amplifier, the complete analysis is performed reflecting offsets and drifts through each stage to the amplifier input.

4.3 Noise Characteristics and Optimum Noise Performance Conditions

Within an operational amplifier are many sources of noise which may be represented by equivalent noise voltage and current generators at the input of each stage in the manner applied to a differential stage in Sec. 2.4. By further reflecting the noise contributions of each stage to the amplifier inputs, the noise characteristics of an operational amplifier may be represented by one set of equivalent input noise voltage and current generators. In this form the noise components of the amplifier can be compared directly with an input signal to determine the signal-to-noise ratio resulting with the amplifier. A signal-to-noise ratio expressed with the input signal provides a measure of noise performance which is independent of the gain or electronic function to be performed by the operational amplifier. When considering the effects of the two equivalent input noise sources, the relative importance of the equivalent input voltage noise as compared with input current noise is determined by the signal source resistance presented to the noise currents. By then considering the source resistance level to be used, the type of amplifier having the most appropriate balance of voltage and current noise can be chosen. When the equivalent input voltage noise and input current noise are defined, some circuit conditions can be optimized for noise performance. Optimum noise performance results in the highest signal-to-noise ratio and not necessarily the minimum noise figure often pursued. As will be discussed, changing circuit conditions to achieve minimum noise figure can degrade signal-to-noise ratio rather than improve it.

Equivalent input noise voltage and input noise currents of an operational amplifier are derived from the individual-stage noise components such as those of Fig. 4.10. The equivalent input noise sources are resolved in the same manner as were the equivalent input offset voltage and input bias currents in the preceding section. To reflect the input noise sources of the second stage in Fig. 4.10 to the first-stage input the noise currents that would be induced in the first-stage outputs by these sources are considered. Input noise currents of the loading stage add directly to the collector or drain currents of the first stage. Additional collector or drain currents reflecting the input noise voltage of the following stage are developed and equal that noise voltage divided by the load resistors. Combining the mean-square noise currents reflected to a differential-stage output from a loading stage results in

$$\overline{i_{no1}^2} = \overline{i_{ni2}^2} + \frac{\overline{e_{ni2}^2}}{(2R_{L1})^2}$$

Considering the above expression for the noise current developed in a

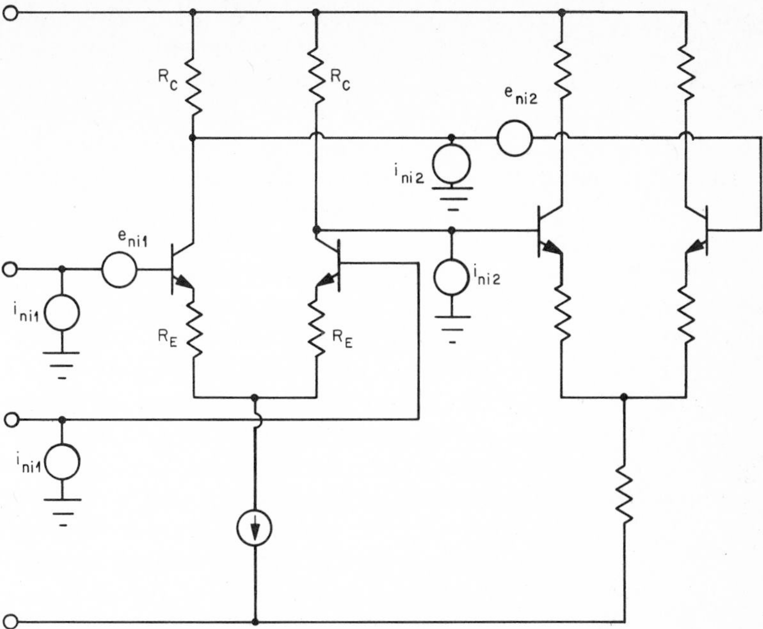

Fig. 4.10 Cascaded differential stages with separate noise representations.

stage by loading circuitry, the net input noise currents of an operational amplifier can be found. For a bipolar transistor input stage, the net equivalent input noise current will be composed of that of the input stage and i_{no1} from above divided by the beta of the stage transistors. Adding the mean-square noise components, this is expressed by

$$\overline{i_{ni}^2} = \overline{i_{ni1}^2} + \frac{\overline{i_{no1}^2}}{\beta_1^2}$$

$$\overline{i_{ni}^2} = \overline{i_{ni1}^2} + \frac{\overline{i_{ni2}^2}}{\beta_1^2} + \frac{\overline{e_{ni2}^2}}{4\beta_1^2 R_{C1}^2} \qquad (4\text{-}27)$$

Using the result of Eqs. (2-45), the individual-stage input noise currents, i_{ni1} and i_{ni2}, are a function of current level as expressed by

$$i_{ni} = \sqrt{2qI_B\,\Delta f}$$

The current levels in a loading bipolar transistor stage are generally set at less than 10 times that of the preceding stage for drift considerations discussed in Sec. 4.2. Input noise currents of the second stage will then compare with those of the first stage as expressed by

$$\overline{i_{ni2}^2} = \frac{I_{B2}}{I_{B1}} \overline{i_{ni1}^2} \leq 10\overline{i_{ni1}^2}$$

Under this condition a first-stage beta of only 30 in the noise current relationship of Eq. (4-27) makes the second term of that equation negligible. For an FET loading stage, i_{ni2} is much smaller than above and may always be neglected in the expression for equivalent input noise current i_{ni}. The relative importance of the third term of this expression is observed by considering typical noise voltage and current levels. Since both the current and voltage noises described in Sec. 2.4 have the same dependence on bandwidth, any given bandwidth may be used for comparison. For a 1-kHz bandwidth the input noise voltage of a bipolar transistor or FET second stage will be between 0.4 and 2 μV rms, for stage currents from 30 to 1 mA, as described by Eqs. (2-44) and (2-48). Considering this range of noise voltage and a beta of 100 in the noise current expressions of Eq. (4-27) the mean-square noise current due to e_{ni2} will be at least a factor of 10 less than that due to the 6 pA rms of i_{ni1} as long as R_{C1} is greater than 5 kΩ.

For all practical cases, then, loading stages do not significantly contribute to the equivalent input noise currents as expressed by Eq. (4-27), and the input noise currents of an operational amplifier having bipolar input transistors are essentially those of the input stage alone. From Eqs. (2-45), the resulting equivalent noise current at each input of an operational amplifier having a bipolar transistor input stage is

$$i_{ni} \doteq i_{ni1} = \sqrt{2qI_{B1}\,\Delta f} \tag{4-28}$$

where I_{B1} is the base current of the input stage. At low current levels the l-f noise greatly increases i_{ni} for the bipolar transistor input stage. With 0.2-μA base current, i_{ni} is typically 20 pA rms for a bandpass from dc to 1 kHz. In the case of FET inputs, all input noise currents result from gate leakage current for which shot noise was described in Chapter 2. Since gate leakage current is not significantly affected by reflected noise of a loading stage, the equivalent input noise currents are essentially independent of load noise. For the FET input operational amplifier from Eqs. (2-49) then

$$i_{ni} \doteq i_{ni1} = \sqrt{2qI_{G1}\,\Delta f} \tag{4-29}$$

where I_{G1} is the input-stage gate leakage current level. With the typical 5-pA gate leakage current, i_{ni} usually has a negligible effect on total input noise in common amplifier applications. The noise voltage created by i_{ni} on the source resistance is generally small compared with the thermal noise of the resistance.

The effect of loading-stage noise upon equivalent input noise voltage is also analyzed by reflecting the noise current i_{no1} defined above to represent the noise effects of a loading stage. A collector or drain noise current such as i_{no1} reflects to the inputs of a differential stage as a noise voltage by the analyses in Sec. 2.4. Using these previous analyses, the equivalent noise at one input of a bipolar transistor input stage due to loading by a second stage will be

$$\overline{e_{ni12}^2} = 2\overline{i_{no1}^2}\left(R_{e1} + \frac{R_G + r'_{b1}}{\beta_1}\right)^2$$

Similarly, for an FET input stage loaded by a seond stage using Fig. 2.18,

$$\overline{e_{ni12}^2} = 2\overline{i_{no1}^2}\left(\frac{1 + g_{fs1}R_{S1}}{g_{fs1}}\right)^2$$

Addition of this input noise to a similar noise at the other input and to the input noise voltage of the first stage itself results in a net equivalent input noise voltage of

$$\overline{e_{ni}^2} = \overline{e_{ni1}^2} + 2\overline{e_{ni12}^2}$$

Replacing the reflected second term above by using the preceding expressions for $\overline{e_{ni12}^2}$ and $\overline{i_{no1}^2}$ resolves this expression in terms of the individual-stage noise sources. For the bipolar transistor stage

$$\overline{e_{ni}^2} = \overline{e_{ni1}^2} + 4\left(\overline{i_{ni2}^2} + \frac{\overline{e_{ni2}^2}}{4R_{C1}^2}\right)\left(R_{e1} + \frac{R_G + r'_{b1}}{\beta_1}\right)^2 \qquad (4\text{-}30)$$

For the loaded FET differential stage

$$\overline{e_{ni}^2} = \overline{e_{ni1}^2} + 4\left(\overline{i_{ni2}^2} + \frac{\overline{e_{ni2}^2}}{4R_{D1}^2}\right)\left(\frac{1 + g_{fs1}R_{S1}}{g_{fs1}}\right)^2 \qquad (4\text{-}31)$$

In both cases above, the input noise voltage of the loading stage reflects to the first-stage input divided by the unloaded gain of the stage, which is approximated in Chapter 1 as

$$A_O \doteq \frac{-R_C}{R_e + R_G/\beta} \qquad \text{or} \qquad A_O \doteq \frac{-g_{fs}R_D}{1 + g_{fs}R_S}$$

For the high gain normally developed in the input stage, the effect of second-stage noise voltage is greatly reduced. In order to compare the relative magnitudes of e_{ni1} and e_{ni2} the differential-stage equivalent input noise voltage expressions from Sec. 2.4 are used. Depending on whether the stages use bipolar transistors or FETs, e_{ni1} and e_{ni2} will be of the form

$$e_{ni} = 2\sqrt{2qI_C\,\Delta f\left(R_e + \frac{r'_b + R_G}{\beta}\right)^2 + 2KT(R_E + r'_b)\,\Delta f}$$

$$e_{ni} = 4\left(\frac{1}{g_{fs}} + R_S\right)\sqrt{2KT\,\Delta f\left(g_{fs} + \frac{1}{R_D}\right)}$$

The greatest difference in noise voltage contributions e_{ni1} and e_{ni2} of two cascaded bipolar transistor stages will result for a second-stage collector current which is much greater than that in the input stage. For drift considerations the ratio of the stage currents is again limited to a factor of 10 for which the bipolar stage input noise expression above predicts

$$\overline{e_{ni2}^2} \le 10\overline{e_{ni1}^2}$$

Such a variation in stage noise voltages does not result between FET stages as indicated by their noise relationship above. From the examples of Chapter 2, the input noise voltages of FET and bipolar input stages are roughly comparable. Thus, for any combination of FET and bipolar stages, $\overline{e_{ni2}^2}$ will reach a level which is only about 10 times $\overline{e_{ni1}^2}$. Then if the effect of $\overline{e_{ni2}^2}$ upon the equivalent input noise voltages of Eqs. (4-30) and (4-31) is divided by the square of a gain of only 10, A_O^2, the equivalent mean-square input noise voltage reflected by $\overline{e_{ni2}^2}$ will be one-tenth that due to $\overline{e_{ni1}^2}$. For the general case, then, $\overline{e_{ni2}^2}$ may be neglected.

The other reflected term of each equivalent input noise voltage expression in Eqs. (4-30) and (4-31) results from the second-stage input current noise reflected by the inverse of the first-stage transconductance. Input noise currents of the loading stage will be greatest when this stage is a bipolar transistor type operated in the 1/f noise frequency range from dc to 1 kHz or when it is operated at high currents. In either case i_{ni2} will typically be limited to no more than 30 pA rms. At this level the resulting component of e_{ni} in Eqs. (4-30) and (4-31) is masked by the 0.4 μV rms level e_{ni} for a first-stage transconductance greater than 240 μmhos. For most input stages the effect of second-stage noise currents, as well as noise voltage, may then be neglected in resolving equivalent input noise sources. Noise due to even later stages is further reduced in effect by the intervening stages. Equivalent input noise voltage of an operational amplifier is then essentially that of the first stage as given by Eqs. (2-44) and (2-48). For the bipolar transistor input operational amplifier

$$e_{ni} \doteq e_{ni1} = 2\sqrt{2qI_{C1}\,\Delta f\left(R_{e1} + \frac{r'_{b1} + R_G}{\beta_1}\right)^2 + 2KT(R_{E1} + r'_{b1})\,\Delta f} \tag{4-32}$$

where

$$R_{e1} = R_{E1} + r_{e1}$$

For the FET input case

$$e_{ni} \doteq e_{ni1} = 4\left(\frac{1}{g_{fs1}} + R_{S1}\right)\sqrt{2KT\,\Delta f\left(g_{fs1} + \frac{1}{R_{D1}}\right)} \tag{4-33}$$

From the examples of Sec. 2.4 using a 1-kHz bandwidth a value of e_{ni} for typical bipolar transistor or FET input operational amplifiers is predicted to be about 0.4 μV rms. Although the bipolar transistor case is fairly well approximated by this prediction from the shot noise and thermal noise considered, FET input operational amplifiers typically display around 1 μV rms equivalent input noise voltage under the conditions described, because of higher l-f noise. When all noise sources of an operational amplifier have been reflected to its inputs, the amplifier can be represented as in Fig. 4.11.

The signal-to-noise ratio resulting with an operational amplifier is determined by the equivalent input noise voltage, the input noise currents, and the source resistance presented to the noise currents. Those noise voltages at the amplifier inputs created by input noise currents in the source resistances combine with the equivalent input noise voltage to produce a total input noise voltage. In general, the source resistances presented to the two inputs are made as nearly equal as possible so that the associated effects of the two input bias currents will cancel. When the two resistances are equal and equal input noise currents are considered, the total input noise voltage is

$$e_{nit} = \sqrt{\overline{e_{ni}^2} + 2\overline{i_{ni}^2}R_G^2} \qquad (4\text{-}34)$$

By considering the relative importance of the individual terms of the expression for e_{nit}, a simplification can be made for the equivalent input noise voltage expression for the bipolar transistor input amplifier. From the noise voltage and current expressions of Eqs. (4-32) and (4-28), the total input noise voltage will be

$$e_{nit} = \sqrt{8qI_{C1}\,\Delta f\left(R_{e1} + \frac{r'_{b1} + R_G}{\beta_1}\right)^2 + 8KT(R_{E1} + r'_{b1})\,\Delta f + 4qI_{B1}\,\Delta f R_G^2}$$

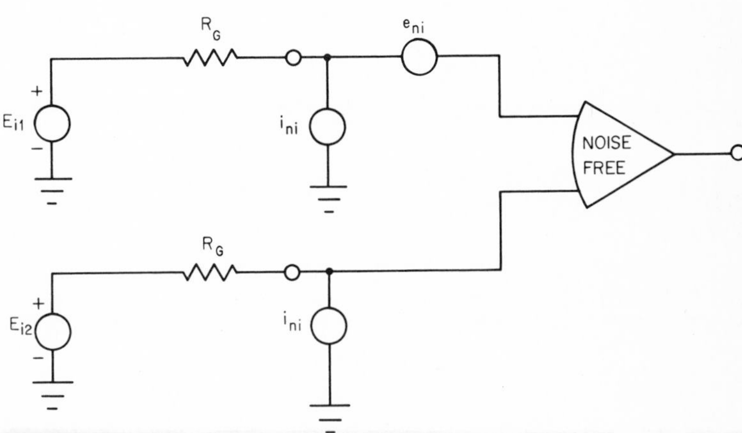

Fig. 4.11 Equivalent noise representation of an operational amplifier.

Combining terms by using $I_B = I_C/\beta$, and simplifying by considering the general case for $R_e \gg r'_b/\beta$, results in

$$e_{nit} = \sqrt{8qI_{C1}\,\Delta f\left(R_{e1}^2 + \frac{2R_{e1}R_G}{\beta_1} + \frac{R_G^2}{\beta_1^2} + \frac{R_G^2}{2\beta_1}\right) + 8KT(R_{E1} + r'_{b1})\,\Delta f}$$

The first series of terms in parentheses above contains the dependency of e_{nit} upon source resistance, and this portion may be simplified. For $\beta \gg 2$, the third term in this series is negligible in comparison with the fourth. At very low values of source resistance the second term of this series is much less than the first and becomes significant by comparison only when R_G reaches $0.05\beta R_e$. At this point the second term is one-tenth the first term. However, for a beta of 100 the fourth term is then one-eighth the size of the first, and beyond this level of source resistance the fourth term increases much faster than does the second. In neglecting the second term of the series discussed, an 8 percent maximum error results at the source resistance level discussed. Omitting the second and third terms as neglected above and writing e_{nit} in its original form give

$$e_{nit} \doteq \sqrt{8qI_{C1}\,\Delta f R_{e1}^2 + 8KT(R_{E1} + r'_{b1})\,\Delta f + 4qI_{B1}\,\Delta f R_G^2}$$

As a result of the approximations made, the dependence of equivalent input noise voltage upon source resistance has been neglected since the associated noise is small in comparison with that resulting from the input noise currents flowing in the source resistances. Comparing the above result with the basic relationship for e_{nit} [Eq. (4-34)], the simplified result for equivalent input noise voltage of a bipolar transistor input operational amplifier is found to be

$$e_{ni} \doteq 2\sqrt{2qI_{C1}\,\Delta f R_{e1}^2 + 2KT(R_{E1} + r'_{b1})\,\Delta f} \qquad (4\text{-}35)$$

Evaluation of operational amplifier signal-to-noise ratio is made by considering the total input noise voltage e_{nit} of Eq. (4-34). With all noise sources reflected to the input and combined in e_{nit} the noise can be compared directly with the input signal independent of the function to be performed by the amplifier. Depending upon the source resistance level, either the equivalent input noise voltage or the input noise currents may have a dominant effect upon the total noise e_{nit}. The various basic types of operational amplifiers have greatly different levels of equivalent input noise voltages and currents. As a result, minimum total noise can be achieved by selecting the type of amplifier which provides the lowest level of e_{nit} at the source resistance level under consideration. Representative curves of e_{nit} for bipolar transistor input, FET input, and chopper-stabilized operational amplifiers are presented

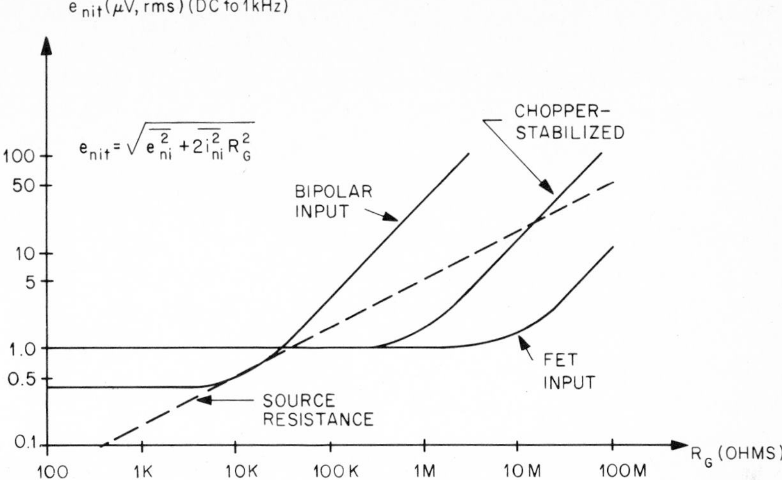

Fig. 4.12 Representative curves of total input noise voltage versus source resistance for various types of operational amplifiers. (*Empirical results from Burr-Brown Models* 3050/01, 3071/25 *and* 3307/12C.)

in Fig. 4.12 for a bandpass of dc to 1 kHz. Compared with these curves is the thermal noise of the combined source resistance, $2R_G$, which is

$$e_{ng} = \sqrt{8KT\,\Delta f R_G}$$

Since the thermal noise of a source resistance shown places a lower limit on the noise added to a signal, a continued improvement of amplifier noise provides diminishing returns. As shown, the source resistance noise will be the dominant source of noise over certain resistance ranges for FET input and chopper-stabilized operational amplifiers.

Each of the amplifier noise curves shown in Fig. 4.12 displays regions over which first its voltage noise and then its current noise component are dominant. The rising portions of each curve result when input noise currents create noise voltages with the source resistances which are significant in comparison with the equivalent input noise voltage. Since most chopper-stabilized operational amplifiers have single-ended rather than differential inputs, only one source resistance is considered for this case. At any given source resistance the type of amplifier having the lowest curve provides the lowest total noise. Below 30 kΩ source resistance, the 0.4 μV rms noise voltage of bipolar transistor input amplifiers makes this type desirable. Above 30 kΩ, the 20 pA rms input noise currents make the total noise with such amplifiers greater than the 1.0 μV rms noise voltage common to FET input or chopper-stabilized operational amplifiers. Beyond the 300 kΩ level the 1.5 pA rms input noise current of the chopper-stabilized amplifier results in greater total noise than that developed by the FET input amplifier which has 0.06 pA rms input noise currents.

In describing the noise characteristics of an amplifier it is convenient to consider noise occurring over a specified frequency range exclusive of all noise outside this range. Measurement of just this noise component would require a rectangular system bandpass having infinite response selectivity such as represented in Fig. 4.13. However, a practical amplifier bandpass which is not rectangular can be related to an equivalent rectangular response and an associated effective noise bandwidth by comparing the resulting noise levels. From Sec. 2.4 the mean-square noise voltage resulting from spectral noise density S_v applied to transfer function $H(j\omega)$ is

$$\overline{e_n{}^2} = \int_0^\infty S_v|H(j\omega)|^2 \, df$$

By equating this mean-square noise voltage developed with the amplifier transfer function and that from an arbitrary rectangular bandpass, an effective noise bandwidth is found which resolves noise characteristics in terms of the convenient rectangular bandpass case. In the simplest case the transfer function of an operational amplifier applied as a voltage amplifier will be a single-pole low-pass function as would be indicated by its gain response curve. For this case the single-pole response of Fig. 4.13 can be compared with the noise equivalent rectangular response shown to determine the effective noise bandwidth of the amplifier. Considering a constant spectral noise density, as holds for shot and thermal noise, the mean-square noise voltages for the two responses are equated below:

$$\overline{e_n{}^2} = S_v \int_0^\infty \frac{A_O{}^2}{|1 + j\omega/\omega_p|^2} \, df = S_v \int_0^{f_e} A_O{}^2 \, df$$

Solving the above provides the effective noise bandwidth of the amplifier response in terms of its pole frequency.

$$\Delta f = f_e = \frac{\pi}{2} f_p \tag{4-36}$$

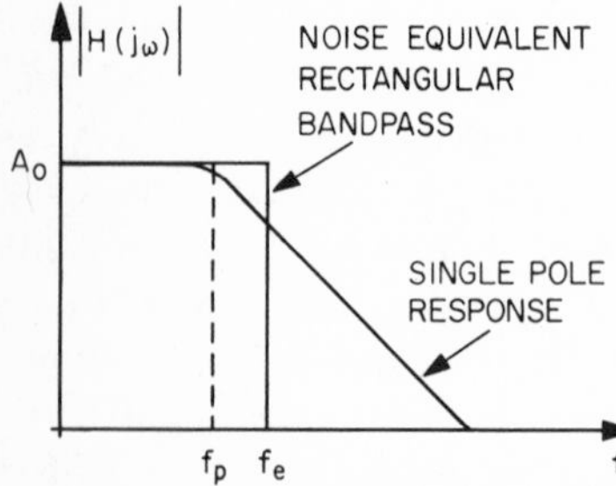

Fig. 4.13 Single-pole response and rectangular response resulting in equivalent noise performance.

Having developed relationships describing operational amplifier noise characteristics in the preceding material of this section, it is now possible to consider optimization of noise performance. Optimum noise performance is achieved with an operational amplifier when circuit conditions are chosen to maximize signal-to-noise ratio and not necessarily when these conditions are chosen to minimize noise figure. This can be seen from a closer examination of the meaning of noise figure. Noise figure (N.F.) is a comparison of the power signal-to-noise ratio at the input to the same ratio at the output as defined by

$$\text{N.F.} = 10 \log \frac{(\text{S/N}) \text{ power, input}}{(\text{S/N}) \text{ power, output}}$$

Note that a lower noise figure can be attained by making the input signal-to-noise ratio worse, but this can only make the noise performance worse. To write the noise-figure expression in terms of the equivalent input noise sources of an operational amplifier, consider

$$\text{N.F.} = 10 \log \frac{(E_i^2/R_I)/(\overline{e_{ng}^2}/R_I)}{(E_o^2/R_L)/(\overline{e_{no}^2}/R_L)}$$

where $\overline{e_{ng}^2}$ and $\overline{e_{no}^2}$ are the mean-square noise voltages from the source and at the output. Since $E_o = AE_i$ and $e_{no} = A\sqrt{e_{nit}^2 + e_{ng}^2}$,

$$\text{N.F.} = 10 \log \left(1 + \frac{\overline{e_{nit}^2}}{\overline{e_{ng}^2}}\right)$$

Considering the noise of the source e_{ng} above as just the thermal noise of $2R_G$ and using Eq. (4-34) for e_{nit}, the noise figure becomes

$$\text{N.F.} = 10 \log \left(1 + \frac{\overline{e_{ni}^2} + 2\overline{i_{ni}^2}R_G^2}{8KTR_G\,\Delta f}\right) \tag{4-37}$$

Traditionally, that source resistance at which the noise figure is minimum is defined as an optimum source resistance. However, the value of source resistance providing optimum noise performance is zero since this provides the lowest total noise added to the signal. Increased source resistance can only increase the noise added to the signal by its interaction with the amplifier noise currents and by the thermal noise of the source resistance itself. As a result, noise figure does not provide an indication of circuit conditions which provide optimum noise performance. Rather, noise figure provides a comparison of noise performance of various amplifiers under given fixed circuit conditions. To compare noise performance of the basic types of operational amplifiers their noise figures are plotted against source resistance in Fig. 4.14 using Eq. (4-37) for a band-

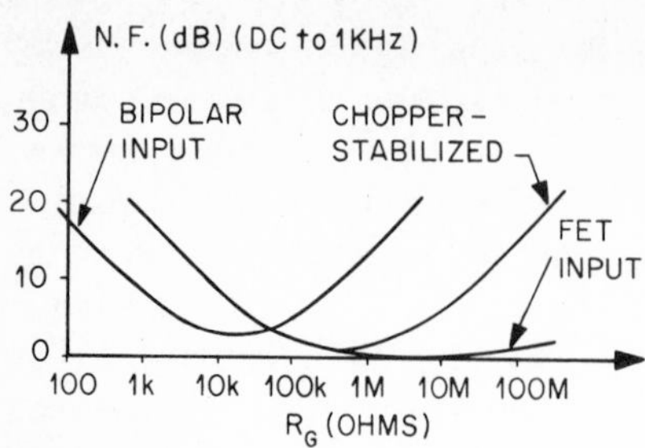

Fig. 4.14 Representative noise figure curves for various types of operational amplifiers.

pass of direct current to 1 kHz. In this form the noise figure indicates which amplifier should be used with a given source resistance for best noise performance as did the curves of Fig. 4.12. It does not, however, indicate that noise performance can be improved by increasing source resistance even though noise figure might be reduced. The output signal-to-noise ratio and not noise figure provides a general measure of noise performance as a function of circuit variables.

Only when increased source resistance also provides a larger input signal does the noise performance of an operational amplifier sometimes improve by the increase. In such cases an optimum source resistance exists for best noise performance as indicated by the highest signal-to-noise ratio. This ratio can be written for the equivalent input noise representation as

$$\frac{S}{N} = \frac{E_i}{\sqrt{\overline{e_{ng}^2} + \overline{e_{nit}^2}}} = \frac{E_i}{\sqrt{\overline{e_{ng}^2} + \overline{e_{ni}^2} + 2\overline{i_{ni}^2}R_G^2}}$$

If e_{ni} is the dominant noise term above, a simultaneous increase of R_G and E_i increases the signal more than the noise. Such a situation can result when the input to the amplifier is a current rather than a voltage,

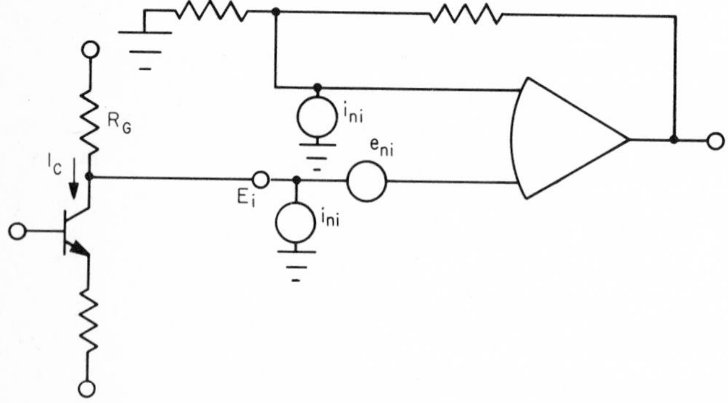

Fig. 4.15 Circuit example for which the signal increases with source resistance.

as is the case for Fig. 4.15. The collector current of the transistor shown flows in R_G to develop the input voltage E_i, and increasing this source resistance also increases the signal. For canceling dc effects from the two input bias currents the resistance presented to the other amplifier input will match R_G and the signal-to-noise ratio will be

$$\frac{S}{N} = \frac{I_c R_G}{\sqrt{8KTR_G\,\Delta f + \overline{e_{ni}^2} + 2\overline{i_{ni}^2}R_G^2}}$$

Maximum signal-to-noise ratio results for $R_G \gg e_{ni}/2i_{ni}$, $R_G \gg 4KT\,\Delta f/\overline{i_{ni}^2}$ and will be

$$\left(\frac{S}{N}\right)_{max} \doteq \frac{I_c}{2i_{ni}}$$

Noise performance can also be improved for ac applications when a transformer-coupled input may be used as in Fig. 4.16a.

For this ac case the effect of input bias current flow in source resistances is not important and the source resistances at the two inputs are

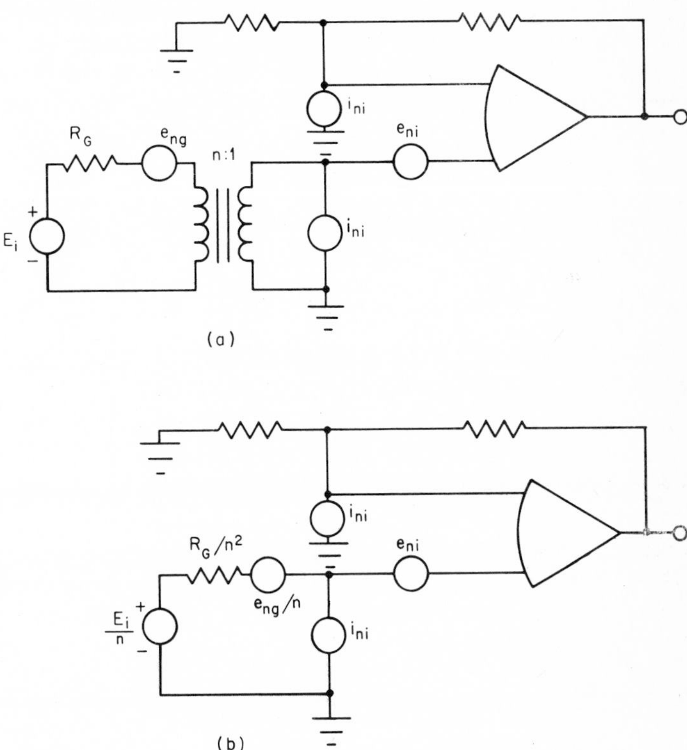

Fig. 4.16 (a) Transformer-coupled input amplifier and (b) equivalent circuit.

not matched. Instead, the feedback and summing resistance levels are minimized to make their effect on total noise small. From the equivalent circuit of Fig. 4.16a the signal-to-noise ratio will be

$$\frac{S}{N} = \frac{E_i}{\sqrt{\overline{e_{ng}^2} + n^2\overline{e_{ni}^2} + \overline{i_{ni}^2}R_G^2/n^2}}$$

Since the signal is varied along with R_G by the transformer, noise performance can be optimized by choosing the best turns ratio n. Differentiating the above it is found that maximum signal-to-noise ratio results for the optimum turns ratio given by

$$n_{opt} = \sqrt{\frac{i_{ni}R_G}{e_{ni}}}$$

Once the equivalent input noise voltage and current are known, the transformer turns ratio providing the best signal-to-noise ratio can be found. In this case alone it can be shown that minimum noise figure also happens to occur for the above turns ratio.

In summary, operational amplifier noise performance can be optimized by using a source resistance level appropriate for the type of input signal applied. For the general use of a voltage input which may extend in frequency to dc the source resistances presented to the amplifier inputs should be as low as possible to minimize the noise voltage created by these resistances with the input noise currents. Typical feedback for an operational amplifier, as in Fig. 4.16, presents a source resistance to the inverting input which equals the parallel combination of the feedback and summing resistors. When the input signal is a current, resulting in a signal voltage proportional to the source resistance, the source resistance should be made large so that signal-to-noise ratio reaches that limit set by input noise currents where effects of e_{ni} and e_{ng} become negligible. With an ac input signal the signal source should be transformer-coupled to the amplifier input. Source resistance is then altered by the square of the turns ratio but the signal is changed by only a factor equal to the turns ratio, and an intermediate level of transformed source resistance provides best noise performance. For any other specific case the level of source resistance providing best noise performance can be resolved by maximizing signal-to-noise ratio. Noise figure, if applied, should be reserved for comparisons of amplifiers under specific circuit conditions.

4.4 Chopper-stabilized and Varactor Diode Carrier Type Operational Amplifiers

In addition to the direct-coupled operational amplifier circuits discussed so far, ac-coupled techniques using modulated carriers to transmit the dc signal are used in chopper-stabilized and varactor DC amplifiers. By preceding the direct-coupled portions of such amplifiers with ac-coupled sections, the dc biasing error voltages and currents considered in Sec. 4.2 are avoided in the critical input portion of these amplifiers. Input bias currents and offset currents are not developed by the dc biasing of the amplifier since ac coupling isolates the inputs from such currents. However, the dc input currents of the direct-coupled stages produce error voltages in these stages in the same manner as considered in Sec. 4.2. The combined dc error voltages, from such currents and from the individual input offset voltages of the direct-coupled stages, are added to signals as before. The effects of these dc errors are again easily represented by an equivalent input offset voltage and its thermal drift. However, for chopper-stabilized and varactor amplifiers the effect of dc biasing errors is reduced since they are not combined with the signal until the signal has been amplified by the gain of the modulated signal section. Thus, in comparison with the amplified signal these dc errors are less significant than in the direct-coupled case, and the equivalent input dc errors of the amplifier are reduced by the gain of the ac-coupled section. Deficiencies of the modulation circuits add new errors to dc signals and generate amplifier input currents as will be described. However, the overall resulting equivalent input offset voltages and input bias or offset currents are greatly reduced by these modulated carrier techniques. The most notable performance improvements result in the thermal drifts of these dc input voltages and currents. Chopper-stabilized operational amplifiers provide around a factor of 50 improvement in input offset voltage drift over the typical bipolar transistor input operational amplifier to result in drifts on the order of 0.1 μV/°C. Although input current drifts are also reduced by using chopper stabilization, the varactor bridge input operational amplifier provides the most dramatic reduction in these drifts. Input bias current drifts of the order of 0.01 pA/°C are attained by using the varactor modulation approach.

Considering first the chopper-stabilized operational amplifier, as represented in Fig. 4.17, the input signal is separated into its high- and low-frequency components, which are then selectively amplified by the two channels of the amplifier. High-frequency signals are passed directly to the main amplifier by the filter formed by R_1 and C_1, and the dc and very low-frequency portion of the input signal is used to modulate a carrier in the chopper channel. Within the chopper channel amplifier the modulated carrier is then amplified, demodulated, and filtered to be recombined with the high-frequency portion of the signal in the main amplifier. Although the general case considered here has a single-ended input, some chopper-stabilized operational amplifiers apply similar techniques to differential input forms. Since both the channels shown are ac-coupled, the input bias currents of the main amplifier cannot

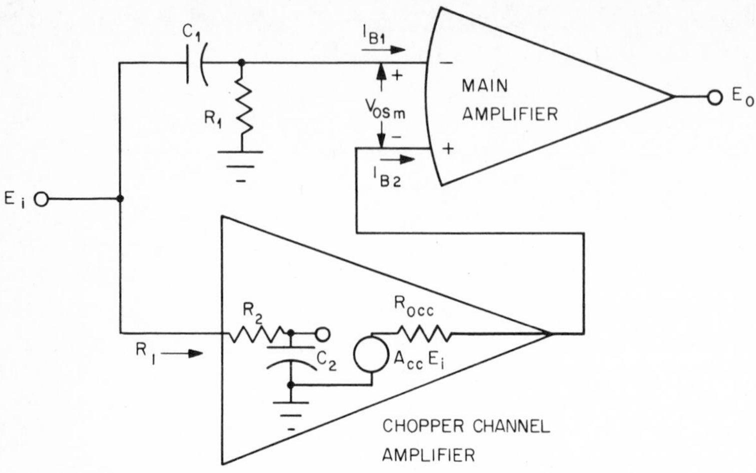

Fig. 4.17 Basic chopper-stabilized operational amplifier representation.

reach the input of the overall amplifier. Also, the error due to the input offset voltage of the main amplifier is reduced since the dc input signal is amplified in the chopper channel before combining with this error voltage. Amplification inaccuracy or nonlinearity from chopper channel modulation and demodulation is corrected by the heavy feedback normally applied around the overall amplifier and presents little error.

To demonstrate the reduction in dc errors provided by chopper stabilization, the input offset voltage and bias currents of the main amplifier represented in Fig. 4.17 can be reflected to the overall amplifier input as an equivalent input offset voltage. This equivalent input error equals the net dc error voltage developed at the chopper channel output divided by the chopper channel dc gain A_{Occ}. Combined with this voltage is an input offset voltage V_{OScc} created by the chopper channel amplifier modulation error. The equivalent input offset voltage of a chopper-stabilized operational amplifier is then

$$V_{OS} = V_{OScc} + \frac{V_{OSm} - I_{B1}R_1 + I_{B2}R_{Occ}}{A_{Occ}} \overset{T}{=} 0.1 \text{ mV}$$

In order to balance out the error voltages produced by the main amplifier input bias currents above, R_1 is commonly made equal to the chopper channel output resistance R_{Occ} so that equal input bias currents will produce canceling voltages. Differences in these resistances or in the currents result in residual error. Representing the difference in input bias currents by the input offset current I_{OSm}, the resulting input offset voltage of a chopper-stabilized operational amplifier becomes

$$V_{OS} = V_{OScc} + \frac{V_{OSm} + I_{B1}(R_{Occ} - R_1) - I_{OSm}R_{Occ}}{A_{Occ}} \qquad (4\text{-}38)$$

where V_{OScc} is the chopper channel amplifier equivalent input offset voltage and R_{Occ} is determined by the demodulation networks considered later. Thermal drift of this input offset voltage is developed by temperature sensitivities of the chopper channel gain A_{Occ} above and of the modulator errors which produce V_{OScc}, as well as the input drifts of the main amplifier. Input bias current of the chopper-stabilized operational amplifier represented in Fig. 4.17 is due primarily to the leakage current of capacitor C_1 along with the leakage current and switching spikes of the modulator network. These sources of chopper channel dc error are discussed in more detail later in conjunction with the modulator. From these factors the input bias currents of the order of 100 pA are produced with drifts typically near 0.5 pA/°C.

Improved operational amplifier input dc errors as indicated above are provided by the ac coupling and gain of the chopper channel amplifier within the limits of the modulation and demodulation circuits. The basic chopper channel amplifier form is shown in Fig. 4.18 from which its operation can be described. As indicated, the major components of the amplifier are an input low-pass filter, an input switch or chopper, an ac amplifier, an output switch, an output low-pass filter, and a drive circuit for the switches. Input filtering passes only the very low-frequency signals, and a dc signal E_I is considered for the associated operating waveforms of Fig. 4.19. When the input switch is closed, this dc signal is shorted to common, and the signal reaching the ac amplifier is the chopped waveform shown as $e_1(t)$ which is amplitude-modulated by the input signal. The amplitude of $e_1(t)$ is not simply that of the input signal since it is affected by the average voltage developed on C_2. This voltage produced by the signal results in a modulator gain, to be discussed later,

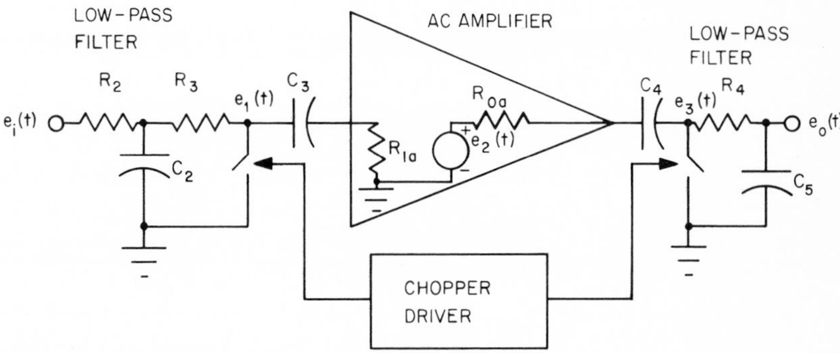

Fig. 4.18 Basic chopper channel amplifier.

which is not unity. The symmetry of the chopper waveform at the ac amplifier input, which will affect the modulator gain, is controlled primarily by that of the chopper channel drive signal. Following amplification in the ac amplifier, the signal is the ac waveform $e_2(t)$. In order to extract a dc signal from $e_2(t)$ its average value must not be left at zero as established by the ac coupling. For this reason the dc reference of the signal is restored, using the output switch, by clamping one side of the waveform to common in synchronism with the input chopping. Resulting from this dc level restoration is the signal $e_3(t)$ whose average value may be extracted by the output low-pass filter to provide the final dc output E_O.

The overall voltage gain of the chopper channel amplifier is determined

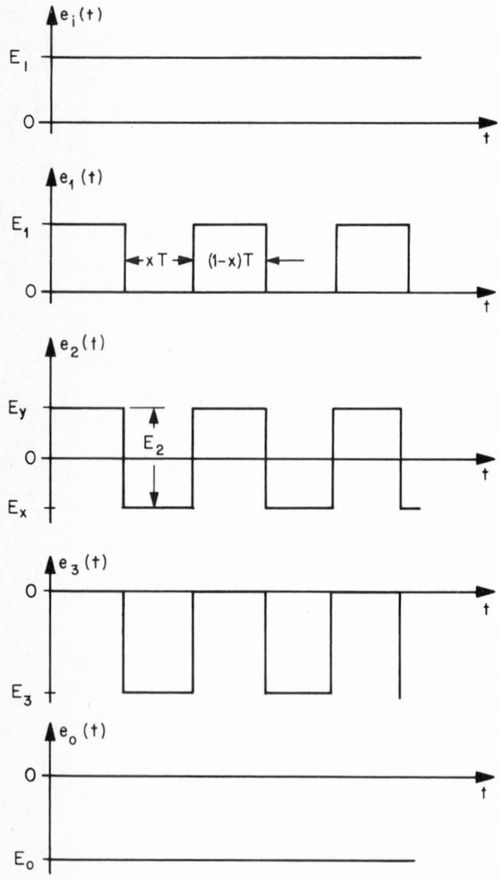

Fig. 4.19 Operating waveforms of the chopper channel amplifier of Fig. 4.18 for a dc input signal.

by the input modulator, the ac amplifier, and the output demodulator. As mentioned, the signal amplitude is not passed by the modulator switch and filter network with unity gain since the circuit capacitors assume average voltages set by the signal. By assuming that the large capacitors in these applications hold their average voltages approximately constant, the gains of the modulator and demodulator circuits are easily resolved. For the demodulator, two circuit states occur as represented in Fig. 4.20 for open and closed output switch conditions and for a signal $e_2(t)$ as presented in Fig. 4.19. When the switch is closed as in Fig. 4.20b, capacitor C_4 will charge through the low amplifier output resistance to essentially the input voltage E_y. In this state the output capacitor discharge current will be

$$I_a = \frac{E_O}{R_4}$$

When the switch opens, resulting in the circuit of Fig. 4.20c, the capacitor voltages will remain at essentially their average values of E_y and E_O, because of the large capacitances and the high resistance of R_4. For this approximation the output capacitor discharge current will be

$$I_b \doteq \frac{E_x - E_y - E_O}{R_4 + R_{Oa}} \qquad \text{for } (R_4 + R_{Oa})(C_4 + C_5) \ll (1 - x)T$$

where $(1 - x)T$ is the portion of the cycle for which the switch is open.

Using the charging and discharging currents found above, the demodulator gain can be expressed. The average value of the output capacitor voltage is that for which the charge deposited by I_b equals that drained by I_a or

$$I_a(xT) = I_b(1 - x)T$$

where xT and $(1 - x)T$ are the intervals defined in Fig. 4.19. Substitution of the previous results for I_a and I_b in the above equation defines the output voltage as

$$E_O = (E_x - E_y)(1 - x)\frac{R_4}{R_4 + R_{Oa}x}$$

Voltage gain of the demodulator is defined as the ratio of this dc output voltage to the peak-to-peak input voltage presented by $e_2(t)$. From the waveform of $e_2(t)$ its peak-to-peak voltage is

$$E_2 = E_y - E_x$$

Thus the demodulator gain is

$$A_{OD} = \frac{E_O}{E_2} = -(1 - x)\frac{R_4}{R_4 + R_{Oa}x}$$

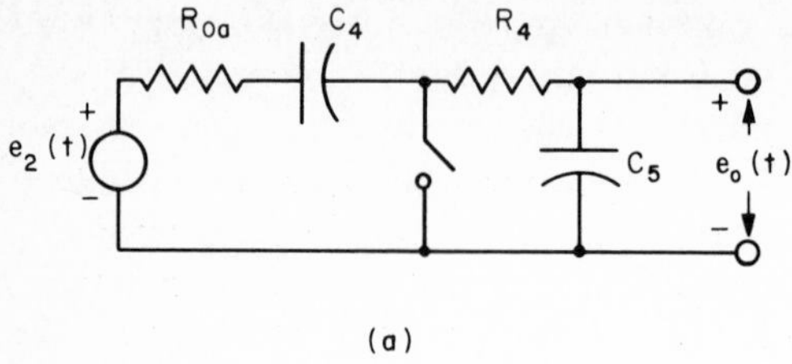

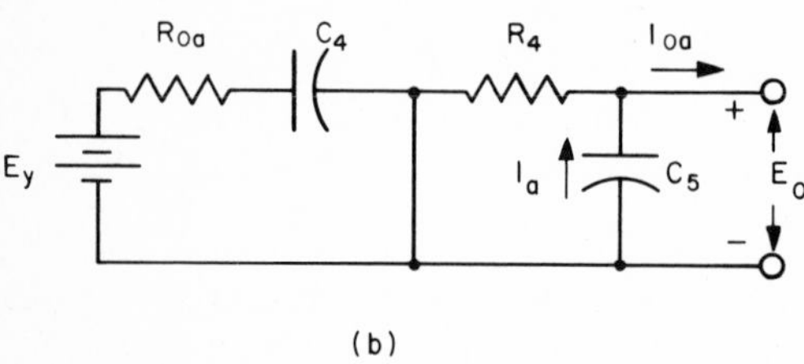

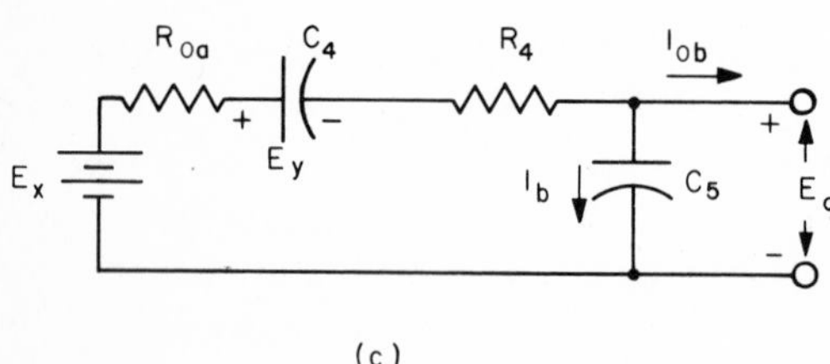

Fig. 4.20 Gain analysis circuits for the output demodulator (a) from Fig. 4.18, (b) with the switch closed, and (c) with the switch open.

In general the output resistance of the ac amplifier is much smaller than the output filter resistor and x is less than 1, so that the demodulator gain is very nearly

$$A_{OD} \doteq -(1 - x) \tag{4-39}$$

Note that for a symmetrically chopped waveform, having a value of one-half for x, the demodulator gain is one-half. Voltage gain of the modulator circuit, consisting of the input filter and switch, is found by a similar analysis[2] to be

$$A_{OM} = \frac{1}{1 + x[R_2/R_3 + (R_2 + R_3)/R_{Ia}]}$$

The input resistance of the ac amplifier, R_{Ia} in the above, is typically much larger than the input filter resistors, R_2 and R_3, and for a balanced filter the latter two resistors are equal, resulting in a modulator gain of

$$A_{OM} \doteq \frac{1}{1+x} \qquad \text{for } R_{Ia} \gg R_2 \qquad R_2 = R_3 \tag{4-40}$$

For a balanced filter having $R_2 = R_3$ and for a symmetrically chopped waveform with $x = \frac{1}{2}$ the modulator voltage gain is two-thirds.

Overall voltage gain of the chopper channel amplifier is

$$A_{Occ} = A_{OM}A_{ac}A_{OD}$$

where A_{ac} is the gain of the chopper channel ac amplifier to the chopped waveform. Substituting the gain results of Eqs. (4-39) and (4-40) in the above, the chopper channel gain is found to be

$$A_{Occ} \doteq -\frac{1-x}{1+x}A_{ac} \qquad \text{for } R_{Ia} \gg R_2 \qquad R_2 = R_3 \qquad R_{Oa} \ll R_4 \tag{4-41}$$

In the symmetrical waveform case

$$A_{Occ} \doteq \frac{A_{ac}}{3} \qquad \text{for } x = \frac{1}{2} \qquad R_2 = R_3$$

Since the chopper channel gain affects input offset voltage, the chopping waveform symmetry needs to be stable. Variations in this symmetry with temperature contribute to input offset voltage drift. The sensitivity of chopper channel gain to chopping symmetry is displayed by differentiating the gain expression of Eq. (4-41) with respect to x. Expressing the result in terms of the chopper channel gain itself yields

$$\frac{dA_{Occ}}{dx} = \frac{-2A_{Occ}}{1-x^2}$$

Rewriting this result in terms of incremental changes rather than differentials,

$$\frac{\Delta A_{Occ}}{A_{Occ}} = \frac{-2x}{1-x^2}\frac{\Delta x}{x}$$

Assuming that small variations in symmetry occur about a nominal of $x = \frac{1}{2}$, the fractional gain change can be related to the fractional symmetry change by

$$\frac{\Delta A_{Occ}}{A_{Occ}} \doteq -\frac{4}{3}\frac{\Delta x}{x} \tag{4-42}$$

The input and output resistances of the chopper channel amplifier are those of the modulator and demodulator circuits, respectively. Although input impedance is determined by both channels, the input resistance of the overall amplifier is that presented by the chopper channel amplifier, which can be found by using an analysis similar to that applied to derive the demodulator gain above. The output resistance of the chopper channel amplifier affects input offset voltage as expressed in Eq. (4-38), and this resistance can be expressed by considering the demodulator models of Fig. 4.20. The average open-circuit output voltage for the two states represented in that figure is simply the output voltage derived previously or

$$E_{Ooc} = E_O = (E_x - E_y)(1 - x)\frac{R_4}{R_4 + R_{Oa}x}$$

To find the average short-circuit output current the currents flowing in shorted outputs for the two states represented are found and will be

$$I_{Oa} = 0$$

$$I_{Ob} = \frac{E_x - E_y}{R_4 + R_{Oa}}$$

Then the average will be

$$I_{Osc} = \frac{I_{Oa}xT + I_{Ob}(1 - x)T}{T} = \frac{(1 - x)(E_x - E_y)}{R_4 + R_{Oa}}$$

The output resistance is then found to be

$$R_{Occ} = \frac{E_{Occ}}{I_{Osc}} \doteq R_4 \qquad \text{for } R_4 \gg R_{Oa} \tag{4-43}$$

Thus, output resistance of the chopper channel amplifier considered is simply equal to the output filter resistor. Input resistance follows from a similar analysis[2] on the modulator circuit and is

$$R_I = R_2 + R_3 + \frac{1 - x}{x}\frac{R_3R_{Ia}}{R_3 + R_{Ia}}$$

Generally the ac amplifier input resistance is much larger than R_3 above, and the input resistance of a chopper-stabilized operational amplifier becomes

$$R_I \doteq R_2 + \frac{R_3}{x} \qquad \text{for } R_{Ia} \gg R_3 \tag{4-44}$$

For a balanced filter, having $R_2 = R_3$ and a symmetrical chopping signal making $x = \frac{1}{2}$, the input resistance is three times the input filter resistor level.

Using the expressions derived above for chopper channel gain and output resistance, the equivalent input offset voltage caused by the main amplifier can be found with Eq. (4-38). However, the remaining term of this equation, V_{OScc}, which represents the offset due to the chopper channel amplifier, is not so well defined. In practice, V_{OScc} represents a major portion of the chopper-stabilized operational amplifier input offset voltage, and the primary sources of this chopper channel error voltage are also the sources of input bias current from this channel. Considering Fig. 4.18, the leakage current flows in the input circuit, creating both input offset voltage and input bias current components. Leakage current of the input switch in its OFF stage has a similar effect. Also, the charging and discharging currents of the input switch capacitances flow in the input circuit, developing input error voltages and currents. For a net change in switch capacitance charge of ΔQ per cycle, an average current is generated:

$$I = \frac{\Delta Q}{T}$$

where T is the period of the chopper signal.

The frequency response of a chopper-stabilized operational amplifier is formed by both cascaded and paralleled amplifier responses. Considering the amplifier diagram of Fig. 4.17, low-frequency signals are amplified by the two individual amplifiers in cascade, and high-frequency signals are amplified by the main amplifier only. However, signals between these frequency limits are amplified by both channels in parallel. This is one case of the feedforward amplifier which will be discussed in the next chapter. The overall frequency response of the amplifier is found by using the block diagram of Fig. 4.21. For the case shown it is assumed that the chopper channel amplifier has essentially a single-pole response set by the very low-frequency output filter. The overall response function is

$$A(s) = \left(\frac{R_1C_1s}{1 + R_1C_1s} + \frac{A_{Occ}}{1 + R_4C_5s}\right) A_m(s)$$

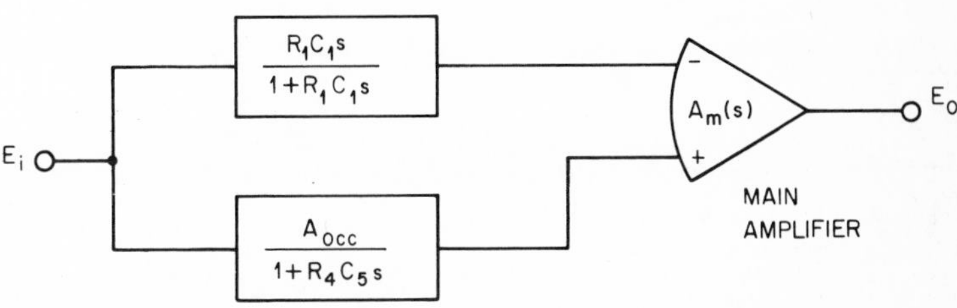

Fig. 4.21 Frequency response block diagram of a chopper-stabilized operational amplifier.

or

$$A(s) = \frac{R_1C_1R_4C_5s^2 + R_1C_1(1 + A_{Occ})s + A_{Occ}}{(1 + R_1C_1s)(1 + R_4C_5s)} A_m(s) \qquad (4\text{-}45)$$

In conjunction with the feedforward amplifier discussion, this type of frequency response is considered in the next chapter.

Considering next the varactor input operational amplifiers, the direct-coupled portion of these amplifiers is also preceded by an ac-coupled section which conducts signals by means of a modulated carrier. In this case the signal varies the capacitances of varactor diodes to modulate the carrier voltage difference between the diodes. The block diagram of Fig. 4.22 displays the basic components of the typical varactor input operational amplifier. Paralleling the chopper channel described previously, the ac portion of the circuit uses the input signal to modulate a carrier which is then amplified and demodulated. Then the amplified signal is supplied to a DC amplifier where the significance of dc input errors is reduced by the gain of the preceding carrier section. To examine the improvement in dc input errors the equivalent input offset voltage can be written in terms of the input error voltage and currents shown for the main amplifier. The result is analogous to that given for a chopper channel in Eq. (4-38), and for a carrier section with gain A_{Oc} and output resistance R_{Oc} it is

$$V_{OS} = V_{OSc} + \frac{V_{OSm} + I_{B1}(R_{Oc} - R_1) - I_{OSm}R_{Oc}}{A_{Oc}} \qquad (4\text{-}46)$$

where V_{OSc} is the equivalent input offset voltage of the carrier section and I_{OSm} is the main amplifier input offset current. In addition to V_{OSc} the modulation errors develop input bias currents, as will be discussed. Extremely small input bias currents of the order of 0.01 pA result with the varactor input section. However, the frequency response of varactor

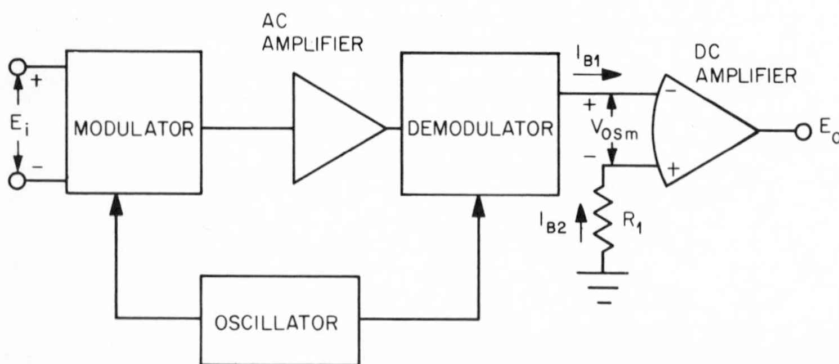

Fig. 4.22 Varactor input operational amplifier block diagram.

input operational amplifiers considered is limited by the modulation technique employed to frequencies which are much less than the carrier frequency.

Varactor modulators utilize the dependence of semiconductor junction diode capacitance upon voltage to modulate a carrier. A typical modulator is shown in Fig. 4.23 where the diode voltages V_{d1} and V_{d2} are determined by the input signal E_i and the carrier voltage E_c. Both the carrier and signal voltages used are small so that the diodes are never under sufficient forward bias to reach a low resistance state. As will be analyzed, the input signal E_i develops a difference in diode voltages, unbalancing their capacitances. From this unbalance a difference in carrier voltages results on the two diodes. This carrier frequency difference voltage is passed by the high-pass filter formed by R_1 and C_2 to the modulator output. Key characteristics of the diodes for the modulator are their voltage-dependent capacitances and their high resistances under small voltages. By considering the expressions governing these characteristics, the diodes can be represented by resistors and capacitors as in Fig. 4.24 to develop an equivalent circuit of the modulator. The resis-

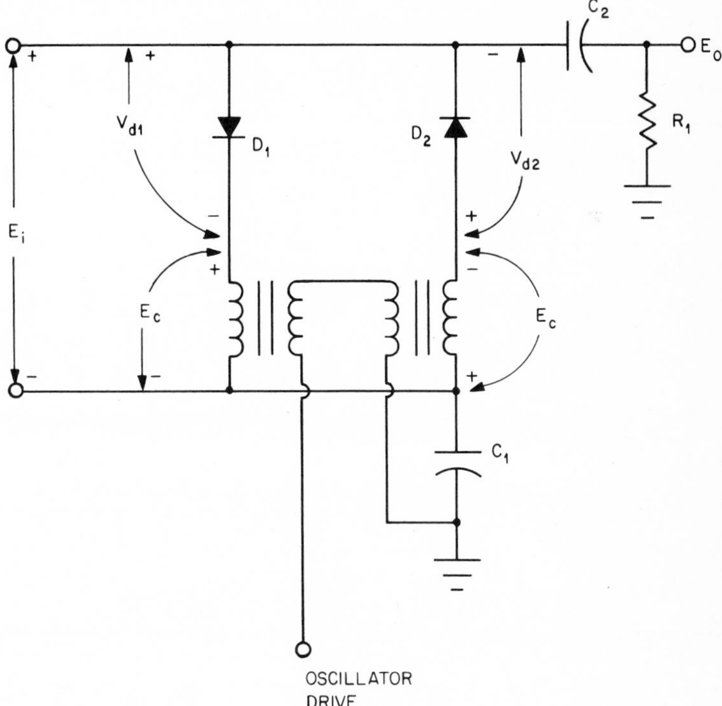

Fig. 4.23 Basic varactor modulator.

tance of a diode follows from the junction equation written from Eq. (2-2) as

$$I_d = I_S(e^{qV_d/KT} - 1)$$

For a small diode voltage V_d the junction equation can be simplified, using the series expansion

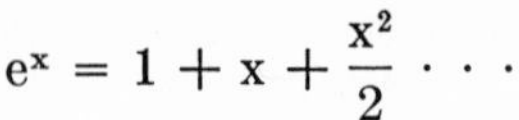

$$e^x = 1 + x + \frac{x^2}{2} \cdots$$

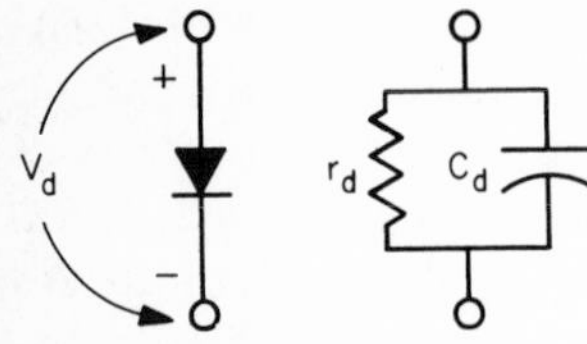

Fig. 4.24 Varactor diode equivalent representation.

Applying the expansion with $kT/q \doteq 25$ mV,

$$I_d \doteq I_S \frac{qV_d}{kT} \qquad \text{for } V_d = E_i + E_c \leq 10 \text{ mV}$$

Then the diode resistance is approximated by

$$r_d = \frac{V_d}{I_d} \doteq \frac{KT}{qI_S} \qquad \text{for } V_d \leq 10 \text{ mV} \tag{4-47}$$

$$r_d \overset{T}{=} 10^9\ \Omega \qquad \text{for } I_S = 10 \text{ pA}$$

Similarly, the capacitance representation can be simplified from its defining relationship of

$$C_d = \frac{C_o}{\sqrt{1 - V_d/\phi}}$$

where C_o is the junction capacitance under zero bias and ϕ is the junction contact potential. For diode voltages small compared with the contact potential of about 0.6 V, C_d may be approximated by using the series expansion

$$\frac{1}{\sqrt{1-x}} \doteq 1 + \frac{x}{2} + \cdots$$

Then

$$C_d \doteq C_o\left(1 + \frac{V_d}{2\phi}\right) \qquad \text{for } V_d \ll 0.6 \text{ V} \tag{4-48}$$

$$C_d \overset{T}{=} 10 \text{ pF}$$

Using the above representation for the varactor diodes, analysis models can be drawn for the modulator. At signal frequencies the diode capacitance presents a negligible shunt to r_d, and the diodes may be represented by their resistances alone. To consider the average diode voltages established by the input signal the carrier signal can be omitted and the modulator equivalent circuit will be that of Fig. 4.25a. From this model, the average diode voltages are seen to be

$$V_{d1} = E_i \qquad V_{d2} = -E_i$$

which establish the average diode capacitances as predicted by Eq. (4-48). Also, the differential input resistance of a varactor input amplifier is found from this model to be

$$R_I = \frac{r_d}{2} \qquad \text{for } r_{d1} \doteq r_d \doteq r_{d2} \tag{4-49}$$

Common-mode input resistance is determined by the leakage resistances of C_1 and C_2 which result in dc return paths to common. In general, this resistance is about $10^{14}\ \Omega$. The voltage ratings of these capacitors and the transformer determine the common-mode voltage limit. At the carrier frequency these capacitors are low impedances, resulting in the model of Fig. 4.25b. For carrier frequencies of 10 kHz or greater the reactance of the diode capacitance is commonly much less than r_d, and the diode resistances are omitted for carrier frequency analysis. From the latter model the modulator output signal can be found in terms of the diode capacitance difference as

$$E_o = \frac{C_{d1} - C_{d2}}{C_{d1} + C_{d2}} E_c$$

The result is expressed in terms of E_i, using Eq. (4-48) for C_d with the previous result of $V_{d1} = E_i = -V_{d2}$. From this operation, using $\Delta C_o = C_{o1} - C_{o2}$ and $C_{oa} = (C_{o1} + C_{o2})/2$, the output is

$$E_o \doteq \left(\frac{E_i}{\phi} + \frac{\Delta C_o}{C_{oa}}\right) E_c \qquad \text{for } C_{oa} \gg \frac{\Delta C_o E_i}{2\phi} \tag{4-50}$$

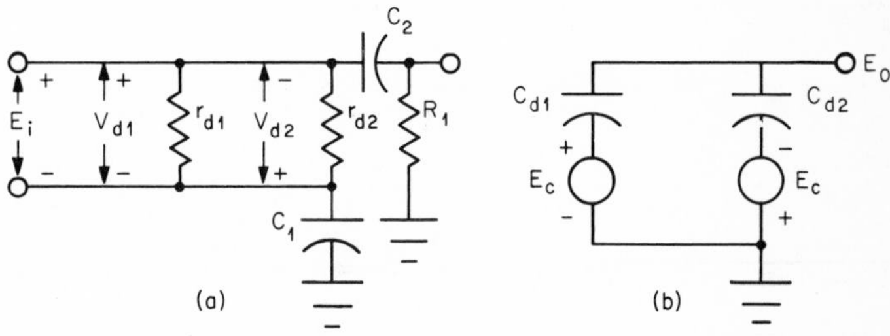

Fig. 4.25 Analyses models of the modulator of Fig. 4.23 at (a) signal frequencies and (b) the carrier frequency.

Note from this result that there is an output signal for zero input, because of mismatched diode capacitances. This is the principal source of input offset voltage in the carrier section. Input bias currents are primarily due to unequal diode leakage currents.

The modulated signal resolved above is amplified and demodulated to provide an output signal in much the same manner as described earlier for the chopper channel amplifier. An output switch chops the signal to ground to restore the output dc level prior to filtering. Heavy filtering removes the remaining carrier frequency component to provide an output which is essentially the amplified signal. Residual carrier frequency output voltage constitutes an added noise compǒnent; however, it is outside the amplifier frequency response and can be further attenuated by lowering the gain of the following DC amplifier at the carrier frequency. Low-frequency noise performance is improved by the varactor input since the 1/f noise of the ac amplifier transistors is not amplified.

REFERENCES

1. E. A. Goldberg, Stabilization of Wideband Amplifiers for Zero and Gain, *RCA Rev.*, June, 1950, p. 298.
2. H. A. Cook, The Contact Modulator, Pt. 2, Modulation Methods and Applications, p. 18, Airpax Electronics, Inc., Cambridge, Md., 1963.

5

PHASE COMPENSATION

Much of the precision provided by operational amplifiers is achieved through negative feedback around these high-gain amplifiers. As with any feedback system, the phase shift around the feedback loop must be controlled to preserve frequency stability. As is well known, if the phase shift around the loop reaches 360° at a frequency for which the loop gain is unity, the feedback becomes effectively positive, resulting in self-sustaining oscillation at that frequency. The phase inversion of the negative feedback produces a stabilizing 180° phase shift, but an additional 180° can be developed in the amplifier or the feedback to induce oscillation. The phase shift developed through an operational amplifier is the combined phase shifts of its several stages, and it can readily develop 180° of phase shift in the feedback loop. To ensure frequency stability under feedback conditions, phase compensation is commonly used with operational amplifiers. Phase compensation reduces the amplifier gain at those frequencies for which phase shift is high, and it reduces high-frequency phase shift by accepting greater phase shift at low frequencies. This is accomplished by adding response poles and zeros.

In the following sections the techniques of phase compensation are discussed, using Bode diagrams, and the associated frequency response and transient response characteristics are described. The straight-line response approximations of the Bode diagrams permit rapid visual evaluation of frequency stability and response characteristics for the majority of operational applications. Using these diagrams, the effects of phase compensation upon gain magnitude and phase responses can be related to frequency stability characteristics. A more precise method of evaluating frequency stability characteristics is also described, which permits accurate determination of maximum response peaking from the gain magnitude and phase responses of an operational amplifier. Concluding the chapter is a consideration of operational amplifier step response. Examination of this transient response quickly reveals any frequency stability deficiencies, and, therefore, this response is commonly used as a criterion in selecting phase compensation networks.

5.1 Frequency Stability and Bode Diagram Analysis

Feedback around an operational amplifier introduces the possibility of an oscillatory state. In general, oscillation will result if the feedback signal itself is sufficient to maintain the output signal independent of any input signal. For this to occur, the phase shift around the feedback loop must make the resultant feedback positive instead of negative, and the loop gain must sustain the output without an input signal. The gain magnitude and phase responses of the feedback loop are determined by the amplifier and the feedback network such as those represented in Fig. 5.1. Represented by $\beta(s)$, the feedback factor indicates the fraction of the output signal which is fed back to the input. For the example shown, this fraction is determined by the voltage divider formed by R_1 and R_2. From the block diagram it can be seen that the gain around the feedback loop or the loop gain A_L is the product of the amplifier open-loop gain $A(s)$ and the feedback factor $\beta(s)$ or $A_L(s) = A(s)\beta(s)$. The phase shift

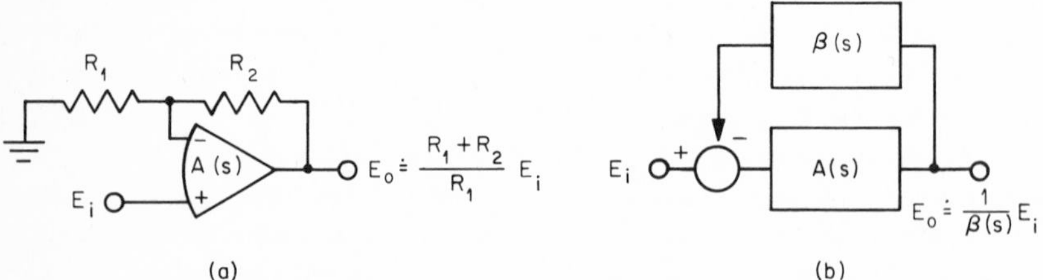

Fig. 5.1 Operational amplifier with negative feedback in (a) a typical configuration and (b) in block diagram form.

around the feedback loop is then the sum of the phase shifts of the amplifier and the feedback network.

For the block diagram, the closed-loop gain is

$$A_{CL}(s) = \frac{E_o}{E_i} = \frac{A(s)}{1 + A(s)\beta(s)} = \frac{A(s)}{1 + A_L(s)}$$

Instability of the amplifier results if the denominator of the above equation becomes zero, as happens when

$$A_L(s) = A(s)\beta(s) = 1 \angle 180° = -1$$

Under the above condition A_{CL} is infinite, indicating that an output results for no input signal. This is characterized by self-sustaining oscillation. Such instability is avoided by limiting the phase shift of the feedback loop to less than 180° when the loop gain is unity. In practice, a second condition can produce oscillation. This is developed by right half plane poles which correspond to a transient response which grows without bound. Such a signal increase, which can be initiated by noise, forces an amplifier into saturation where a noise signal will then drive the output to the opposite saturation limit, and so forth. The result is an oscillation between the two saturation limits. A common feedback system stability test is performed by plotting the loop gain magnitude versus its phase to ensure that the curve does not intersect or circle the point $1 \angle 180°$. Known as a Nyquist diagram, this test is well described in other treatments of frequency stability.[1] Of greater interest in the operational amplifier case is the actual amount of phase shift present when the loop gain reaches unity. In addition to answering the stability question, for the general case of no right half plane poles, the amount of phase shift at this point also provides an indication of other response characteristics. Although 180° of phase shift must result in producing self-sustaining oscillations in the general case, smaller phase shifts in the feedback loop can result in gain response peaking and a transient response having overshoot and peaking.

The phase responses of most operational amplifiers can be adequately approximated from their gain magnitude response curves by making Bode diagrams. Bode diagrams are basically straight-line approximations to gain magnitude and phase response curves, as will be outlined. For operational amplifiers these responses are convenient approximations since operational amplifiers are essentially minimum-phase systems. By combining the Bode diagrams of the feedback factor $\beta(j\omega)$ with that of the amplifier open-loop gain $A(j\omega)$ it is generally possible to approximate the magnitude and phase responses of the loop gain $A_L(j\omega) = A(j\omega)\beta(j\omega)$. The Bode diagrams approximate the gain magnitude response curve plotted in decibels and the phase response curve plotted in degrees, both versus frequency on a log scale. For a single-pole response the gain magnitude is

$$|A|(dB) = 20 \log \left| \frac{E_o}{E_i} \right| = 20 \log \left| \frac{A_O}{1 + j(f/f_p)} \right|$$

$$|A|(dB) = 20 \log \frac{A_O}{\sqrt{1 + (f/f_p)^2}} \tag{5-1}$$

Below the pole frequency the gain magnitude is asymptotic to the line

$$|A|(dB) \doteq 20 \log A_O \qquad \text{for } f \ll f_p$$

Above the pole frequency the asymptote is the line

$$|A|(dB) \doteq 20 \log \left(\frac{f_p}{f} A_O \right) \qquad \text{for } f \gg f_p$$

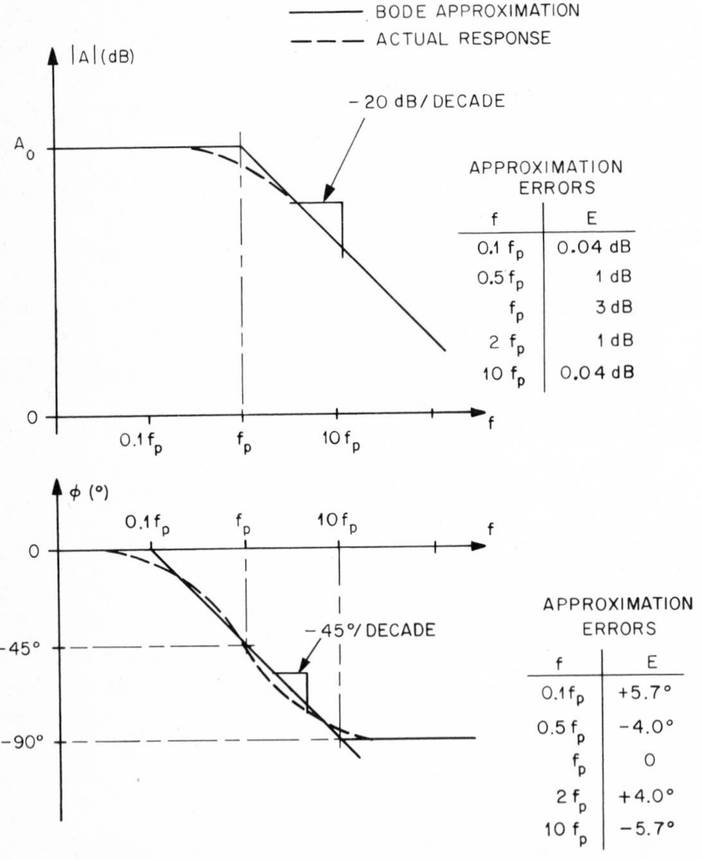

Fig. 5.2 Single-pole gain magnitude and phase responses and their Bode approximations.

Note that the slope of this line is −20 dB per decade of frequency or equivalently −6 dB per octave. In a similar manner it can be shown that the asymptotic slope of a multiple-pole responses is −20 dB per decade times the number of poles. For response zeros the slopes become positive. The Bode diagram approximates the gain magnitude response curves by these straight-line asymptotes which intersect at the pole or zero frequency, as shown in Figs. 5.2 and 5.3. Note that a maximum gain error of 3 dB results for the Bode-diagram approximation and it occurs at the pole frequency. One decade of frequency above or below the pole the error is negligible for most operational amplifier applications.

Accompanying each magnitude plot in Figs. 5.2 and 5.3 are the associated phase response and its Bode diagram approximation aligned with the magnitude plot for comparison. The phase shift of the single-pole response expression considered above is

$$\phi = -\arctan\frac{f}{f_p} \tag{5-2}$$

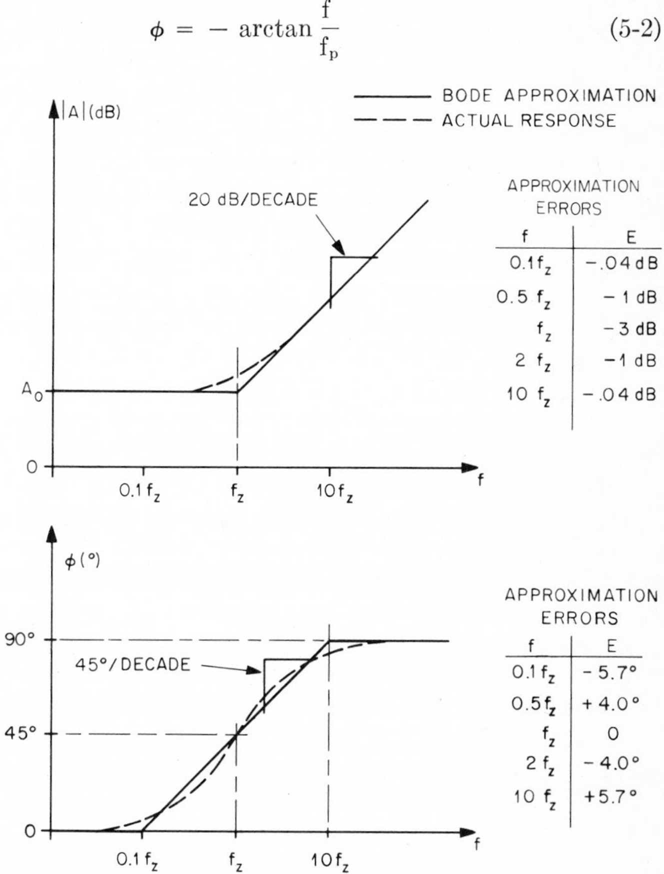

Fig. 5.3 Single zero-gain magnitude and phase response curves and their Bode approximations.

At frequencies far below the pole frequency, o asymptotically approaches zero, and far above f_p the phase shift approaches $-90°$. These asymptotes are connected in the Bode diagram by a line whose slope is $-45°$ per decade as shown. Note that the phase shift at the pole frequency is $-45°$, and the error of approximation is zero at this point. The Bode diagram approximates the phase shift by the asymptotic limits of 0 and $-90°$ for frequencies a decade below and above f_p, respectively. For most operational amplifier stability and response considerations the 5.7° maximum errors of the Bode diagram are not serious. In the case of a response zero the Bode phase diagram varies between asymptotes of 0 and 90° with a slope of 45° per decade and reaches 45° at f_z. Additional poles or zeros increase the phase response slope and final value by a factor equal to the number of poles or zeros.

Using Bode diagrams the gain magnitude and phase response curves of operational amplifiers can be drawn to permit rapid stability evaluation. When these curves are considered together with those of the feedback network the overall responses of the feedback loop result, providing a stability indicator to guide the selection of phase compensation. In general, each stage of an operational amplifier develops a response pole, and the individual stage responses are determined as described in Sec. 4.1. By linearly adding the Bode diagrams of the separate stages, the overall amplifier response curves are formed as in Fig. 5.4. In this and following curves f_I is an arbitrary frequency. As shown, each pole adds -20 dB per decade to the slope of the magnitude response curve, resulting in a -60 dB per decade slope for the three-stage example. Each pole also increases the slope of the phase shift response by $-45°$ per decade for a decade of frequency above and below the pole frequency, and each pole adds $-90°$ to the final overall phase shift. As can be seen from these curves, the amplifier phase shift exceeds 180° over much of the response. Since the phase shift of the feedback loop includes that of the amplifier, unstable operation can potentially occur for some feedback conditions.

The feedback conditions determining stability discussed earlier can be related to operational amplifier Bode diagrams which are readily obtained. By experimentally deriving the gain magnitude response curve, the Bode diagrams of any operational amplifier can be drawn. Since the straight-line portions of this actual response curve represent the asymptotes of the response curve, continuation of these straight-line portions produces the gain magnitude Bode diagram and identifies the pole and zero frequencies at the intersections. From these pole frequencies the phase response curve can be drawn directly. The open-loop gain of the amplifier combines with the feedback factor to form the loop gain described earlier by $A_L(j\omega) = A(j\omega)\beta(j\omega)$, and stability depends generally upon the phase shift of $A_L(j\omega)$ when its magnitude is unity. As expressed before, the

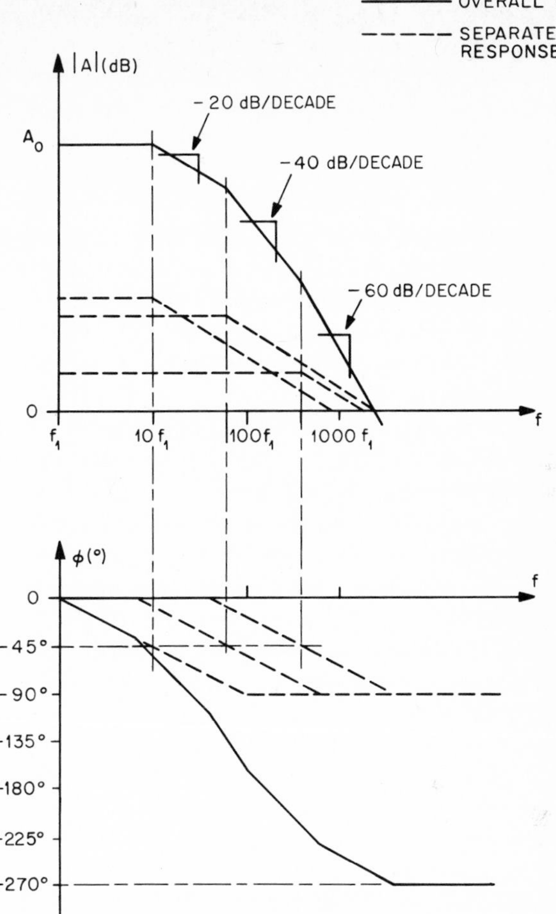

Fig. 5.4 Gain magnitude and phase response curves of a three-stage operational amplifier.

critical condition is

$$A_L(j\omega) = A(j\omega)\beta(j\omega) = 1 \angle 180°$$

which can be expressed as

$$A(j\omega) = \frac{1}{\beta(j\omega)} \angle 180°$$

From this expression the point of interest is seen to be that at which the reciprocal of the feedback factor equals the open-loop gain in magnitude and the phase difference between the two is 180°. From this expression the condition resulting in instability can be expressed in terms of the magnitude and phase characteristics of $A(j\omega)$ and $1/\beta(j\omega)$. Oscillation will result at any frequency for which

$$|A(j\omega)| = \frac{1}{|\beta(j\omega)|} \tag{5-3a}$$

and

$$\phi_a - \phi_b = 180° \tag{5-3b}$$

where ϕ_a is the amplifier open-loop phase shift and ϕ_b is the phase shift of $1/\beta(j\omega)$. By plotting the Bode magnitude response of $1/\beta(j\omega)$ on the operational amplifier open-loop gain magnitude Bode diagram, the frequency at which the first condition above is met is found at the intersection of the two curves. The difference in the two phase response curves is then examined to evaluate frequency stability.

For many common operational amplifier applications the preceding frequency stability evaluation is facilitated by relying on the relationship between $1/\beta(j\omega)$ and the ideal approximation to the closed-loop amplifier gain $A_{CL}(j\omega)$. Frequently the two functions are nearly equal and stability can be evaluated by simply considering the intercept of the open-loop and ideal closed-loop responses. Such is the case for the basic amplifier configurations of Fig. 5.5 whose gain expressions are derived in Appendix A. In each case the feedback factor determining that portion of the output fed back to the input is simply determined by the voltage divider of R_1 and R_2, giving

$$\beta(j\omega) = \frac{R_1}{R_1 + R_2} \tag{5-4}$$

Then $1/\beta(j\omega)$ equals the ideal closed-loop gain expression shown for the noninverting amplifier. Frequency stability can then be evaluated by examining the phase-shift difference between the open-loop gain and ideal closed-loop gain diagrams at the frequency of intersection of the two gain magnitude diagrams.

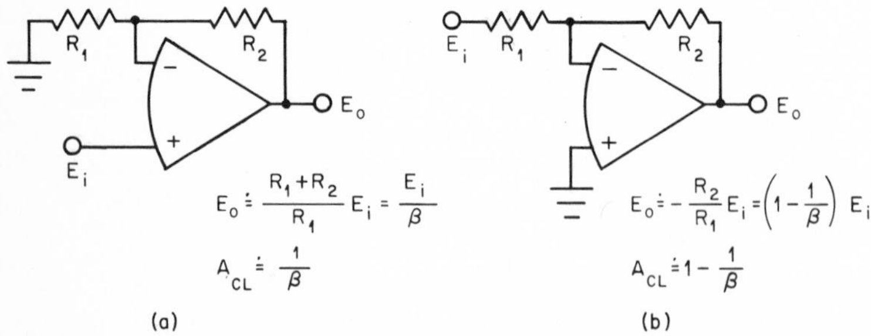

Fig. 5.5 Basic operational amplifier configurations and their gain and feedback factor relationships for (a) noninverting gain and (b) inverting gain.

Using this relationship for the noninverting amplifier, a stability evaluation is shown in Fig. 5.6 for two levels of feedback factor or ideal closed-loop gain. For $|A_{CL}| = |1/\beta| = A_1$ the phase shift of the amplifier is 90°. This is also the phase shift of the feedback loop in this case, since the resistive feedback network has no phase shift, and the amplifier will be stable at this gain level. However, for closed-loop gains equal to A_2 or less the phase shift is 180°, and oscillation will result at the frequency of the intersection of the open-loop and closed-loop gain magnitude plots. Operation at closed-loop gain levels between A_1 and A_2 results in feedback loop phase shifts between 90 and 180° which develops varying degrees of response peaking and overshoot. These characteristics will be described in Sec. 5.3. A similar analysis can be performed for the inverting amplifier of Fig. 5.5b for which $1/\beta = -A_{CL} + 1$. For high closed-loop gains the $1/\beta$ magnitude and phase responses are essentially those of

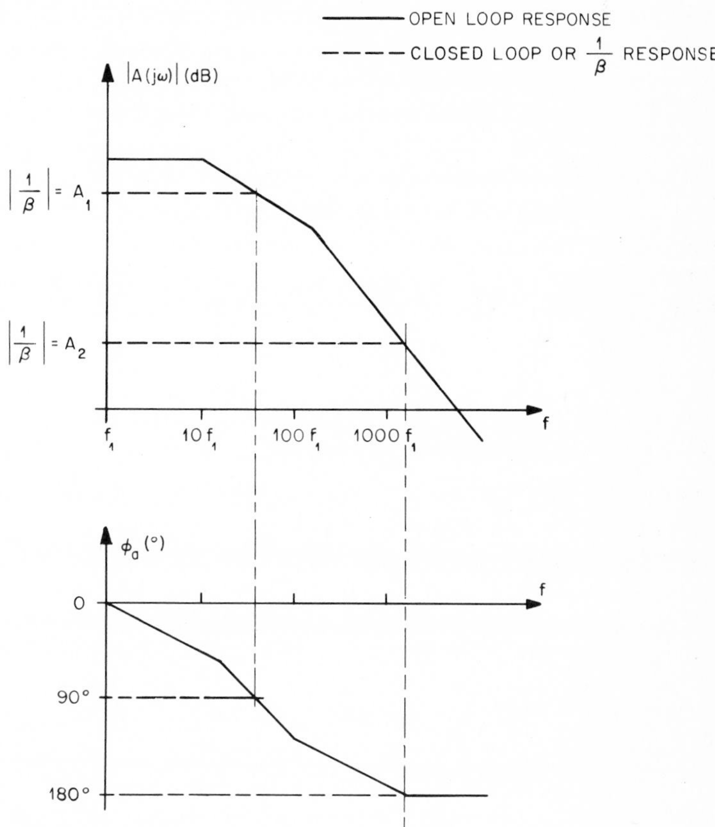

Fig. 5.6 Frequency stability evaluation for two levels of feedback factor or noninverting amplifier gain.

$-A_{CL} \doteq R_2/R_1$, and the stability evaluation can be made in the same manner as for the noninverting case. When lower gains are involved for the inverting amplifier, the $|1/\beta|$ curve is drawn by adding 1 to the closed-loop gain magnitude curve of $-A_{CL}$.

For the resistive feedback cases considered above, the phase shift of the feedback loop is solely that of the amplifier. Control of amplifier phase shift is then sufficient to ensure frequency stability. However, when capacitive or inductive elements are used for feedback, it may also be necessary to control the phase shift of the feedback network. Such is the case for the differentiator circuit of Fig. 5.7 having

$$\frac{1}{\beta(j\omega)} = 1 + j\omega R_2 C = -A_{CL}(j\omega) + 1$$

Plotting the magnitude of $1/\beta(j\omega)$ on the gain magnitude diagram of the amplifier results in Fig. 5.8. From these magnitude curves the phase shifts of $A(j\omega)$ and $1/\beta(j\omega)$ are known from the Bode diagram techniques, and the phase difference $\phi_a - \phi_b$ is plotted directly. The latter curve indicates that the phase difference is 180° at the frequency of intersection of the two magnitude diagrams. The differentiator will then oscillate at this frequency unless the phase difference is reduced. Note that reduced phase difference between $A(j\omega)$ and $1/\beta(j\omega)$ corresponds to a smaller difference in slopes of the magnitude curves. By adding a response pole to $1/\beta$, as shown for the compensated response, stability is restored. A pole is added to the differentiator circuit by inserting a resistor in series with C. Denoting this resistor as R_1, the responses are modified as shown to result in 90° phase difference at the intercept frequency.

5.2 Phase Compensation Techniques

By adding poles and zeros to the frequency response of an operational amplifier, its phase is compensated to ensure stability and limit response peaking, overshoot, and ringing. For a specific amplifier application the phase compensation is chosen to accommodate the feedback conditions involved, but for a general-purpose amplifier the compensation is selected

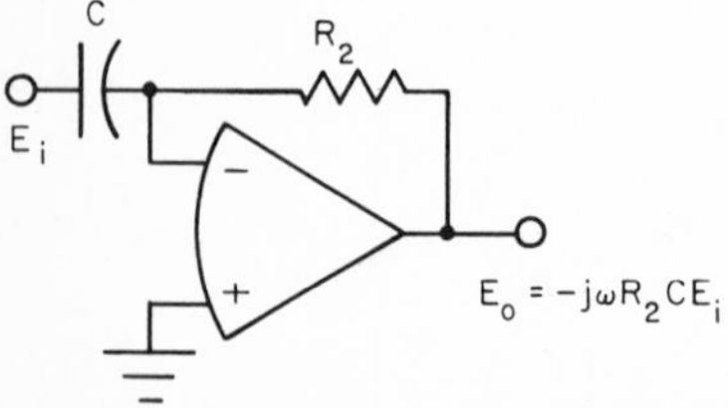

Fig. 5.7 Differentiator circuit.

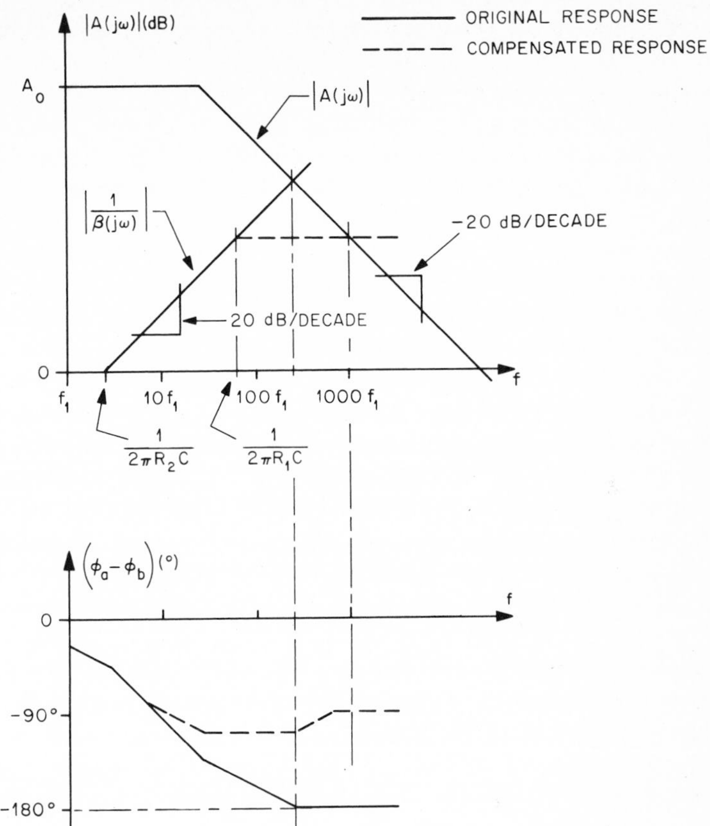

Fig. 5.8 Stability evaluation of the differentiator of Fig. 5.7.

for operation under a wide range of feedback conditions. Typically the phase shift of the general-purpose amplifier is controlled to be around 135° or less over the entire frequency range for which the amplifier open-loop gain is unity or greater. Beyond the amplifier unity-gain crossover frequency, phase shift does not have to be limited since unity loop gain $A_L = A\beta$ cannot be developed beyond this point with passive feedback networks which limit β to a maximum of 1. If the phase shift of $1/\beta(j\omega)$ at the intercept frequency considered in Sec. 5.1 is small, the phase shift of the feedback loop at this critical point will be essentially that of the amplifier. Feedback loop phase shifts limited to 135° at the intersection frequency of the $A(j\omega)$ and $1/\beta(j\omega)$ magnitude curves ensure phase margins of at least 45° from the 180° critical phase-shift level. A 45° phase margin is typically chosen to limit frequency response peaking and to control transient response overshoot and ringing. The phase compensation of a general-purpose operational amplifier is then chosen to provide a phase margin of 45° or more for any closed-loop gain of unity or greater, assuming that the phase shift of the feedback network is small at the critical intercept frequency.

With more than one stage in an operational amplifier the phase shift through the amplifier will readily exceed 135°, as discussed in the previous section. To control frequency response characteristics the phase compensation poles and zeros are added to the amplifier response, lowering the gain at frequencies for which phase shift is high and decreasing high-frequency phase shift. Several techniques of phase compensation will be discussed involving the use of an RC shunt, Miller-effect multiplication of a feedback capacitance, and the feedforward approach to wide-band response. The basic series RC phase compensation network shown as R_x and C_x in Fig. 5.9 shunts the output of one differential stage of an operational amplifier to create a pole and a zero and to shift the original pole of the stage to a higher frequency. In the figure R_{I2} and C_{I2} represent the loading effects of the following stage. Since C_{I2} is controlled by stage interaction it is best approximated by using the technique described in Sec. 4.1. Generally, the RC shunt phase compensation is connected at the first stage of an operational amplifier to optimize slewing rate. This optimum results because the low voltage swing at the first-stage output can be developed on C_x with far less current than required for the larger voltage swings on later stages.

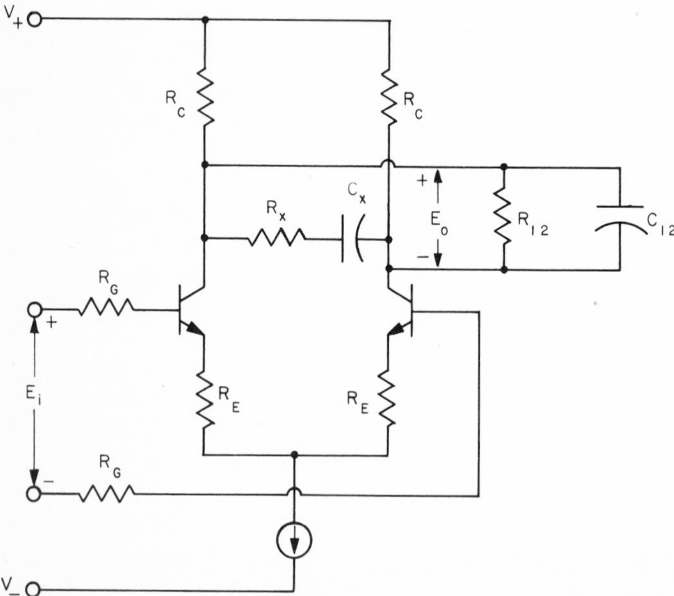

Fig. 5.9 Series RC phase compensation applied to a differential stage.

To develop expressions describing response the differential stage is replaced by the model of Fig. 1.12, or of Fig. 1.15 for an FET stage, resulting in an equivalent circuit such as that of Fig. 5.10a. Combining the parallel resistive and capacitive elements, the simplified model of Fig. 5.10b is the result for

$$R_L = R'_O \| 2R_C \| R_{I2}$$
$$C_L = C'_o + C_{I2}$$

From the simplified model the uncompensated response pole is at

$$f_{p1} = \frac{1}{2\pi R_L C_L}$$

The addition of R_x and C_x results in a compensated response of

$$A = \frac{A_O(1 + R_x C_x s)}{R_L C_L R_x C_x s^2 + (R_L C_L + R_x C_x + R_L C_x)s + 1}$$

For typical phase compensation networks the resistor R_x is much smaller than the stage load resistance R_L and the term $R_x C_x$ can be neglected in comparison with $R_L C_x$ above. Similarly $C_L \ll C_x$, and the term $R_L C_L$ can be omitted. With these simplifications the quadratic equation is used to solve for the roots of the denominator which are the response poles. This gives

$$s = \frac{-1}{2R_x C_L}\left(1 \pm \sqrt{1 - \frac{4R_x C_L}{R_L C_x}}\right)$$

The second term under the square-root sign is small compared with unity since $R_x \ll R_L$ and $C_L \ll C_x$ for the general case. Thus, the square-root term may be simplified with the binomial expansion

$$\sqrt{1 - x} \doteq 1 - \frac{x}{2}$$

to get

$$s \doteq \frac{-1}{2R_x C_L}\left[1 \pm \left(1 - \frac{2R_x C_L}{R_L C_x}\right)\right]$$

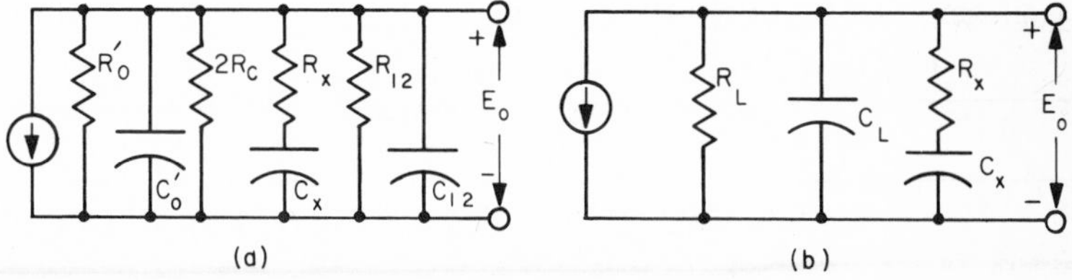

Fig. 5.10 Equivalent circuits of the phase-compensated stage of Fig. 5.9.

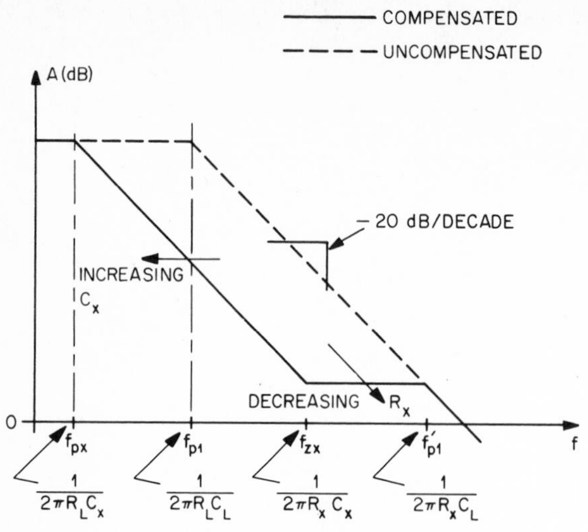

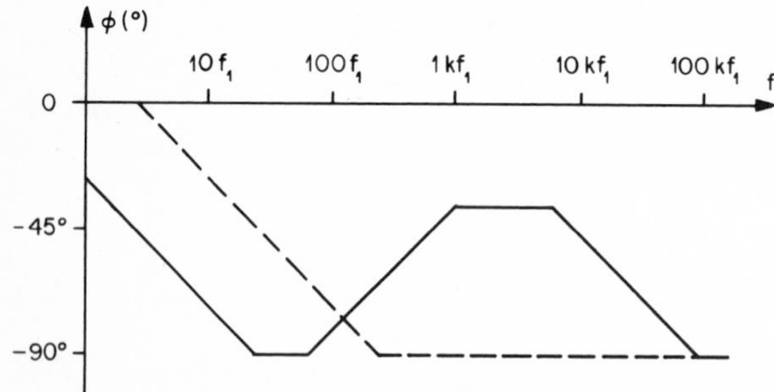

Fig. 5.11 Bode diagram representation of the effects of series RC phase compensation on a differential stage.

From this result the pole frequencies are found to be

$$\left.\begin{aligned} f_{px} &= \frac{1}{2\pi R_L C_x} \\ f'_{p1} &= \frac{1}{2\pi R_x C_L} \end{aligned}\right\} \text{for } R_x \ll R_L \qquad C_x \gg C_L \qquad \begin{aligned} (5\text{-}5) \\ (5\text{-}6) \end{aligned}$$

The response zero from the numerator of the gain expression is at

$$f_{zx} = \frac{1}{2\pi R_x C_x} \tag{5-7}$$

For the preceding response approximation the gain of a phase-compensated differential stage is described by

$$A(j\omega) \doteq \frac{A_O(1 + j\omega R_x C_x)}{(1 + j\omega R_L C_x)(1 + j\omega R_x C_L)}$$

The effect of the phase compensation is displayed by comparing the Bode diagrams of this response with that of the uncompensated case, as in Fig. 5.11. Note that the phase compensation reduces both gain and phase shift at high frequencies. Both effects tend to improve the stability of the typical amplifier whose phase-shift buildup becomes large at high frequencies. Note that increasing C_x lowers the first pole frequency and that decreasing R_x increases the second pole frequency and the zero frequency. Phase compensation changes in these directions improve frequency stability by the associated greater shunting of the stage output.

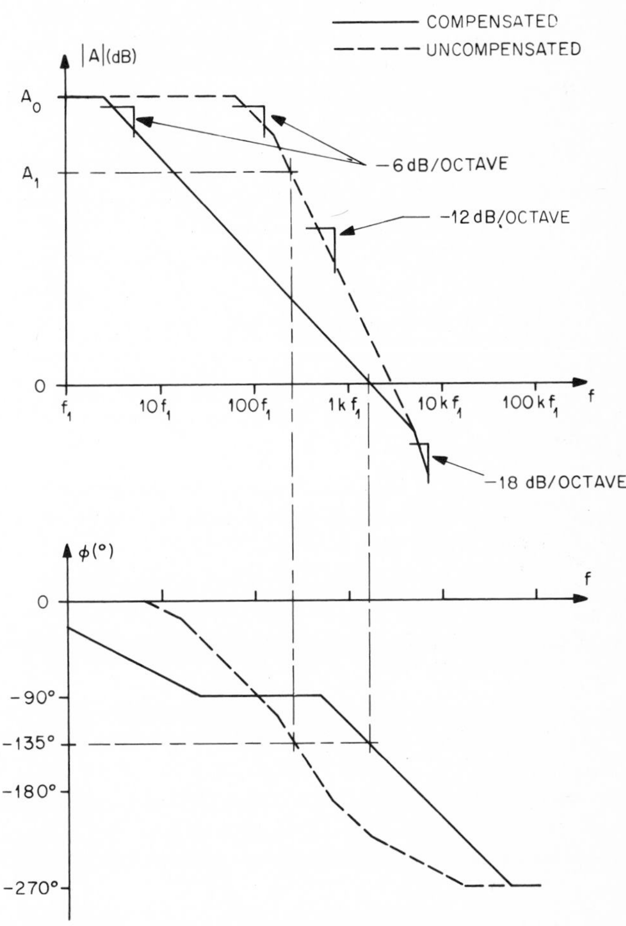

Fig. 5.12 Three-pole amplifier response compensated to provide a −6 dB per octave magnitude slope and 45° phase margin.

Through the addition of response poles and zeros as described above, the response of an operational amplifier can be tailored to provide 45° phase margin in a variety of ways. The most straightforward response shaping limits the slope of the gain magnitude response curve to no more than −20 dB per decade, or equivalently −6 dB per octave, until the gain magnitude is below unity. Such a compensated response resembles that of a single-pole network and limits phase shift to 90° over most of the frequency range for which gain is unity or greater. From the compensated response curve of Fig. 5.12, the −6 dB per octave compensated response has an associated 45° phase margin at the unity-gain crossover frequency if the remaining poles occur at sufficiently higher frequency. For all higher gain levels the phase margin would be greater than 45°. Without the phase compensation such phase margin would result only over a small range of closed-loop gains as bounded by the level A_1 shown. The above compensation can be achieved by the series RC shunt just described. With this technique the phase compensation capacitance C_x would be used to produce the added low-frequency pole, and the compensation resistor R_x would be chosen to create a response zero with C_x which coincides with and cancels one of the original low-frequency poles. The other low-frequency pole is moved to a much higher frequency beyond the unity-gain crossover by the low impedance shunting of the compensation network. Each of these response changes occurs as previously analyzed for the RC shunt network. The remaining pole of the amplifier which originally occurred below crossover is unaffected by the compensation, and two poles result beyond crossover for this compensated response.

A simple modification of this technique is frequently made to achieve additional gain and better slewing rate with the compensated response. By decreasing the phase compensation capacitance from that chosen above, the frequency of the added pole and zero are increased, resulting in the Bode diagrams of Fig. 5.13. In this case the zero no longer coincides with one of the original poles, and the slope of the magnitude response curve reaches −12 dB per octave at a low frequency. However, the frequency range over which this slope extends is limited to avoid phase shifts above 135° as indicated. For this compensation the phase margin is lower for some high-gain levels, but it never falls below 45°. It can be seen that, with stability thus preserved, higher open-loop gain is achieved for much of the mid-frequency range by using this compensation technique. In addition, the reduced phase compensation capacitance raises the slewing rate limit imposed by the capacitance.

Departing from the RC shunt compensation, another phase compensation technique ensures exact coincidence of a compensation zero with an amplifier pole to provide an accurate −6 dB per octave slope. Such a response may be desired when the feedback network contributes enough phase shift to the loop that the amplifier phase shift of the last case is not acceptable. As will be described, feedback capacitance around an intermediate stage provides phase compensation through Miller-effect multiplication of the capacitance and develops a response zero coincident with the pole of that stage. Such compensation is illustrated in Fig. 5.14 for a two-stage amplifier. For a response analysis of this amplifier the output circuit model of Fig. 1.12 is used for the first stage, and the second-stage loading is represented by the stage input resistance, input capacitance, and the Miller-effect capacitance C_m as shown. Upon combining the various resistive elements and adding the two stage capacitances, this

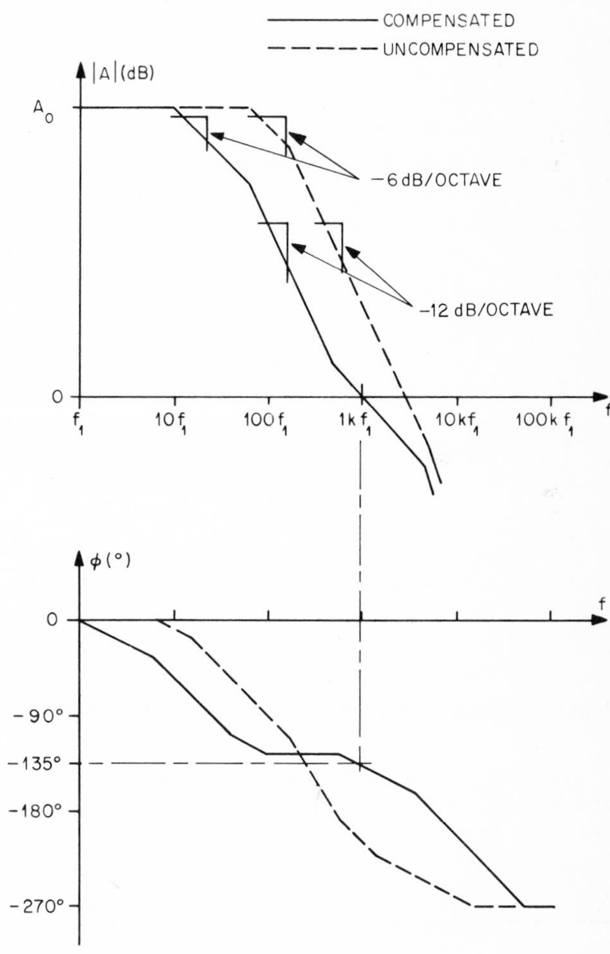

Fig. 5.13 Three-pole amplifier response of Fig. 5.12 compensated for 45° phase margin with −6 dB per octave and −12 dB per octave magnitude slopes.

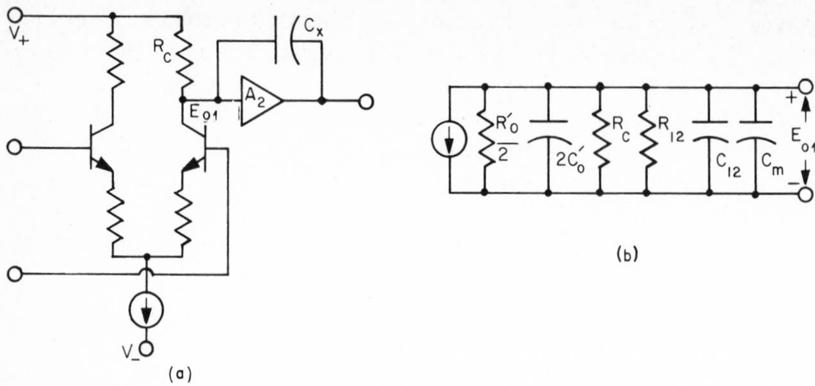

Fig. 5.14 (a) Miller-effect phase-compensated operational amplifier and (b) first-stage output circuit model.

analysis model is simplified to that of Fig. 5.15a, where

$$R_L = \frac{R_O'}{2} \| R_C \| R_{I2}$$

$$C_L = 2C_o' + C_{I2}$$

From this model a single-pole frequency response would be expected for the first stage except that the Miller capacitance C_m is a function of frequency through its dependence on the second-stage gain. This capacitance is expressed by

$$C_m = (1 - A_2)C_x = \left(1 - \frac{A_{O2}}{1 + jf/f_{p2}}\right) C_x \qquad A_{O2} < 0$$

where A_{O2} and f_{p2} are the second-stage dc gain and pole frequency.

As a result of this frequency-dependent phase compensation capacitance, the compensation adds more than a single pole, and the response approximation technique of Sec. 4.1 will be used to analyze the compensation effects. At low frequencies C_m is essentially determined by the dc gain of the second stage and can be approximated by its low-frequency

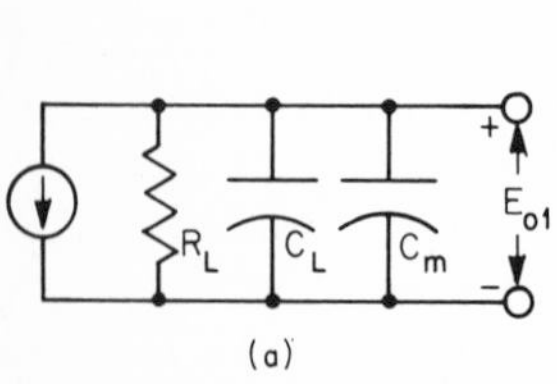

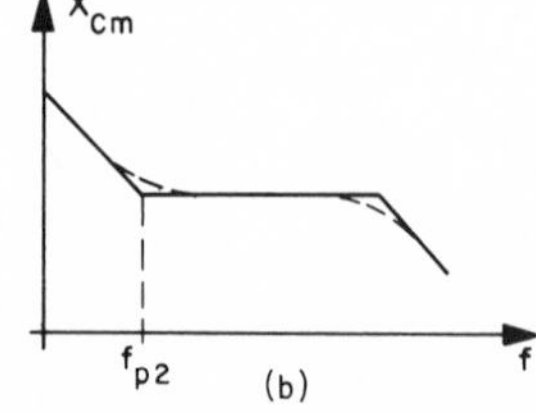

Fig. 5.15 (a) Simplified response analysis model of Fig. 5.14 and (b) the response of its Miller-effect capacitance C_m.

limit as

$$C_{mL} = (1 - A_{O2})C_x$$

In general, this capacitance is chosen to create a phase compensation pole at a frequency below that of A_2 so that C_m will be at essentially this higher limit when it develops the first pole. Then, with the model of Fig. 5.15a the first pole is found to be at

$$f_{px} \doteq \frac{1}{-2\pi R_L A_{O2} C_x} \qquad \text{for } -A_{O2} \gg 1 \qquad -A_{O2}C_x \gg C_L \tag{5-8}$$

However, the capacitive shunting effect which created the above pole is interrupted when the second-stage pole is reached. At f_{p2} the gain which multiplied the effect of C_x becomes essentially a constant or simply resistive for a range of frequencies, as shown in Fig. 5.15b. When its reactance becomes constant, C_m does not further shunt the first-stage load to produce continued gain decrease. In other words, a response zero has been added to the first-stage output circuit which coincides with the second-stage output circuit pole, as shown in Fig. 5.16. At this frequency no overall response change occurs. Following this portion of its reactance curve, C_m reaches its high-frequency limit of C_x, and the reactance again decreases to develop a response pole at

$$f'_{p1} = \frac{1}{2\pi R_L(C_L + C_x)} \tag{5-9}$$

Note from the gain magnitude plots that this shifted first-stage pole occurs at that frequency at which the compensated stage response returns to the uncompensated response. By adding the gain magnitude plots of the two stages, the overall amplifier response results as shown with the compensated response slope at -6 dB per octave prior to the unity-gain crossover frequency.

An additional common phase compensation technique is that of feedforward which achieves wideband response. With this approach the frequency limitations of the high-gain portion of an operational amplifier are partly avoided by feeding a signal around that portion of the amplifier. As indicated in Fig. 5.17 the input signal is amplified by both a high-gain, low-frequency channel and a low-gain, high-frequency channel, as was the case considered for the chopper-stabilized amplifier. The high-gain channel provides the high accuracy under feedback at low frequencies normally achieved with operational amplifiers. The high-frequency channel provides a low-phase-shift path through the amplifier at high frequencies to maintain frequency stability. Even though the high-gain channel might develop very high phase shifts at higher frequencies, the gain of the high-frequency path will dominate at these frequencies and

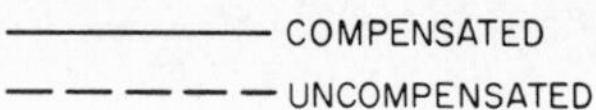

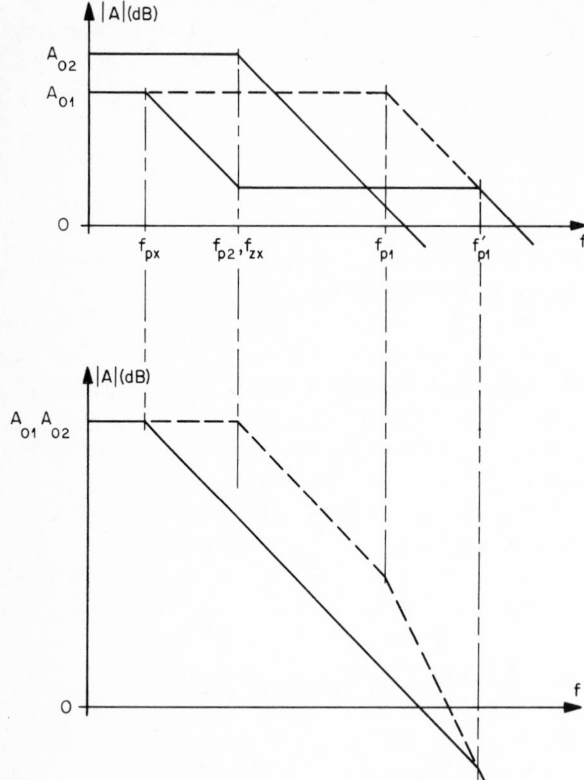

Fig. 5.16 Gain magnitude Bode diagrams of the separate stages and overall amplifier for the Miller-effect phase-compensated amplifier of Fig. 5.14a.

control the phase shift of the overall amplifier. From the block diagram, the overall amplifier gain will be the algebraic sum of the gains of the two channels as expressed by

$$A = \left(A_1 + \frac{j\omega RC}{1 + j\omega RC}\right) A_2 \tag{5-10}$$

where R is the resistance at the feedforward capacitor connection points. At low frequencies the first term above dominates, and at high frequencies the second term controls the gain when the magnitude of A_1 becomes much less than unity.

These characteristics are displayed by the associated Bode diagram of

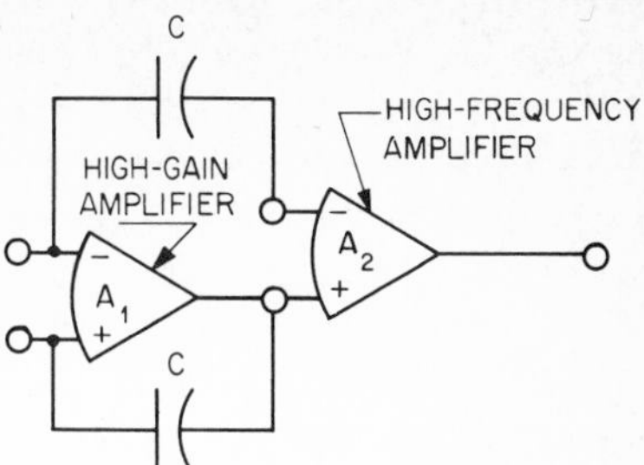

Fig. 5.17 Block diagram of a feedforward operational amplifier.

Fig. 5.18 which also shows the responses of the two separate channels. Note that the conventional response A_1A_2 can now have more than one pole before its unity-gain crossover since the high-frequency channel controls stability over part of its response. This fact permits use of less severe phase compensation in the high-gain channel. By easing this phase compensation and adding high-frequency gain in the other channel, the response of the feedforward amplifier can be increased an order of magnitude. In some cases unity-gain crossover frequencies of 100 MHz and slewing rates of 1,000 V/μs are achieved. Of critica importance in attaining such response with frequency stability is the manner in which the responses of the two channels add at the transition from one response to the other. If phase shift in the high-gain channel approaches $-180°$ at the transition, the two signals can be of opposite phase and would cancel to develop a response notch. Also, if this phase shift reaches $-180°$ before the transition, oscillation can occur for operation in the affected range.

When the phase compensation is chosen for an operational amplifier, the external impedances presented to its terminals during actual use should be considered. Both the source impedance presented to the inputs and the load impedance connected to the output can alter the frequency stability characteristics of the amplifier. High source resistances lower the frequency of the first-stage response pole, as described by the differential-stage response analysis of Sec. 1.2. In this case the

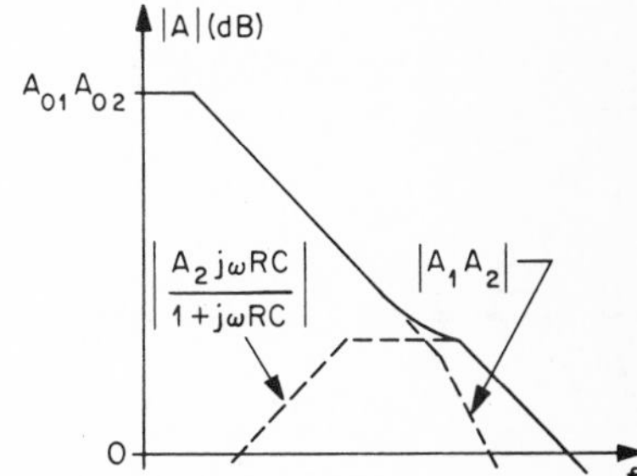

Fig. 5.18 Feedforward amplifier separate channel and overall Bode diagrams.

feedback loop will acquire greater phase shift because of the low-pass filter action of the feedback resistance and the input capacitance. Correction for this added phase shift is commonly achieved with a capacitive bypass of the feedback resistor. Also, the output load impedance affects the responses of the later stages. For the common case of an emitter-follower output stage the effect of a capacitive load will primarily be reflected as increased input capacitance to this stage. From this increase, the response pole frequency of the preceding-stage output circuit will drop and the phase shift below the crossover frequency will similarly increase.

5.3 Frequency Response Peaking and Step Response

In the preceding sections of this chapter the gain magnitude and phase response curves of an operational amplifier were used as a test for frequency stability and as a guide in choosing phase compensation. Phase compensation was used to limit the phase shift of the feedback loop at its unity-gain point to less than 180°, thereby providing a phase margin. In this section a closer examination is made of the dependence of amplifier response characteristics upon the feedback loop response. The response of the feedback loop gain determines frequency response peaking along with the overshoot, ringing, and settling time of the step response. For response peaking analysis a simple graphical technique is developed which permits determination of the response peak level without plotting the entire response. Characteristics of the amplifier step response are described by relating them to the response of a second-order system.

Application of negative feedback around an operational amplifier constrains the amplifier response to some closed-loop response as represented in Fig. 5.19. However, at high frequencies the decrease in loop gain diminishes the feedback control and the phase shift added by the second response pole shown raises the feedback loop phase lag above 90°.

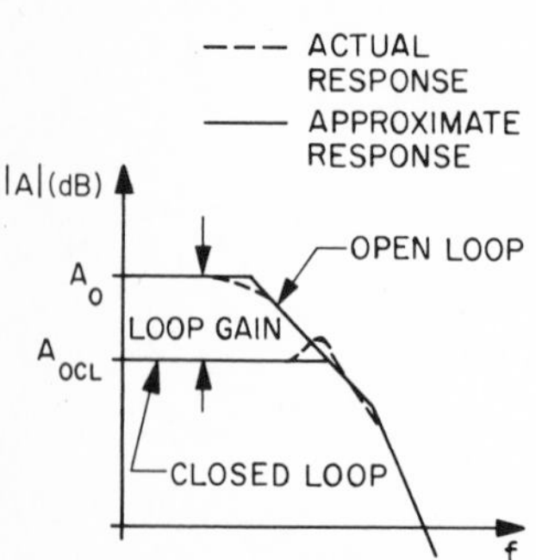

Fig. 5.19 Closed-loop response peaking.

As a result, the closed-loop response deviates from its straight-line approximation and peaks near the intercept of this approximate curve with the open-loop response. Since this peaking represents closed-loop gain error, it is desirable to know the magnitude of the response more accurately, as can be found from the closed-loop gain expression. First for the noninverting case the closed-loop gain from Appendix A can be expressed as

$$A_{CL} = \frac{A}{1 + A\beta} = \frac{1/\beta}{1 + 1/A\beta}$$

Separating the magnitude and phase of the open-loop gain term A, it is written as

$$A = |A|e^{j\phi}$$

By substituting this in the preceding expression, A_{CL} becomes

$$A_{CL} = \frac{1/\beta}{1 + e^{-j\phi}/|A|\beta}$$

or

$$A_{CL} = \frac{1/\beta}{1 + (\cos\phi)/|A|\beta - j[(\sin\phi)/|A|\beta]}$$

When β is a constant, as provided by resistive feedback components, the magnitude of the closed-loop gain will be

$$|A_{CL}| = \frac{1/\beta}{\sqrt{1 + (2\cos\phi)/|A|\beta + 1/|A|^2\beta^2}} \quad (5\text{-}11)$$

where

$$\phi = \arg A$$

Using this result, the closed-loop gain magnitude response can be drawn from the open-loop gain magnitude and phase responses.

As a measure of maximum gain error due to peaking, only the maximum closed-loop gain need be found. This approach permits a much simpler evaluation of response peaking. To find the condition for maximum closed-loop gain magnitude the expression for $|A_{CL}|$ in Eq. (5-11) is differentiated with respect to $|A|$ and equated to zero. In this way the maximum is found to occur when the loop gain magnitude and amplifier phase shift are related by

$$(|A|\beta)_P = \frac{-1}{\cos\phi_P} \quad (5\text{-}12)$$

where $(|A|\beta)_P$ and ϕ_P occur at the closed-loop response peak. The maximum closed-loop gain is resolved by substituting this relationship in

Eq. (5-11) to get

$$|A_{CL}|_{max} = \frac{1/\beta}{\sin \phi_P} \tag{5-13}$$

To simplify analysis, this result may be expressed in terms of the gain of the peak alone, giving a direct measure of the amount of peaking. In decibel form the gain of the peak will be

$$|A_P| = 20 \log \frac{|A_{CL}|_{max}}{A_{OCL}}$$

where A_{OCL} is the dc or ideal closed-loop gain provided by the feedback. This gain is defined by Eq. (5-11) for $\phi = 0°$ and $|A|\beta \gg 1$ to be

$$A_{OCL} = \frac{1}{\beta}$$

By substituting this expression and that of Eq. (5-13) in the expression for $|A_P|$, the magnitude of the gain peak is found to be

$$|A_P| = 20 \log \frac{1}{\sin \phi_P} \tag{5-14}$$

With this result the maximum amount of peaking resulting for a given closed-loop gain can be found, once ϕ_P has been found, by using the condition for maximum $|A_{CL}|$ expressed by Eq. (5-12).

To find the value of amplifier phase shift ϕ_P at which the peak occurs, a graphical solution can be used to avoid trial-and-error evaluation. The closed-loop gain peaks at that point for which the loop gain magnitude $|A|\beta$ curve versus phase shift ϕ intersects the curve defined by the condition of Eq. (5-12). From the amplifier open-loop gain magnitude and phase curves the $|A|\beta$ versus ϕ curve for a given resistive feedback condition can be drawn on a curve of $(|A|\beta)_P$ versus ϕ_P to find the point of interest. The latter curve is plotted along with $|A_P|$ of Eq. (5-14) in Fig. 5.20 for peaking evaluation in the constant β case. To use these curves, $|A|\beta$ versus ϕ is plotted as shown by the dashed-line example. The intersection of this curve with $(|A|\beta)_P$ identifies ϕ_P or the phase shift at the gain peak. At this value of ϕ_P the peaking maximum is found from the $|A_P|$ curve and is 5.5 dB for the example used. The frequency of the peak can be found from the open-loop phase plot.

In addition to defining peaking for the noninverting case above, it can be shown that the same curves apply for the inverting connection. From Appendix A the gain relationship for the inverting configuration can be expressed as

$$A_{CL} = \frac{1 - 1/\beta}{1 + 1/A\beta}$$

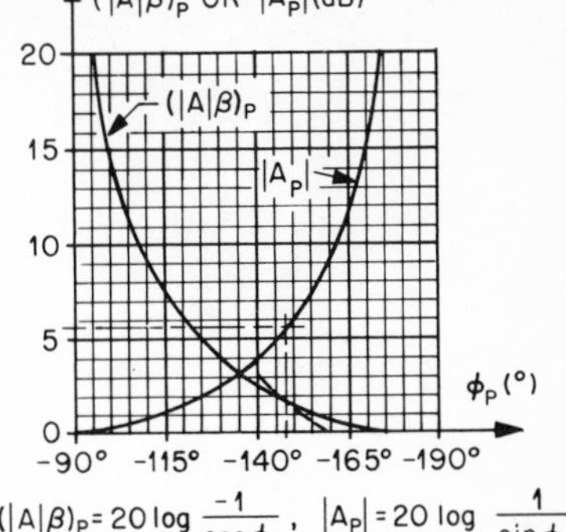

Fig. 5.20 Curves for determination of closed-loop peaking.

Following the previous analysis, the condition for maximum $|A_{CL}|$ is again found to be

$$(|A|\beta)_P = \frac{-1}{\cos \phi_P}$$

Since this condition is the same as that expressed by Eq. (5-12) for the noninverting case, the $(|A|\beta)_P$ curve of Fig. 5.20 applies to the inverting configuration. The gain maximum is found to be

$$|A_{CL}|_{max} = \frac{1 - 1/\beta}{\sin \phi_P} \tag{5-15}$$

In this case A_{OCL} is found to be $1 - 1/\beta$ and the peaking gain is again

$$|A_P| = 20 \log \frac{|A_{CL}|_{max}}{A_{OCL}} = 20 \log \frac{1}{\sin \phi_P}$$

Thus, the peaking gain for the inverting case is identical to that of the noninverting connection, and the second curve of Fig. 5.20 also applies to both cases.

One additional step can be made to approximately relate peaking to the phase margin used previously as the guide in selection of phase compensation. This is achieved by relating the gain and phase at each point of the $(|A|\beta)_P$ curve to an associated phase shift at the $|A|\beta = 1$ point. For a given loop response there is a unique value of ϕ and, thereby, of phase margin ϕ_m which will result at the $|A|\beta = 1$ point for a closed-loop response which passes through a given point of the $(|A|\beta)_P$ curve. Associated with each point on the $(|A|\beta)_P$ curve is a value of peaking gain found from the $|A_P|$ curve. For the most general case an open-loop response curve will be considered for which only one pole significantly influences the phase response slope in the region of interest, as in Fig. 5.21. As long as the point of interest is at least a decade in frequency away from

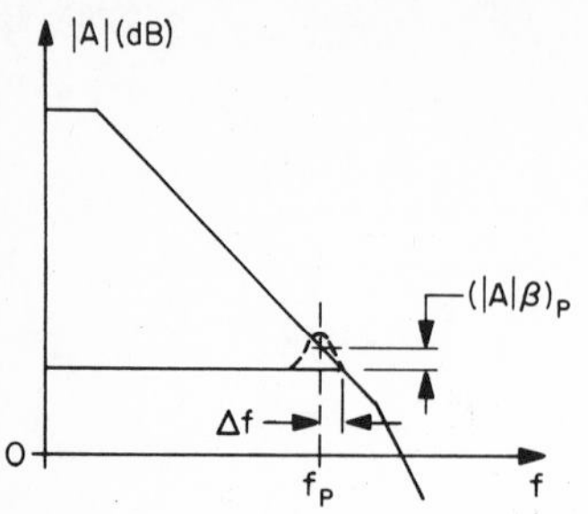

Fig. 5.21 Response curve used to define peaking versus phase margin.

all pole locations, except one, only this nearest pole influences the variation of ϕ with frequency. Then the frequency location of the peak f_P with respect to this pole is defined for a given ϕ_P as indicated. The associated value of $(|A|\beta)_P$ defines a Δf between f_P and the intercept as shown, and Δf determines the added phase shift. In this way the phase shift at the intercept and the associated phase margin are found to result in the curve of peaking versus phase margin in Fig. 5.22. Similar curves can be drawn for other more specialized open-loop response conditions.

The step response characteristics of a multiple-pole operational amplifier have complex relationships to the poles and are not readily expressed mathematically except for specific cases. However, an approximation to the amplifier frequency response can be made to simplify the step response analysis. For the general case in which the frequency response is primarily controlled by only two poles, the well-established second-order system response characteristics can be applied to an operational amplifier. This approximation is fairly accurate as long as additional response poles or zeros occur at no less than a decade higher frequency than the $|A|\beta = 1$ intercept frequency of interest. Such is the case for the previous amplifier response of Fig. 5.21. In order to relate the amplifier response to the second-order system results, the closed-loop gain is rewritten, assuming a two-pole open-loop gain response of

$$A = \frac{A_O}{(1 + j\omega/\omega_{p1})(1 + j\omega/\omega_{p2})}$$

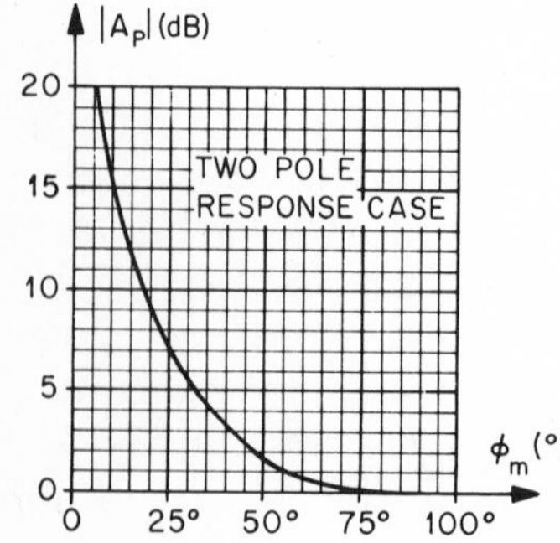

Fig. 5.22 Closed-loop gain peaking defined by phase margin for a two-pole response such as Fig. 5.21.

From the previous analysis the closed-loop gain is

$$A_{CL} = \frac{A_{OCL}}{1 + 1/A\beta}$$

where $A_{OCL} = 1/\beta$ for the noninverting case and $A_{OCL} = 1 - 1/\beta$ in the inverting configuration. By combining the last two expressions, the closed-loop gain can be written as

$$A_{CL} = \frac{A_{OCL}}{1 - \omega^2/A_O\beta\omega_{p1}\omega_{p2} + j(\omega/A_O\beta)(1/\omega_{p1} + 1/\omega_{p2})}$$

In second-order system format the closed-loop gain is

$$A_{CL} = \frac{A_{OCL}}{1 - \omega^2/\omega_n{}^2 + 2\zeta j(\omega/\omega_n)} \qquad (5\text{-}16)$$

which has a step response described by[2]

$$e_o(t) = A_{OCL}e_i(t)\left[1 - \frac{e^{-\zeta\omega_n t}}{\sqrt{1-\zeta^2}} \sin\left(\omega_n\sqrt{1-\zeta^2}\,t + \cos^{-1}\zeta\right)\right]$$

The characteristics of this response can be related to amplifier parameters by comparing the last three equations. First it is seen that

$$\omega_n = \sqrt{A_O\beta\omega_{p1}\omega_{p2}} \qquad (5\text{-}17)$$

where ω_n is the natural frequency of oscillation. This frequency and the damping ratio ζ determine the response overshoot and ringing illustrated in Fig. 5.23. By again comparing the two previous expressions for A_{CL}, the damping ratio is expressed also in terms of amplifier characteristics as

$$\zeta = \frac{\omega_{p1} + \omega_{p2}}{2\sqrt{A_O\beta\omega_{p1}\omega_{p2}}} \qquad (5\text{-}18)$$

The frequency of the ringing shown is defined by the sine term of the step response equation and is the damped frequency of oscillation as expressed by

$$\omega_d = \omega_n\sqrt{1-\zeta^2} \qquad (5\text{-}19)$$

Both the initial overshoot and the continued ringing of the step response represent transient errors of the amplifier. To describe this error the maxima and minima or peaks of the step response function are found by differentiating the response equation with respect to time and setting it equal to zero. From this operation the times at which the peaks occur are found to be

$$t_P = \frac{n\pi}{\omega_d} \qquad n = 1, 2, 3, \ldots \qquad (5\text{-}20)$$

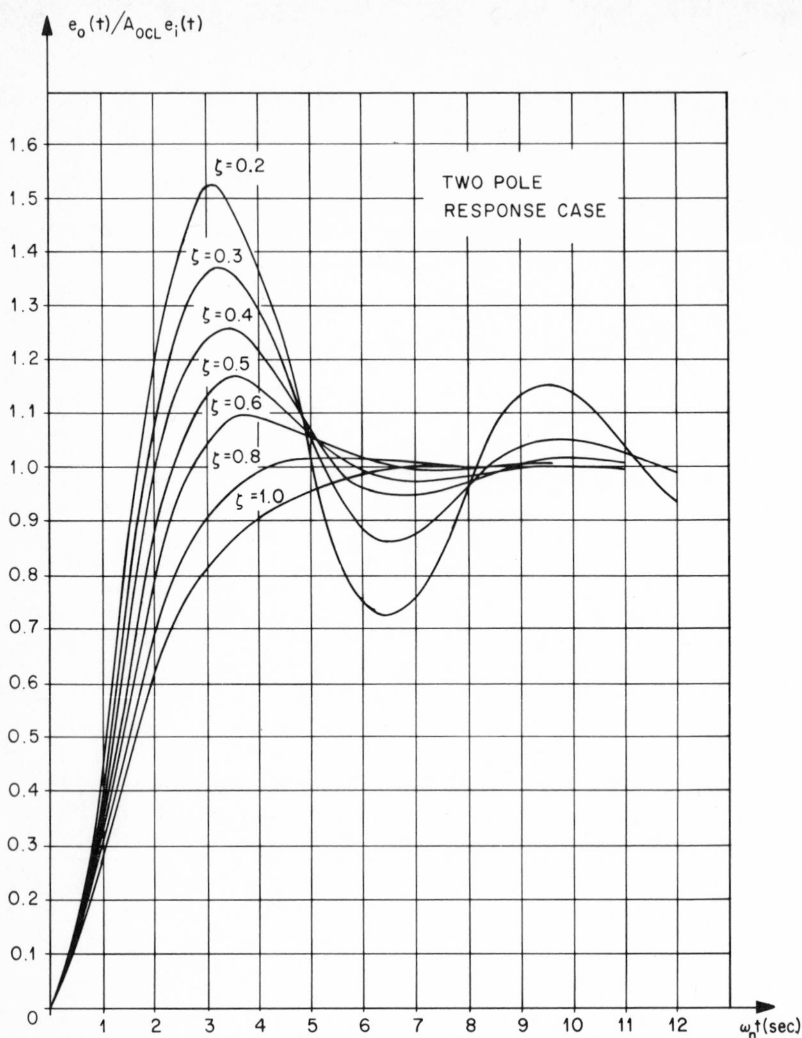

Fig. 5.23 Step response for the two-pole response approximation for various levels of damping ratio.

By substituting this result into the step response equation, the expression for the peaks is found to be

$$\left[\frac{e_o(t)}{A_{OCL}e_i(t)}\right]_{max} = 1 + \exp\frac{-\zeta n\pi}{\sqrt{1-\zeta^2}} \tag{5-21}$$

The overshoot or the amount by which the last expression exceeds unity for n = 1 is expressed as a percentage by

$$\text{Overshoot} = 100\% \exp \frac{-\zeta\pi}{\sqrt{1 - \zeta^2}} \tag{5-22}$$

Although the overshoot identifies the maximum transient response error following the initial rise or fall, the characteristic more often of interest is the time required for the amplifier output to settle to within a certain accuracy following a transient. This characteristic is called *settling time* and is commonly specified as the time required for response settling to within 0.1 or 0.01 percent of final value. Within this time the magnitudes of the ringing and other transient responses have dropped to the specified percentage of the output. For small-signal settling time, as governed by the above analysis, the first peak within the error band approximately defines settling time. For an error band of x percent the appropriate peak is that for the smallest value of n satisfying

$$100\% \exp \frac{-\zeta n\pi}{\sqrt{1 - \zeta^2}} \leq x\%$$

With the value of n found in this way, a conservative estimate of small-signal settling time is provided by Eq. (5-20) for t_P.

In addition to the small-signal step response described above, the analogous characteristics under large-signal conditions are critical aspects of operational amplifier behavior. Output signals in most applications exceed the several-hundred-millivolt range for which transient response is governed by small-signal conditions. Generally large-signal transient response is greatly altered by nonlinear operation and bias disturbances developed by the large transients. Nonlinear amplification results as the amplifier output swing becomes rate-limited, and the swing can no longer respond within the small-signal rise time. Settling time then increases. As outlined in Sec. 3.1, output response speed is bounded by the slewing rate limit imposed by circuit capacitances. Slewing rate is limited by the ability of the circuit to provide charging current to such capacitances, and the dominant limit is commonly the phase compensation capacitance. When the phase compensation is connected to a differential stage, the stage will unbalance under feedback during rate limit to supply the total stage current to the capacitance. Under this condition the slewing rate across the capacitor is

$$\frac{de_{Cx}}{dt} = \frac{2I_C}{C_x} \tag{5-23}$$

where I_C is the quiescent collector current of one side of the stage. This slewing rate is amplified by the gain following C_x. Because of the imposed rate of change limit, the output signal is distorted from the input signal shape as indicated in Fig. 5.24. Following the output signal rise in the large-signal step response a further increase in settling time frequently results from associated bias disturbances. Circuit capacitances which were rapidly charged by the input signal rise may affect bias voltages, and the slow discharging of these capacitances can produce output errors with long settling times.

Also available from the second-order system characteristics used for the small-signal step response is a prediction of peaking from the damping ratio ζ. Although this representation of peaking is limited to the two-pole response approximation, it does provide a simple peaking evaluation. As before, the amount of peaking is derived by finding the maximum of the closed-loop gain magnitude. From the gain expression of Eq. (5-16), the gain magnitude is

$$|A_{CL}| = \frac{A_{OCL}}{\sqrt{(1 - \omega^2/\omega_n{}^2)^2 + 4(\zeta^2\omega^2/\omega_n{}^2)}}$$

By equating the derivative of this equation with respect to ω to zero, the condition at the peak is found to be

$$\omega = \omega_n \sqrt{1 - 2\zeta^2} \tag{5-24}$$

Substitution of the result in the above gain magnitude result yields the maximum closed-loop gain magnitude of

$$|A_{CL}|_{max} = \frac{A_{OCL}}{2\zeta \sqrt{1 - \zeta^2}}$$

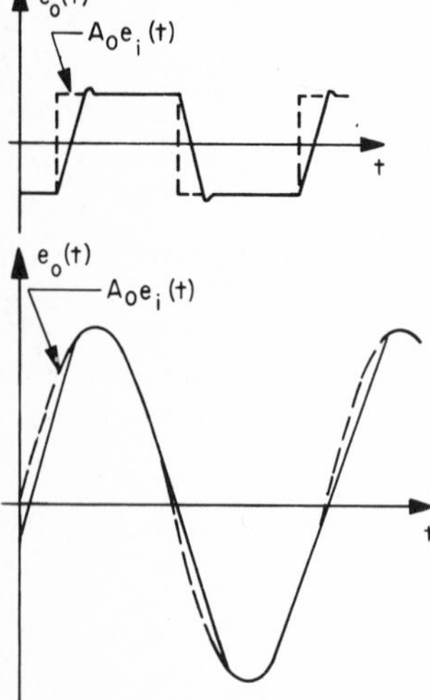

Fig. 5.24 Output signal distortion produced by slewing rate limiting.

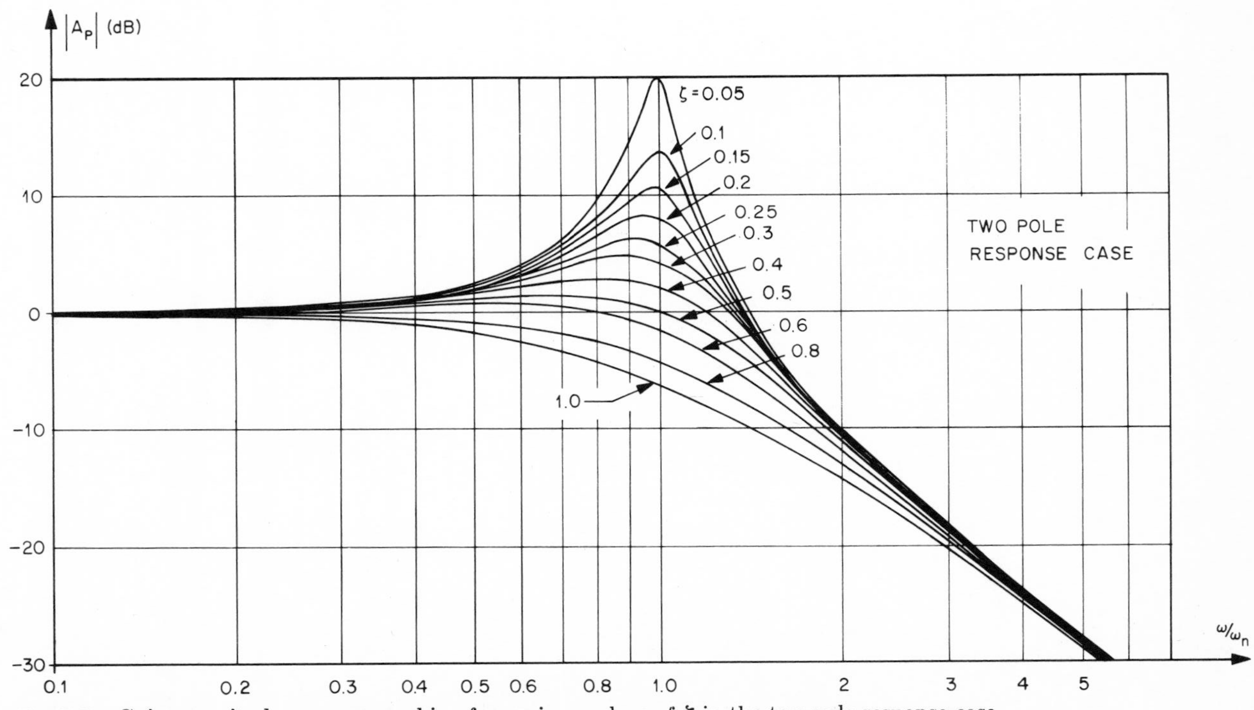

Fig. 5.25 Gain magnitude response peaking for various values of ζ in the two-pole response case.

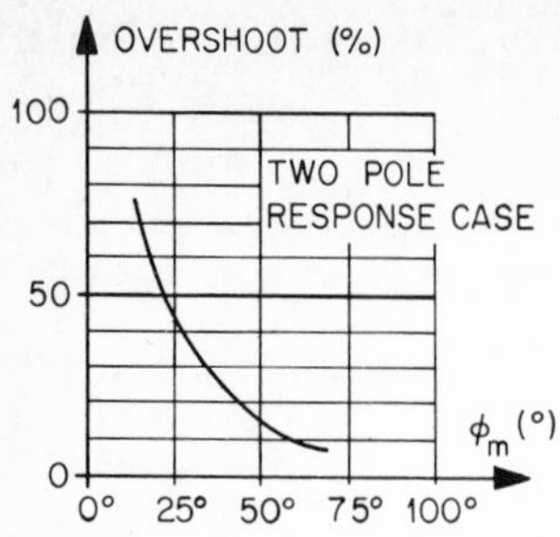

Fig. 5.26 Approximate overshoot versus phase margin for closed-loop gain intercepts at least a decade above the first open-loop pole frequency.

The peaking gain A_P then has a magnitude of

$$|A_P| = 20 \log \frac{|A_{CL}|_{max}}{A_{OCL}}$$

$$|A_P| = 20 \log \frac{1}{2\zeta\sqrt{1-\zeta^2}} \tag{5-25}$$

Using this result, the amount of peaking under the second-order system approximation is rapidly found. Alternatively, the approximate peaking can be found from a family of curves of $|A_{CL}|$ for various values of ζ as presented in Fig. 5.25.

In practice the phase compensation is chosen by considering the various response characteristics to minimize gain error. Gain errors result from the magnitude peaking, overshoot, ringing, settling time, and slewing rate limit of an operational amplifier. The control of the small-signal response characteristics is achieved by maintaining phase margin. In Sec. 5.2 the phase margin ϕ_m desired as a result of phase compensation was defined as approximately 45°. To relate small-signal response characteristics to phase margin the frequently encountered two-pole response case depicted earlier in Fig. 5.21 is considered. When the intercept of the closed-loop and open-loop responses is at least a decade in frequency above the first pole frequency, the peaking magnitude is related to ϕ_m by the curve of Fig. 5.22. The overshoot may also be related to ϕ_m by associating the peaking of this curve with that of the general second-order system of Eq. (5-25). From this comparison the damping ratio ζ for a given phase margin is found. The damping ratio in turn defines overshoot from Eq. (5-22) to result in the plot of overshoot versus ϕ_m in Fig. 5.26.

In practice, the selection of operational amplifier phase compensation is greatly aided by experimental evaluation of response characteristics. Phase compensation deficiencies are rapidly detected by observing the amplifier step response with a square-wave input signal. The analytical relationships serve as a guide in the experimental process by associating response deficiencies with their sources and directing their resolution.

Those response characteristics not well approximated by the analytical expressions such as large-signal settling time must be evaluated solely by experiment. To select phase compensation its component values can be varied while initially observing small-signal overshoot and ringing to find optimum values. By repeating this procedure at various closed-loop gain levels a phase compensation can be chosen for general-purpose applications of any gain level. Specific closed-loop gains at which the square-wave test should be performed in choosing this general-purpose compensation can be identified from the slope of the open-loop gain magnitude response curve. Tests should be made in the vicinity of those gain levels for which the response slope suggests phase margins approaching 45° or less. Generally, the worst case results for the unity-gain voltage follower configuration as the feedback factor β is then at its maximum of unity, and the open-loop phase shift is commonly highest at the associated high-frequency intercept. Once the step response has been optimized, the magnitude peaking, slewing rate, and settling time are examined.

REFERENCES

1. G. J. Thaler and R. G. Brown, *Servomechanism Analysis,* pp. 148–176, McGraw-Hill Book Company, New York, 1953.
2. J. J. DiStefano, A. R. Stubberd, and I. J. Williams, *Feedback and Control Systems,* Schaum's Outline, p. 39–40, McGraw-Hill Book Company, New York, 1967.

Part 2
APPLICATION

6
LINEAR CIRCUIT APPLICATIONS

In this chapter we discuss some of the most frequently encountered linear circuit applications of operational amplifiers. These include differential DC amplifiers, bridge amplifiers, analog integrators, differentiators, line-driving amplifiers, ac-coupled feedback amplifiers, current-to-voltage converters, reference voltage sources, voltage regulators, current amplifiers, and charge amplifiers. The details of these applications are given in the sections which follow. Inverting, noninverting, and summing amplifiers are discussed in the basic theory of Appendix A.

6.1 Differential DC Amplifiers

The amplifiers to be discussed in this section are most descriptively known as differential DC amplifiers, denoting the fact that they amplify the difference between two signals and that the inputs are direct-coupled. Other common terms used for this basic type of amplifier are transducer amplifier, bridge amplifier, data amplifier, instrumentation amplifier, difference amplifier, and error amplifier. Such amplifiers are easily realized through the use of one or more operational amplifiers with linear feedback. The idealized characteristics of these amplifiers are infinite input impedance, zero output impedance, no dc offsets or drift, zero amplifier noise, a constant gain factor with no gain error, and complete rejection of signals common to both inputs (infinite common-mode rejection). Inputs are typically from transducers which convert a physical parameter and its variations to electrical signals. Examples of such transducers are thermocouples, strain-gage bridges, etc. Several types of such differential DC amplifiers, of varying complexity and performance characteristics, are discussed in the following paragraphs.

6.1.1 Differential DC amplifiers using one operational amplifier[2,6,7]

The circuit of Fig. 6.1a has the virtue of simplicity, using only one operational amplifier and four matched resistors. The presence of a common-mode voltage e_{cm} and a differential voltage $e_1 - e_2$ are characteristic of most transducers. The common-mode voltage may represent a dc level, as in a bridge, or may be noise pickup. If an ideal operational amplifier is assumed, the following equations apply:

$$e_3 = (e_{cm} + e_2)\frac{R_4}{R_3 + R_4}$$

$$\frac{e_{cm} + e_1 - e_3}{R_1} = \frac{e_3 - e_o}{R_2}$$

Combining these gives the resulting equation for output voltage,

$$e_o = e_{cm}\frac{R_4R_2 + R_4R_1 - R_2R_3 - R_2R_4}{R_1(R_3 + R_4)} - \frac{R_2}{R_1}e_1 + \frac{R_4}{R_3}\frac{1 + R_2/R_1}{1 + R_4/R_3}e_2$$

If $R_2/R_1 = R_4/R_3$, the above equation reduces to $e_o = (R_2/R_1)(e_2 - e_1)$. The resistor ratios R_2/R_1 and R_4/R_3 must be carefully matched in order to ensure the rejection of common-mode signals. The value of these resistor ratios sets the gain for differential signals. These equations illustrate the performance of the circuit when one is dealing with zero source impedances and nonzero common-mode signals. For zero source impedance the gain is determined solely by the feedback resistors and, if these resistors are matched in pairs as indicated, common-mode signals are rejected completely. Actually, of course, the operational amplifier has been assumed ideal in having infinite input impedance, infinite gain, and infinite common-mode rejection. If these factors are given real values and their effects evaluated, it will be found that the finite input impedance of the operational amplifier and its inherent finite common-mode rejection will place limits on the overall common-mode rejection of the closed-loop differential amplifier. The finite open-loop gain will limit the gain accuracy of the overall circuit.

Figure 6.1b illustrates a model for unbalanced source impedances and their interactions with the finite resistances of the amplifier feedback network. An analysis similar to that for the circuit of Fig. 6.1a yields

$$e_o = e_{cm} \frac{R_2(R_{S_1} - R_{S_2})}{(R_1 + R_{S_1})(R_3 + R_{S_2} + R_4)} + \frac{R_2}{R_1 + R_{S_1}} \left[\frac{1 + (R_1 + R_{S_1})/R_2}{1 + (R_3 + R_{S_2})/R_4} e_2 - e_1 \right]$$

Note that, if the source impedances are nonzero but equal, the only effect is a gain error due to the source loading. However, if the source impedances are also unequal, the common-mode rejection is degraded. Input bias currents (I_{B1}, I_{B2}) and input voltage offset (V_{OS}) of the operational amplifier will cause dc offset errors at the output of the differential amplifier circuit. Bias current (I_{B2}) from the noninverting side of the operational amplifier flows through the parallel combination of R_4 and R_3 to create a dc error voltage at the noninverting input terminal. This dc voltage effectively adds to the offset voltage of the operational amplifier and is amplified by the factor $(R_2 + R_1)/R_1$. Bias current (I_{B1}) from the inverting input of the operational amplifier flows principally through resistor R_2 and causes an output offset adding to the other two components to give the total dc offset error of

$$E_{OS} = V_{OS} \frac{R_2 + R_1}{R_1} + I_{B1} \frac{R_3 R_4}{R_3 + R_4} \frac{R_2 + R_1}{R_1} - I_{B2} R_2$$

Tracking between the two bias currents reduces the bias-current-induced error term by as much as a factor of 10. The principal limitations of this circuit are its low input impedance and the difficulty of varying the gain. The input impedance, of course, is determined by the feedback and input resistors. If these resistors are made large in order to increase the input impedance the dc errors due to bias currents will be proportionately increased, thus placing an upper limit on the feasible values of input impedance. The gain of the differential amplifier can be changed only by varying the ratios of the feedback resistors. Because of the necessity of maintaining the equality of the resistive ratios, it is quite difficult to continuously vary the gain. Gain steps can be achieved if the common-mode rejection is carefully adjusted at each gain setting. The differential amplifier circuit of Fig. 6.2 is a similar type of circuit with the added feature of a gain vernier which allows the gain to be continuously varied without affecting the common-mode rejection of the circuit. The output voltage is

$$e_o = 2\left(1 + \frac{1}{K}\right) \frac{R_2}{R_1} (e_2 - e_1)$$

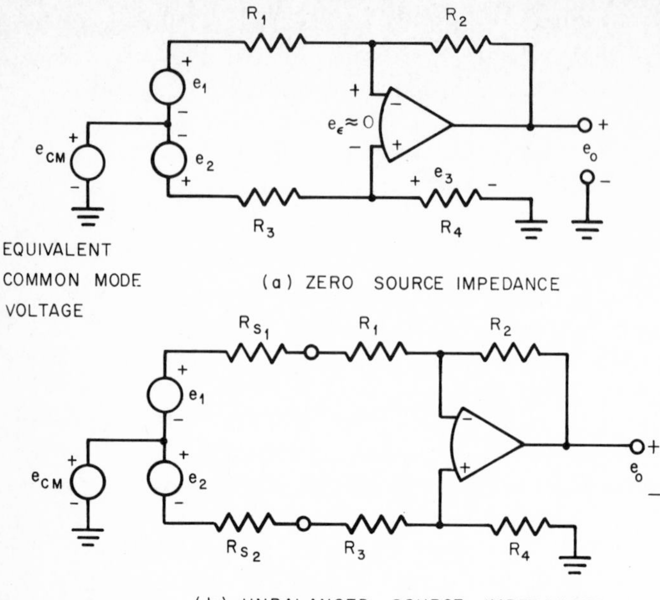

Fig. 6.1 Simple differential amplifier.

Note, however, that this circuit requires four matched resistors of value R_2 and two matched resistors of value R_1. The gain is an inverse function of the setting of the vernier potentiometer and as such is highly nonlinear. The potentiometer can, however, provide approximate linearity over limited ranges. The circuit still suffers from the limitations of low input impedance. The dc offset errors are much the same as those for the circuit of Fig. 6.1.

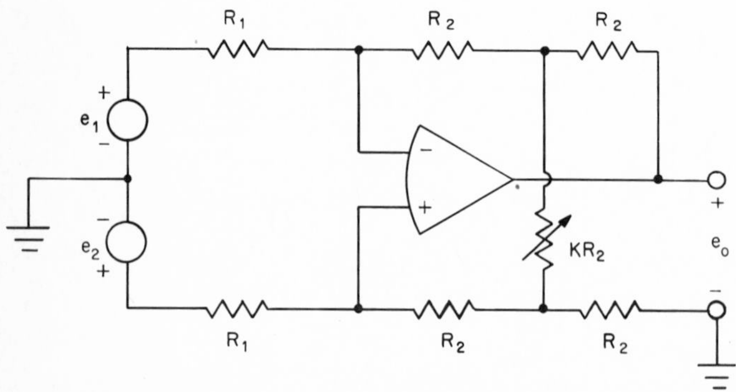

Fig. 6.2 Simple adjustable-gain differential amplifier.

6.1.2 Differential DC amplifiers using more than one operational amplifier[2,7] The circuit of Fig. 6.3 provides another low impedance alternative to those of Figs. 6.1 and 6.2. The two amplifiers required operate in the inverting mode and need not have a noninverting capability. Thus they can be chopper-stabilized amplifiers for low drift, or they may be FET input types which may have rather poor linearity when used noninverting. The output voltage is

$$e_o = \frac{R_2}{R_1}(e_2 - e_1)$$

The gain can be easily varied, in steps or continuously, by changing the value of R_2, without affecting the common-mode rejection properties. Good common-mode rejection requires four closely matched resistors of value R_1. Note that the dc offset error is approximately four times that of a single amplifier, if it is assumed that the offset errors add, as given by the expression

$$E_{os} = \left(1 + 2\frac{R_2}{R_1}\right) V_{OS2} + 2\frac{R_2}{R_1} V_{OS1}$$

$$= \left(1 + 4\frac{R_2}{R_1}\right) V_{OS} \qquad \text{(worst case)}$$

Since the common-mode rejection of the operational amplifiers is not a factor, the common-mode rejection of the closed-loop amplifier can be trimmed to quite high values simply by allowing a small amount of adjustability of one of the R_1 resistors. The common-mode voltage capability of the circuit is limited only by the output voltage capability of the unity-gain inverter. This capability can be increased by making the gain of amplifier A_1 less than unity. The gain of amplifier A_2 must then be increased accordingly, however, which increases the output offset

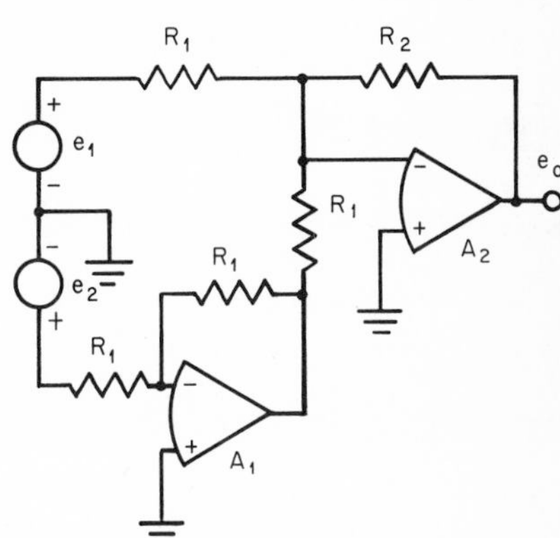

Fig. 6.3 Differential DC amplifier using inverting operational amplifiers.

error. Another differential DC amplifier circuit using two operational amplifiers is shown in Fig. 6.4. This circuit provides the high input impedance lacking in the circuits discussed up to now. For this circuit

$$e_o = \left(1 + \frac{R_4}{R_3}\right)(e_2 - e_1) \quad , \quad \text{if } \frac{R_1}{R_2} = \frac{R_4}{R_3}$$

Again, equality of the two resistor ratios is required in order for the circuit to reject common-mode signals. The operational amplifiers, since they operate in the noninverting mode, must have good common-mode properties. The input impedance at each terminal of the differential amplifier is simply the common-mode input impedance of the operational amplifiers. This can be quite large (10 MΩ and up), depending on the type of operational amplifier used. For fixed gains, or gain steps, the circuit is quite useful, but it is not feasible for continuously variable gain. Also, since the input voltage of the upper amplifier must be less than $R_1/(R_1 + R_2)$ times the output saturation voltage, the common-mode voltage range is very limited at low values of overall gain. This is not considered a serious limitation since such amplifiers are usually used at gains of 10 or greater. The differential DC amplifier circuit of Fig. 6.5 overcomes most of the weaknesses of the circuits discussed up to this point. Analysis of the circuit yields the following equations:

$$e_3 = \left(1 + \frac{R_2}{R_1}\right)e_1 - \frac{R_2}{R_1}e_2 + e_{cm}$$

$$e_4 = \left(1 + \frac{R_3}{R_1}\right)e_2 - \frac{R_3}{R_1}e_1 + e_{cm}$$

$$e_o = e_4 - e_3$$

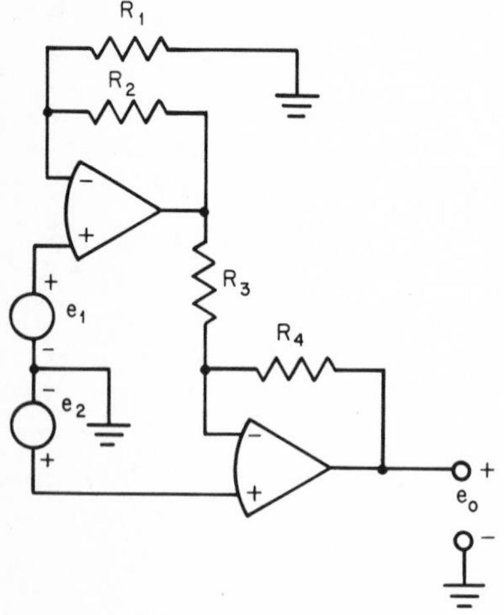

Fig. 6.4 High input impedance differential amplifier.

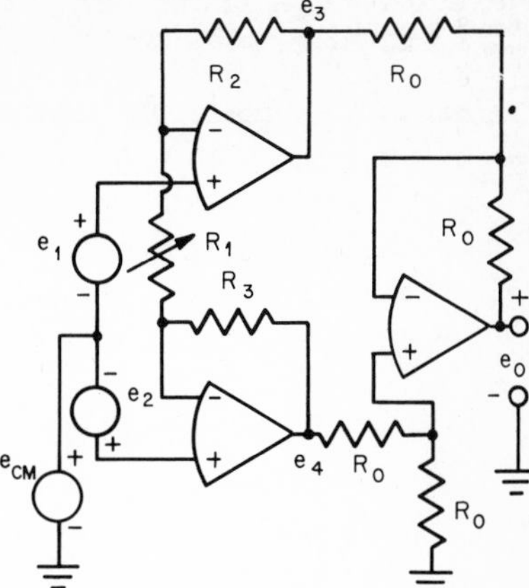

Fig. 6.5 High input impedance adjustable-gain differential amplifier.

If $R_2 = R_3$, the output voltage is

$$e_o = \left(1 + \frac{2R_2}{R_1}\right)(e_2 - e_1)$$

The two input amplifiers constitute a differential buffer amplifier with a gain of $1 + 2R_2/R_1$ for differential signals, and unity gain for common-mode signals. The noninverting configuration of these input amplifiers ensures high input impedance at both inputs. The gain is easily varied by a single resistor R_1. The effects of mismatch in resistors R_2 and R_3 is simply to create a gain error without affecting the common-mode rejection of the circuit. The resistors R_o of the output amplifier must be accurately matched, or trimmed, to ensure the rejection of common-mode signals at this point. This final amplifier acts simply as a differential-input to single-ended-output converter. Feedback impedances in both stages can be relatively low in value to minimize the effects of bias current, since these feedback elements do not affect the input impedance of the differential amplifier. Usually, all the gain of this differential amplifier is in the input stage, thus ensuring that only the offset voltages of these two operational amplifiers are significant in determining the output offset. Since the output voltage offset is proportional to the difference of the voltage offsets of these two amplifiers, it is desirable to use amplifiers whose voltage offsets tend to track with temperature. Such techniques are the basis for some low-drift differential amplifier modules. The bias currents of these input amplifiers will flow through the impedance of the source and will thus generate additional offset voltage which will appear at the output of the differential amplifier amplified by the differential gain factor. The use of amplifiers with FET input stages will greatly reduce this effect.

6.2 Bridge Amplifiers[2]

Probably the most common use for a differential DC amplifier is in amplifying the output signal from a transducer bridge, such as a strain gage. The most straightforward way of doing this is with one of the high impedance amplifiers discussed in the preceding section. Such a strain-gage bridge, with one active bridge arm, is shown in Fig. 6.6. The following equations describe its operation:

$$e_2 = V \frac{R}{2R + \Delta R}$$

$$e_1 = \frac{V}{2}$$

$$e_2 - e_1 = -\frac{V}{4}\frac{\delta}{1 + \delta/2}$$

where

$$\delta = \frac{\Delta R}{R}$$

$$e_o = K(e_2 - e_1) = -\frac{KV}{4}\frac{\delta}{1 + \delta/2}$$

$$e_o \approx -KV\frac{\delta}{4}, \quad \text{if } \delta \ll 1$$

The output signal is a linear function of the variation of the active element only for small percentage changes in the element. If larger changes are to be measured, the exact equation must be used and a conversion or linearization performed at some point in the data-gathering process.

It is sometimes desirable to use an amplifier less complex than the fully developed differential instrumentation amplifier for amplifying the output signal from a bridge. There are several such circuits which use only a single operational amplifier, such as the one shown in Fig. 6.7. This

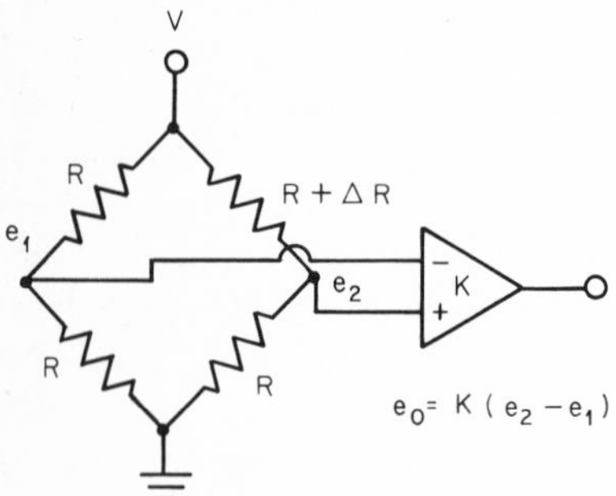

Fig. 6.6 Bridge amplifier.

circuit forces the differential output voltage of the bridge to be zero since opposite sides are connected directly to the inputs of an operational amplifier with feedback. Thus the amplifier is used to measure the current flowing into the bridge under short-circuit conditions. The resulting output voltage is

$$e_o = \frac{R_F}{R} \frac{\delta}{1 + \delta} \frac{V}{(2 + \delta)/(1 + \delta) + R/R_F}$$

If $\delta \ll 1$ and $R_F \gg R$, this equation reduces to the approximate form

$$e_o \approx V \frac{\delta}{2} \frac{R_F}{R}$$

Note that here again the equation for the output voltage of the bridge amplifier is a nonlinear function of the variation of the active bridge element, but for small deviations the nonlinearity is negligible. For the simplified, approximate form of the equation, it has also been assumed that the values of resistance in the bridge are much smaller than the resistors R_F. The bridge resistance appears in the gain equation, thus requiring that the values of the bridge elements be insensitive to temperature in order that the gain of the amplifier be stable with temperature. If the assumption that R_F is much greater than the nominal bridge resistance applies, there is no loading effect.

The dc offset voltage generated at the output of the bridge amplifier as a result of the input offset voltage and bias currents of the operational amplifier is given by

$$E_{OS} = V_{OS} \frac{2R_F + R}{R} + (I_{B2} - I_{B1})R_F$$

where I_{B1} and I_{B2} are input bias currents. The main advantage of this circuit is its simplicity. It does require an amplifier which has reasonably good common-mode rejection.

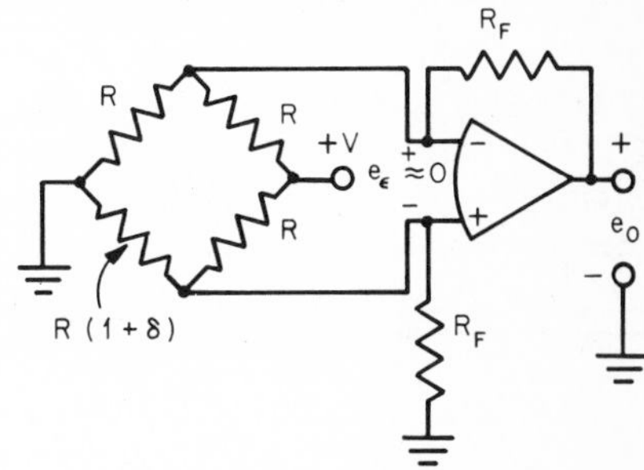

Fig. 6.7 Bridge current amplifier.

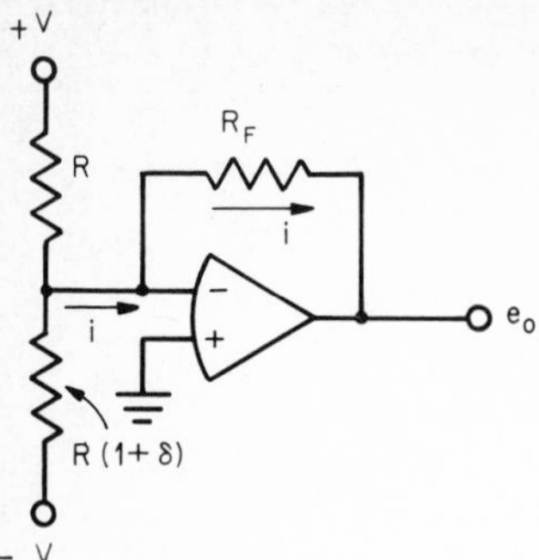

Fig. 6.8 Half-bridge current amplifier.

Where the rejection of common-mode noise signals is not a problem, the half-bridge measuring circuit of Fig. 6.8 is sometimes used. Here also the output of the bridge is connected directly to the input terminal of the operational amplifier, as is the feedback through R_F. Since the other input of the operational amplifier is held at ground potential, the output of the half-bridge is held at zero voltage, and the amplifier responds to the short-circuit output current

$$e_o = -iR_F = -V \frac{R_F}{R} \frac{\delta}{1 + \delta}$$

If $\delta \ll 1$

$$e_o \approx -V \frac{R_F}{R} \delta$$

Because the amplifier operates single-ended, the amplifier used can be chopper-stabilized for lowest possible drift and dc offset errors. Also, the maximum voltage supplied to the bridge, or half-bridge, is not limited by common-mode voltage limitations of the operational amplifier, as it is in those circuits which use the noninverting input of the operational amplifier. Thus it is possible to increase the sensitivity of the bridge by increasing the supply voltage within the limitations of the bridge elements and the ability of the amplifier to supply the current flowing through the feedback resistor.

The major drawback of the half-bridge circuit is its inability to reject noise pickup, as is normally accomplished by the differential type of bridge amplifier. Consequently, the noise and ripple of the half-bridge supply must be very low, and all wiring must be kept short and well shielded. As in the previous bridge amplifier, the gain is a function of the bridge elements. This can be a serious drawback if the bridge elements are sensitive to environmental factors other than the one that it is desired to measure. The output dc offset voltage of the half-bridge

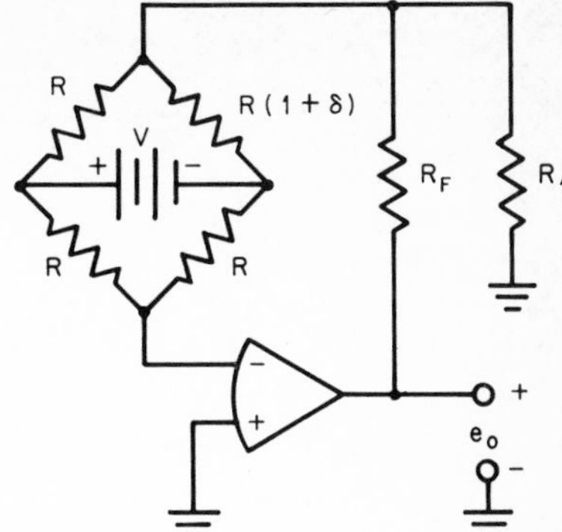

Fig. 6.9 Inverting bridge amplifier.

amplifier is given by the expression

$$E_{OS} = V_{OS}\left(1 + \frac{2R_F}{R}\right) - I_{B1}R_F$$

where I_{B1} is the input bias current.

Figure 6.9 illustrates another bridge amplifier using a single operational amplifier in the inverting mode. Thus it is once again possible to use a single-ended chopper-stabilized amplifier with its attendant low drift. The amplifier output voltage is

$$e_o = V\left(1 + \frac{R_F}{R_1}\right)\frac{\delta}{4(1 + \delta/2)}$$

which, for $\delta \ll 1$, reduces to

$$e_o \approx V\left(1 + \frac{R_F}{R_1}\right)\frac{\delta}{4}$$

Another advantage of this circuit, not shared by the preceding two, is that the gain is not dependent upon the absolute value of the bridge resistors. The output voltage is proportional to the open-circuit coltage of the bridge since the input to the amplifier draws negligible signal current. The inverting input of the operational amplifier is maintained at virtual ground by the high open-loop gain. Since the gain is a function of R_F and R_1, it can be varied easily with either resistor. A small-valued potentiometer can be added in series with either resistor for calibration purposes. This type of bridge amplifier can be very accurate and is recommended when it is necessary to detect very small bridge signals. The primary disadvantage is that a floating bridge supply is required. Since it uses a single-ended amplifier it does not have the common-mode rejection capabilities of the true differential amplifier. However, careful shielding and filtering to remove noise can help to eliminate this problem.

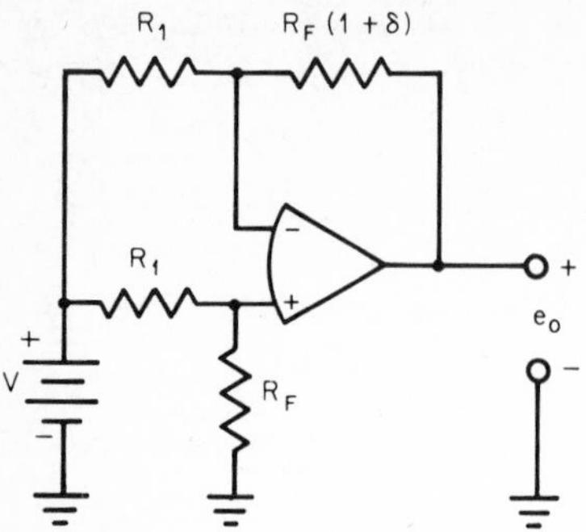

Fig. 6.10 Wide-deviation bridge amplifier.

The output voltage offset as a function of input offset voltage and bias currents is similar to that of the inverting amplifier circuit. That is,

$$E_{OS} = \frac{R_F + R_1}{R_1}(V_{OS} - I_{B1}R) - I_{B1}R_F$$

where I_{B1} is the amplifier input bias current.

The final bridge amplifier circuit to be discussed is that given in Fig. 6.10 where the output voltage is directly proportional to the transducer deviation even for large fractional changes in the active element:

$$e_o = -V\frac{\delta R_F}{R_1 + R_F}$$

This particular circuit should be used whenever the deviation of the active element is large enough so that the linear approximations made in the previous bridge equations are no longer valid. Examples are semiconductor strain gages that have high gage factors, thermistors, etc. The bridge elements must be so matched that the two input resistors are equal and the active element is equal to the value of R_F when the bridge is at null. Calibration is somewhat difficult since it requires the trimming of two values of resistance to maintain null while varying sensitivity.

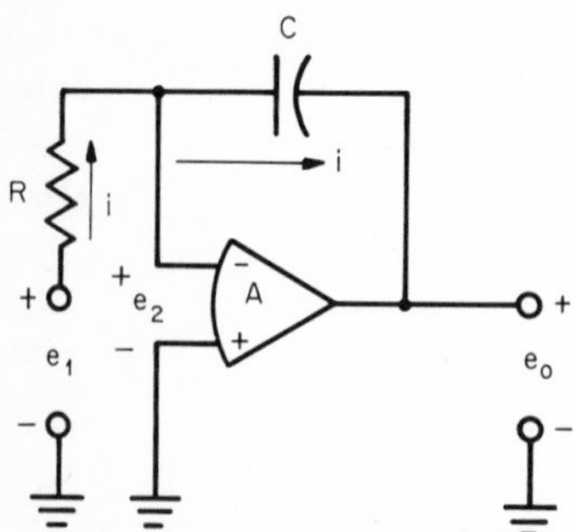

Fig. 6.11 Analog integrator.

6.3 Analog Integrators[1,5,8]

The analog integrator is extremely useful in computing, signal processing, and signal generating applications. It uses an operational amplifier in the inverting configuration, as shown in Fig. 6.11. The equations of operation are derived by assuming an ideal operational amplifier of gain A. These are

$$\frac{e_1 - e_2}{R} = i$$

$$e_2 - e_o = \frac{1}{C}\int_0^t i\,dt = \frac{1}{RC}\int_0^t (e_1 - e_2)\,dt$$

$$e_2 = -\frac{e_o}{A}$$

If $A \rightarrow \infty$, then $e_2 \rightarrow 0$ and $e_o = -(1/RC)\int e_1\,dt$. As in the inverting amplifier, the summing point is held at a virtual ground by the high gain of the amplifier and its feedback network. Since no current flows into the input terminal of the operational amplifier, all the input current $i = e_1/R_1$ is forced to flow into the feedback capacitor, causing a charge voltage to appear across this element. Because one end of the capacitor is tied to the virtual ground point, the output voltage of the amplifier equals the capacitor charging voltage. The overall integrator circuit has the low output impedance normally associated with a feedback amplifier.

The dc offset and bias current of the analog integrator are taken into account in the more realistic model of Fig. 6.12. Because these dc errors exist, the output of the integrator now consists of two components: the integrated signal term and an error term

$$e_o = -\frac{1}{RC}\int e_1\,dt + \frac{1}{RC}\int V_{OS}\,dt + \frac{1}{C}\int I_B\,dt + V_{OS}$$

The error term itself is made up of a component due to the input offset

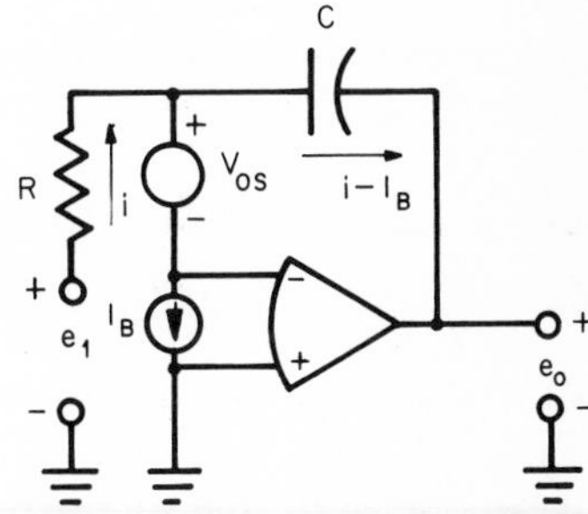

Fig. 6.12 Effect of offset voltage and bias current in an integrator circuit.

voltage and another due to the input bias current. The integral of the dc offset voltage results in a ramp voltage, a linearly increasing term whose polarity is determined by the polarity of the input offset voltage. In addition to this ramp voltage error, the input offset voltage creates an output offset voltage equal to it in value. The bias current flows almost entirely through the feedback capacitor, charging it in ramp fashion, similar to the ramp voltage due to input offset voltage. These two ramp voltage errors will continue to increase until the amplifier reaches its saturation voltage or some limit set by external circuitry. These error components usually set the upper limit on feasible length of integration time. The error component due to bias current can be minimized by increasing the capacitance of the feedback element. This can be done only by decreasing the value of the input resistor, if a specific value of the RC time constant is to be achieved. A lower limit usually exists on R, because of current limitations and loading of the input signal source.

The effects of bias current can be reduced by inserting a resistance R between the noninverting input of the amplifier and ground. This equalizes the resistances at the two inputs and changes the effects of bias current to that of offset (difference) current. Thus, in the equation for output voltage, I_B, the bias current, should be replaced by I_{OS}, the offset current, if the compensating resistor is used. The error ramp due to voltage offset is fixed by the chosen value of RC time constant.

In order to realize the performance possibilities of an operational amplifier as an integrator, a feedback capacitor must be selected with a dielectric leakage current which is less than the bias current of the amplifier. Polystyrene and Teflon are usually the best choices for the ultimate in long-term integrating accuracy. If shorter integration times are required, the requirements on capacitor quality can accordingly be relaxed. Mylar capacitors may then prove satisfactory, or silver-mica types, if small values of capacitance, corresponding to high-speed integration, are to be used.

The choice of the type of amplifier is also governed by the length of computing time and the desired accuracy. Chopper-stabilized amplifiers are usually used for long-term integrators because of their superior long-term dc stability. FET amplifiers are used for medium-length integration because of their low bias current. Amplifiers with bipolar transistor input stages may be used in very short-term integration such as in signal generation (sweep generation, triangle waves, etc.).

If the finite gain and bandwidth are taken into account, their effects on the integrator response function may be evaluated. The open-loop frequency response of the amplifier is approximated by a single pole located at $1/\tau_o$, and a low-frequency gain of A_o.

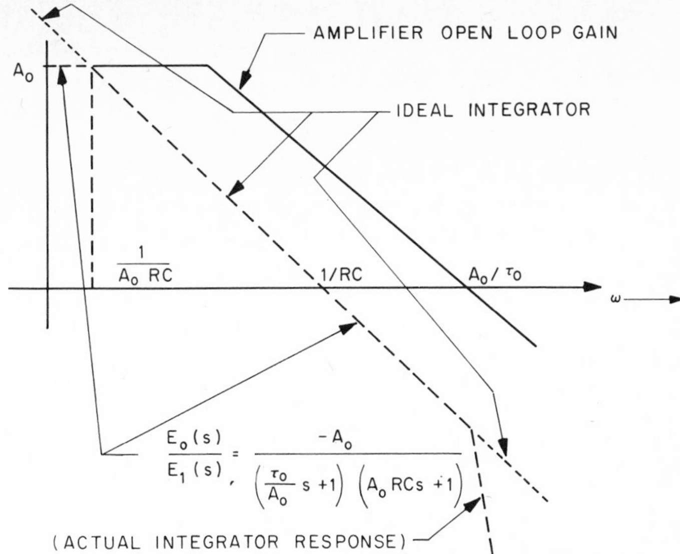

Fig. 6.13 Bode plots; amplifier, integrator.

The resulting integrator response function is

$$\frac{E_o}{E_1}(s) \approx \frac{-A_o}{(\tau_o/A_o s + 1)(A_o RCs + 1)}$$

if $A_o \gg 1$ and $A_oRC \gg \tau_o$. This function has two poles on the real axis, as opposed to the ideal integrator function which has a single pole at the origin. In Fig. 6.13 the frequency response of this approximate integrator is compared with the response of an ideal integrator, along with an open-loop frequency response of the operational amplifier. Note that the response of the real integrator departs from the ideal response only at the extremes of frequency. At low frequencies the departure is due to the finite gain of the operational amplifier. At high frequencies, it is due to the finite amplifier bandwidth.

The transient response of the integrator is studied in Fig. 6.14 by calculating the response to a step function. The response of an ideal integrator to a step function $-E/s$ would be a linear ramp voltage increasing to infinity. The step response of the practical integrator is a close approximation of this ramp throughout most of the signal range:

$$e_o(t) = A_oE\left(1 - \frac{e^{-t/A_oRC}}{1 - \tau_o/A_oRC} + \frac{e^{-t/\tau_o}}{A_oRC/\tau_o - 1}\right)$$

In order to compare the ideal and real responses, it is necessary to examine the responses for very small and very large time. For small values of

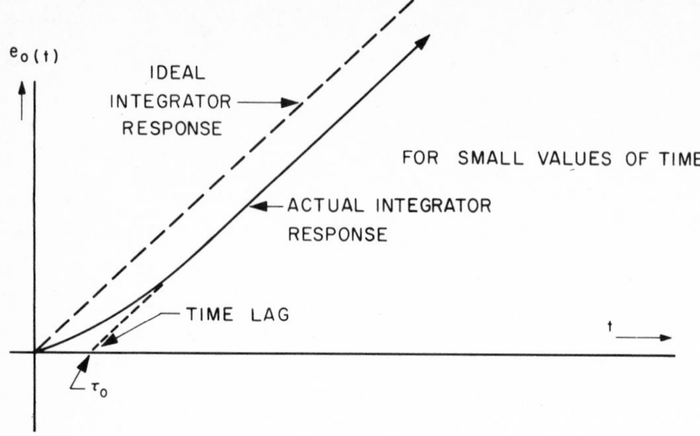

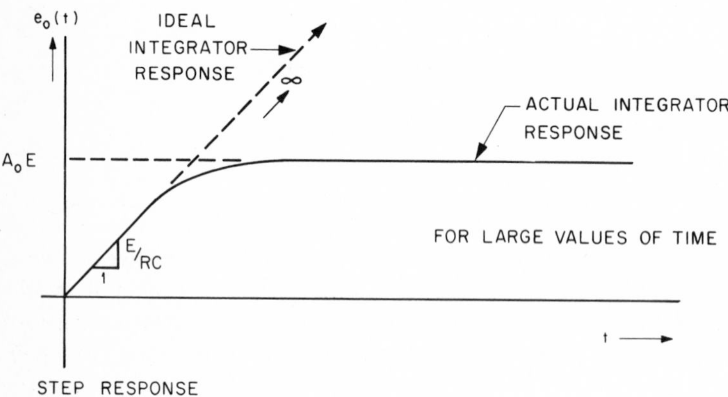

Fig. 6.14 Step response of integrator.

time the response is approximately

$$e_o(t) \approx E\left(\frac{t}{RC} + \frac{\tau_o}{RC} + \frac{e^{-t/\tau_o}}{RC/\tau_o}\right)$$

For large values of time the response is approximately

$$e_o(t) = A_oE(1 - e^{-t/A_oRC})$$

For small values of time the principal error effect is caused by the finite bandwidth, which causes a time lag error in the actual response. For large values of time the output signal would approach an exponential with time constant A_oRC and final value A_oE. For accurate computation, the

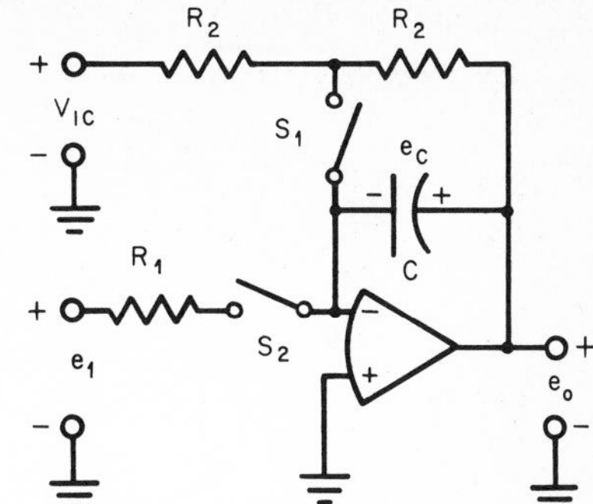

Fig. 6.15 Three-mode integrator.

integration should be terminated at a time much less than A_oRC and an output amplitude much less than A_oE.

Figure 6.15 illustrates the switching techniques used to initiate and terminate the period of the integration. This integrator circuit has three modes. The first of these is RESET, in which the initial conditions are established by placing an initial charge on the capacitor. This is done by closing switch S_1 to allow the output voltage to rise to the negative of V_{IC}. If switch S_1 is then opened and S_2 is closed, the circuit begins integration of the input signal e_1 beginning at the value $-V_{IC}$. This is the second or INTEGRATE mode. If both switches are held open, the output voltage will hold its latest value and will not respond to input or initial condition voltages. During this HOLD mode, the only discharge of the capacitor is due to the bias current of the amplifier and dielectric leakage in the capacitor. Since electronic switch modules are commonly used for the mode control function in place of the simple switches shown, any leakage current flowing from these switches must be added to amplifier bias current in calculating the decay of the capacitor voltage during HOLD, or during the INTEGRATE mode.

Although the analog integrator is a linear device, its maximum rate of change of output signal can lead to slew rate distortion for signals of relatively high frequency and large amplitude. The inherent slew rate limit of the operational amplifier places one of these limitations on the operation. However, another limitation, usually much more restrictive, is that placed on the rate of change of capacitor voltage by the output current limits of the amplifier. The expression for this is

$$\left(\frac{de_c}{dt}\right)_{max} = \left(\frac{de_o}{dt}\right)_{max} = \frac{I_{lim}}{C}$$

where

$$I_{lim} = \text{output current limit}$$

The time required for the amplifier to RESET to initial conditions is

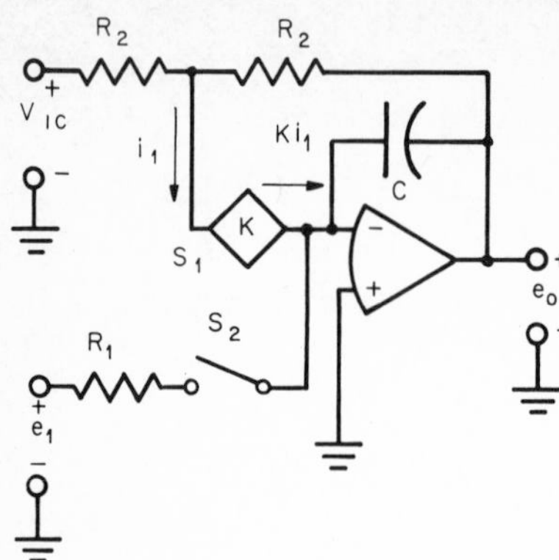

Fig. 6.16 Current amplifying switch used for integrator reset.

limited by the RC time constant of the RESET network and also by the slew rate achievable in the closed-loop circuit. If a reset switch which has a large current gain factor is used, the reset time can be considerably reduced. The use of such a switch is illustrated in Fig. 6.16, where the circuit is shown in the RESET state. Analysis of the circuit yields the equation for the output voltage,

$$e_o = -e_1(1 - e^{-Kt/R_2C})$$

This is again the equation of an exponentially increasing voltage. Here, however, the time constant is R_2C/K, reduced by a factor equal to the current gain of the switch. The RESET time can potentially be reduced by the factor K, if the operational amplifier and switched current amplifier do not reach their current limits, thus limiting the slew rate. Maximum current is required at $t = 0$, the initiation of the RESET mode.

6.4 Differentiators[2,3,6]

By interchanging the resistor and capacitor of an integrator circuit we obtain the inverse function, differentiation. However, as will be shown, the differentiator circuit (Fig. 6.17a) has some troublesome properties. If the usual single-pole open-loop gain function is assumed for the amplifier the transfer function of the differentiator circuit may be reduced to

$$\frac{e_o}{e_1} = \frac{-RCs}{1 + (1/A_o)(\tau_o + RC)s + (RC\tau_o/A_o)s^2}$$

This transfer function has the form

$$H(s) = \frac{-H_os}{1 + (\alpha/\omega_n)s + s^2/\omega_n^2}$$

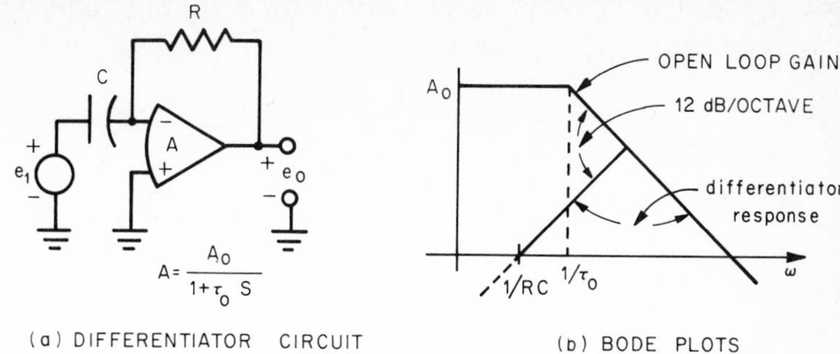

Fig. 6.17 Differentiator using an operational amplifier. (a) Differentiator circuit; (b) Bode plots.

where

$$\omega_n{}^2 = \frac{A_o}{RC\tau_o}$$

and

$$\alpha \text{ (damping factor)} = \sqrt{\frac{(\tau_o + RC)^2}{RC\tau_o} \frac{1}{A_o}} \ll 1$$

Thus the damping factor α is very small, indicating a lightly damped circuit response and complex poles near the $j\omega$ axis. Such a response would also be indicated by the 12 dB per octave rate of closure of the Bode plots (Fig. 6.17b). Thus the differentiator circuit as shown has a tendency toward instability. If the amplifier open-loop gain has an attenuation rate of greater than 6 dB per octave over a portion of its Bode plot, the circuit may well oscillate. Another problem with this differentiator circuit is its high gain at high frequencies. This allows the high-frequency components of amplifier noise to be amplified even though the signal may not have high-frequency components. Thus the high-frequency output noise may obscure the differentiated signal.

The modified differentiator circuit of Fig. 6.18a is usually preferred as a means of eliminating the problems of the simpler circuit. Two additional real poles are introduced by use of R_1 and C_F. This creates a very stable system and reduces the high-frequency noise. The poles are placed sufficiently high in frequency to prevent significant phase-shift error in the signal frequency range. The modified frequency response is shown in Fig. 6.18b.

6.5 Line-driving Amplifiers

One of the primary areas of application for the operational amplifier is that of buffering between a signal source and the desired load. Usually

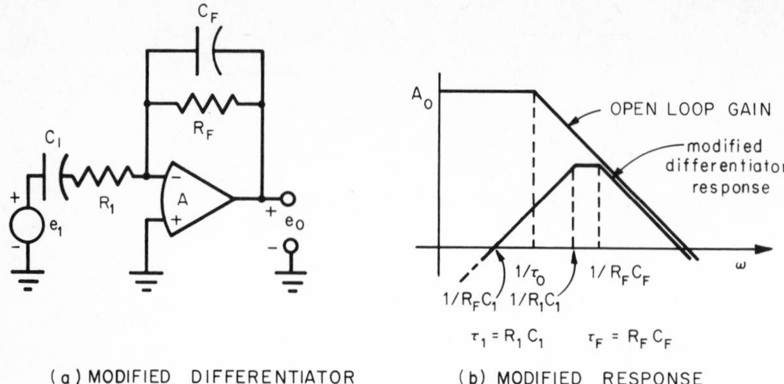

Fig. 6.18 Modified differentiator with improved noise and stability. (a) Modified differentiator; (b) Modified response.

the signal source is very limited in power, has relatively high internal impedance, and is low level. The load is relatively low in impedance (possibly capacitive) and requires high-level signals. Thus the amplifier must provide impedance buffering, signal scaling, and power gain. Needless to say, it must be stable under the desired conditions of loading and feedback and must have sufficient gain and bandwidth to ensure accurate response to input signals. A typical example of such an application is the line-driving amplifier.

When data signals must be transmitted over long signal lines from a remote measuring station, the line-driving amplifier is usually required. Figure 6.19 illustrates a simulated load of this type. The capacitance is that of a shielded cable and may be as little as a few picofarads or as much as several microfarads. If the output impedance of the amplifier is considered, the equation for effective open-loop gain, A′(s), becomes

$$A'(s) = A(s)\,\frac{R_p}{R_p + R_o}\,\frac{1}{1 + R_qC_Ls}$$

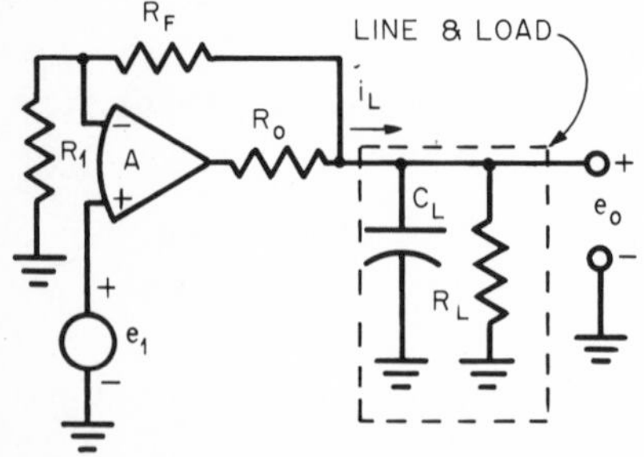

Fig. 6.19 Line-driving amplifier.

where

$$R_p = \frac{1}{1/R_F + 1/R_L} \qquad R_q = \frac{1}{1/R_F + 1/R_L + 1/R_o}$$

where A(s) is the unloaded open-loop gain, and R_o is the dynamic output impedance of the operational amplifier. If A(s) is approximated by a single-pole transfer function

$$A(s) = \frac{A_o}{1 + s/\omega_o}$$

then the effective (loaded) open-loop gain becomes

$$A'(s) = \frac{R_p}{R_p + R_o} \frac{A_o}{1 + s/\omega_o} \frac{1}{1 + R_q C_L s}$$

A Bode plot of this transfer function, for $s = j\omega$, is shown in Fig. 6.20, along with a plot of the unloaded open-loop gain. Note that the effect of the resistive loading is to reduce the open-loop gain, lowering the entire curve. Thus resistive loading alone reduces the unity-gain bandwidth and will consequently reduce closed-loop bandwidth by the same factor. This bandwidth reduction factor is extremely important for fast line drivers since the very low impedance of the line can severely degrade the bandwidth unless the operational amplifier has very low output impedance. The capacitive component of load impedance introduces another pole in the gain function at $s = -1/R_q C_L$. This causes an additional "break" in the frequency response and a rolloff of −12 dB per octave above the frequency $\omega = 1/R_q C_L$. If the closed-loop gain curve intersects this section of the effective open-loop gain curve, the amplifier will be marginally stable with unacceptable transient response.

There are a number of techniques for dealing with the problems of loading. The most satisfactory of these is to choose an amplifier with very low open-loop output impedance or to create one by adding a power booster stage to an available operational amplifier. This will reduce the

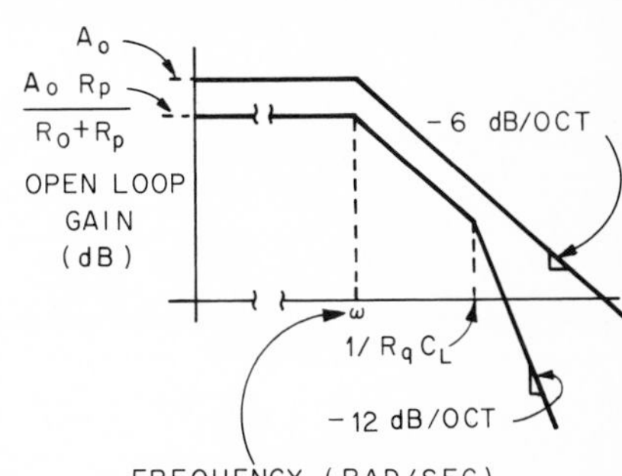

Fig. 6.20 Effect of loading on open-loop gain.

gain and bandwidth loading factors caused by the load resistance and will increase the frequency at which the additional pole occurs. The higher in frequency this pole occurs, the more stable the closed-loop response will be. The power output stage also supplies the current necessary to meet the condition

$$i_{L\,max} = C_L \left(\frac{de_o}{dt}\right)_{max}$$

As an example, the amplifier must be capable of supplying 63 mA to the capacitive load if $C_L = 10{,}000$ pF and the output voltage is a 10-V sine wave at 100 kHz.

6.6 AC-coupled Feedback Amplifiers[2,6]

Although the operational amplifier is designed to amplify dc signals, it has a rather broad frequency response and is consequently quite useful for strictly ac signals. The feedback network can be tailored for exactly the desired passband. One of the simplest ac amplifiers is that shown in Fig. 6.21a, where the closed-loop gain is given by

$$\frac{E_o}{E_1}(s) = -\frac{R_F}{R_1}\frac{s}{s + 1/R_1C_1}$$

The dc gain is zero, and the high-frequency gain approaches $-R_F/R_1$. The lower cutoff frequency is

$$f_c = \frac{1}{2\pi R_1C_1}$$

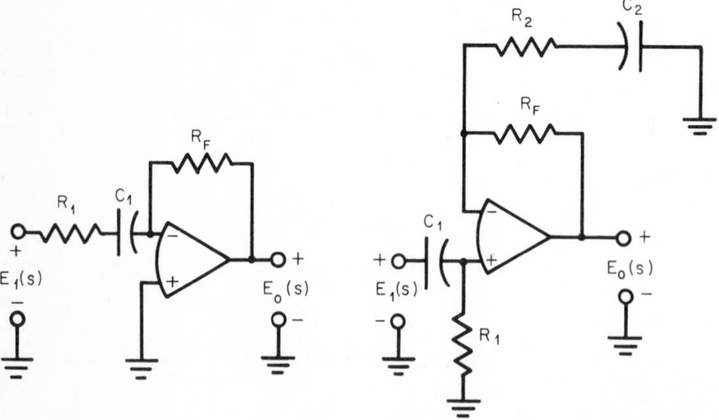

Fig. 6.21 Ac-coupled feedback amplifiers.

The dc output offset voltage E_{OS} is equal to the dc input offset voltage, plus the dc offset voltage generated by the input bias current flowing through R_F.

$$E_{OS} = V_{OS} \times 1.0 + I_{B1}R_F$$

A noninverting ac amplifier is shown in Fig. 6.21b. The response is given by

$$\frac{E_o}{E_1}(s) = \frac{s}{s + 1/R_1C_1} \frac{(R_2 + R_F)C_2s + 1}{R_2C_2s + 1}$$

Both of the circuits of Fig. 6.21 have relatively low input impedance above the cutoff frequency, determined by the resistors denoted R_1 in both cases.

The circuit of Fig. 6.22 is an ac amplifier whose input impedance is "bootstrapped" to a high value. Resistor R_2 provides a decoupling for dc input signals. However, for high-frequency signals the voltage across R_2 becomes very small. Consequently, very little current flows through R_2, and the effective input impedance is very high.

The analysis of the circuit is greatly simplified if it is assumed that $e_2 = e_4$ $(A \to \infty)$. Then we may write the equations

$$\frac{e_1 - e_2}{X_2} = \frac{e_2 - e_3}{R_2}$$

$$\frac{e_o - e_2}{R_F} = \frac{e_2 - e_3}{X_1}$$

$$\frac{e_2 - e_3}{R_2} + \frac{e_2 - e_3}{X_1} = \frac{e_3}{R_1}$$

where

$$X_1 = \frac{1}{j\omega C_1} \qquad \text{and} \qquad X_2 = \frac{1}{j\omega C_2}$$

If these equations are solved for e_2, the input impedance may be calculated from

$$Z_{in} = \frac{e_1}{i_{in}} = \frac{e_1X_2}{e_1 - e_2}$$

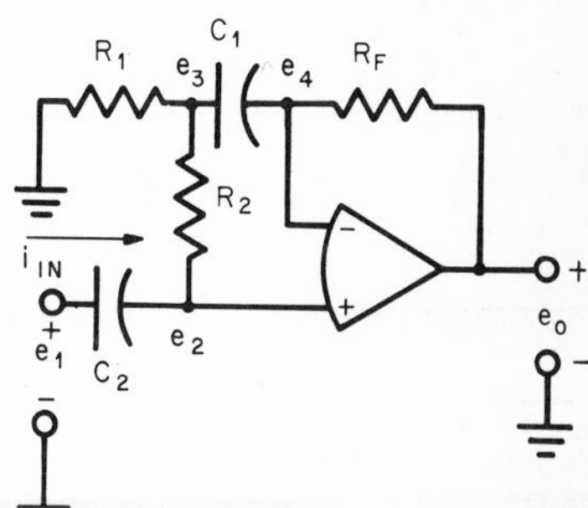

Fig. 6.22 Bootstrapped ac amplifier.

which yields

$$Z_{in} = X_2 + R_2 + R_1 + \frac{R_1R_2}{X_1}$$

As the frequency increases, X_1 and X_2 approach zero and the input

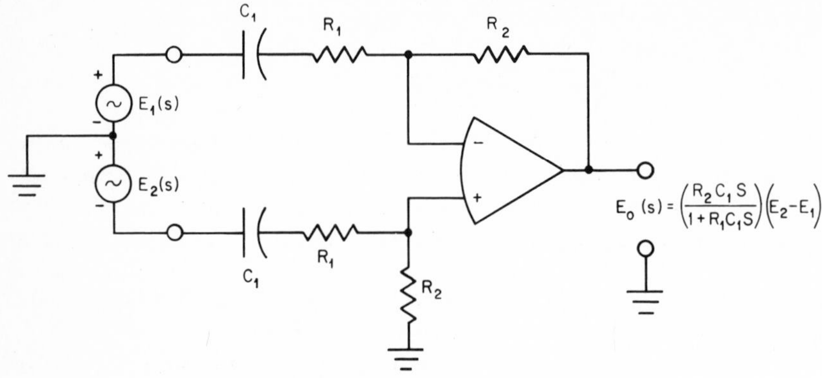

(a) SIMPLE ONE-AMPLIFIER CIRCUIT

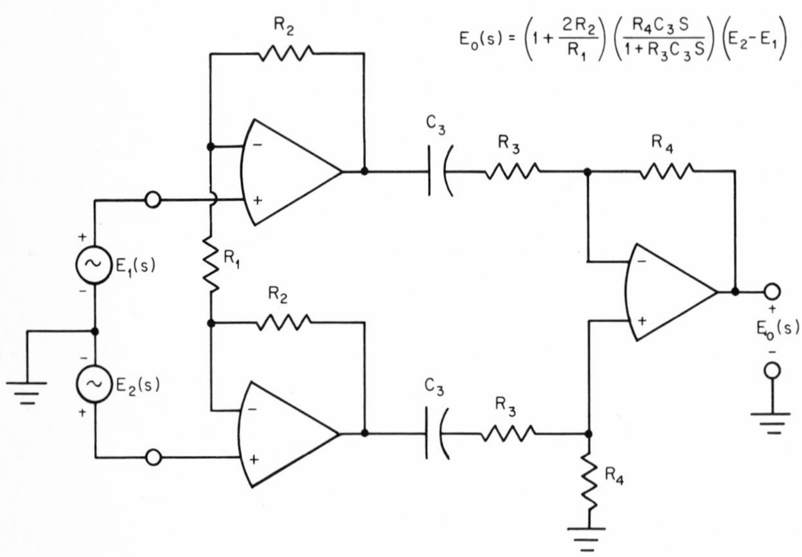

(b) HIGH INPUT IMPEDANCE CIRCUIT

Fig. 6.23 Differential ac amplifiers. (a) Simple one-amplifier circuit; (b) high input impedance circuit.

impedance becomes very large. As frequency increases still further, the open-loop gain decreases and the condition $e_2 = e_4$ is no longer enforced. The input impedance then decreases.

Differential ac amplifiers are also easily realized through the use of operational amplifiers. Two examples are shown in Fig. 6.23. That of Fig. 6.23a introduces simple dc decoupling into the familiar differential DC amplifier circuit. The circuit of Fig. 6.23b provides high input impedance while decoupling dc signals in the second stage. The dc offset voltages of the first-stage amplifiers are removed by the capacitive coupling. The dc offset voltage of the second-stage amplifier is multiplied by the dc gain, 1.0.

6.7 Voltage-to-Current Converters[2,7]

In applications such as coil driving and transmission of signals over long lines, it is sometimes desirable to convert a voltage to an output current. With operational amplifiers this is quite easily done. Several realizations of the voltage-to-current converter (VIC) will be examined in this section.

The simplest VICs are those for floating loads. The circuits of Fig. 6.24a and b are the prime examples of this type. The circuit of Fig. 6.24a is a simple inverting circuit. The input current is given by

$$i_1 = \frac{e_1}{R_1}$$

since R_1 is terminated at the virtual ground of the summing junction. This same current flows through the feedback load impedance Z_L in the feedback loop. The current i_1 is independent of the value of Z_L. Both the signal source and the operational amplifier must be capable of supplying the desired amount of load current. The circuit of Fig. 6.24b

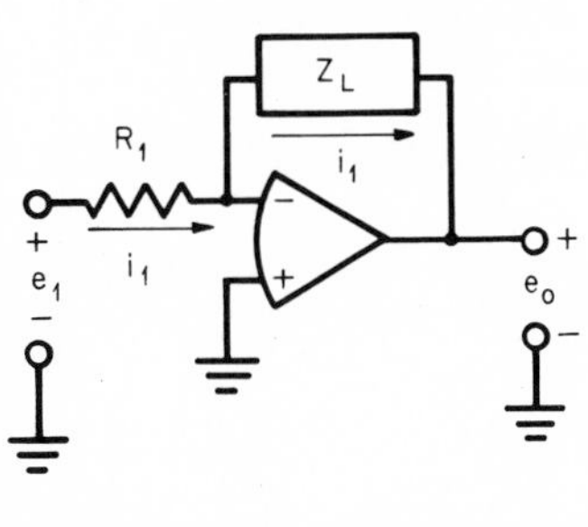

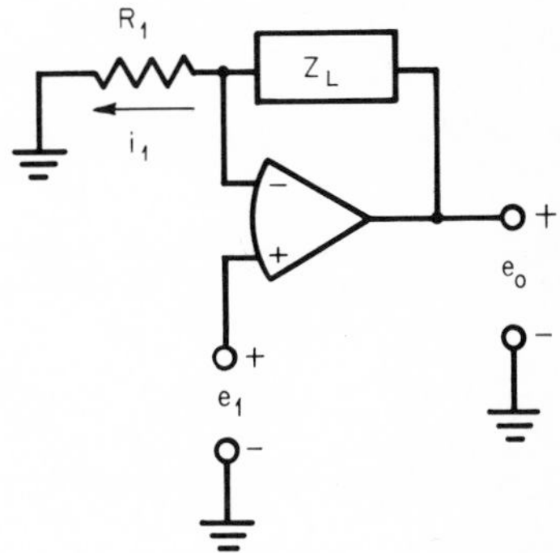

(a) INVERTING AMPLIFIER TYPE (b) NONINVERTING AMPLIFIER TYPE

Fig. 6.24 Voltage-to-current converters, floating load.

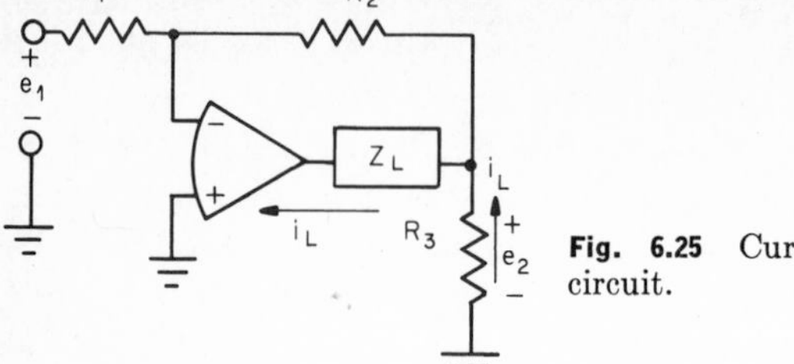

Fig. 6.25 Current amplifying circuit.

operates in the noninverting mode and, hence, presents a high impedance to the driving source. The current is again given by the equation

$$i_1 = \frac{e_1}{R_1}$$

and, again, i_1 is the load current. Very little current, however, is required from the signal source, because of the high input impedance of the noninverting amplifier.

Another VIC for a floating load is shown in Fig. 6.25. Here, most of the current is provided by the amplifier and only a small portion by the signal source. Analysis of the circuit yields the following equation for load current:

$$i_L = \frac{e_1}{R_1}\left(1 + \frac{R_2}{R_3}\right)$$

The resistor R_3 provides a convenient means for scaling the current. The resistor R_1 can be made relatively large to minimize the loading of the signal source. The amplifier must be capable of providing all the current to the load and must also be capable of output voltage equal to

$$e_{o\,max} \approx i_{L\,max}(Z_L + R_3)$$

For loads which are grounded on one side, there are also circuits which give voltage-to-current conversion. The single amplifier circuit of Fig. 6.26 acts as a current source controlled by e_1,

$$i_L = -\frac{e_1}{R_2}$$

if

$$\frac{R_3}{R_2} = \frac{R_F}{R_1}$$

If these ratios of resistances are matched, the circuit will function as a true source of current with very high internal impedance. A mismatch of the ratios will be seen as a decreased internal impedance of the current source. Fluctuations in effective load impedance will then cause fluctua-

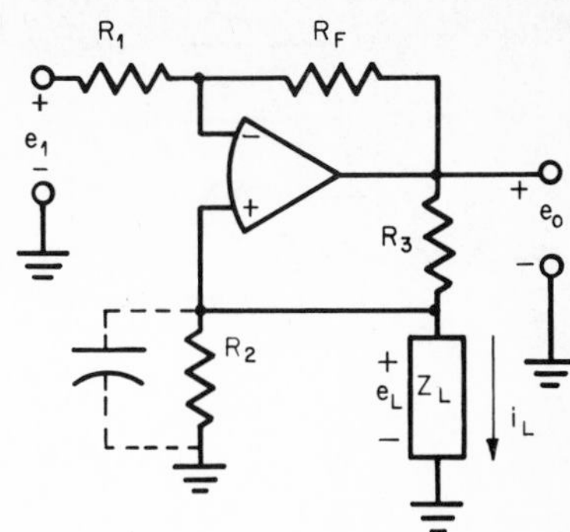

Fig. 6.26 VIC, grounded load.

tions of the output current. The operational amplifier for the circuit of Fig. 6.26 must have an output voltage range sufficient to provide the maximum load voltage plus the voltage drop across R_3. Normally, R_1 and R_2 will be chosen to draw small currents, and R_F and R_3 will be made small to minimize voltage drops.

The circuit of Fig. 6.27 utilizes two inverting amplifiers to drive a current into a grounded load. This current is given by the expression

$$I_L = e_1 \frac{R_5R_F/R_4R_1}{R_3 + Z_L[1 + R_3/R_2 - (R_5/R_4)(R_F/R_2)]}$$

If resistors are selected so that

$$1 + \frac{R_3}{R_2} = \frac{R_5R_F}{R_4R_2}$$

then

$$i_L = \frac{e_1}{R_3}\frac{R_5R_F}{R_4R_1}$$

In particular, if

$$R_1 = R_F = R_4 = R_5$$

then

$$i_L = \frac{e_1}{R_3}$$

and

$$R_2 = R_F - R_3$$

If R_1 is large, very little current is drawn from the signal source and very little flows through the feedback elements. Then the output voltage is given by

$$e_{o\,max} \approx I_{L\,max}(Z_L + R_3)$$

Note that, in the circuits of Figs. 6.26 and 6.27, when the load is open-circuited the positive feedback is equal to the negative feedback. This is

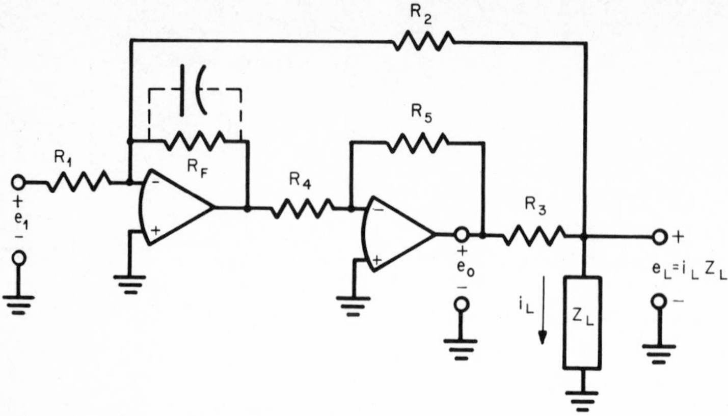

Fig. 6.27 Two-amplifier VIC, grounded load.

equivalent to an open-loop condition. The stabilizing capacitors shown by dotted lines are therefore desirable to prevent excessive noise and possible oscillations. Figure 6.28 illustrates a modified form of the two-amplifier VIC which provides the additional feature of very high input impedance. The expression for output current as a function of input voltage is

$$i_L = \frac{e_1(R_5/R_4)(1 + R_F/R_2 + R_3/R_2)}{R_3 + Z_L(1 + R_3/R_2 - R_5R_F/R_2R_4)}$$

If we again select resistors such that

$$1 + \frac{R_3}{R_2} = \frac{R_5R_F}{R_2R_4}$$

and

$$R_F = R_4 = R_5$$

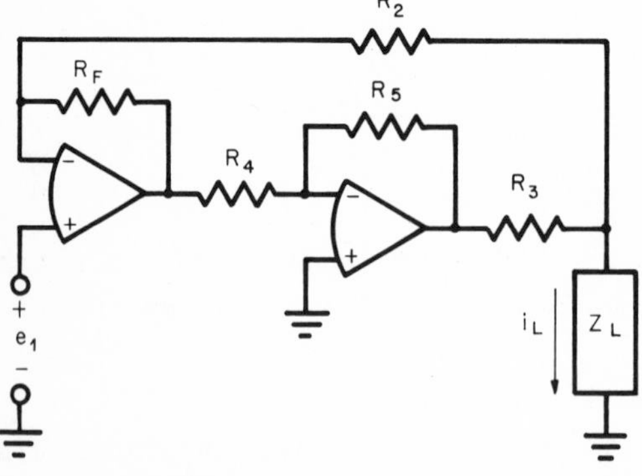

Fig. 6.28 Buffered VIC, grounded load.

then

$$i_L = \frac{2e_1R_F}{R_2R_3}$$

and

$$R_2 = R_F - R_3$$

6.8 Reference Voltage Sources and Regulators[2,6,7]

Because of its high input impedance and easily adjustable gain, the operational amplifier may be used as a reference voltage source with very low output impedance and substantial output current capability. Two circuits for use with standard cells are shown in Fig. 6.29. In both instances the output voltage is given by

$$E_o = V_{REF}\left(1 + \frac{R_F}{R_1}\right)$$

The circuit of Fig. 6.29a can be used with single-ended amplifiers (such as chopper-stabilized types), as well as those with differential input. The circuit of Fig. 6.29b is used if the reference source or cell must be grounded on one side. The only current drawn from the cell is the input bias current of the amplifier plus a term given by

$$I_i = \frac{E_o}{AR_i} = \frac{V_{REF}(1 + R_F/R_1)}{AR_i}$$

where R_i is the differential input impedance of the operational amplifier. This component of current is negligible in comparison with bias current for most amplifiers. The reference voltage cell is, for all practical purposes, isolated from any load being driven. The effective output

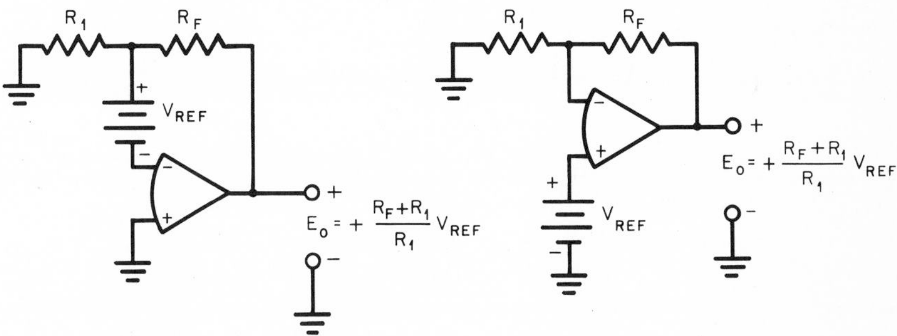

Fig. 6.29 Reference voltage sources.

impedance R_{OUT} is given by

$$R_{OUT} = \frac{R_o}{A\beta}$$

where

$$\beta = \frac{R_1}{R_1 + R_F}$$

and

$$R_o = \text{open-loop output impedance}$$

The load regulation is therefore given by

$$\text{Regulation } \% = \frac{R_o}{A\beta R_L} \times 100$$

where R_L is the minimum load impedance.

Similar circuits for use with zener diodes are shown in Fig. 6.30a and b. The loading conditions on the zener diodes are constant and the load regulation is the same as derived for the circuits of Fig. 6.29. Regulation with respect to the input voltage V_s depends upon the dynamic resistance of the reference zener diode Z_1. The circuit of Fig. 6.29c further reduces this regulation due to input voltage by providing the output reference voltage as the source for the zener diode current. The dc voltage V_s now functions only as a "startup" voltage through the network of R_2, R_3, and D_1.

6.9 Voltage Regulators

Any one of the voltage references described in the preceding section may be considered a voltage regulator, with extremely tight regulation characteristics. Where higher output currents are required, a power booster can be added, inside the feedback loop. However, in speaking of voltage regulators, it is more usual to consider operation from a single source of unregulated dc voltage, rather than the dual supplies tacitly assumed in the reference voltage circuits. Figure 6.31 shows such a regulator. The amplifier, which normally operates on dual power supplies of opposite polarity, is biased for operation on a single unregulated power supply. The negative supply terminal is grounded and the noninverting input is biased at the zener voltage. The zener diode Z_1 operates at constant load current, since the output current is provided by the transistor Q_1. If the amplifier has a minimum (balanced) supply rating of $\pm V_m$, then V_s must be larger than $2V_m$. Similarly, if $\pm V_M$ is the maximum (balanced) supply rating, V_s must not exceed $2V_M$. The amplifier will saturate as the output voltage approaches either supply voltage. This determines

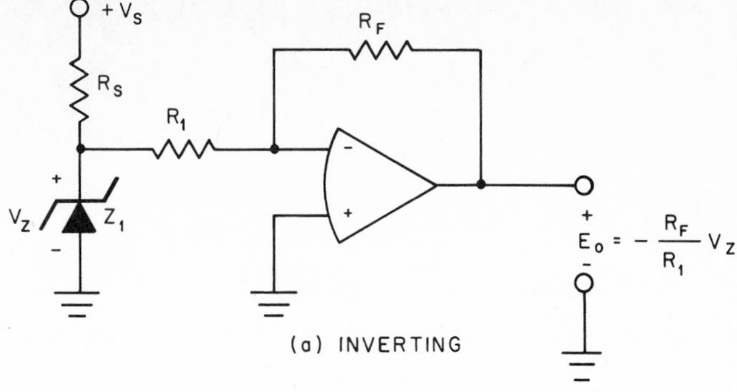

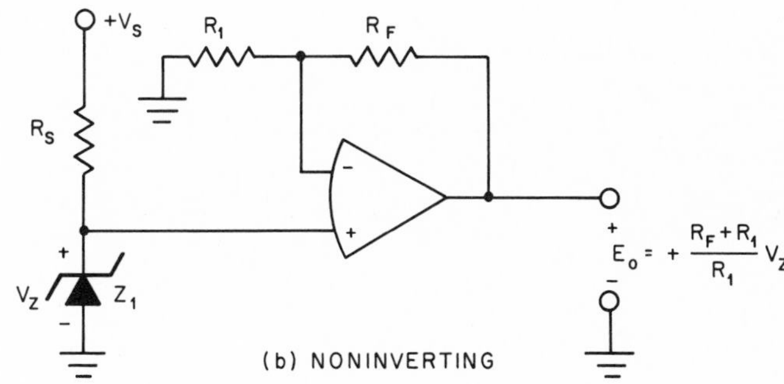

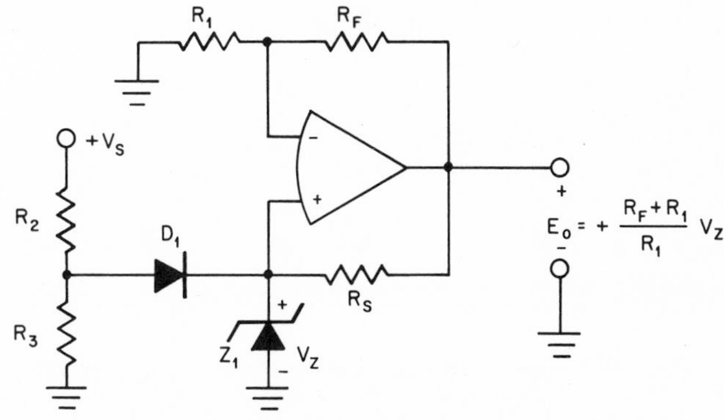

Fig. 6.30 Zener reference sources.

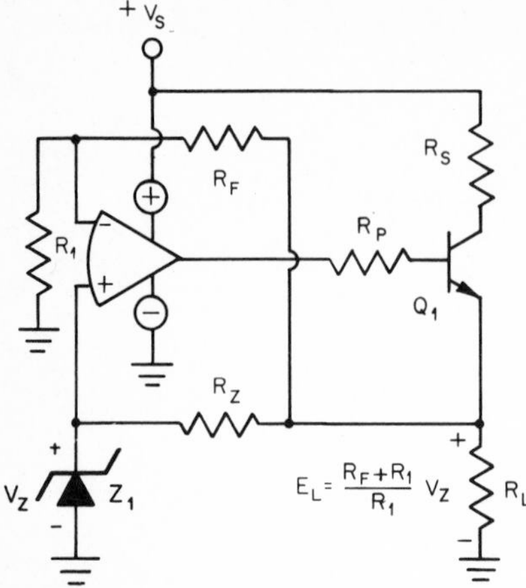

Fig. 6.31 Voltage regulator.

the limit on output; the common-mode voltage range sets the lower limit on zener voltage.

Although the amplifier may have an internal current limit, the resistor R_p is required to protect against short circuit in this type of regulator. This is because a short circuit to ground is equivalent to a short circuit to negative supply. This causes a power dissipation equal to twice that of a short circuit to ground when operating on balanced dual supplies. Thus the internal protection may not be sufficient. The value of R_p should be chosen to limit the amplifier short-circuit current to approximately one-half the internal current limit value when the output is at positive saturation voltage. The resistor R_s provides current limiting to protect Q_1.

The load regulation of this type of regulator can exceed 0.01 percent, since the effective output impedance is very low. The line regulation is increased beyond that of the zener by using the output voltage as excitation for the zener.

6.10 Current Amplifiers

Current amplifiers, or current-to-voltage converters, are realized very simply by using operational amplifiers. An ideal current source has infinite output impedance and output current which is independent of load. Photocells and photomultiplier tubes are basically current sources with output impedance which is finite but very large. For small load impedances, the output impedance may be considered infinite.

The current-to-voltage converter of Fig. 6.32 presents almost zero load impedance to ground because the inverting input appears as a virtual ground. The input current, however, flows through the feedback resistor,

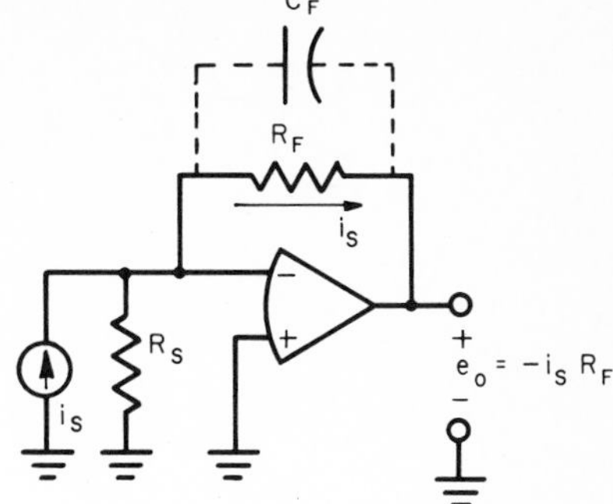

Fig. 6.32 Current amplifier (current-to-voltage converter).

generating an output voltage

$$e_o = -i_s R_F$$

The actual input impedance of the current-to-voltage converter, Z_{in}, taking into account the finite gain A and differential (open-loop) input impedance Z_{id}, is

$$Z_{in} = \frac{Z_{id}}{1 + (Z_{id}/R_F)(1 + A)} \approx \frac{R_F}{1 + A}$$

The lower limit on measurement of current input is determined by the bias current of the inverting input. For greatest resolution, FET or varactor bridge amplifiers are usually used.

The gain of the amplifier for dc offset voltage and noise voltage is given by

$$\frac{R_F + R_s}{R_s} \approx 1.0 \qquad \text{since } R_s \gg R_F$$

Thus errors due to these parameters are very small. However, current noise can be a factor because of the very large impedances. Since most such measuring circuits are used for very low-frequency signals, it is usual to parallel R_F with a capacitor C_F to reduce the high-frequency current noise. Output impedance of the current-to-voltage converter is very low because of the nearly 100 percent feedback.

6.11 Charge Amplifiers

Some transducers, such as capacitance microphones and some types of accelerometers, operate on the principle of conversion of the measurement variable into an equivalent charge. The equivalent circuit of such a transducer may be represented by a battery and capacitor in series, as shown in Fig. 6.33a. As the capacitance varies, the charge also changes

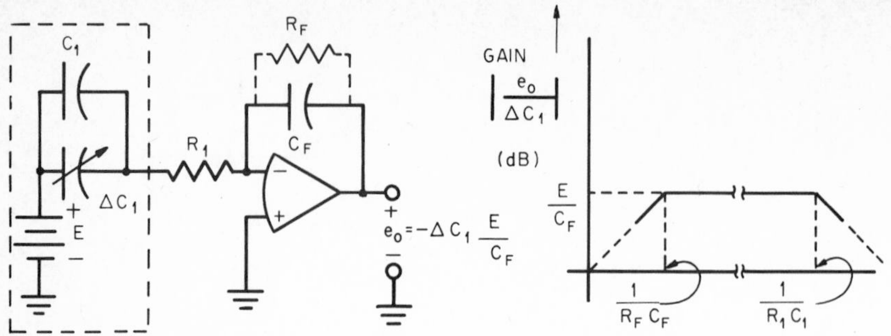

(a) CHARGE AMPLIFIER CIRCUIT (b) FREQUENCY RESPONSE

Fig. 6.33 Charge amplifier operation.

according to the equation

$$\Delta q = \Delta C_1 E$$

When the transducer is connected to the inverting input of an operational amplifier as in Fig. 6.33a, this charge flows into the feedback capacitor C_F. The resultant change in charge on C_F generates an output voltage,

$$e_o = -\Delta C_1 \frac{E}{C_F}$$

Since the operational amplifier requires a dc path from each input to common (for bias current flow) it is necessary to insert the resistor R_F. In the absence of this resistor, the capacitors will build up a dc charge until the output voltage reaches saturation. This resistor limits the lower cutoff frequency of the charge amplifier. For stabilization purposes, and sometimes for protection of the amplifier input stage, it is also desirable to insert the series resistor R_1. This resistor limits the upper response frequency as shown in Fig. 6.33b.

The gain, or sensitivity of the charge amplifier, in its passband is given by

$$\frac{e_o}{\Delta C_1} = -\frac{E}{C_F}$$

and can be varied only by changes in C_F. It is usually desirable to use a small value of C_F consistent with the desired frequency response and a reasonable value of R_F. FET amplifiers are usually the first choice for charge amplification, because of their high input impedance, low bias current, and wide bandwidth.

Another common form of charge amplifier is shown in Fig. 6.34. Here the amplifier operates as a noninverting buffer with gain. Charge

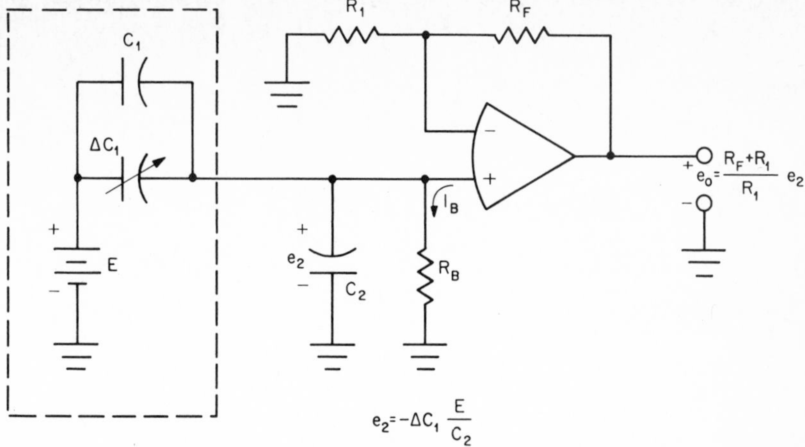

Fig. 6.34 Alternative charge amplifier circuit.

flows into, and out of, the capacitor C_2 as the capacitance of the transducer varies. Once again these capacitance variations are converted into voltage variations at the amplifier output. An amplifier with FET input stage is usually also required in this circuit to minimize the bias and noise currents. The resistor R_B provides the dc path for this bias current and limits the low-frequency response of the circuit.

REFERENCES

1. G. A. Korn, and T. M. Korn, *Electronic Analog and Hybrid Computers,* McGraw-Hill Book Company, New York, 1964.
2. *Applications Manual for Operational Amplifiers,* Philbrick/Nexus Research, Dedham, Mass., 1965.
3. N. D. Diamantides, Improved Electronic Differentiator, *Electronics,* July 27, 1962.
4. G. A. Korn, Exact Design Equations for Operational Amplifiers with Four-terminal Computing Networks, *IRE Trans. Electron. Computers,* February, 1962.
5. T. Miura, et al., On Computing Errors of an Integrator, *Proc. 2nd AICA Conf.,* Strasbourg, France, 1958, Presses Académiques Européennes, Brussels.
6. *Handbook of Operational Amplifier Applications,* Burr-Brown Research Corporation, Tucson, Ariz., 1961. (Out of print.)
7. *Handbook and Catalog of Operational Amplifiers,* Burr-Brown Research Corporation, Tucson, Ariz., 1969. (Out of print.)
8. G. Tobey, Analog Integration, *Instrum. Control Syst.,* January, 1969.

7

OPERATIONAL AMPLIFIERS IN NONLINEAR CIRCUITS

Some of the more interesting applications of operational amplifiers require the use of nonlinear feedback networks. By the use of such networks the amplifier with feedback can be made to approximate transfer curves, linearize transducers, limit the amplitude of signals, perform mathematical operations, and do a variety of other tasks. Basic to most of these nonlinear feedback networks is the use of the voltage-to-current characteristics of semiconductor junctions: diodes, zener diodes, and transistors. In some applications, the large-signal switching properties of such elements are used, whereas in others the nonlinearity of the junction itself is utilized. In this chapter we present a discussion of such circuits and their applications. Since the operation of diode limiter networks is basic to a great many of the circuits considered in this chapter, the first section is devoted to a brief discussion of the operation of these simple circuits. The remainder of the chapter treats feedback limiters, diode function generators, logarithmic amplifiers, and analog multipliers. Each of the sections concludes with a brief discussion of the primary areas of application for each functional circuit.

7.1 Diode Limiter Networks

In this section we present a discussion of idealized series and shunt limiter networks. The operation of these may be best understood by first considering some basic models for limiters.

7.1.1 Basic limiter models[1,4,6] The idealized model for the limiting element consists of an ideal diode in series with a floating bias source, as shown in Fig. 7.1a. When the signal voltage equals the bias voltage, the ideal diode conducts, with the amount of current dependent upon the resistance of the circuit containing the limiter. A series limiter is shown in Fig. 7.1b along with its transfer curve. For input voltage less than V_B the diode D is nonconducting and output voltage is zero. When e_i exceeds V_B, the output voltage follows the input. The shunt limiter of Fig. 7.1c provides an alternative means of obtaining an abrupt transition in the slope of the transfer curve. For output voltage e_o less than V_B the diode is nonconducting, and the circuit acts as a simple resistive divider. As input voltage is increased, however, the output eventually reaches the value V_B, and the diode begins to conduct, thus preventing further increases in e_o.

Both the series limiter and the shunt limiter find useful application as a part of the feedback network of an operational amplifier. Figure 7.2 illustrates a simple inverting amplifier circuit in which a diode and series bias source are used to provide a limit on the output voltage of an operational amplifier. For output voltage less than V_B, the output is a simple linear function of the input voltage with gain equal to the ratio $-R_F/R_1$. When the output reaches V_B the diode conducts, preventing further increase in e_o. If the input voltage increases still further, the additional input current passes through the limiting elements, generating no additional voltage at the output. The summing point remains at a virtual ground. Actually, of course, all practical limiting circuits will have some internal impedance, usually nonlinear, which modifies the ideal behavior described here. Also, floating bias sources are rather impractical in most cases and must be approximated by other means. These more practical limiters are discussed in the following sections.

7.1.2 Series limiters A practical and very close approximation to the ideal series limiter discussed above is achieved through the use of a silicon diode and a zener diode as shown in Fig. 7.3a. When the input voltage exceeds the sum of the zener voltage and the forward voltage of the silicon diode, the combination conducts. The output voltage then approximately follows the input. Because the two diodes have finite

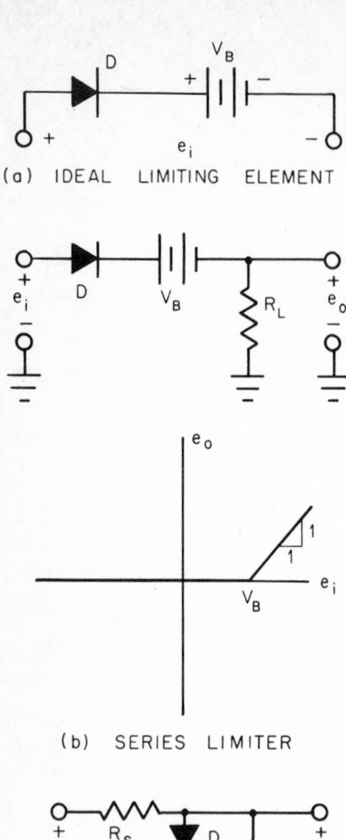

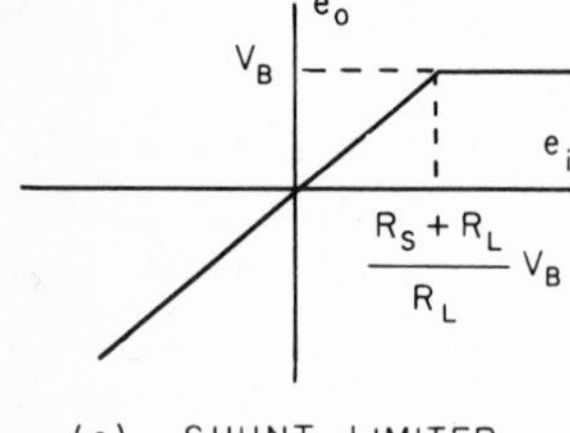

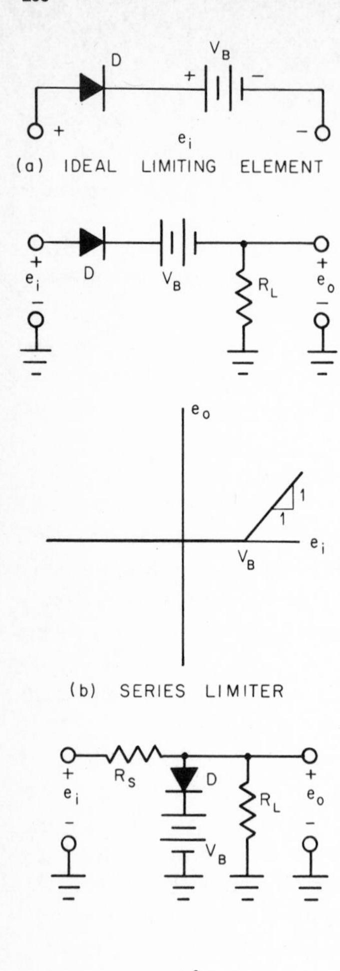

Fig. 7.1 Limiter concepts: (a) ideal limiting element; (b) series limiter; (c) shunt limiter.

ON resistances, the output does not follow the input exactly but is attenuated slightly by this series resistance. A double series limiter can be formed as shown in Fig. 7.3b by paralleling two such combinations of diodes in opposite polarities.

An even simpler method of obtaining a double series limiter is to use

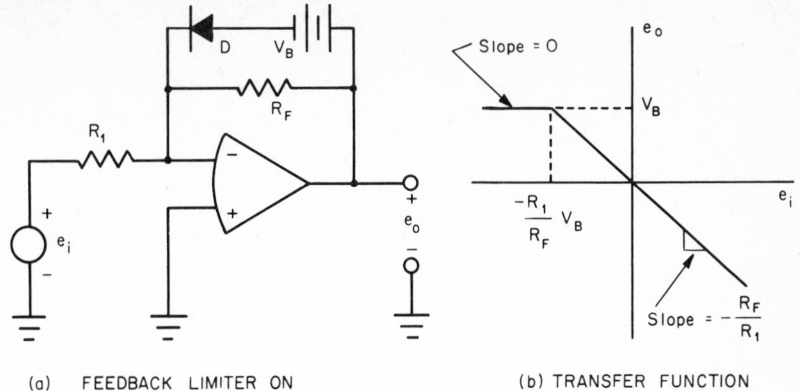

Fig. 7.2 Operation of a feedback limiter. (a) Feedback limiter on an inverting amplifier; (b) transfer function.

r_f = FORWARD RESISTANCE OF D_1
r_z = ZENER RESISTANCE OF Z_1 .

SLOPE = $\frac{R_L}{r_f + r_z + R_L}$

D_1 Z_1 $+V_f-$ $+V_z-$ R_L e_i e_o (V_z+V_f)

(a) SERIES ZENER LIMITER

D_1 Z_1 D_2 Z_2 R_L e_i e_o $-(V_{z_2}+V_{f_2})$ $(V_{z_1}+V_{f_1})$

(b) DOUBLE ENDED SERIES ZENER LIMITER

Z_1 Z_2 R_L e_i e_o $-(V_{z_1}+V_{f_2})$ $(V_{z_2}+V_{f_1})$

(c) BACK-TO-BACK ZENER LIMITER

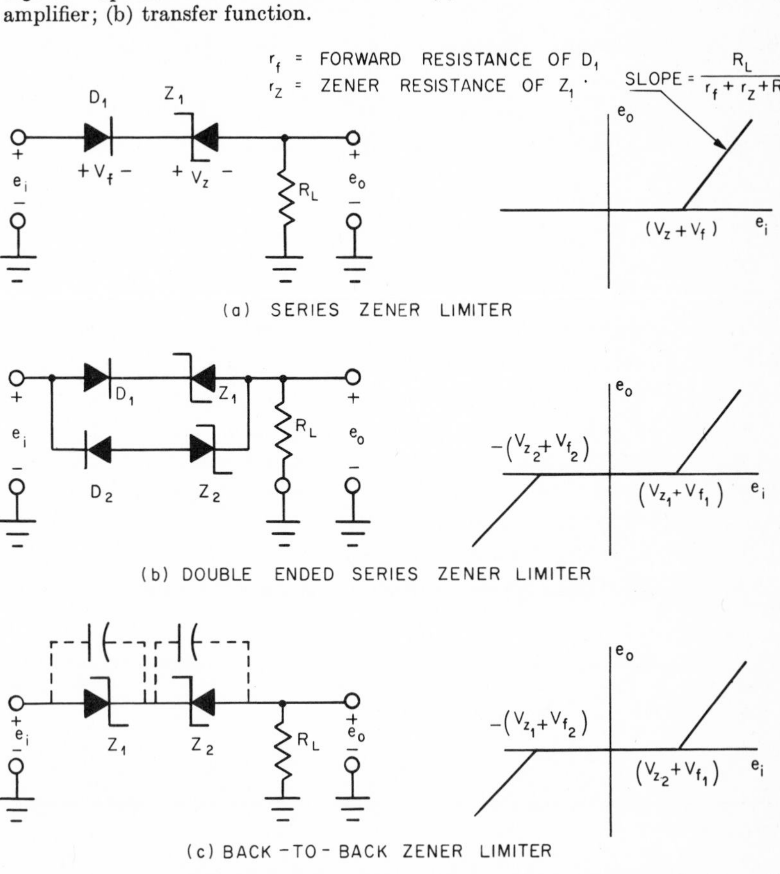

Fig. 7.3 Series limiters using zener diodes: (a) series zener limiter; (b) double-ended series zener limiter; (c) back-to-back zener limiter.

back-to-back zeners, as shown in Fig. 7.3c. Here the function of the silicon diodes is served by the zener diodes in their forward-conducting region. This circuit suffers from the high junction capacitance of the zeners and may present feedthrough problems at high frequency. The method of Fig. 7.3b is somewhat superior in this respect, especially if low-capacitance silicon diodes are used for D_1 and D_2. The transition between the ON and OFF states will actually not be a sharp one, as shown in the figure, but will have a degree of rounding determined by the diode and zener characteristics and by the value of the load resistor R_L. This resistor could equally well be the summing resistor of an operational amplifier network. In this case, it would be terminated in a virtual ground instead of true ground.

Another type of series limiter is shown in Fig. 7.4. Here the biasing is accomplished through the use of an external reference voltage and a shunt resistor. The diode begins conducting when the junction voltage e_j exceeds the forward voltage drop of the diode. The breakpoint voltage V_B is given by the expression $V_B = V_f(1 + R_1/R_2) + V_R(R_1/R_2)$ and can easily be varied by adjustment of R_2. Such limiters are useful in the piecewise approximation of functions, a topic to be discussed later in the chapter.

7.1.3 Shunt limiters A simple means of realizing a shunt limiter is shown in Fig. 7.5a where, again, the combined silicon diode and zener diode are used as the actual limiting elements. The circuits of Fig. 7.5b and c are actually the duals of the double series limiting elements of Fig. 7.3. The circuit of Fig. 7.5b achieves lower shunt capacitance than that of 7.5c and is therefore preferable for high-frequency applications.

As another approach to shunt limiting, the resistive divider shunt limiter of Fig. 7.6a is quite useful where it is necessary to accurately adjust the breakpoint voltage V_B. When the output voltage e_o equals the reference voltage V_R, plus the diode forward voltage, the diode conducts and prevents further increase in e_o. The actual value of input voltage at which the breakpoint occurs is determined by the ratio of R_1

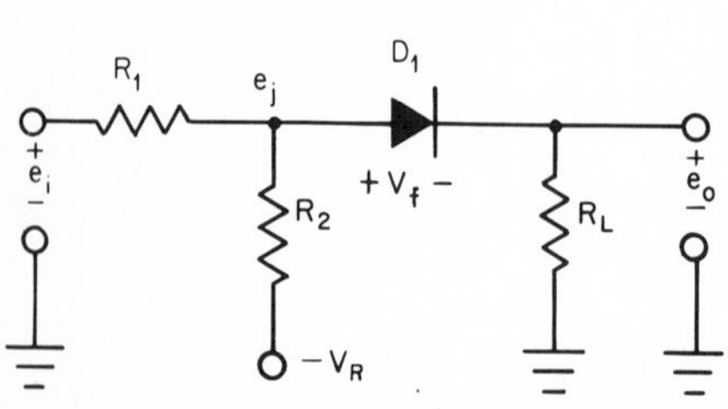

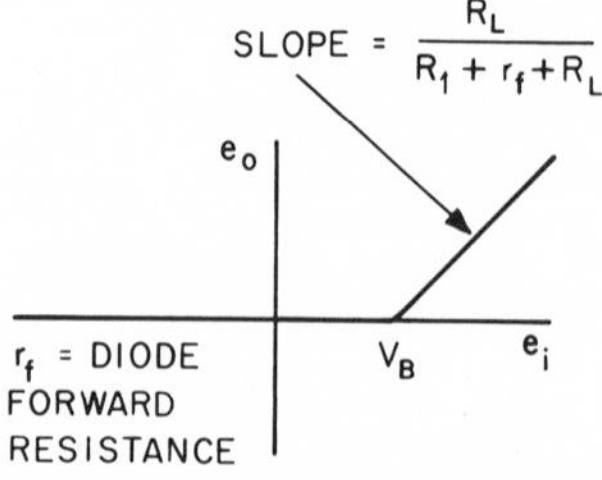

Fig. 7.4 Externally biased series limiter.

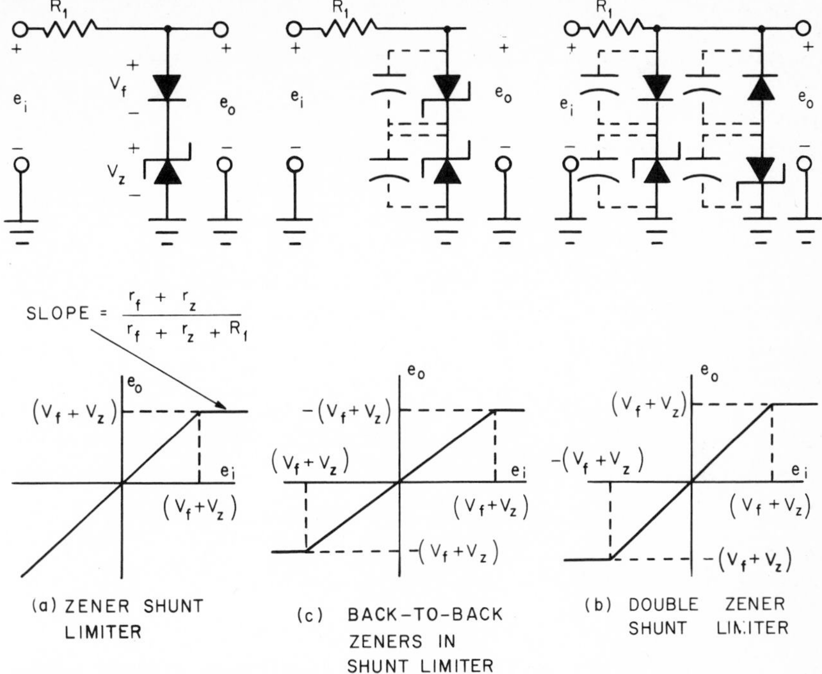

Fig. 7.5 Practical shunt limiters: (a) zener shunt limiter; (b) double zener shunt limiter; (c) back-to-back zeners in a shunt limiter.

and R_L and by the value of V_R.

$$V_B = V_R \frac{R_L + R_1}{R_L}$$

An additional, negative breakpoint can easily be achieved, as in Fig. 7.6b, by adding another diode and reference source.

Another practical limiter circuit is the bridge limiter circuit of Fig. 7.7. This is actually a form of shunt limiter which provides a double limit and partial compensation of the temperature-sensitive characteristics of the diodes. The breakpoints, of course, still exhibit rounding because of the gradual turnoff of the diodes. The breakpoint voltages, or limits, are easily varied through the bias resistors R_1 and R_2, or by varying $+V_c$ and $-V_c$.

7.2 Feedback Limiters[1,2,4,6]

In the preceding section we discussed the operation of several series and shunt limiting networks. In this section we will illustrate some circuits which use networks of this type to obtain feedback limiting. Three dif-

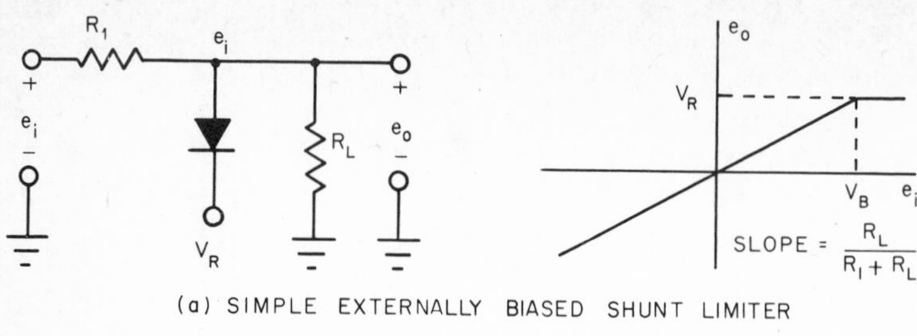

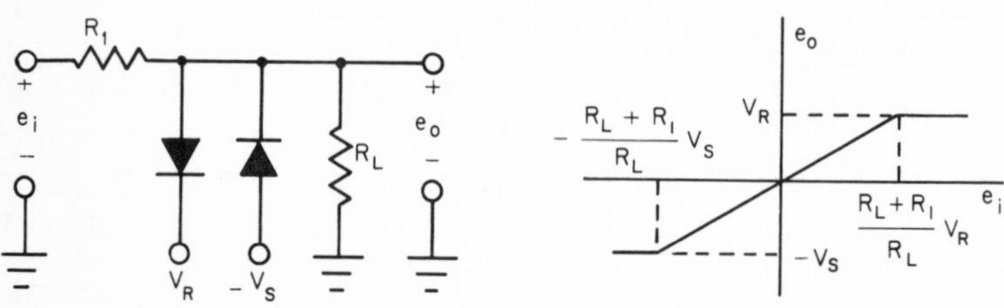

Fig. 7.6 Externally biased shunt limiters: (a) simple externally biased shunt limiter; (b) double shunt limiter with external biasing.

ferent approaches will be considered. They are resistive ratio methods, zener diode feedback limiters, and precision limiters.

7.2.1 Resistive ratio methods In the feedback limiter, series or shunt limiting networks provide an abrupt change in the feedback ratio, and hence the closed-loop gain, of the operational amplifier. The resistive divider feedback circuit of Fig. 7.8a makes use of a simple series limiting

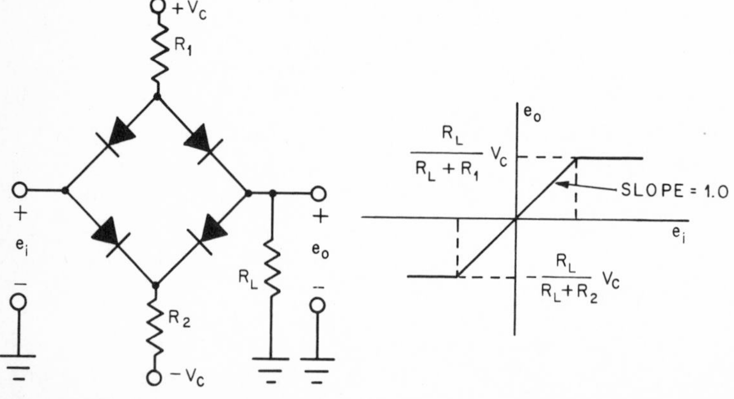

Fig. 7.7 Diode bridge limiter.

circuit. The diode begins conduction when the voltage e_j exceeds the forward voltage V_f. The output voltage is then limited at the value

$$V_L = \frac{R_3}{R_2} V_R + \left(1 + \frac{R_3}{R_2}\right) V_f$$

The gain before limiting is $-R_F/R_1$ and, after the limit occurs, is $-R_F R_3/(R_F + R_3)R_1$. The slope, or gain, in the limit region can be made to approach zero if $R_3 \ll R_1$. Since such small values of R_3 may be impractical, the circuit of Fig. 7.8b may be used to obtain slope ≈ 0. This is done by adding a transistor to the circuit. This transistor then provides the necessary current to the summing junction while drawing only relatively small base current through R_3. Total current into the summing point remains zero both before and after limiting occurs. The diode to ground protects the transistor from reverse breakdown of the base emitter junction. The limit voltage in this case is given by the

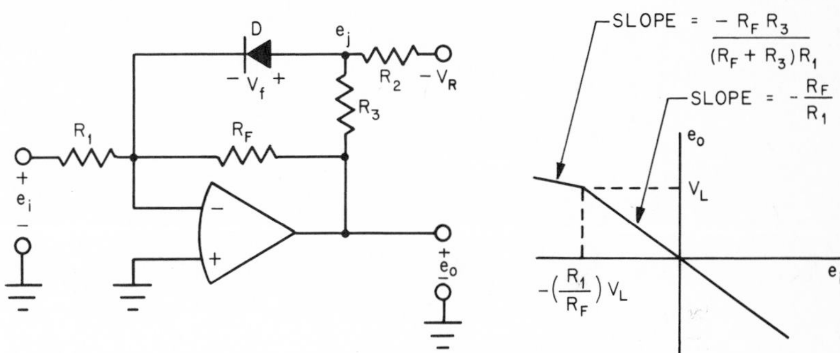

(a) SIMPLE FEEDBACK LIMITER

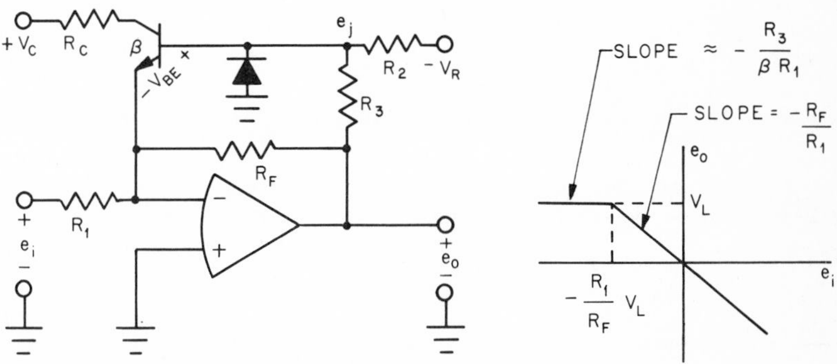

(b) FEEDBACK LIMITER WITH TRANSISTOR

Fig. 7.8 Resistive divider feedback limiters: (a) simple feedback limiter; (b) feedback limiter with a transistor.

expression

$$V_L = \frac{R_3}{R_2} V_R + \left(1 + \frac{R_3}{R_2}\right) V_{BE}$$

The circuits of Fig. 7.8 are quite useful because of the ease with which the limiting level can be varied. If a "soft" limit is sufficient, the circuit of Fig. 7.8a is used. If "hard" limiting is necessary, the transistor limiter of Fig. 7.8b is preferable. Both circuits suffer from the temperature sensitivity of the diode and transistor forward voltage drops, V_f and V_{BE}. Also, they cannot limit at voltages smaller than V_f or V_{BE}. The capacitance of these limiters is rather low; consequently they perform well in high-frequency applications.

7.2.2 Zener diode feedback limiters Two circuits which may be categorized as zener diode feedback limiters are shown in Fig. 7.9.

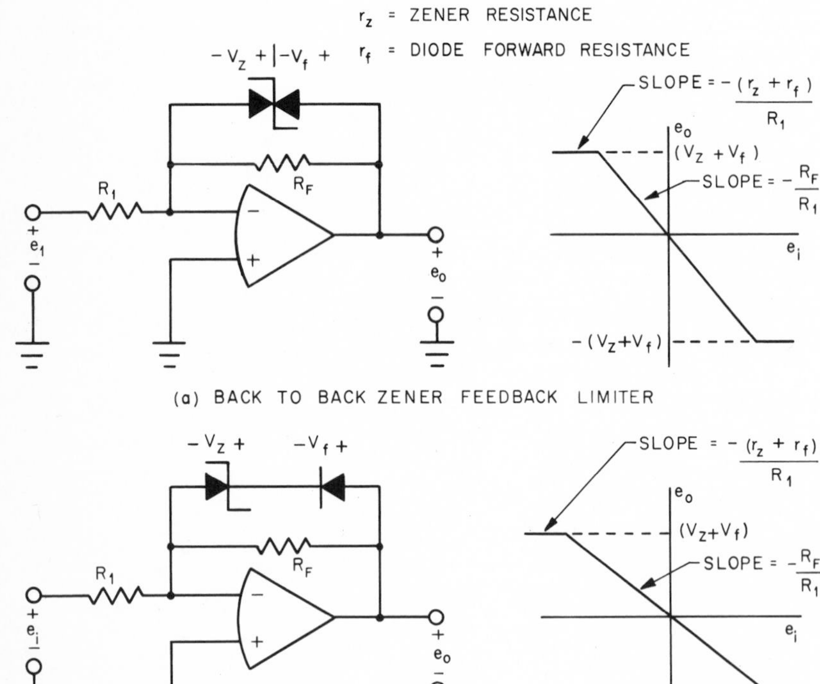

Fig. 7.9 Zener diode feedback limiters: (a) back-to-back zener feedback limiter; (b) low capacitance zener feedback limiter.

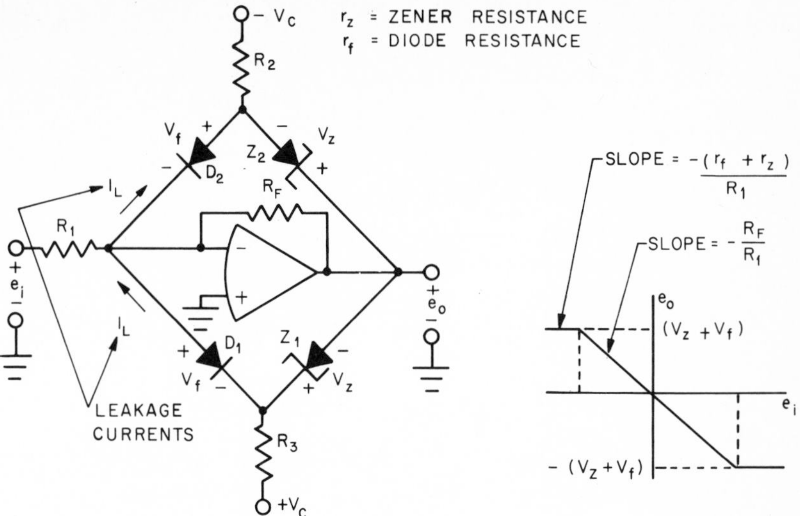

Fig. 7.10 Zener diode bridge feedback limiter.

Either of these limiters works satisfactorily at low frequency. The circuit of Fig. 7.9b provides lower capacitance than that of 7.9a and therefore is better at high frequency. The limits are set by the zener voltage and forward voltage drop of the diode. For good limiting action, the ON value of resistance of the diodes must be negligible in comparison with R_F and the OFF resistance must be much larger than R_F. Zener diodes selected for such limiters should have a sharp "knee" to avoid distortion of the transfer curve as the output voltage approaches its limit. Another zener diode feedback limiter is the bridge circuit shown in Fig. 7.10. This is a double-ended version of the zener limiter of Fig. 7.9b with the addition of external biasing to obtain sharper transition between the ON and OFF regions of the limiter. The small leakage currents through the silicon diodes in the OFF state will tend to cancel at the summing junction. In applying the circuits shown in Figs. 7.9 and 7.10, it should be noted that zener diode limiters are useful mainly for protection against overvoltage and not as a means of obtaining precisely known limits for signal processing or computation purposes.

7.2.3 Precision limiters As discussed in earlier sections, the actual limiting elements (diodes, zener diodes, and transistors) have finite resistance and nonlinear temperature-sensitive switching characteristics. These characteristics contribute to a "rounding" of the breakpoint area of the limiter characteristics as illustrated in Fig. 7.11. Ideally, the breakpoint would be sharp and well defined, plus being insensitive to

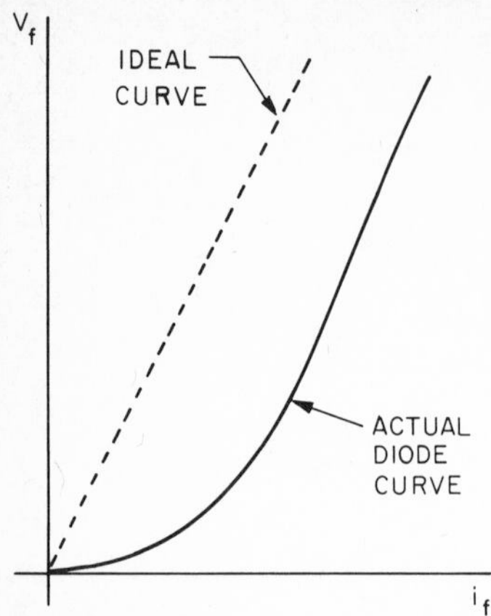

Fig. 7.11 Practical limiter characteristic near the breakpoint.

temperature. Circuits which achieve such characteristics are referred to as precision limiters. In the precision limiter of Fig. 7.12, the high open-loop gain of the operational amplifier is used to reduce the effect of the diode nonlinearity and temperature sensitivity. The operation of the circuit can be analyzed by the usual techniques except that the relationship of current and voltage in the diodes must be taken into account. This relationship is given by

$$v_f = \frac{nkT}{q} [\ln (i_f - I_o) - \ln I_o] = f(i_f)$$

For silicon diodes this voltage has a maximum value of approximately +0.6 V for full-conduction. For $e_3 > 0$ ($e_1 < 0$) the current i_3 will be zero because $e_2 \approx 0$ and D_2 is back-biased. Essentially all input current

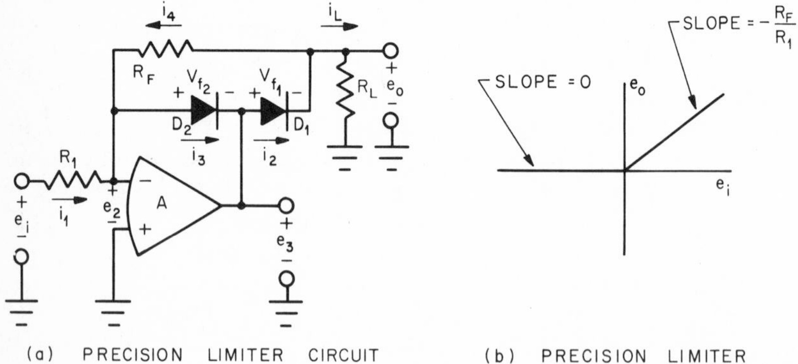

Fig. 7.12 Precision limiter operation. (a) Precision limiter circuit; (b) precision limiter transfer curve.

i_1 flows through R_F, generating an output voltage

$$e_o = -R_F i_1 = -\frac{R_F}{R_1} e_i, \qquad e_i < 0$$

If the finite amplifier gain and diode nonlinearity are taken into account, the expression becomes

$$e_o = \frac{(-R_F/R_1)e_i}{1 - (1/A\beta)[1 + f(i_2)/e_o]}$$

where

$$\beta = \frac{R_1}{R_1 + R_F}$$

Note that the effect of the diode forward voltage $f(i_2)$ is reduced by the loop gain $A\beta$ of the closed-loop circuit. Thus, the "rounding" of the turn-on region virtually disappears.

For $e_3 < 0$, diode D_1 no longer conducts and all the input current i_1 flows through D_2. Theoretically, the output voltage is then exactly equal to zero. The expression for e_o, considering finite gain and the diode nonlinearity, is

$$e_o \approx -\frac{f(i_1)}{A}\frac{R_L}{R_F + R_L}, \qquad e_i > 0$$

This is an extremely small voltage, probably less than the dc offset voltage of the amplifier. Thus the precision limiter provides a good approximation of ideal diode behavior, reducing the diode nonlinearity, temperature sensitivity, and forward voltage drop by a factor equal to the loop gain of the amplifier. A simpler analysis of the precision limiter can be made where the diode is represented by the linear model of a resistance, a bias source, and an ideal diode. Using this analysis, it is seen that the effects of diode resistance and internal bias voltage are reduced by the same factor, $A\beta$.

Nonzero precision limits can also be achieved, using the circuit shown in Fig. 7.13. Here, a diode bridge limiter gives both positive and negative limits. Because the diode bridge is inside the feedback loop, the nonlinearity, temperature sensitivity, and forward resistance of the bridge are all reduced by the loop-gain factor. Thus the limits are sharp and relatively independent of diode parameters.

7.2.4 Applications of limiters The simple series and shunt limiters described earlier in this section are used extensively in diode function generators which are discussed in later sections of this chapter. The limiter applications to be discussed here are principally those for feed-

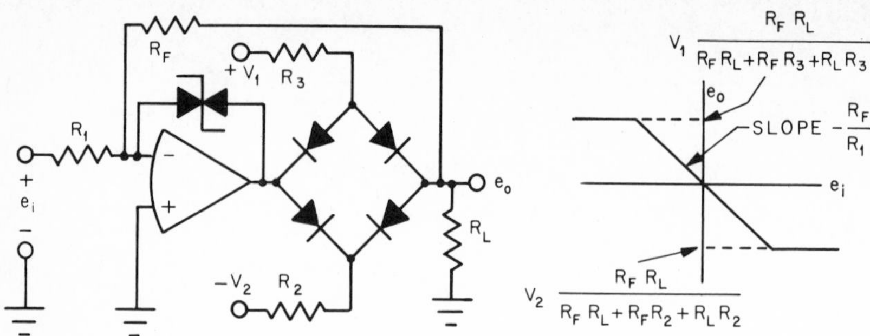

Fig. 7.13 Precision bridge limiter.

back limiters. As mentioned earlier, one of the primary reasons for using a feedback limiter is to prevent overload of the amplifier output stage. In many amplifiers, particularly chopper-stabilized types, a considerable time is required to recover from saturation of the output stage. The feedback limiter, by preventing such saturation, ensures fast recovery when the output voltage reaches the preselected limit. In circuits where input bias current is to be kept to a minimum, the bias current decoupling technique shown in Fig. 7.14 may be necessary. Here the resistor R_D shunts the bias current to ground since the diodes D_1 and D_2 are operating with zero voltage drop and zero current when the limiter is OFF. When the circuit is limiting, of course, the diodes conduct the feedback current.

Another of the basic applications for a feedback limiter is in comparator circuits. The limiter determines the ON and OFF voltage levels for the comparator output (see Chapter 9 for more details of comparators). As another application, limiters are often used with operational amplifiers for signal generation. Usually such use is in conjunction with a comparator for generation of square waves. Triangle and ramp wave-

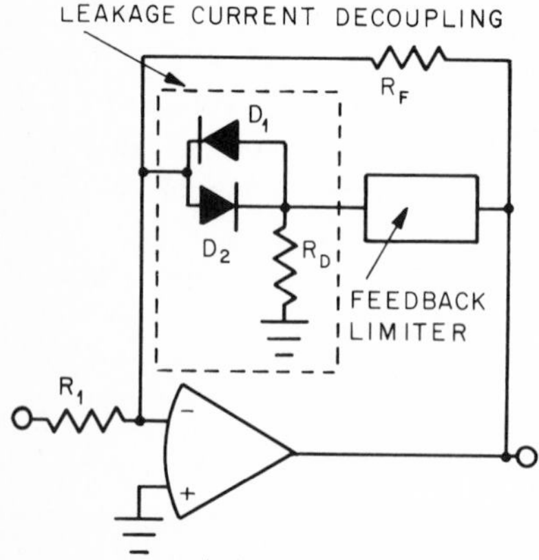

Fig. 7.14 Use of a leakage current decoupling circuit.

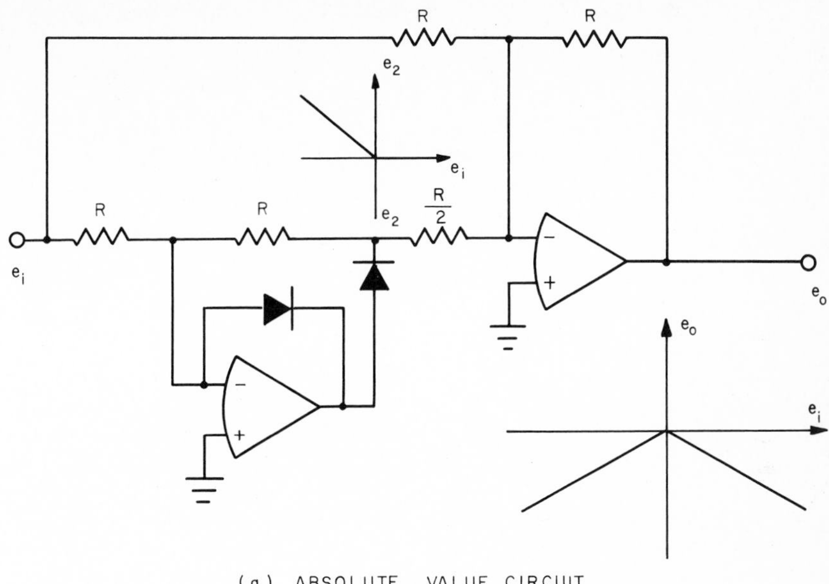

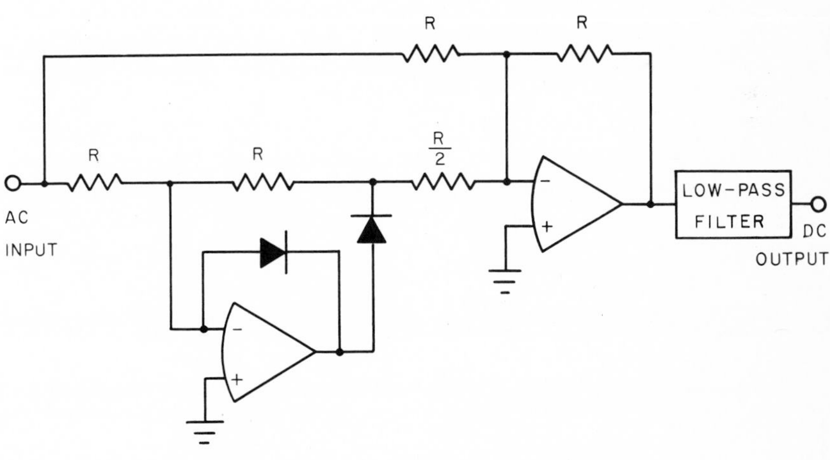

Fig. 7.15 Precision rectification. (a) Absolute-value circuit; (b) precision ac to dc conversion.

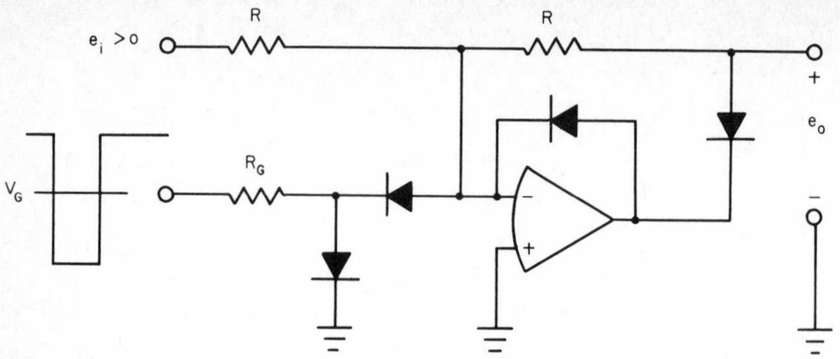

Fig. 7.16 Precision gate circuit.

forms can also be generated if an integrator follows the limiter or comparator (see Chapter 10 for more detail on signal generation).

Some limiters have breakpoints determined by an external reference voltage. Thus these levels can be easily changed by making the reference a variable or programmed voltage. This feature also makes possible the use of limiters for modulation of pulses and square waves (see Chapter 11).

The precision rectifier circuit discussed in Sec. 7.2.3 is useful in a variety of applications. As an example, the absolute-value circuit of Fig. 7.15a performs the function of precision full-wave rectification. With the addition of a low-pass filter as shown in Fig. 7.15b, the circuit achieves precision ac to dc conversion. Another interesting use of the precision rectifier circuit is shown in Fig. 7.16. Here the circuit functions as

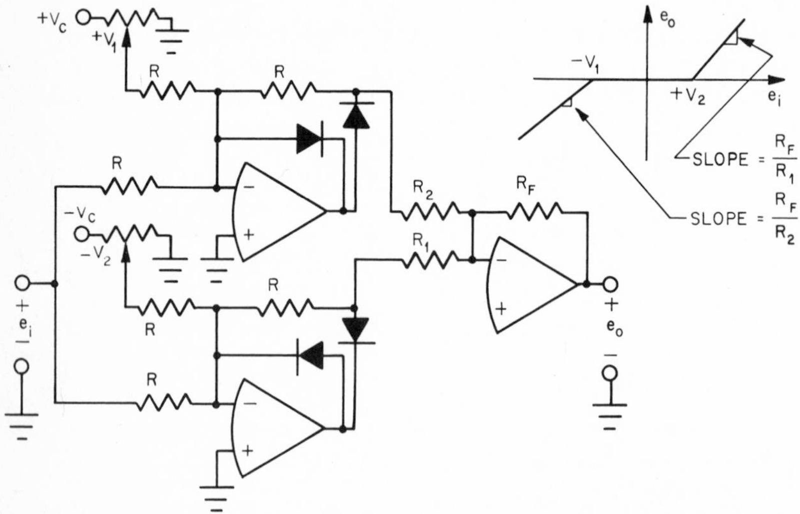

Fig. 7.17 Precision deadspace circuit.

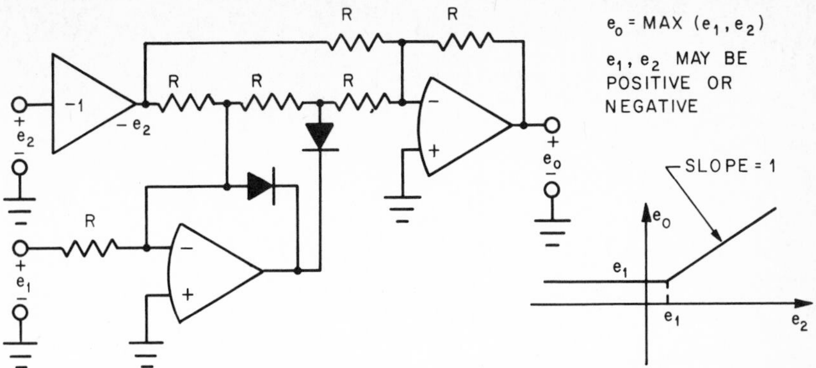

Fig. 7.18 Precision maximum selector.

a precision gate for positive signals. The negative gating signal V_G drives the limiter into its OFF region. If V_G is larger than the largest expected signal level, no signal can pass through the precision gate. Other uses of the precision rectifier principle are shown in Fig. 7.17 (precision deadspace circuit), and Fig. 7.18 (precision maximum selector).

7.3 Diode Function Generators[1,5,8–10]

The approximation of nonlinear functions is achieved with operational amplifiers by use of appropriate nonlinear feedback networks. The most general way of generating such functions is through the use of piecewise linear approximation, as shown in Fig. 7.19. The accuracy of such an approximation is determined by the number of line segments used. The complete piecewise curve is obtained by the summation of individual line segments whose "breakpoint" voltages and slopes are determined separately for each segment. Figure 7.20 illustrates how such segments may

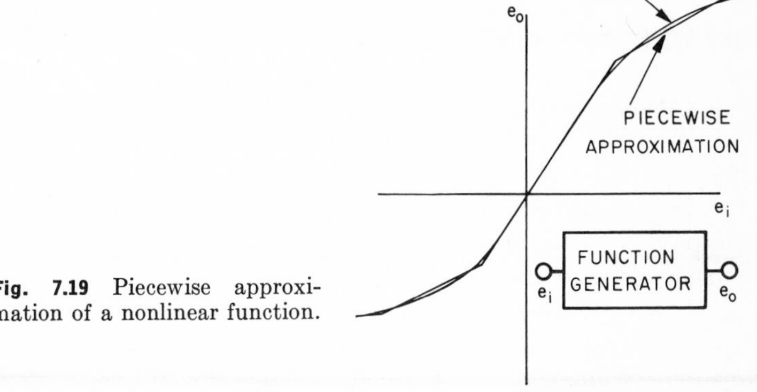

Fig. 7.19 Piecewise approximation of a nonlinear function.

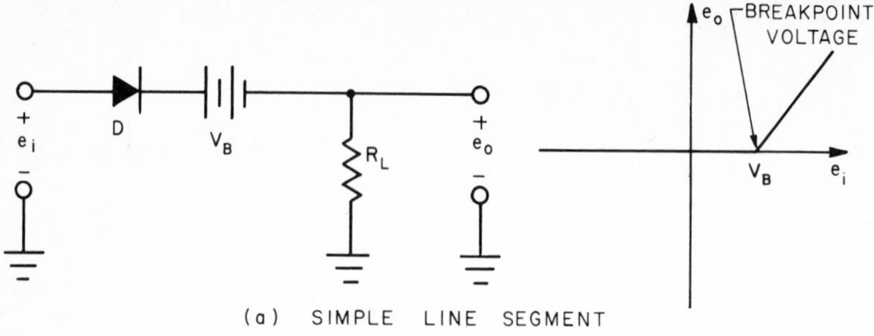

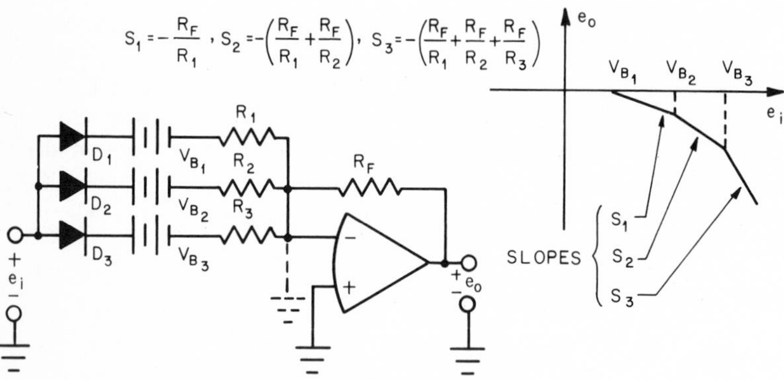

Fig. 7.20 Generation of piecewise approximation: (a) simple line segment; (b) summation of line segments.

be generated with simple limiter circuits and summed by the operational amplifier. The amplifier summing junction is a summation point for the currents from the breakpoint networks and the resistor R_F provides the scaling function. A more practical means of obtaining the desired line segments is through the use of series and shunt limiters as shown in Fig. 7.21. Note that each of the diode breakpoint circuits of Fig. 7.21 can be represented as a nonlinear transconductance. By using such networks as feedback elements, as shown in Fig. 7.22b, we obtain the inverse function.

As discussed earlier, the forward conduction characteristics of the silicon diode are somewhat temperature-sensitive and can cause changes in the breakpoints of the curve. This effect can be compensated partially by the methods shown in Fig. 7.23. In both cases the forward voltage drop of the breakpoint diode is compensated by a similar voltage drop in series with the biasing source. In the shunt limiter, the base-to-emitter voltage drop offsets much of the temperature sensitivity of the diode forward

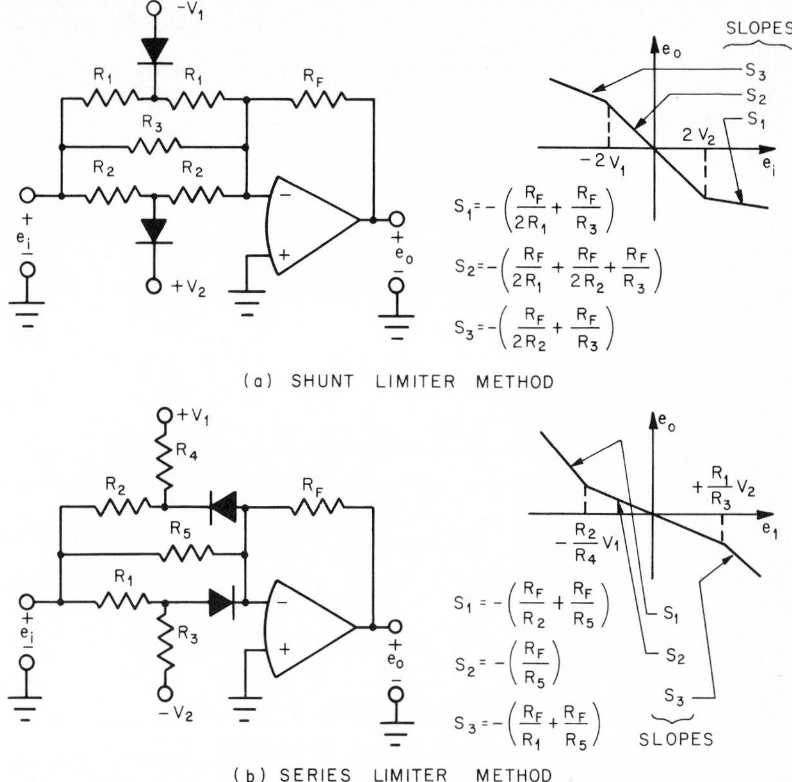

Fig. 7.21 Practical diode function generator circuits: (a) shunt limiter method; (b) series limiter method.

voltage. In the series limiter, a second diode acts as a temperature-compensating element for the breakpoint diode.

A more flexible approach to the approximation of nonlinear functions is illustrated in Fig. 7.24. The figure shows a variable diode function generator (VDFG) wherein both the locations of the breakpoints and the slopes of the line segments are individually adjustable. Note that the slopes can be positive, negative, or zero. The breakpoints can easily be made variable or can be temperature-compensated as in Fig. 7.23. This particular version of the VDFG is of the shunt type. A series-type VDFG is shown in Fig. 7.24b. As still another approach, the precision limiter, with its ability to simulate ideal diodes, can be used to generate line segments whose breakpoints are precisely known and which are temperature-insensitive. A simple version of such a function generator is shown in Fig. 7.25. Each breakpoint requires an operational amplifier, which made this approach prohibitive in cost before the advent of the

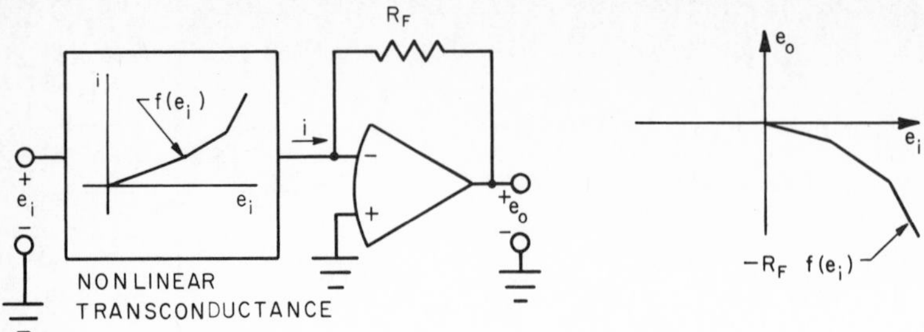

(a) USE OF NONLINEAR TRANSCONDUCTANCE

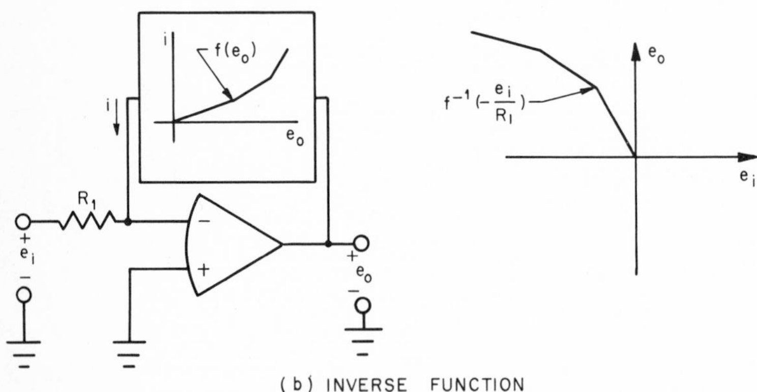

(b) INVERSE FUNCTION

Fig. 7.22 Use of nonlinear transconductance to obtain an inverse function. (a) Use of nonlinear transconductance; (b) inverse function.

integrated-circuit operational amplifier. The principal drift factor in this circuit is not the diode junction but is actually the input voltage drift of the operational amplifiers, which are at least two orders of magnitude better than an uncompensated diode. The breakpoints are sharp, rather than rounded—a fact which may be a disadvantage of this technique. The technique is easily adapted to arbitrary function generation where both the breakpoints and the slopes are adjustable.

7.3.1 Applications of diode function generators One of the more obvious uses of the function generators which have been described above is in the linearization of response curves. Primary examples are the linearizing of thermocouples, thermistors, and pressure transducers. The nonlinearity of the transducer is balanced by a compensating nonlinearity of the function generator, thus achieving a composite function which is linear. The general procedure is illustrated in Fig. 7.26. As another approach the diode function generator can be used for waveform generation when its input voltage is a linear sweep such as a triangle wave.

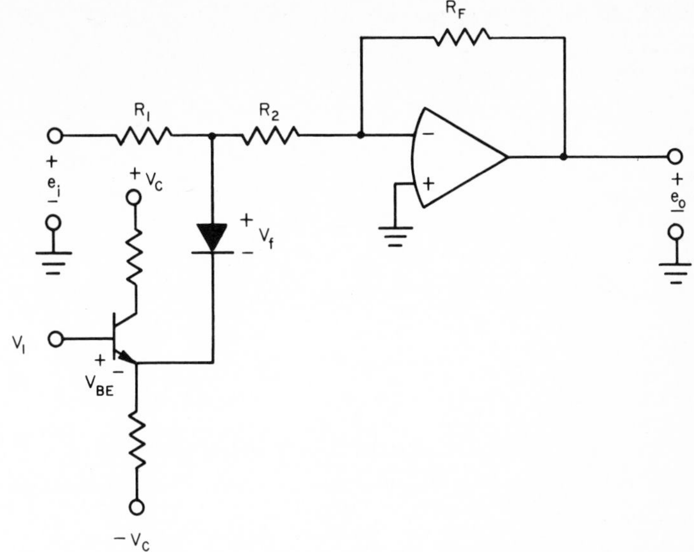

(a) TEMPERATURE COMPENSATED BREAKPOINT – SHUNT LIMITER

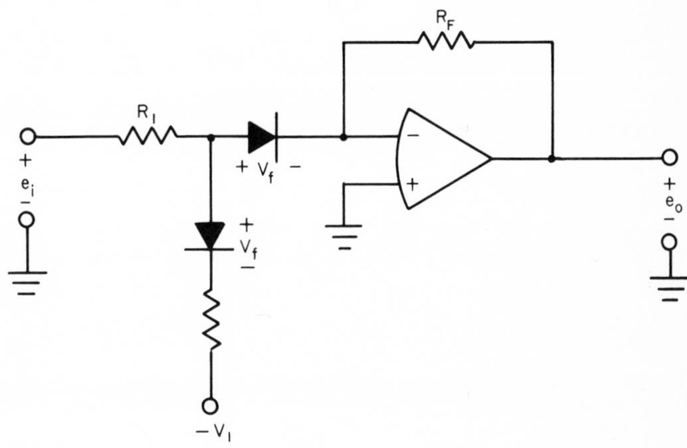

(b) TEMPERATURE COMPENSATED BREAKPOINT – SERIES LIMITER

Fig. 7.23 Compensation of breakpoint temperature drift. (a) Temperature-compensated breakpoint shunt limiter; (b) temperature-compensated breakpoint series limiter.

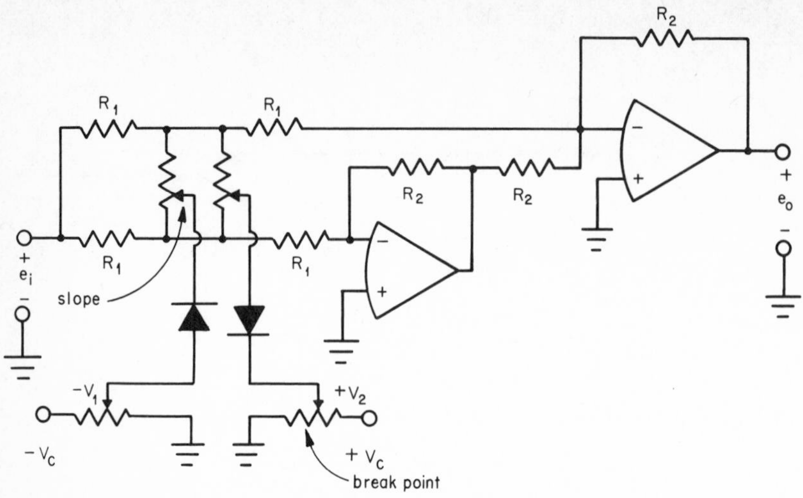

(a) SHUNT LIMITER DIODE FUNCTION GENERATOR

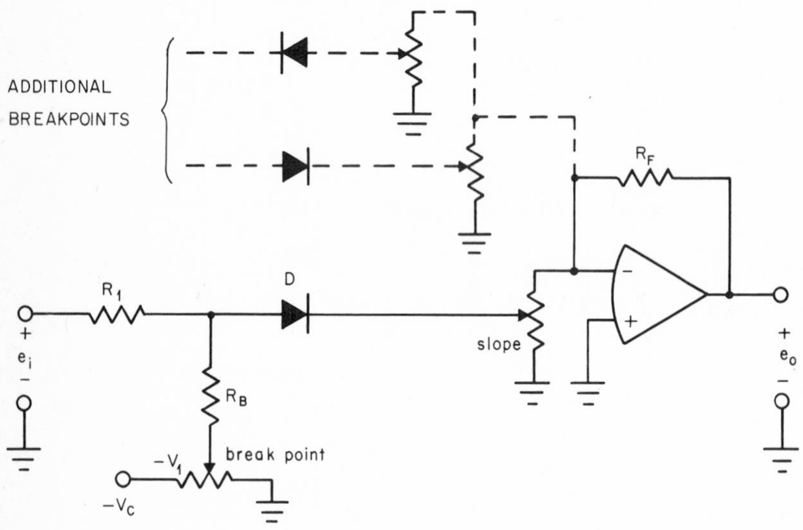

(b) SERIES LIMITER DIODE FUNCTION GENERATORS

Fig. 7.24 Variable diode function generators: (a) shunt limiter diode function generator; (b) series limiter diode function generator.

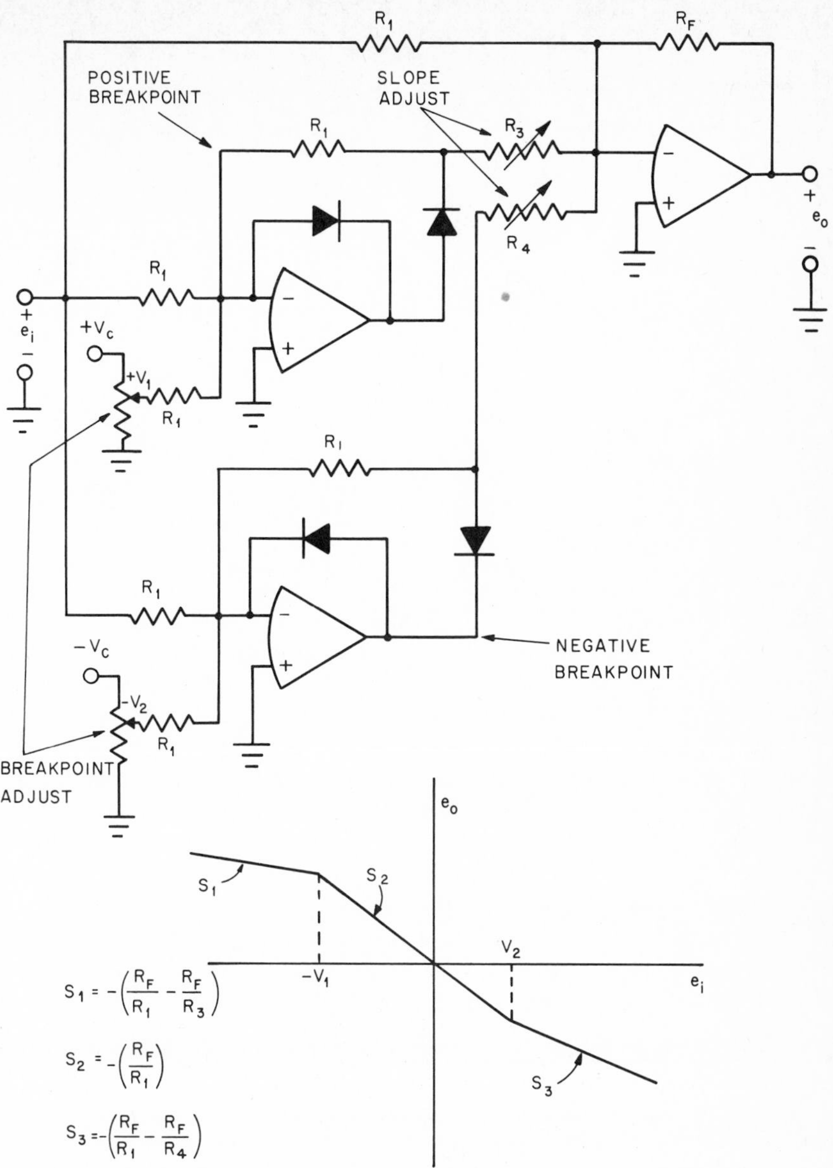

Fig. 7.25 Precision limiter diode function generator.

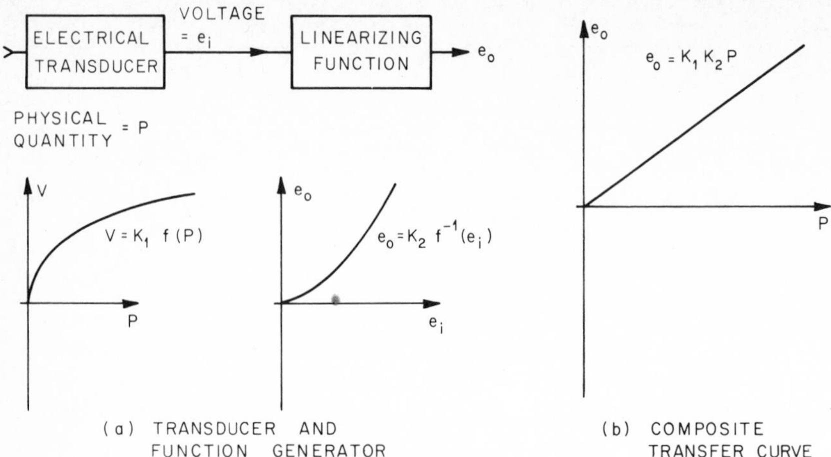

Fig. 7.26 Linearization of a transducer using a function generator. (a) Transducer and function generator; (b) composite transfer curve.

Either arbitrary functions, as shown in Fig. 7.27a, or common functions, such as the sine wave shown in Fig. 7.27b, can be generated. This method is especially attractive for the generation of very low-frequency waveforms. Other well-known functions such as the square and square root of an input voltage can be accurately approximated by the diode breakpoint method. Such function generators are extremely time-consuming to design for good accuracy, because of the large number of breakpoints and the interaction of all adjustments. However, these devices are available commercially and can be used for a variety of functions. Some examples are shown in Fig. 7.28. Other interesting areas for the application of function generator techniques are the simulation of physical effects in computation, the realization of nonlinear sensitivities for control systems, and the compression of signals having wide dynamic range. These applications are illustrated in Fig. 7.29.

7.4 Logarithmic Amplifiers[2,15]

In the preceding sections of this chapter we have discussed the use of nonlinear feedback networks to provide limiting and function generation. The diode function generators (DFG) which were discussed used the large-signal switching properties of the diode. In this section we shall present a treatment of logarithmic amplifiers. Such amplifiers use the nonlinear volt-ampere relationship of the p-n junction itself. This relationship is given by

$$i_f = I_o(e^{v_f/\eta V_T} - 1)$$

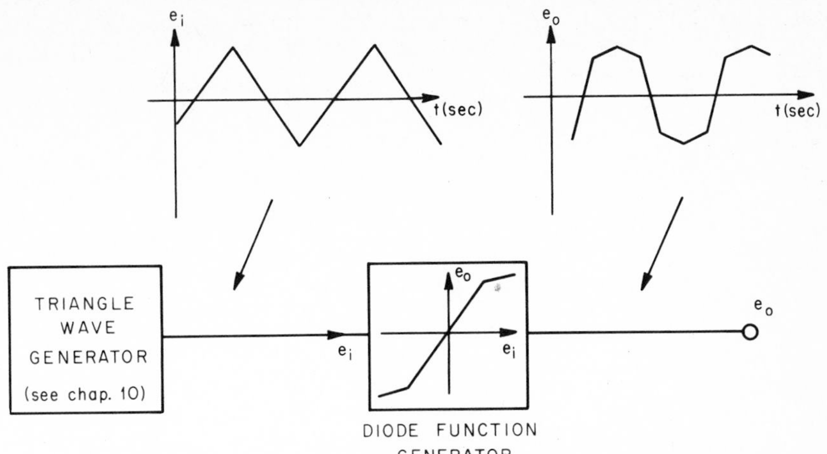

(a) GENERATION OF ARBITRARY WAVESHAPES

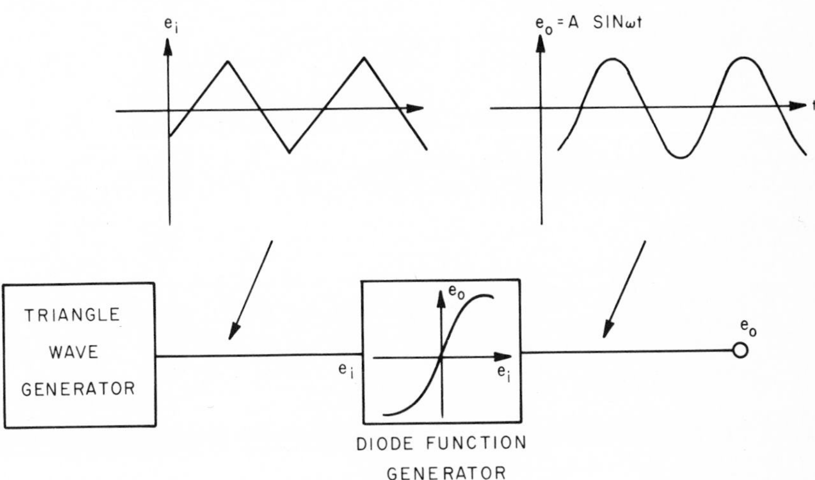

(b) SINE WAVE GENERATION

Fig. 7.27 Use of diode function generators for waveform generation. (a) Generation of arbitrary waveshapes; (b) sine-wave generation.

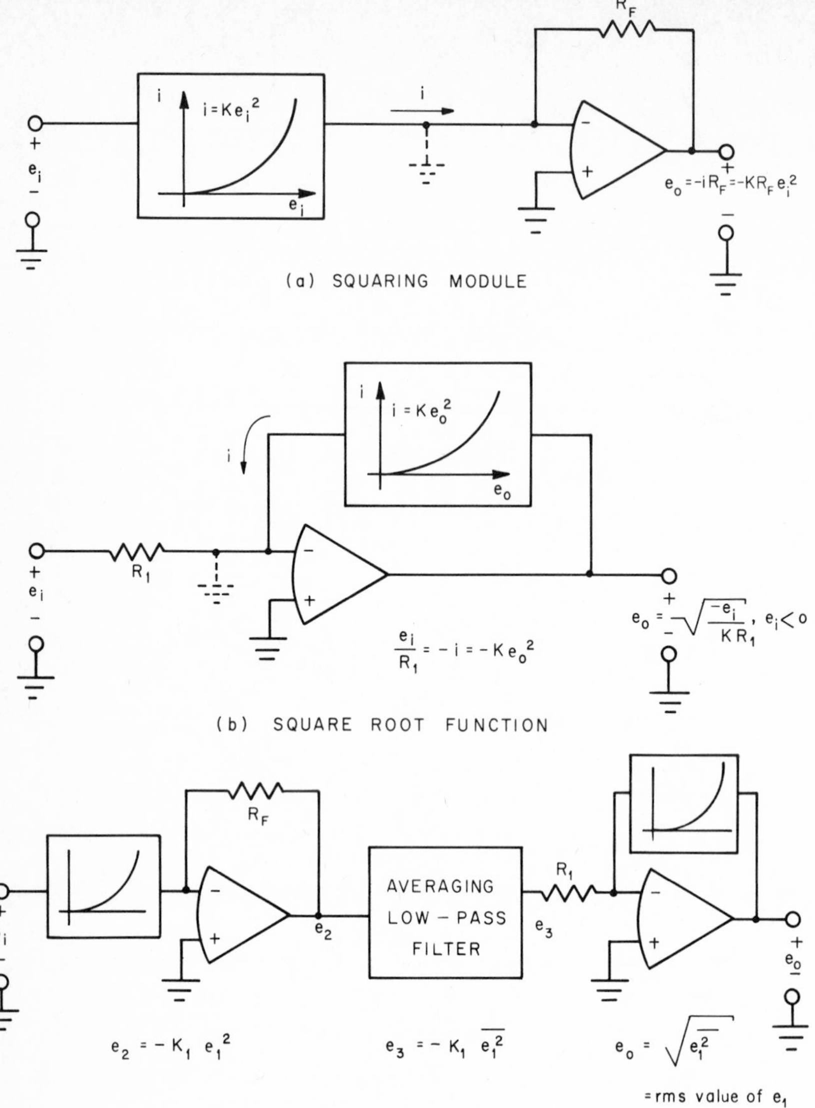

Fig. 7.28 Use of squaring function. (a) Squaring module; (b) square-root function; (c) rms circuit.

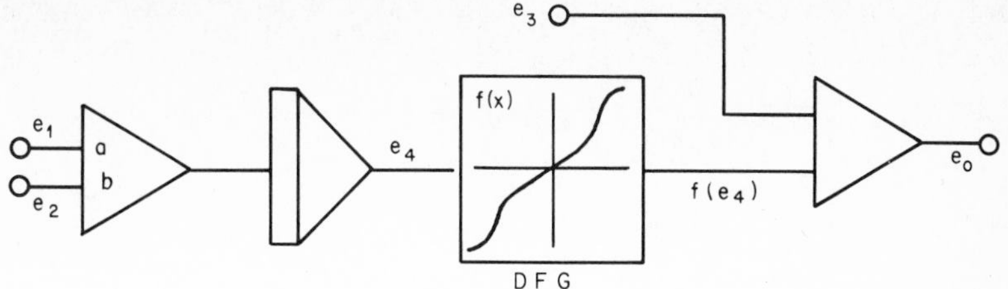

(a) USE OF DFG IN ANALOG SIMULATION

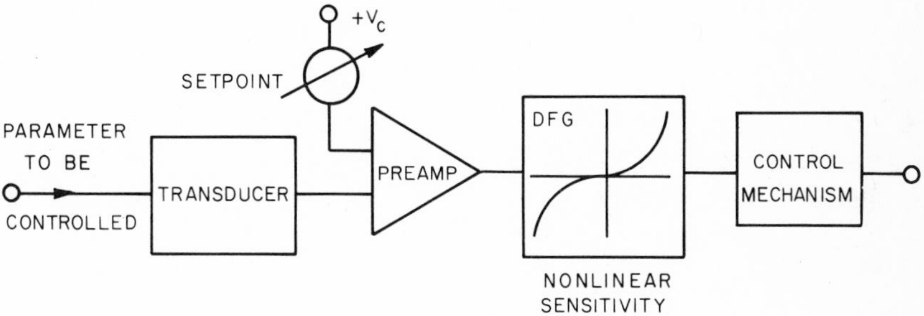

(b) USE OF DFG IN CONTROL SYSTEM

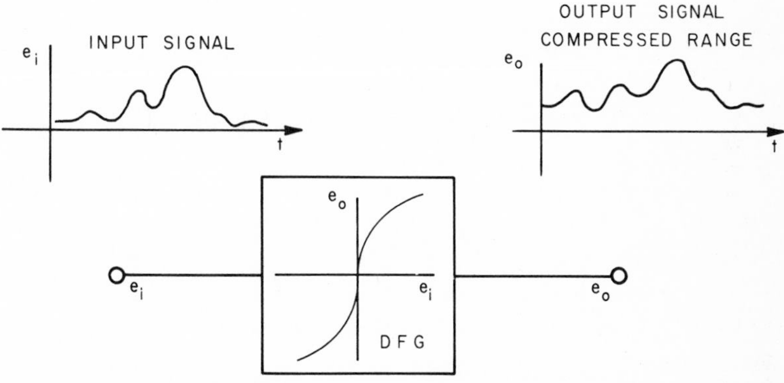

(c) USE OF DFG FOR SIGNAL COMPRESSION

Fig. 7.29 Applications of a diode function generator (DFG). (a) Use of a DFG in an analog simulation; (b) use of a DFG in a control system; (c) use of a DFG for signal compression.

where I_o = reverse saturation current

$I \approx 2$ for small currents in silicon devices

$$V_T = \frac{k}{q} T \approx \frac{T}{11{,}000} \text{ volts; T in °K}$$

If we restrict the operating region of v_f so that $e^{v_f/\eta V_T} \gg 1$, the logarithmic relationship may be expressed as

$$\ln i_f = \ln I_o + \frac{v_f}{\eta V_T}$$

or

$$v_f = \eta V_T (\ln i_f - \ln I_o)$$

Ignoring temperature effects for the moment, η, V_T, and I_o may be considered as constants. If the diode is connected in the feedback path of an operational amplifier, as shown in Fig. 7.30, the output voltage of the amplifier is a logarithmic function of the input voltage. The derivation of the logarithmic relation proceeds as follows:

$$i_1 = \frac{e_i}{R_1} \qquad i_f = i_1$$

$$v_f = \eta V_T \left(\ln \frac{e_i}{R_1} - \ln I_o \right)$$

$$e_o = -v_f = -\eta V_T \left(\ln \frac{e_i}{R_1} - \ln I_o \right)$$

In considering the temperature compensation of such an amplifier, it should be noted that there are actually two separate temperature effects to be compensated: a temperature-sensitive scale factor, ηV_T, and a temperature-sensitive offset term, $\eta V_T \ln I_o$. The saturation current term can be removed or reduced by the use of a current source and a second, matched diode, D_2, as shown in Fig. 7.31. The current source forces a constant current I_R through D_2, which in turn generates the voltage V_f'. If the two diodes are perfectly matched, the V_T and I_o terms for the two diodes will be equal and the $\ln I_o$ term will be absent from e_3.

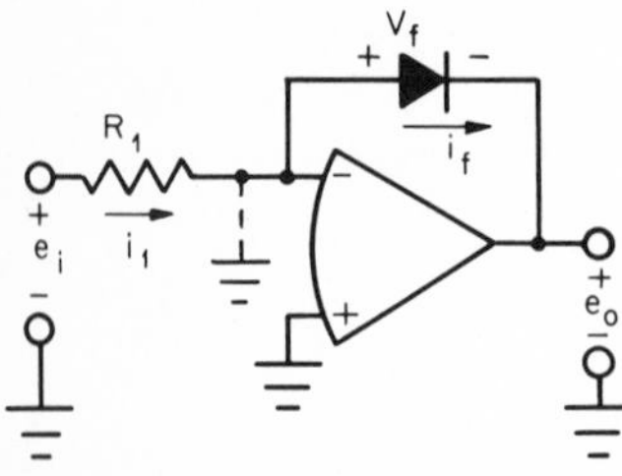

Fig. 7.30 Simple logarithmic amplifier.

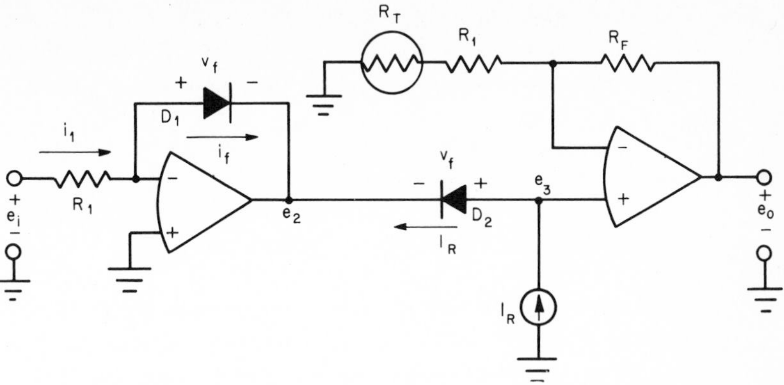

Fig. 7.31 Temperature-compensated logarithmic amplifier.

Thus we may write

$$e_3 = e_2 + V_f = -\eta V_T \left(\ln \frac{e_i}{R_1} - \ln I_o - \ln I_R + \ln I_o \right)$$

$$e_3 = -\eta V_T \ln \frac{e_i}{R_1 I_R}$$

The only remaining temperature sensitivity in e_3 is that of the scale factor term. This can be compensated in the output amplifier by making its gain temperature-sensitive and compensating for the V_T factor. This is most easily done by using a temperature-sensitive resistor R_T in the feedback network as shown in Fig. 7.31. The output voltage is then given by

$$e_o = -\frac{R_F + R_1 + R_T}{R_1 + R_T} \eta V_T \ln \frac{e_i}{R_1 I_R} = K_1 \ln (K_2 e_i)$$

The gain of the output amplifier determines the constant K_1, and the values of I_R and R_1 determine the constant K_2. Together, they determine the range of input voltage which will drive the output amplifier through its rated range.

The dynamic range of a logarithmic amplifier of the type described in this section is limited by several independent factors. The diode itself follows the logarithmic relationship between v_f and i_f rather closely over as much as 6 decades of i_f. However, i_f contains not only the input signal current i_1 but also the input bias current and the noise current of the operational amplifier, plus currents generated by the input offset voltage and input noise voltage applied across R_1. If the maximum current allowed to flow through the diode (for accurate logging) is 1 mA, then R_1 must be 10 kΩ, if the maximum input voltage is to be 10 V. If we assume an amplifier which has an input bias current of 10 nA and an input offset voltage of 1.0 mV, the dynamic range of input signal for percent accuracy is 100 mV to 10 V—a range of 40 dB. The limitation is provided by the offset voltage. In order to achieve a dynamic range of 80 dB at 1 percent accuracy we must have a total error current of less than 1 nA. An amplifier with FET input will have a bias current of the order of a few picoamperes. However the input offset voltage must then be less than 10^{-9} A $\times$ 10^4 Ω = 10 μV. This is a very difficult figure to maintain, particularly over a long period of time and over a range of temperatures. The noise voltage of the amplifier e_n, also generates an error current, $i_n = e_n/R_1$, which must be considered as a part of the total error if transient signals with a broad frequency spectrum are being measured. These problems are not as formidable if the signal source is a high impedance source of current, as shown in Fig. 7.32. In such a case the input signal current flows through the feedback diode, generating a voltage proportional to the logarithm of the input current. Since the source resistance is extremely large, the effective voltage gain of the circuit is small, and the voltage offset and noise of the amplifier are not so critical as error sources. The principal sources of error are bias current and noise current, which can be made very small if an FET operational amplifier is used. Note that in the log amplifiers discussed the input signal must be unipolar. It may, however, be negative or positive, depending on the orientation of the diodes.

Another variation of the logarithmic amplifier is the log-ratio circuit shown in Fig. 7.33. Here there are two input signals (e_1,e_2) which are converted to temperature-sensitive logarithmic voltages (e_3,e_4) by the diodes D_1 and D_2 and amplifiers A_1 and A_2. The relations are

$$e_3 = -\eta V_T \left(\ln \frac{e_1}{R_1} - \ln I_o \right)$$

$$e_4 = -\eta V_T \left(\ln \frac{e_2}{R_1} - \ln I_o \right)$$

Amplifier A_3 acts as a difference amplifier with gain. By subtracting e_3 from e_4, the temperature-sensitive offset terms $\eta V_T \ln I_o$ tend to cancel

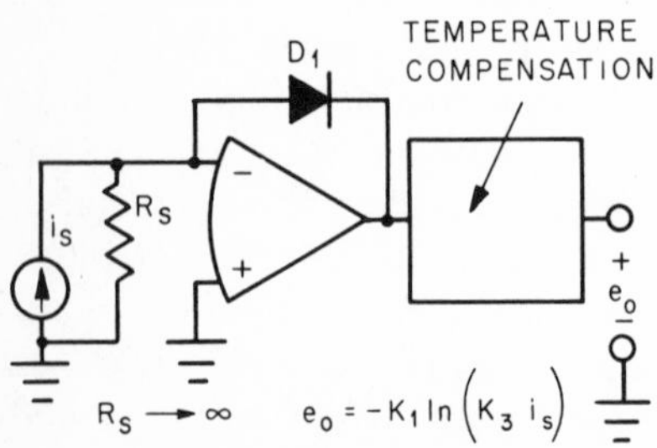

Fig. 7.32 Logarithmic current-to-voltage converter.

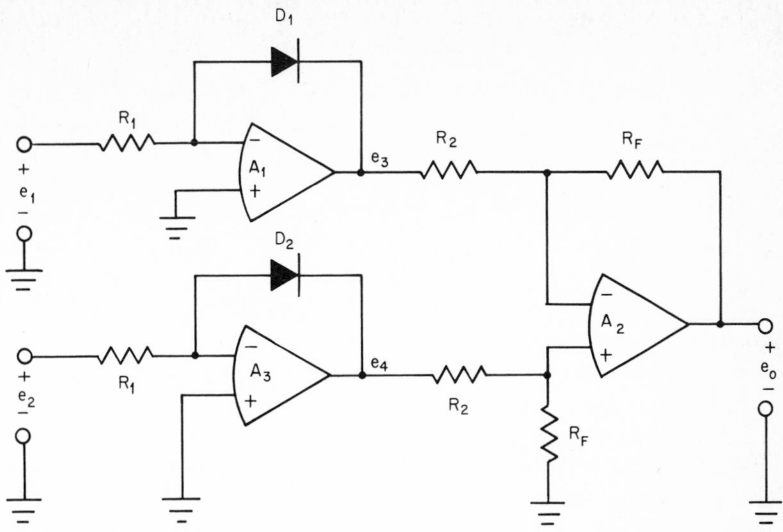

Fig. 7.33 Log-ratio amplifier.

each other. The output voltage is

$$e_o = \frac{R_F}{R_2}(e_4 - e_3) = \frac{R_F}{R_2}\eta V_T \ln \frac{e_1}{e_2}$$

The temperature-sensitive gain term ηV_T can be canceled from the expression for e_o if R_F/R_2 is made to have a compensating temperature sensitivity.

The antilog function also can be obtained by using a diode with an operational amplifier as shown in Fig. 7.34. The following relationships apply,

$$e_2 = e_i \frac{R_1}{R_1 + R_2}$$

$$e_3 = e_i \frac{R_1}{R_1 + R_2} - \eta V_T(\ln I_f - \ln I_o)$$

but, also,

$$e_3 = -\eta V_T(\ln i_2 - \ln I_o)$$

Therefore

$$e_i \frac{R_1}{R_1 + R_2} = \eta V_T \ln \frac{I_f}{i_2}$$

$$e_o = +i_2 R_F$$

$$-e_i \frac{R_1}{R_1 + R_2} = \eta V_T \ln \frac{e_o}{R_F I_f}$$

$$e_o = R_F I_f \ln^{-1}\left[-e_i \frac{R_1}{(R_1 + R_2)\eta V_T}\right]$$

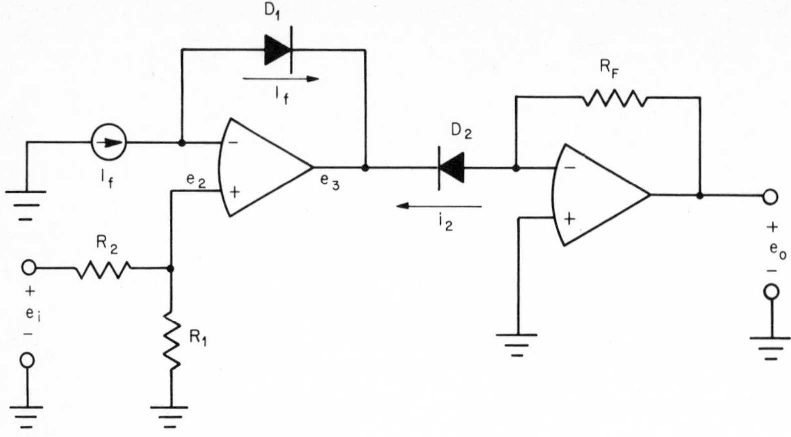

Fig. 7.34 Antilog amplifier.

Once again the temperature-sensitive offset terms, $\eta V_T \ln I_o$, cancel if the diodes are well matched. The temperature sensitivity of the proportionality factor $R_1/(R_1 + R_2)\eta V_T$ can be eliminated if the resistive divider (R_1,R_2) can be made to have a compensating sensitivity. The current generator I_f and feedback resistor R_o are adjusted for proper scaling. The dynamic range of the antilog amplifier is determined by the noise, voltage offset, and bias current of both amplifiers. This means that for wide dynamic range, such as 80 dB, both amplifiers must have voltage offset and noise less than about 10 μV. This is extremely difficult to achieve with normal operational amplifier designs.

7.4.1 Applications of log amplifiers The log and antilog amplifiers described above can be used in combination for the generation of arbitrary functions by raising the input to a power, as shown in Fig. 7.35. The exponent α is obtained by simply multiplying $\ln e_1$ by a constant through a coefficient network. The voltage e_2 is proportional to $\ln e_i$; thus

$$e_2 = \alpha K_1 \ln K_2 e_i$$

The output voltage is proportional to the antilog of e_2:

$$e_o = K_3 \ln^{-1} K_4 e_2 = K_3 e^{K_4 e_2}$$
$$e_o = K_3 (K_2 e_i)^{\alpha K_1 K_4}$$

The value of the coefficient can be greater or less than 1.0, if a scaling amplifier is used, thus allowing a variety of functions. For special situations where the function is to remain fixed, the scaling of the log and antilog amplifiers can include the coefficient α.

Another of the primary uses of the logarithmic amplifier is the com-

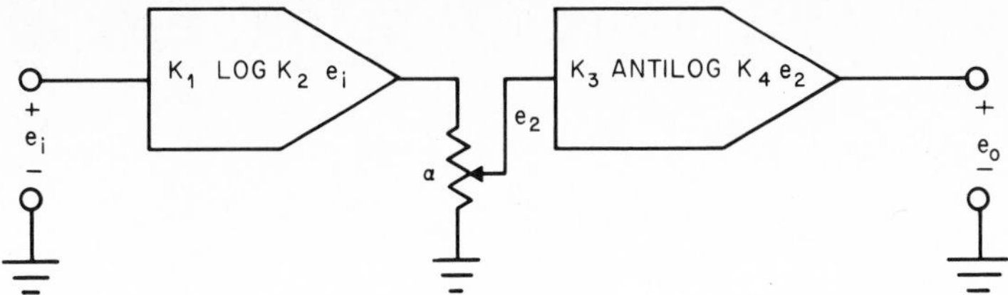

Fig. 7.35 Raising variables to a power with log techniques.

pression/expansion of signals having wide dynamic range. Consider as an example the logarithmic amplifier with characteristic

$$e_o = -10 \log e_i$$

where $-10 \text{ V} \leq e_o \leq +10 \text{ V}$. A plot of this characteristic is shown in Fig. 7.36. The input signal range is from +100 mV to +10 V, which corresponds to an output range of $-10 \leq e_o \leq +10$ V. Input signals in the range $+0.1 \text{ V} \leq e_i \leq +10.$ are expanded to an output range of $0 \text{ V} \leq e_o \leq +10 \text{ V}$. Input signals in the range $+1.0 \text{ V} \leq e_i \leq +10 \text{ V}$ correspond to an output range of $-10 \text{ V} \leq e_o \leq 0 \text{ V}$. Thus each decade of input range is represented by equal increments of output voltage. This makes somewhat easier the reading and recording of input data having a wide dynamic range. An ac compression amplifier with a psuedo-logarithmic response can be achieved with the circuit of Fig. 7.37a. The diodes D_1 and D_2 generate the logarithmic response for positive and negative output voltages, respectively. The resistor R_F is required because of the discontinuity in the log curve at zero. This feedback resistor modifies the curve near zero as shown in Fig. 7.37b. The transfer curve of this compression amplifier will vary with temperature and cannot be effectively temperature-compensated in the output scaling amplifier. Thus it is useful only under temperature-controlled conditions or for very "rough" signal compression.

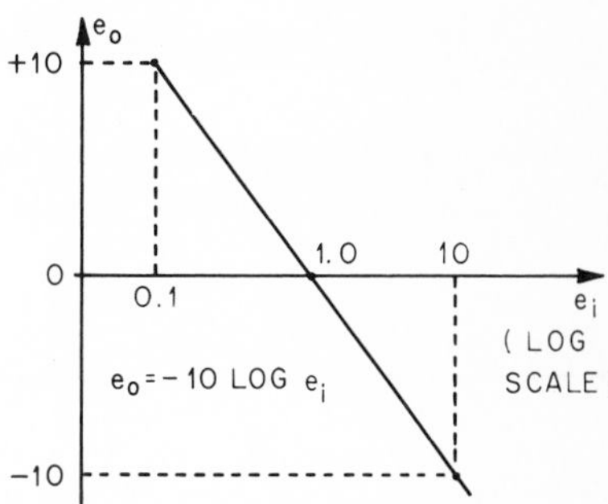

Fig. 7.36 Two-decade log amplifier gain curve.

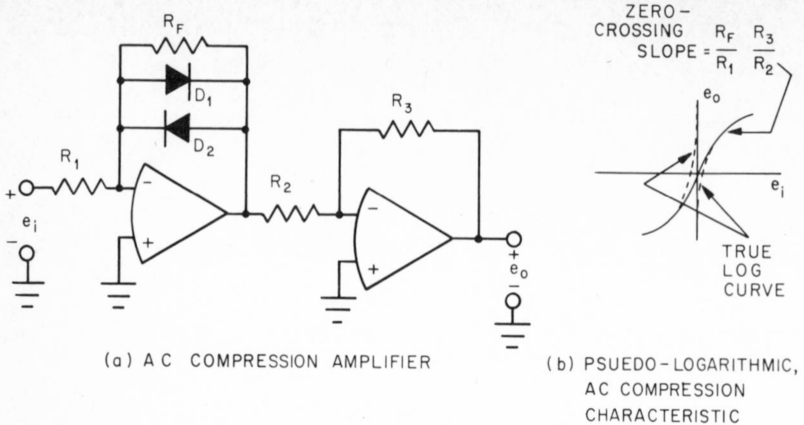

Fig. 7.37 Ac log compression technique. (a) Ac compression amplifier; (b) psuedo-logarithmic, ac compression characteristics.

7.5 Analag Multiplication and Division[1,11–14,17]

In the preceding sections of this chapter we have discussed operational amplifier nonlinear circuits which provide limiting, function generation, and logarithmic amplification. Another frequently encountered non-linear application of operational amplifiers is for accurate multiplication and division of analog signals. The six most common solid-state methods are logarithmic, quarter-square, triangle averaging, time division, variable transconductance, and current ratioing. There are other techniques for multiplying, but these six are the most suitable for all-solid-state instrumentation. Together, they span a wide spectrum of accuracy, speed, and cost.

7.5.1 Logarithmic multiplier The first multiplier to be discussed is the logarithmic type shown in Fig. 7.38a. The log and antilog amplifier techniques discussed in the preceding section are used in this circuit. It is only necessary to take the log of each input, sum these inputs, and then take the antilog of the sum. The result is the product of the two inputs. In terms of the variables shown in Fig. 7.38a,

$$e_3 = K_1 (\ln e_1 + \ln e_2) = K_1 \ln e_1e_2$$

and

$$e_o = K_2 \ln^{-1} \frac{e_3}{K_1} = K_2e_1e_2$$

Division can be accomplished by subtracting the logarithms of the two inputs and then taking the antilog. This can most easily be accom-

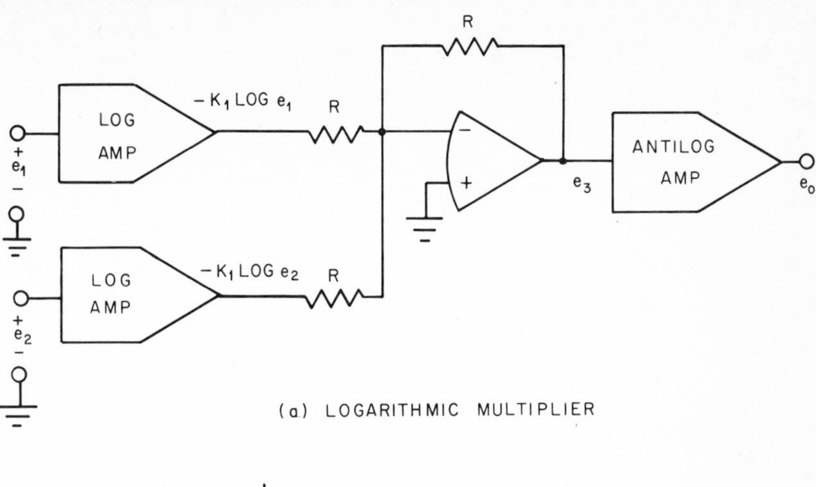

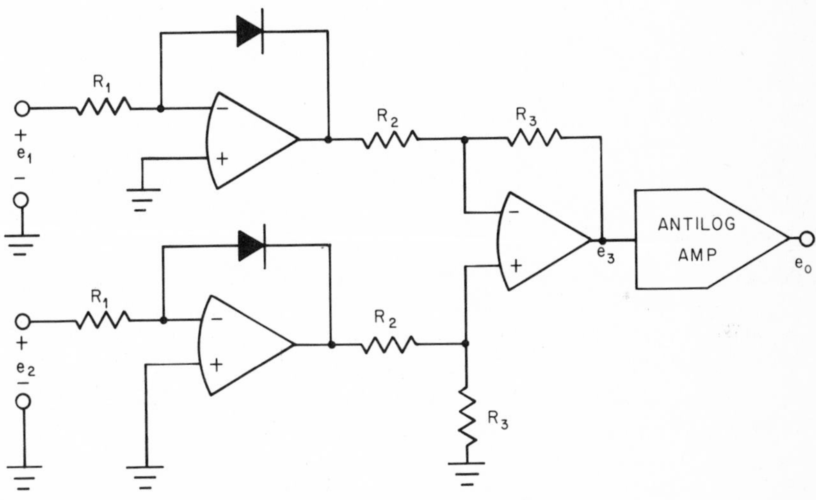

Fig. 7.38 Multiplication and division by log techniques. (a) Logarithmic multiplier; (b) logarithmic divider.

plished with the log-ratio circuit of Fig. 7.38b. Here,

$$e_3 = K_1 \ln \frac{e_1}{e_2}$$

$$e_o = K_2 \ln^{-1} \frac{e_3}{K_1}$$

$$e_o = K_2 \frac{e_1}{e_2}$$

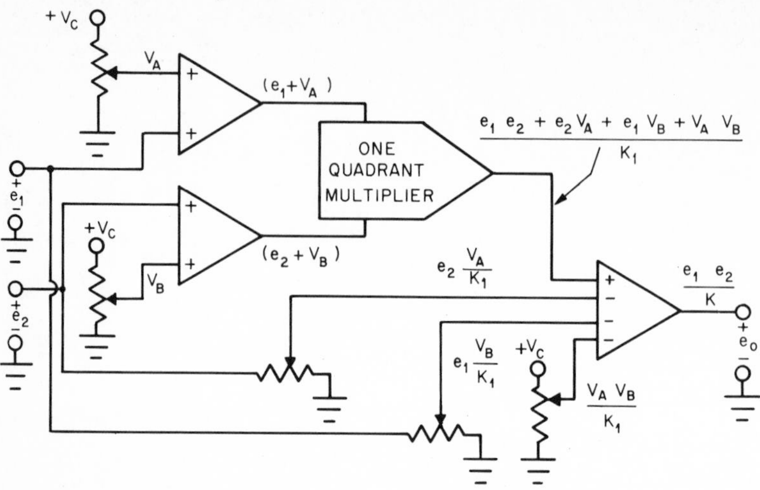

Fig. 7.39 Two- and four-quadrant multiplication obtained from a single-quadrant multiplier.

The logarithmic technique of multiplication and division is, of course, useful only for unipolar inputs, or one-quadrant operation, as it is sometimes described. Actually, however, any single-quadrant multiplier can be converted to two- or four-quadrant operation by the technique shown in Fig. 7.39. It is only necessary to subtract from the output all unwanted terms in e_1 and e_2, and the constant term. The addition of V_A and V_B to the input variables ensures that the inputs to the multiplier remain unipolar.

The logarithmic approach, unfortunately, suffers from rather strong temperature sensitivity. This can be compensated to a certain extent by methods previously discussed. However it is difficult to achieve better than 1 percent overall accuracy even for a moderate temperature range. Because of its basic simplicity, however, the logarithmic method may be attractive where accuracies of 1 to 5 percent are satisfactory and where careful temperature compensation is not required.

7.5.2 Quarter-square multiplier The quarter-square multiplier makes use of the equation

$$\frac{(X + Y)^2 - (X - Y)^2}{4} = \frac{(X^2 - X^2) + (Y^2 - Y^2) + 2XY + 2XY}{4}$$

$$= XY$$

to obtain the product. The squared terms are usually obtained through the use of special diode function generators, using the piecewise linear

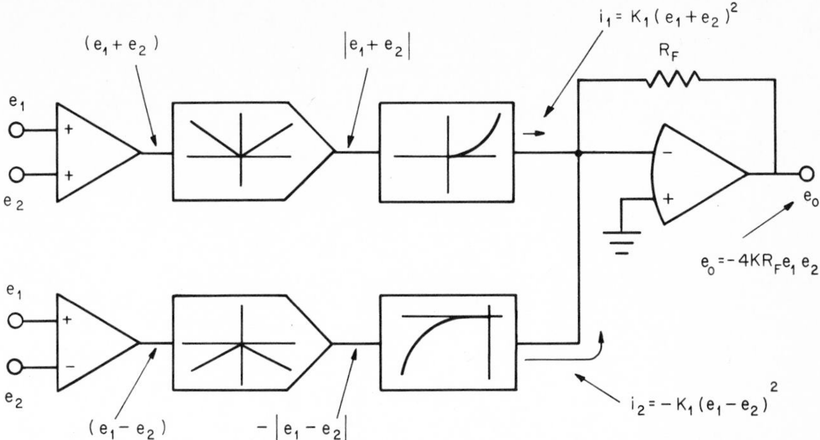

Fig. 7.40 Quarter-square multiplier.

techniques outlined earlier in this chapter. Squaring modules with 10 breakpoints can approximate the function

$$i_{sc} = Ke_1^2$$

to within ± 0.1 percent of full scale. The term i_{sc} is the output short-circuit current to ground, or virtual ground. The circuit of Fig. 7.40 illustrates one means of obtaining multiplication through the quarter-square relationship. This method of multiplication is useful over a wide frequency range, which is its most attractive feature. Its principal disadvantages are the complexity and cost and the fact that the maximum error voltage, although small as a percentage of full scale, may exist at low input levels. This statement is illustrated by the typical error curve of Fig. 7.41. The "ripple" in the error curve arises from the piecewise linear approximation used in the squaring modules. Additional errors are introduced by dc offset shift as a function of temperature.

7.5.3 Triangle-averaging multiplier The method of multiplication known as triangle averaging is illustrated in Fig. 7.42. The voltage e_3 is

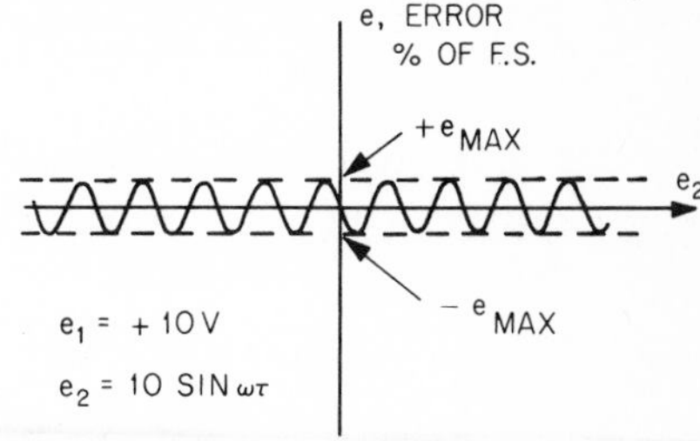

Fig. 7.41 Typical error curve of a quarter-square multiplier.

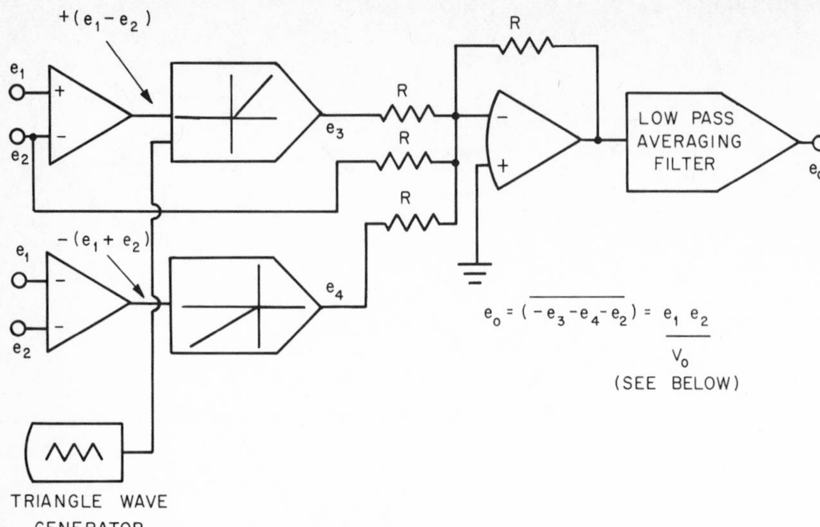

(a) TRIANGLE AVERAGING MULTIPLIER

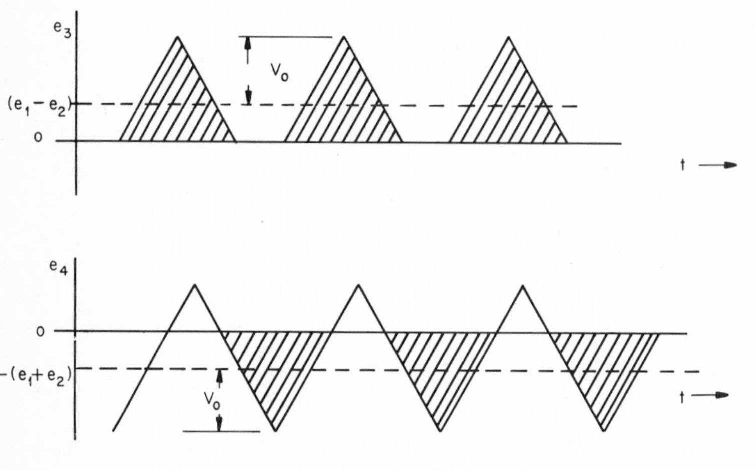

(b) MULTIPLIER WAVE FORMS

Fig. 7.42 Illustration of the triangle-averaging multiplier: (a) triangle-averaging multiplier; (b) multiplier waveforms.

the half-wave rectified sum of the triangle wave and $e_1 - e_2$. Only the positive part of the waveform is retained, and this is time-averaged by a low-pass filter. The resulting average value is

$$\bar{e}_3 = \frac{1}{2}\left(\frac{1}{2} + \frac{e_1 - e_2}{2V_o}\right)(V_o + e_1 - e_2)$$

Similarly

$$\bar{e}_4 = -\frac{1}{2}\left(\frac{1}{2} + \frac{e_1 + e_2}{2V_o}\right)(V_o + e_1 + e_2)$$

The sum of the two voltages is

$$\bar{e}_3 + \bar{e}_4 = -e_2 - \frac{e_1 e_2}{V_o}$$

If the e_2 term is removed in a summing amplifier, the resulting voltage is the desired product. The frequency response of such multipliers is necessarily quite restricted because of the low-pass averaging filter at the output. This filter must effectively eliminate the carrier frequency and must therefore have a cutoff frequency well below the fundamental of the triangle wave. Increasing the frequency of the triangle wave to improve the overall frequency response leads to problems of capacitive coupling of carrier frequencies to the output as well as linearity problems in the triangle waveform. The linearity of the triangle wave and the sharpness of the peaks of the waveform are the principal limitations on the accuracy of this method of multiplication. (See Chapter 10 for triangle-wave generators.)

7.5.4 Time-division multiplier Another carrier technique of multiplication is the so-called time-division multiplier illustrated in Fig. 7.43. It is necessary to generate a square wave whose average value depends upon both of the input signals. In this method of multiplication a triangle wave is once again used. However, instead of clipping and averaging as in the triangle-averaging multiplier, the triangle wave is used to control an electronic switch. The triangle wave is summed with one of the input signals, e_2, and the sum is applied to a zero-biased comparator. The resulting asymmetric square wave has a duty cycle determined by the magnitude and polarity of e_2. That is,

$$T_2 = \frac{e_2 + V_o}{2V_o} T$$

$$T_1 = \frac{V_o - e_2}{2V_o} T$$

This square wave in turn controls the electronic switch. Amplifier A_2 transmits $+e_1$ when the switch is at ON, and $-e_1$ when the switch is at OFF. Since the duty cycle of e_4 is proportional to e_2 and the magnitude is $\pm e_1$, the resulting average value is proportional to the product. When

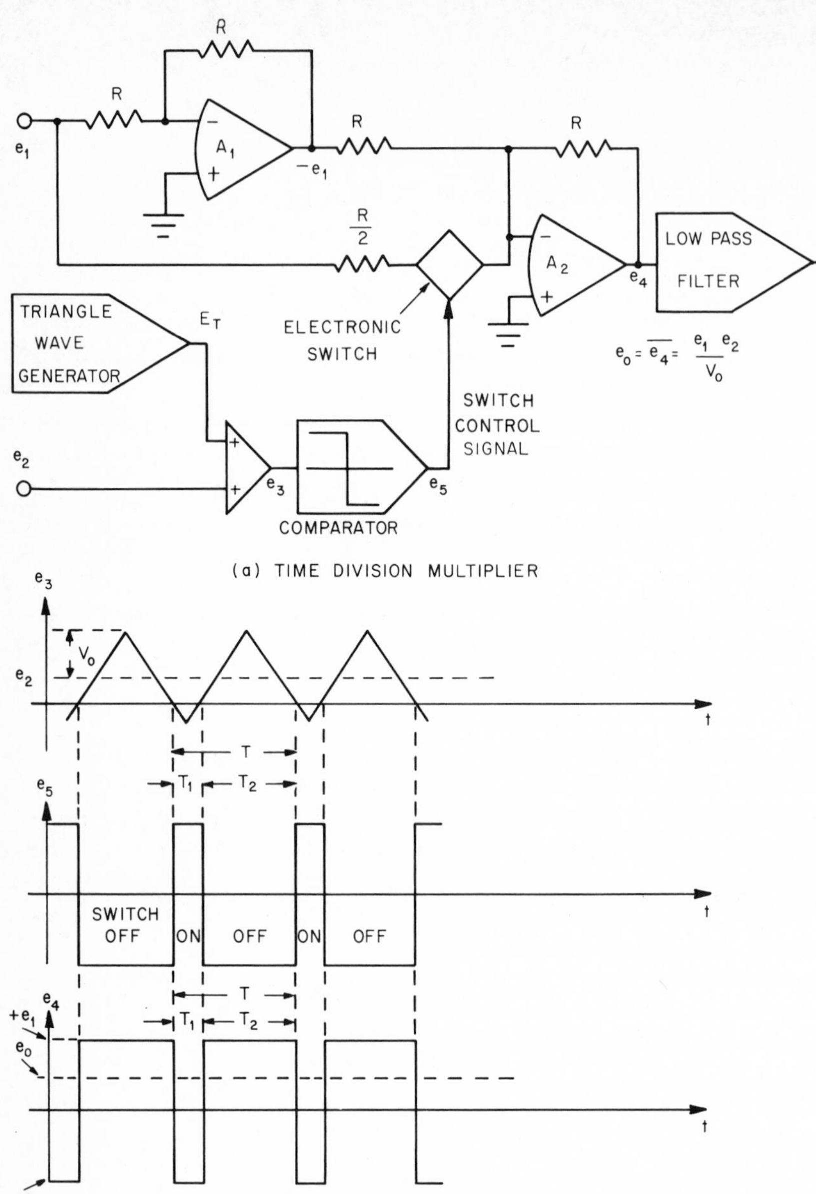

Fig. 7.43 Illustration of the time-division multiplier. (a) Time-division multiplier; (b) time-division multiplier waveforms.

this waveform is averaged by a low-pass filter, the result is equal to the product, which a scaling factor.

$$e_o = \bar{e}_4 = e_1 \frac{e_2 + V_o}{2V_o} - e_1 \frac{V_o - e_2}{2V_o}$$

$$e_o = \frac{e_1 e_2}{V_o}$$

The time-division multiplier suffers from much the same problems as the triangle-averaging type. Accuracy depends strongly upon the linearity, symmetry, and "sharpness" of the triangle wave. The resistors used in the feedback networks of A_1 and A_2 must be precisely matched, taking into account the series resistance of the switch. Offset voltage of the comparator will appear as an error term added to e_2. The switching time for e_4 to change from $+e_1$ to $-e_1$ is a critical error factor and must be small compared with the period T. This places a stringent limit on the upper frequency of the carrier and thus on the frequency response of

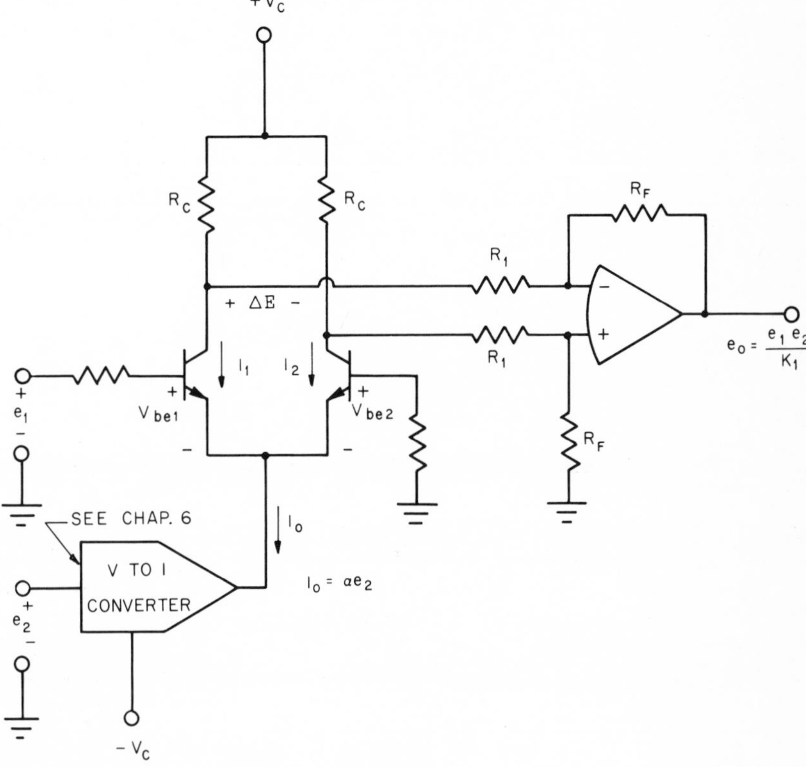

Fig. 7.44 Variable transconductance multiplier.

the multiplier. There are numerous variations of the triangle-averaging and time-division multiplier circuits. However, the examples given illustrate the techniques and some of the design limitations.

7.5.5 Variable transconductance multiplier Perhaps the simplest multiplication technique is the variable transconductance method illustrated in Fig. 7.44. This method depends upon the current through the matched pair of transistors being proportional to one of the input signals, e_2. Assuming that the transistors are a perfectly matched pair, the differential collector current (and consequently the differential collector voltage) is proportional to the product of e_1 and e_2. The result is derived as follows:

$$I_1 = I_S e^{\frac{qV_{be1}}{kT}}$$

$$\frac{\Delta I_1}{\Delta V_{be1}} = \frac{q}{kT} I_1$$

$$I_o = I_1 + I_2 = 2I_S e^{\frac{qV_{be1}}{kT}}$$

$$\Delta I_1 = \frac{q}{2kT} I_o \, \Delta V_{be1}$$

$$\Delta I_2 = \frac{q}{2kT} I_o \, \Delta V_{be2}$$

$$\Delta I_1 - \Delta I_2 = \frac{q}{2kT} I_o \overbrace{(\Delta V_{be1} - \Delta V_{be2})}^{e_1}$$

$$\Delta E = R(\Delta I_1 - \Delta I_2) = R_c \frac{q}{2kT} \alpha e_2 e_1$$

$$e_o = \frac{R_o}{R_1} R_c \frac{q}{2kT} \alpha e_1 e_2 = \frac{e_1 e_2}{K_1}$$

The differential input operational amplifier provides proper scaling and conversion to a single-ended output. Because of the extreme temperature sensitivity of this method of multiplication, it is of limited usefulness. Both the scale factor and the dc level will tend to drift, the latter because of unavoidable mismatch between the multiplying transistors. The linearity is also rather poor and ac feedthrough is appreciable. The ac feedthrough is measured by grounding one of the inputs and applying a sine wave to the other input. The output should be zero but actually contains a component of the input sine wave. This is particularly true when e_2 is grounded and the ac signal is applied at e_1. The variable transconductance method is important chiefly because of its relation to the current-rationing method to be described next. This method, the basic principles of which were first described by Gilbert,[17] is a vast improvement over the variable transconductance method although it uses the same basic relationships of the semiconductor junction.

7.5.6 Current ratioing multiplier One realization of the *current ratioing multiplier* is shown in Fig. 7.45a. The heart of this multiplier is the gain cell shown in Fig. 7.45b. This device ensures that the currents i_3, i_4 in the collectors of transistors Q_3, Q_4 remain in a constant ratio equal to the ratio of the external currents I_7 and I_8. The currents I_1, I_7, and I_8 are generated by constant current sources. The currents and voltages of the gain cell are related by the equations

$$I_7 = K_1 e^{\alpha_1 V_{d1}}$$
$$I_8 = K_2 e^{\alpha_2 V_{d2}}$$
$$I_3 = K_3 e^{\alpha_3 V_{be3}}$$
$$I_4 = K_4 e^{\alpha_4 V_{be4}}$$

where

$$\alpha = \frac{q}{kT}$$

If the transistors and diodes are matched to make α's equal and K's equal, then

$$\frac{I_7}{I_8} = e^{\alpha(V_{d1} - V_{d2})}$$

and

$$\frac{I_4}{I_3} = e^{\alpha(V_{be4} - V_{be3})}$$

The loop equation can be written

$$V_{d1} + V_{be3} = V_{be4} + V_{d2}$$

or

$$V_{d1} - V_{d2} = V_{be4} - V_{be3}$$

and, if this is substituted into the expression for I_4/I_3, the result is

$$\frac{I_7}{I_8} = \frac{I_3}{I_4}$$

In the multiplier circuit of Fig. 7.45a, the gain cell concept is used to enforce the conditions

$$\frac{I_4}{I_3} = \frac{I_8}{I_7} \quad \text{and} \quad \frac{I_6}{I_5} = \frac{I_7}{I_8}$$

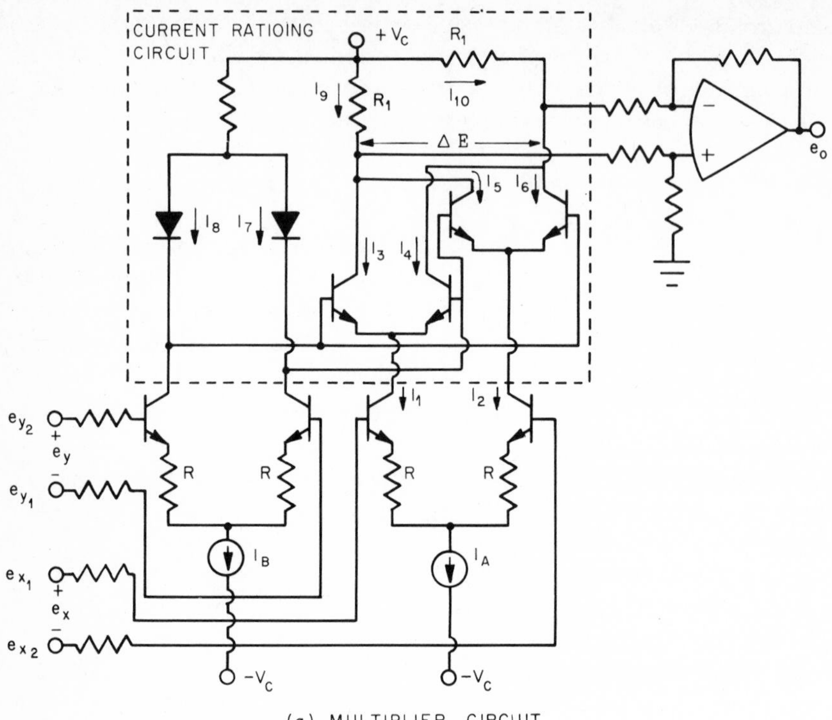

(a) MULTIPLIER CIRCUIT

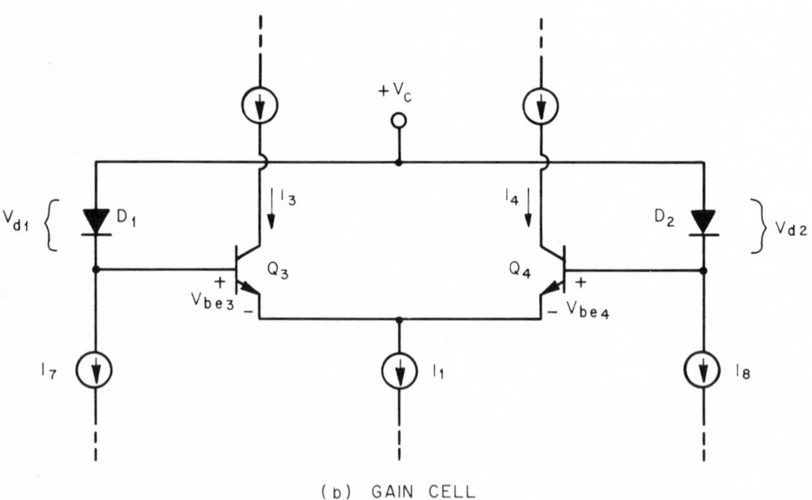

(b) GAIN CELL

Fig. 7.45 Current ratioing multiplier. (a) Multiplier circuit; (b) gain cell.

Other necessary relations are

$$I_1 = I_3 + I_4 \qquad I_9 = I_3 + I_5$$
$$I_2 = I_5 + I_6 \qquad I_{10} = I_6 + I_4$$
$$I_1 + I_2 = I_A \qquad I_7 + I_8 = I_B$$
$$e_X = R(I_1 - I_2) \qquad e_Y = R(I_8 - I_7)$$

Combining the above equations and using a considerable amount of simple algebra, the relationship for ΔE is obtained:

$$\Delta E = R_1(I_9 - I_{10}) = \frac{(-e_Y/R)(+e_X/R)}{I_B} R_1$$

With constant I_B and proper scaling the output voltage is

$$E_o = \frac{(e_{X_1} - e_{X_2})(e_{Y_1} - e_{Y_2})}{10}$$

Accurate multiplication requires that the transistors used be dynamically matched, a requirement that makes monolithic construction attractive for this type of multiplier. However, it has been found possible to achieve 1 percent accuracy of multiplication through the use of carefully matched discrete transistors.

The current ratioing multiplier has several desirable features which give it the potential for widespread use. These are:

1. Good linearity
2. Wide bandwidth
3. Differential input
4. Stability with temperature
5. Low ac feedthrough
6. Low cost

7.5.7 Analog dividers Any of the multipliers discussed in the preceding pages can be used as analog dividers by using the feedback circuits of Fig.

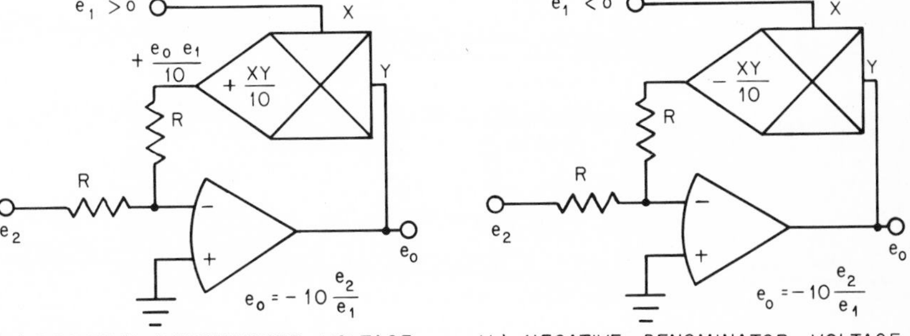

Fig. 7.46 Analog division techniques. (a) Positive denominator voltage; (b) negative denominator voltage.

7.46a and b. Note that only two-quadrant operation is possible because the voltage e_4 must be of opposite polarity to e_1. For $e_1 < 0$, the multiplier must provide polarity reversal whereas for $e_1 > 0$ the multiplier must generate $+ e_1e_o/10 = -e_2$ to ensure stable operation (negative feedback). The principal limitation of such feedback dividers is the large error term as $e_2 \to 0$. This error term severely limits the dynamic range of the divider, especially where the error of the multiplier may have its largest value when the input signals are small (such as in the quarter-square multiplier). The best multipliers for use in feedback division are those whose error curve passes smoothly through the origin (such as the triangle-averaging and current-ratioing types).

7.5.8 Squarers and square rooters One of the most obvious applications of an analog multiplier is for computing the square of a signal voltage. Such calculations are quite common in power measurement, rms level measurement, and in computations of vector magnitude. Figure 7.47 illustrates such an application. The square-root function is obtained by using the multiplier as the feedback element, such as in the case of the divider circuit. The operational amplifier enforces the conditions

$$\frac{e_1^2}{10}\frac{1}{2R} + \frac{e_2^2}{10}\frac{1}{2R} = \frac{e_o^2}{10}\frac{1}{R}$$

$$e_o^2 = \frac{e_1^2}{2} + \frac{e_2^2}{2}$$

$$e_o = \sqrt{e_1^2 + e_2^2}$$

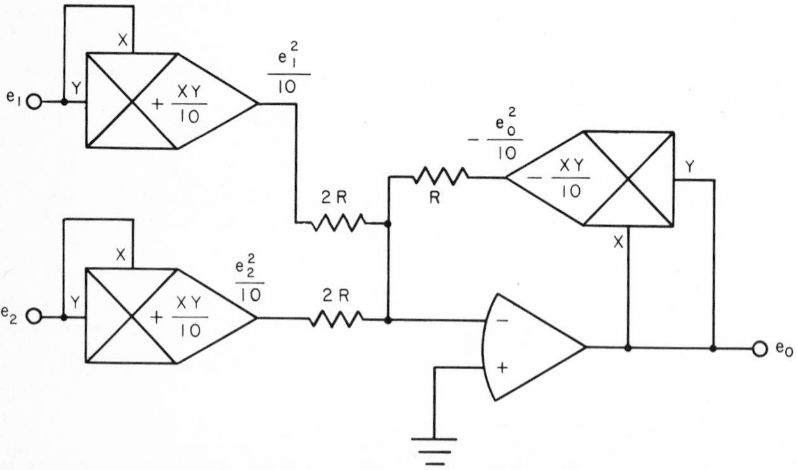

Fig. 7.47 Multipliers used in vector magnitude computation.

REFERENCES

1. G. A. Korn and T. M. Korn, *Electronic Analog and Hybrid Computers*, McGraw-Hill Book Company, New York, 1964.
2. *Applications Manual for Operational Amplifiers*, Philbrick/Nexus Research, Dedham, Mass., 1965.
3. V. V. Abrahanian, An Improved Square Root Circuit, *Instrum. Control Syst.*, April, 1963.
4. F. R. Bradley, and R. P. McCoy, Voltage-limiting Circuit, *Electronics*, May, 1955.
5. R. M. Howe, Representation of Nonlinear Functions by Means of Operational Amplifiers, *IRE, Trans. Electron. Computers*, December, 1956.
6. R. J. Medkeff, and R. J. Parent, A Diode Bridge Limiter for Use with Electronic Analog Computers, *Proc. AIEE*, vol. 70. sec. T-1-70, 1951.
7. S. Godet, The Gated Amplifier Computer Technique, *Proc. Natl. Simultation Conf.*, Dallas, Texas, 1956.
8. The Use of Silicon-junction Diodes in Analog Simulation, *Instrum. Control Sys.*, August, 1961.
9. H. E. Koerner and G. A. Korn, Function Generation with Operational Amplifiers, *Electronics*, Nov. 6, 1959.
10. H. Hamer, Optimum Linear-segment Function Generator, *Trans. AIEE*, vol. 75, sec. 1, p. 518, 1956.
11. E. A. Goldberg, A High Accuracy Time-division Multiplier, *RCA Rev.*, vol. 23, p. 265, September 1952.
12. C. D. Morrill and R. V. Baum, A Stabilized Electronic Multiplier, *IRE Trans. Electron. Computers*, December, 1952.
13. S. Giser, All-electronic High-speed Multiplier, *MIT Instrumentation Lab. Rep. R*-67, November, 1953.
14. R. L. Mills, A New Electronic Multiplication Method Involving Only Simple Conventional Circuits, *Magnolia Petroleum Co. Field Res. Lab. Rep.* 680 (00)-4, Dallas, Texas, Nov. 9, 1953.
15. T. S. Gray and H. B. Frey, Acorn Diode Has Logarithmic Range of 10^9, *Rev. Sci. Instrum.*, February, 1951.
16. C. D. Morrill and R. V. Baum, Diode Limiters Simulate Mechanical Phenomena, *Electronics*, November, 1952.
17. B. Gilbert, A Precise Four-quadrant Multiplier with Subnanosecond Response, *IEEE J. Solid State Circuits*, December, 1968, p. 210.
18. T. Cate, Designing with Nonlinear Function Modules, *EEE*, September, 1969.
19. T. Cate, Designing with Packaged Analog Multipliers, *EEE*, May, 1969.

8

ACTIVE FILTERS

The operational amplifier, especially the integrated-circuit operational amplifier, proves to be an extremely useful active device in the realization of active RC networks. Operational amplifiers have high input impedance, low output impedance, large open-loop gain, and low cost. These qualities are used to advantage in the circuits to be discussed in this chapter. Enough has been discussed about operational amplifiers in the first chapters of this book so that we will not dwell on their properties here.

We will begin our discussion of active filters in Sec. 8.1 by making statements about active filters in general. Then in Sec. 8.2 we will discuss transfer functions and their parameters. Useful formulas are presented to help evaluate the effects of tolerances and temperature coefficients of resistors and capacitors. In Sec. 8.3 we will then describe several realizations and provide design equations and sensitivity equations. The basic sensitivity relations are derived and discussed in Appendix C. After we have become familiar with the circuit realizations, we will discuss tuning (Sec. 8.4), how operational amplifier characteristics affect filter performance (Sec. 8.5), and, briefly, the characteristics of resistors and capacitors (Sec. 8.6). The chapter concludes with a set of filter design and tuning tables (Sec. 8.7).

8.1 Active Filter Characteristics

The active element, in this case the operational amplifier, in active networks is necessary to permit the realization of complex left hand plane poles using only resistors and capacitors for the passive elements. The operational amplifier permits the use of reasonable-valued resistors and capacitors even at frequencies as low as 10^{-3} Hz. An added bonus is the isolation afforded by the low output impedances of individual stages so that network stages can be designed and tuned independently with minimal interaction. Other active elements, the negative immittance converter and the gyrator, can be implemented with operational amplifiers but for practical reasons are not widely used.

Active filters have some characteristics of their own that make them sufficiently different from passive filters that one who uses them must be aware of these differences. For example, active filters usually have single-ended inputs and outputs and thus do not "float" with respect to the system power supply or common as a passive RLC network can. Amplifiers used for active elements have a limited input and output voltage range (± 10 V for most operational amplifier circuits) and an output current capability of a few milliamperes.

The outputs of active filters built with operational amplifiers have a dc voltage offset which drifts with ambient temperature changes. The voltage offset might range from a few microvolts up to several hundred millivolts. Drifts may range from 1 to 100 μV/°C or even more from a multiple-pole low-pass filter built up of many pole-pair stages. The inputs of active filters may have a bias current; this would be true for low-pass and band-reject filters and may be true for bandpass and high-pass filters, depending upon the particular circuit realization. The bias current may range from a few picoamperes for field-effect transistor operational amplifiers to a few microamperes for bipolar transistor and integrated-circuit amplifiers.

Active filters can provide excellent isolation capabilities, that is, a high input impedance ranging from a few kilohms to several thousand megohms if input buffer amplifiers are used, and a low output impedance ranging from a few hundred ohms down to less than 1 Ω. Unity-gain bandwidths as high as 100 MHz are available in operational amplifiers and permit filter designs in the vicinity of 1 Mc. Slewing rate, which is related to full-power response, is the limiting factor for large-signal characteristics. Frequencies as low as 10^{-3} Hz are possible, but filters at these frequencies can become rather bulky because of capacitor sizes. Active filters can have voltage gain, as much as 40 dB in low-frequency low-Q filters.

The primary advantage of active filters is their small size and weight for low-frequency applications and their ruggedness.

All types of responses are possible: the old standards, Butterworth, Chebyshev, and Bessel (Thompson), single-tuned and stagger-tuned bandpass as well as other responses that meet special needs.

The range of Q's possible for active filters extends up to Q's of a few hundred. However, high-Q networks capable of maintaining stability of their characteristics, in the face of element changes with time, temperature, voltage, and frequency, require more expensive (and usually larger size) resistors and capacitors and generally more operational amplifiers than low-Q (less than 10) filters. These facts will become apparent as individual circuits are discussed.

8.2 Pole Pairs, Network Functions, and Parameters

The circuits to be described realize a single-pole or a single complex pole pair. More complicated filters are then built up from these individual building blocks. This approach permits ease of design and tuning of a complex filter, an important practical matter, by reducing interactions between elements. This approach also permits a single systematic approach to answering the question: What happens to the response of a filter if the network element values are not accurate and if they drift with time and temperature?

The filter network functions that are of most interest are magnitude, phase, and group delay. The network parameters that are important are some characteristic frequency, Q, and passband gain. In this section these functions and parameters will be briefly examined for single-pole and complex-pole-pair networks for low-pass, high-pass, and bandpass networks. These relations will all be useful in the next section for deriving the sensitivity functions of these network realizations.

8.2.1 Low-pass network functions

Single Pole. The single-pole low-pass transfer function in the complex frequency variables is

$$H(s) = \frac{H_o \omega_o}{s + \omega_o}$$

The magnitude of the transfer function for the response to sinusoidal steady-state excitation is

$$|H(j\omega)| = G(\omega) = \left(\frac{H_o{}^2 \omega_o{}^2}{\omega^2 + \omega_o{}^2}\right)^{1/2}$$

The phase is

$$\phi(\omega) = -\arctan\frac{\omega}{\omega_o}$$

and the group delay is

$$\tau(\omega) = -\frac{d\phi(\omega)}{d\omega} = \frac{\cos^2\phi}{\omega_o}$$

Complex Conjugate Pole Pair. The complex-conjugate-pole-pair low-pass transfer function and the sinusoidal steady-state magnitude and phase functions are

$$H(s) = \frac{H_o\omega_o^2}{s^2 + \alpha\omega_o s + \omega_o^2}$$

$$|H(j\omega)| = G(\omega) = \left[\frac{H_o^2\omega_o^4}{\omega^4 + \omega^2\omega_o^2(\alpha^2 - 2) + \omega_o^4}\right]^{1/2}$$

$$\phi(\omega) = -\arctan\left[\frac{1}{\alpha}\left(2\frac{\omega}{\omega_o} + \sqrt{4-\alpha^2}\right)\right] - \arctan\left[\frac{1}{\alpha}\left(2\frac{\omega}{\omega_o} - \sqrt{4-\alpha^2}\right)\right]$$

The relation for phase given above is expressed in a form suitable for general computer use since, on many computers, the arctan function can be determined only for the principal angle. Note that α^2 is usually never greater than 4. If it is, the poles will no longer be complex. The Q of a complex pole pair equals $1/\alpha$.

The group delay for a complex conjugate pole pair is

$$\tau(\omega) = \frac{2\sin^2\phi}{\alpha\omega_o} - \frac{\sin 2\phi}{2\omega}$$

8.2.2 High-pass network functions

Single Pole. The single-pole high-pass transfer function and the sinusoidal steady-state magnitude, phase, and delay functions are

$$H(s) = \frac{H_o s}{s + \omega_o}$$

$$G(\omega) = \left(\frac{H_o^2\omega^2}{\omega^2 + \omega_o^2}\right)^{1/2}$$

$$\phi(\omega) = \frac{\pi}{2} - \arctan\frac{\omega}{\omega_o}$$

$$\tau(\omega) = \frac{\sin^2\phi}{\omega_o}$$

Complex Conjugate Pole Pair. The complex-conjugate-pole-pair high-pass transfer function and the sinusoidal steady-state magnitude, phase, and delay functions are

$$H(s) = \frac{H_o s^2}{s^2 + \alpha\omega_o s + \omega_o{}^2}$$

$$G(\omega) = \left[\frac{H_o{}^2\omega^4}{\omega^4 + \omega^2\omega_o{}^2(\alpha^2 - 2) + \omega_o{}^4}\right]^{1/2}$$

$$\phi(\omega) = \pi - \arctan\left[\frac{1}{\alpha}\left(2\frac{\omega}{\omega_o} + \sqrt{4 - \alpha^2}\right)\right] - \arctan\left[\frac{1}{\alpha}\left(2\frac{\omega}{\omega_o} - \sqrt{4 - \alpha^2}\right)\right]$$

$$\tau(\omega) = \frac{2\sin^2\phi}{\alpha\omega_o} - \frac{\sin 2\phi}{2\omega}$$

8.2.3 Bandpass network function The complex-conjugate-pole-pair bandpass transfer function is

$$H(s) = \frac{H_o\alpha\omega_o s}{s^2 + \alpha\omega_o s + \omega_o{}^2}$$

where

$$\alpha = \frac{1}{Q} \qquad \text{and} \qquad Q = \frac{\omega_o}{\omega_2 - \omega_1} = \frac{f_o}{f_2 - f_1}$$

and where f_2 and f_1 are the frequencies where the magnitude response is -3 dB from H_o, the passband gain which occurs at $\omega_o = 2\pi f_o$. The sinusoidal steady-state transfer function may be written in the form

$$H(j\omega) = \frac{H_o}{1 + jQ(\omega/\omega_o - \omega_o/\omega)}$$

Thus, the magnitude, phase, and delay functions are

$$G(\omega) = \left[\frac{H_o{}^2}{1 + Q^2(\omega/\omega_o - \omega_o/\omega)^2}\right]^{1/2} = \left[\frac{H_o{}^2\alpha^2\omega_o{}^2\omega^2}{\omega^4 + \omega^2\omega_o{}^2(\alpha^2 - 2) + \omega_o{}^4}\right]^{1/2}$$

$$\phi(\omega) = \frac{\pi}{2} - \arctan\left(\frac{2Q\omega}{\omega_o} + \sqrt{4Q^2 - 1}\right) - \arctan\left(2Q\frac{\omega}{\omega_o} - \sqrt{4Q^2 - 1}\right)$$

$$\tau(\omega) = \frac{2Q\cos^2\phi}{\omega_o} + \frac{\sin 2\phi}{2\omega}$$

8.2.4 Band-reject network function A band-reject filter can be realized by performing the operation $1 - H_{BP}(s)$, where $H_{BP}(s)$ is a bandpass transfer function (Fig. 8.1). If $R' = RH_o$ (bandpass)

$$H(s) = -\frac{s^2 + \omega_o^2}{s^2 + \alpha\omega_o s + \omega_o^2}\frac{R_F}{R}$$

Since this filter is very closely related to the bandpass filter, its properties will not be discussed.

8.3 Filter Realizations[1,2]

In this section we shall present some realizations for active filters. The operational amplifier filter circuits to be analyzed, and for which design procedures and sensitivity equations are given, are the infinite-gain multiple feedback, controlled source, infinite-gain state-variable feedback, and negative immittance converter realizations. Another realization sometimes used is the infinite-gain single-feedback type. This type involves the use of bridged-T and twin-T networks as well as requiring the cancellation of unwanted zeros and poles. Thus this type requires many elements to realize a transfer function with complex poles and is therefore uneconomical. Trimming and adjustment of bridged-T or twin-T networks is difficult since the passive elements interact to a high degree in such networks. For these reasons this circuit will not be discussed. Single real pole realizations will not be shown since these are rather trivial and easy to design. In addition, single-operational-amplifier single-pole circuits are rather uneconomical.

Design procedures given in this section are only suggested procedures. Other choices are possible, and as one gains experience with these circuits, it becomes desirable to design procedures for minimizing sensitivity in certain network parameters or to ensure a convenient spread of element values. In the design procedures given, the capacitors are always made equal. In addition, one usually starts the design process by selecting the capacitor value because there are fewer standard values of capacitors than there are resistors. Resistors are less expensive than capacitors and are more easily used in trimming schemes. In some cases the passband gain

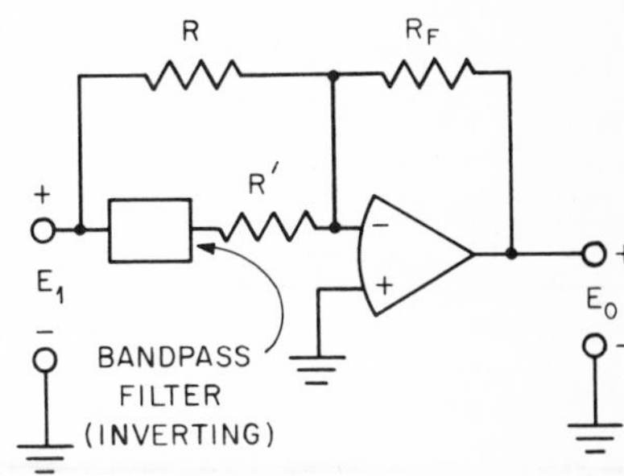

Fig. 8.1 Band-reject filter.

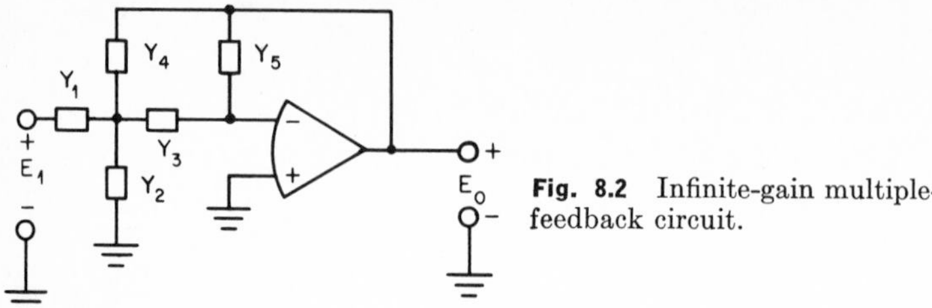

Fig. 8.2 Infinite-gain multiple-feedback circuit.

H_o is a free parameter. It is often convenient to let H_o be a variable which can be used as a parameter for determining optimum, or at least small, sensitivities of certain parameters. An example of this is given in the controlled-source circuit designs. Also, letting H_o be a free parameter simplifies complicated design equations.

8.3.1 Infinite-gain multiple-feedback circuits Figure 8.2 illustrates the infinite-gain multiple-feedback connection for a pair of complex conjugate s-plane poles with zeros restricted to the origin or infinity. The amplifier is used in its inverting configuration, with the + input grounded. Each element Y_i represents a single resistor or capacitor. The voltage transfer function is

$$\frac{E_o}{E_1}(s) = \frac{-Y_1Y_3}{Y_5(Y_1+Y_2+Y_3+Y_4)+Y_3Y_4 + (1/A_{OL})[(Y_3+Y_5)(Y_1+Y_2+Y_4)+Y_3Y_5]}$$

In the limiting case as A_{OL} approaches infinity we obtain

$$\frac{E_o}{E_1}(s) = \frac{-Y_1Y_3}{Y_5(Y_1+Y_2+Y_3+Y_4)+Y_3Y_4}$$

Examples that follow show how these five elements may be chosen so as to realize low-pass, high-pass and bandpass network functions.

Low Pass. The infinite-gain multiple-feedback circuit for a low-pass network function is shown in Fig. 8.3. The voltage transfer function is

$$\frac{E_o}{E_1}(s) = \frac{-1/R_1R_3C_2C_5}{s^2+(s/C_2)(1/R_1+1/R_3+1/R_4)+1/R_3R_4C_2C_5}$$

Note that this circuit produces a signal inversion, as will all circuits realized by this technique.

For this circuit, following the notation of the low-pass network function,

$$H_o = \frac{R_4}{R_1}$$

$$\omega_o = \left(\frac{1}{R_3R_4C_2C_5}\right)^{1/2}$$

$$\alpha = \sqrt{\frac{C_5}{C_2}}\left(\sqrt{\frac{R_3}{R_4}} + \sqrt{\frac{R_4}{R_3}} + \frac{\sqrt{R_3R_4}}{R_1}\right)$$

$$\phi = \pi + \phi_{LP}$$

$$\tau = \tau_{LP}$$

Note that the phase inversion has been incorporated into the phase function. A tuning procedure for this circuit would be first to adjust ω_o with R_3 at a frequency of 10ω, as outlined in the section on tuning. Then adjust α with R_1 at the α peaking frequency.

The sensitivities of the network parameters to circuit element changes follow. Remember that the open-loop gain of the operational amplifier is assumed to be infinite (at least very large), and so sensitivity functions for open-loop gain changes are not considered.

$$S_{R_3}^{\omega_o} = S_{R_4}^{\omega_o} = S_{C_2}^{\omega_o} = S_{C_5}^{\omega_o} = -\frac{1}{2}$$

$$S_{C_5}^{\alpha} = -S_{C_2}^{\alpha} = \frac{1}{2}$$

$$S_{R_1}^{\alpha} = \frac{1}{\alpha\omega_o R_1 C_2}$$

$$S_{R_3}^{\alpha} = \frac{1}{2} - \frac{1}{\alpha\omega_o R_3 C_2}$$

$$S_{R_4}^{\alpha} = \frac{1}{2} - \frac{1}{\alpha\omega_o R_4 C_2}$$

$$S_{R_4}^{H_o} = -S_{R_1}^{H_o} = 1$$

Note that $S_{C_2}^{\alpha}$ and $S_{C_5}^{\alpha}$ are constant and opposite in sign and so are $S_{R_1}^{H_o}$ and $S_{R_4}^{H_o}$.

DESIGN PROCEDURE

Given: H_o, α, $\omega_o = 2\pi f_o$

Choose: $C_2 = C$, a convenient value

$C_5 = KC$

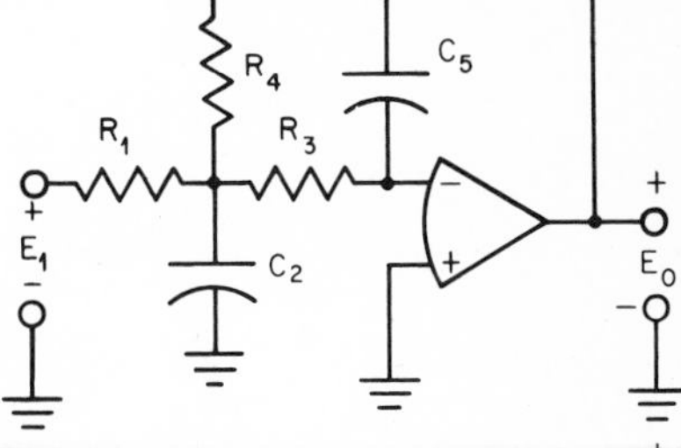

Fig. 8.3 Multiple-feedback low-pass filter.

$$\text{Calculate: } R_4 = \frac{\alpha}{2\omega_o C}\left[1 \pm \sqrt{1 - \frac{4(H_o + 1)}{K\alpha^2}}\right]$$

$$R_1 = \frac{R_4}{H_o}$$

$$R_3 = \frac{1}{\omega_o{}^2 C^2 R_4 K}$$

For best results H_o should be less than 10 for circuits with an α of about 0.1 ($Q = 10$) and can be as high as 100 for an α of about 1 ($Q = 1$) or less. These extreme limits assume that the operational amplifier has an open-loop gain of at least 80 dB at the frequency of interest. The effects of finite open-loop gain for multiple feedback circuits will be discussed later.

High Pass. A high-pass realization is illustrated in Fig. 8.4. The voltage transfer function is

$$\frac{E_o}{E_1}(s) = \frac{-(C_1/C_4)s^2}{s^2 + s(1/R_5)(C_1/C_3C_4 + 1/C_4 + 1/C_3) + 1/R_2R_5C_3C_4}$$

In terms of our high-pass network function

$$H_o = \frac{C_1}{C_4}$$

$$\omega_o = \left(\frac{1}{R_2R_5C_3C_4}\right)^{1/2}$$

$$\alpha = \sqrt{\frac{R_2}{R_5}}\left(\frac{C_1}{\sqrt{C_3C_4}} + \sqrt{\frac{C_3}{C_4}} + \sqrt{\frac{C_4}{C_3}}\right)$$

$$\phi = \pi + \phi_{HP}$$

$$\tau = \tau_{HP}$$

Tuning this high-pass filter will have to be done in the reverse order to that of the low-pass filter. First, adjust α with R_2 or R_5 at the frequency where the α peak occurs (the ω_α frequency is not known

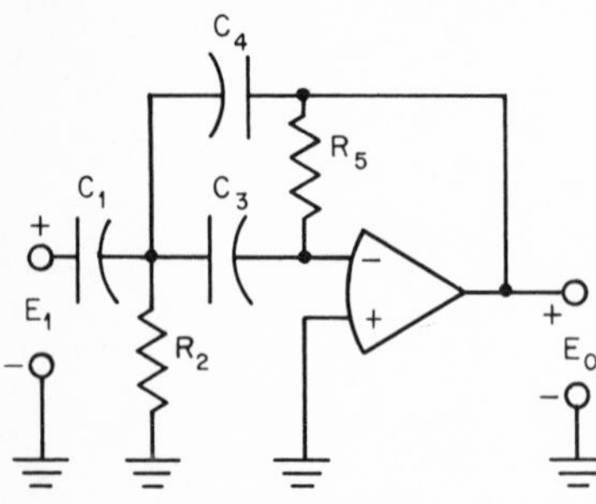

Fig. 8.4 Multiple-feedback high-pass filter.

because ω_o has not been set yet). Then adjust ω_o by adjusting R_2 and R_5 simultaneously by the *same* percentage: α will remain constant. A trimming scheme involving C_1 would be simpler. The sensitivities to element value changes are

$$S_{R_2}^{\omega_o} = S_{R_5}^{\omega_o} = S_{C_3}^{\omega_o} = S_{C_4}^{\omega_o} = -\frac{1}{2}$$

$$S_{R_2}^{\alpha} = -S_{R_5}^{\alpha} = \frac{1}{2}$$

$$S_{C_3}^{\alpha} = \frac{1}{2} - \frac{1}{\alpha\omega_o R_5 C_3}\left(\frac{C_1}{C_3} + 1\right)$$

$$S_{C_4}^{\alpha} = \frac{1}{2} - \frac{1}{\alpha\omega_o R_5 C_4}\left(\frac{C_1}{C_3} + 1\right)$$

$$S_{C_1}^{\alpha} = \frac{1}{\alpha\omega_o R_5}\frac{C_1}{C_3 C_4}$$

$$S_{C_1}^{H_o} = -S_{C_4}^{H_o} = 1$$

DESIGN PROCEDURE

Given: H_o, α, $\omega_o = 2\pi f_o$

Choose: $C = C_1 = C_3$, a convenient value

Calculate: $R_5 = \dfrac{1}{\alpha\omega_o C}(2H_o + 1)$

$$R_2 = \frac{\alpha}{\omega_o C(2H_o + 1)}$$

$$C_4 = \frac{C_1}{H_o}$$

Again, restrictions on H_o are the same as those for the low-pass case. Note that this realization requires three capacitors, a feature which might make it undesirable when compared with other circuits.

Bandpass 1. There are several configurations of the five elements which may be used to realize a bandpass function. One of the more practical configurations is the one shown in Fig. 8.5. The voltage transfer function is

$$\frac{E_o}{E_1}(s) = \frac{-s(1/R_1C_4)}{s^2 + s(1/R_5)(1/C_3 + 1/C_4) + (1/R_5C_3C_4)(1/R_1 + 1/R_2)}$$

In terms of our bandpass network function

$$H_o = \frac{1}{(R_1/R_5)(1 + C_4/C_3)}$$

$$\omega_o = \left[\frac{1}{R_5C_3C_4}\left(\frac{1}{R_1} + \frac{1}{R_2}\right)\right]^{1/2}$$

$$\frac{1}{Q} = \alpha = \sqrt{\frac{1}{R_5(1/R_1 + 1/R_2)}}\left[\sqrt{\frac{C_3}{C_4}} + \sqrt{\frac{C_4}{C_3}}\right]$$

$$\phi = \pi + \phi_{BP}$$

$$\tau = \tau_{BP}$$

Tuning this filter appears rather formidable. In practice $R_1 \gg R_2$ and so R_2 can be used to trim the Q. Then, to adjust the center frequency, R_2 and R_5 can be simultaneously adjusted by the same percentage with negligible effect on the Q.

The sensitivities of the network parameters with respect to the elements are

$$S_{R_5}^{\omega_o} = S_{C_3}^{\omega_o} = S_{C_4}^{\omega_o} = -\frac{1}{2}$$

$$S_{R_1}^{\omega_o} = \frac{-1}{2\omega_o^2R_1R_5C_3C_4}$$

$$S_{R_2}^{\omega_o} = \frac{-1}{2\omega_o^2R_2R_5C_3C_4}$$

$$S_{R_1}^{Q} = \frac{R_1}{2(R_1 + R_2)} - \frac{1}{2}$$

$$S_{R_2}^{Q} = \frac{R_2}{2(R_1 + R_2)} - \frac{1}{2}$$

$$S_{R_5}^{Q} = \frac{1}{2}$$

$$S_{C_3}^{Q} = \frac{Q}{\omega_oR_5C_3} - \frac{1}{2}$$

$$S_{C_4}^{Q} = \frac{Q}{\omega_oR_5C_4} - \frac{1}{2}$$

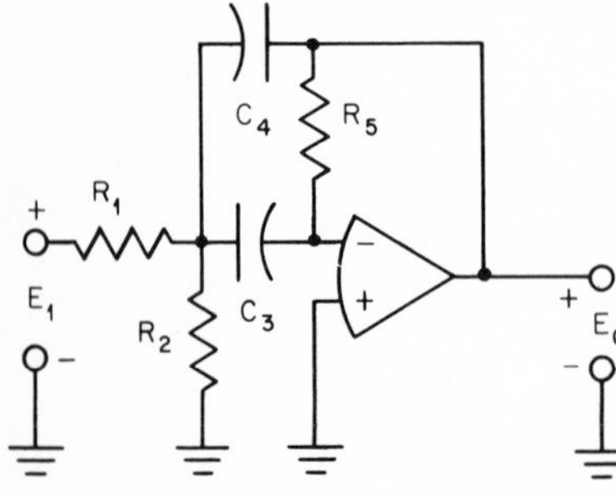

Fig. 8.5 Multiple-feedback bandpass filter.

DESIGN PROCEDURE

Given: H_o, $Q = \frac{1}{\alpha}$, $\omega_o = 2\pi f_o$

Choose: $C = C_3 = C_4$

Calculate: $Q = \frac{1}{\alpha}$

$$R_1 = \frac{Q}{H_o \omega_o C}$$

$$R_2 = \frac{Q}{(2Q^2 - H_o)\omega_o C}$$

$$R_5 = \frac{2Q}{\omega_o C}$$

Again, restrictions on H_o apply to guarantee that the design equations give fairly accurate results.

Bandpass 2. Another multiple-feedback circuit uses an additional active element to overcome some of the disadvantages of the single-amplifier circuit, especially the bandpass realization for Q's roughly between 10 and 50. High Q's realized with bandpass 1 have large spreads of element values and high Q sensitivities to element value changes. The multiple-feedback circuit with positive feedback is shown in Fig. 8.6. The voltage transfer function is

$$\frac{E_o}{E_1}(s) = \frac{s(K/R_1C_4)}{s^2 + (s/R_5C_4)(1 + C_4/C_3 - KR_5/R_6) + (1/C_3C_4R_5)(1/R_1 + 1/R_2 + 1/R_6)}$$

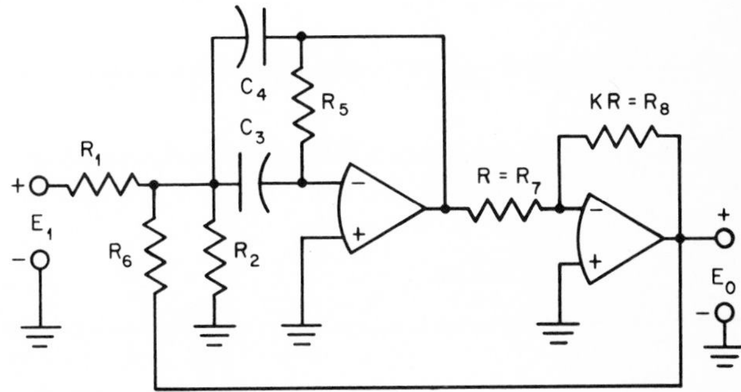

Fig. 8.6 Multiple-feedback bandpass circuit with positive feedback.

Note that the output is taken at the second amplifier. The overall signal transfer is noninverting. The circuit parameters are

$$H_o = \frac{1}{R_1}\frac{1}{(1/KR_5)(1 + C_4/C_3) - 1/R_6}$$

$$\omega_o = \left[\frac{1}{R_5C_3C_4}\left(\frac{1}{R_1} + \frac{1}{R_2} + \frac{1}{R_6}\right)\right]^{1/2}$$

$$\frac{1}{Q} = \alpha = \sqrt{\frac{1}{R_5(1/R_1 + 1/R_2 + 1/R_6)}}\sqrt{\frac{C_3}{C_4}}\left(1 + \frac{C_4}{C_3} - \frac{KR_5}{R_6}\right)$$

$$\phi = \phi_{BP}$$

$$\tau = \tau_{BP}$$

Since R_1 and R_6 are larger than R_2, R_2 is used to trim the center frequency. Note that in this circuit Q can be adjusted with K without influencing ω_o. The sensitivity of the network parameters to element value changes are

$$S_{R_1}^{H_o} = -1$$

$$S_{C_3}^{H_o} = -S_{C_4}^{H_o} = \frac{H_o}{K}\frac{R_1}{R_5}\frac{C_4}{C_3}$$

$$S_K^{H_o} = S_{R_5}^{H_o} = \frac{H_o}{K}\frac{R_1}{R_5}\left(1 + \frac{C_4}{C_3}\right)$$

$$S_{R_6}^{H_o} = -H_o\frac{R_1}{R_6}$$

$$S_{C_3}^{\omega_o} = S_{C_4}^{\omega_o} = S_{R_5}^{\omega_o} = -\frac{1}{2}$$

$$S_{R_1}^{\omega_o} = \frac{-1}{2\omega_o^2R_1R_5C_3C_4}$$

$$S_{R_2}^{\omega_o} = \frac{-1}{2\omega_o^2R_2R_5C_3C_4}$$

$$S_{R_6}^{\omega_o} = \frac{-1}{2\omega_o^2R_5R_6C_3C_4}$$

$$S_{R_1}^{Q} = \frac{-1}{2\omega_o^2R_1R_5C_3C_4}$$

$$S_{R_2}^{Q} = \frac{-1}{2\omega_o^2R_2R_5C_3C_4}$$

$$S_{R_6}^{Q} = -\frac{1}{2}\frac{1}{(1 + R_6/R_1 + R_6/R_2)} - \frac{1}{(R_6/KR_5)(1 + C_4/C_3) - 1}$$

$$S_{C_3}^{Q} = \frac{Q}{\omega_oR_5C_3} - \frac{1}{2}$$

$$S_{C_4}{}^{Q} = \frac{Q}{\omega_o R_5 C_4}\left(1 - \frac{KR_5}{R_6}\right) - \frac{1}{2}$$

$$S_{R_5}{}^{Q} = \frac{Q}{\omega_o R_5 C_4}\left(1 + \frac{C_4}{C_3}\right) - \frac{1}{2}$$

$$S_K{}^{Q} = \frac{-KQ}{\omega_o R_6 C_4}$$

$$S_{R_8}K = -S_{R_9}K = 1$$

DESIGN PROCEDURE

Given: $Q = 1/\alpha$, $\omega_o = 2\pi f_o$

H_o must be a free parameter.

Choose: $C = C_3 = C_4$, $R = R_1 = R_5$

K is chosen to reduce the spread of element values or to optimize sensitivity. It might typically be between 1 and 10.

Calculate: $R = \dfrac{Q}{\omega_o C}$

$$R_6 = R\,\frac{KQ}{2Q - 1}$$

$$G_2 = \frac{1}{R_2} = \frac{1}{R}\left(Q - 1 - \frac{2}{K} + \frac{1}{KQ}\right)$$

For this procedure, $H_o = \sqrt{Q}\,K$.

This completes the section on infinite-gain multiple-feedback realizations. A few general comments are in order. An advantage of this realization is that the output impedance is low; thus networks may be cascaded with negligible interaction. A disadvantage is that it is not possible to obtain high Q without resorting to large spreads of element values and also incurring large Q sensitivities. The multiple-feedback realization with positive feedback can overcome this and allow reasonable sensitivities up to a Q of 50.

8.3.2 Controlled-source circuits A noninverting voltage-controlled voltage source (VCVS) implemented with an operational amplifier is illustrated in Fig. 8.7. The input impedance is very large, tens to hundreds of thousands of megohms, depending upon the type of operational amplifier, and the output impedance is very low, usually less than 1 Ω for K between 1 and 10. The voltage transfer function is

$$\frac{E_o}{E_1}(s) = 1 + \frac{R_b}{R_a} = K$$

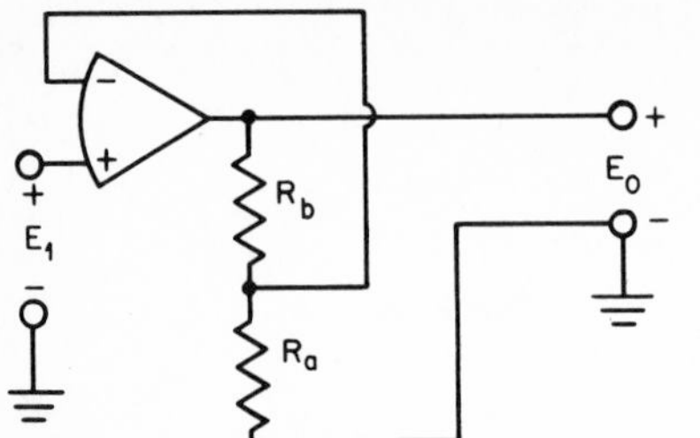

Fig. 8.7 Noninverting operational amplifier VCVS.

The sensitivities of K to the two resistors are

$$S_{R_b}{}^{K} = 1$$
$$S_{R_a}{}^{K} = -1$$

Figure 8.8 shows the controlled-source connection for a circuit which may be used to realize voltage transfer functions with a single pair of complex conjugate s-plane poles with zeros restricted to the origin or infinity. The Y_i are restricted to be single elements, R's and C's. These five elements may be chosen so as to realize low-pass, high-pass, and bandpass network functions. Realizations are possible with $K < 0$; but, since this operational amplifier circuit always has K greater than +1, these will not be discussed. The voltage transfer function is

$$\frac{E_o}{E_1}(s) = \frac{KY_1Y_4}{Y_5(Y_1 + Y_2 + Y_3 + Y_4) + [Y_1 + Y_2(1 - K) + Y_3]}$$

Low Pass. A VCVS circuit for a low-pass network function is shown in Fig. 8.9. The voltage transfer function is

$$\frac{E_o}{E_1}(s) = \frac{K/R_1R_2C_1C_2}{s^2 + s[1/R_1C_1 + 1/R_2C_1 + (1 - K)/R_2C_2] + 1/R_1R_2C_1C_2}$$

The network parameters are

$$H_o = K$$
$$\omega_{o'} = \left(\frac{1}{R_1R_2C_1C_2}\right)^{1/2}$$

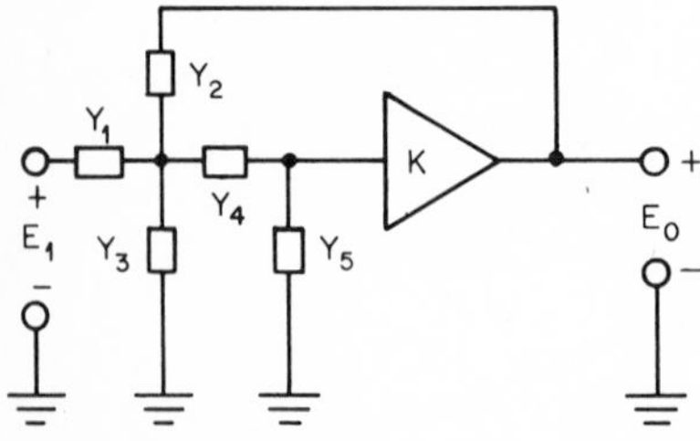

Fig. 8.8 VCVS configuration for a second-degree voltage transfer function.

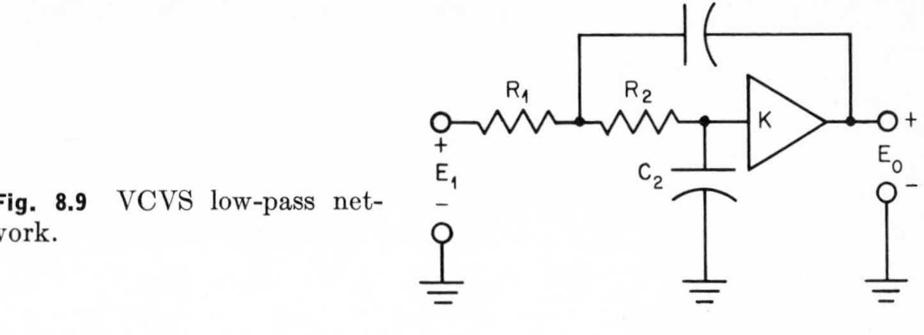

Fig. 8.9 VCVS low-pass network.

$$\alpha = \left(\frac{R_2C_2}{R_1C_1}\right)^{1/2} + \left(\frac{R_1C_2}{R_2C_1}\right)^{1/2} + \left(\frac{R_1C_1}{R_2C_2}\right)^{1/2} - K\left(\frac{R_1C_1}{R_2C_2}\right)^{1/2}$$

$$\phi = \phi_{LP}$$

$$\tau = \tau_{LP}$$

Controlled-source circuits are easier to tune than other circuit realizations. In fact, they can be adjusted over wide ranges without interaction of the network parameters. ω_o is tuned by adjusting R_1 and R_2 by equal percentages: α will not be affected. Capacitance C_1 and C_2 can be adjusted in the same way for the same result. α is trimmed by adjusting K. The sensitivities of the network parameters to element value changes are

$$S_{R_1}^{\omega_o} = S_{R_2}^{\omega_o} = S_{C_1}^{\omega_o} = S_{C_2}^{\omega_o} = -\frac{1}{2}$$

$$S_K^{H_o} = 1$$

$$S_{R_1}^{\alpha} = \frac{1}{2} - \frac{1}{\alpha\omega_o R_1 C_1}$$

$$S_{R_2}^{\alpha} = \frac{1}{2} - \frac{1}{\alpha\omega_o R_2}\left(\frac{1}{C_1} + \frac{1-K}{C_2}\right)$$

$$S_{C_1}^{\alpha} = \frac{1}{2} - \frac{1}{\alpha\omega_o C_1}\left(\frac{1}{R_1} + \frac{1}{R_2}\right)$$

$$S_{C_2}^{\alpha} = \frac{1}{2} - \frac{1-K}{\alpha\omega_o R_2 C_2}$$

$$S_K^{\alpha} = \frac{-K}{\alpha\omega_o R_2 C_2}$$

DESIGN PROCEDURE

Given: H_o, α, $\omega_o = 2\pi f_o$
Choose: $C_1 = C_2 = C$, a convenient value
Calculate: $K = H_o > 2$

$$R_2 = \frac{\alpha}{2\omega_o C}\left[1 + \sqrt{1 + \frac{4(H_o - 2)}{\alpha^2}}\right]$$

$$R_1 = \frac{1}{\omega_o{}^2 C^2 R_2}$$

If H_o is large, say greater than 10, there will be large spreads in element values and high sensitivities. An interesting design procedure is to use K to vary the sensitivities of circuit parameters.

Capacitors are often the components which have the largest temperature coefficients. It is possible to set K such that the overall α sensitivity is minimum, assuming that the capacitors drift equally. In this case we set $S_{C_1}{}^{\alpha} = -S_{C_2}{}^{\alpha}$.

Choose $C = C_1 = C_2$ and let $R_1 = R_2 = R$; then $K = 3 - \alpha$ and $R = 1/\omega_o C$.

High Pass. A VCVS circuit realization of a high-pass network function is shown in Fig. 8.10. The voltage transfer function is

$$\frac{E_o}{E_1}(s) = \frac{Ks^2}{s^2 + s[1/R_2C_1 + 1/R_2C_2 + (1 - K)/R_1C_1] + 1/R_1R_2C_1C_2}$$

The network parameters are

$$H_o = K$$

$$\omega_o = \left(\frac{1}{R_1R_2C_1C_2}\right)^{1/2}$$

$$\alpha = \left(\frac{R_1C_1}{R_2C_2}\right)^{1/2} + \left(\frac{R_1C_2}{R_2C_1}\right)^{1/2} + \left(\frac{R_2C_2}{R_1C_1}\right)^{1/2} - K\left(\frac{R_2C_2}{R_1C_1}\right)^{1/2}$$

The same comments about frequency adjustment and tuning that we mentioned in the low-pass case apply for the high-pass case also. The network parameter sensitivities with respect to element value changes are

$$S_{R_1}{}^{\omega_o} = S_{R_2}{}^{\omega_o} = S_{C_1}{}^{\omega_o} = S_{C_2}{}^{\omega_o} = -\frac{1}{2}$$

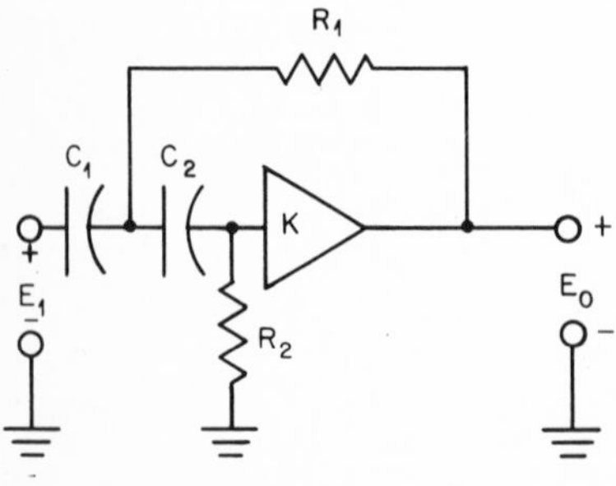

Fig. 8.10 VCVS high-pass network.

$$S_{R_1}{}^{\alpha} = \frac{1}{2} - \frac{1-K}{R_1C_1\alpha\omega_o}$$

$$S_{R_2}{}^{\alpha} = \frac{1}{2} - \frac{1}{R_2}\left(\frac{1}{C_1} + \frac{1}{C_2}\right)\frac{1}{\alpha\omega_o}$$

$$S_{C_1}{}^{\alpha} = \frac{1}{2} - \frac{1}{\alpha\omega_o C_1}\left(\frac{1-K}{R_1} + \frac{1}{R_2}\right)$$

$$S_{C_2}{}^{\alpha} = \frac{1}{2} - \frac{1}{\alpha\omega_o C_2 R_2}$$

$$S_K{}^{\alpha} = \frac{-K}{\alpha\omega_o R_1 C_1}$$

$$S_K{}^{H_o} = 1$$

DESIGN PROCEDURE

Given H_o, α, $\omega_o = 2\pi f_o$

Choose $C_1 = C_2 = C$

$$\text{Calculate: } R_1 = \frac{\alpha + \sqrt{\alpha^2 + 8(H_o - 1)}}{4\omega_o C}$$

$$R_2 = \frac{4}{\omega_o C}\frac{1}{\sqrt{\alpha^2 + 8(H_o - 1)}}$$

Naturally, $H_o = K$ must be such that R_1 and R_2 are positive-valued resistors. Again, a large H_o will result in a large spread of element values and high sensitivities. We can use the same scheme for making $S_{C_1}{}^{\alpha} = -S_{C_2}{}^{\alpha}$ as in the low-pass case.

Choose $C_1 = C_2 = C$; let $R_1 = R_2 = R$. Then $K = 3 - \alpha$ and $R = 1/\omega_o C$.

Bandpass 1. A VCVS realization for the bandpass network function is shown in Fig. 8.11. The voltage transfer function is

$$\frac{E_o}{E_1}(s) = \frac{Ks/R_1C_2}{s^2 + s\left[\frac{1}{R_3C_2} + \frac{1}{R_1C_2} + \frac{1}{R_1C_1} + \frac{1}{R_2C_1} + \frac{1-K}{R_2C_2}\right] + \frac{1}{R_3}\left(\frac{1}{R_1} + \frac{1}{R_2}\right)\frac{1}{C_1C_2}}$$

The network parameters are

$$H_o = \frac{K}{1 + R_1/R_3 + C_2/C_1(1 + R_1/R_2) + (1 - K)(R_1/R_2)}$$

$$\omega_o = \left[\frac{1}{R_3}\left(\frac{1}{R_1} + \frac{1}{R_2}\right)\frac{1}{C_1C_2}\right]^{1/2}$$

$$\frac{1}{Q} = \alpha = \sqrt{\frac{R_3}{(1/R_1 + 1/R_2)}}\left[\sqrt{\frac{C_1}{C_2}}\left(\frac{1}{R_1} + \frac{1}{R_3} + \frac{1-K}{R_2}\right) + \sqrt{\frac{C_2}{C_1}}\left(\frac{1}{R_1} + \frac{1}{R_2}\right)\right]$$

The sensitivities of the network parameters to element changes are

$$S_{C_1}{}^{\omega_o} = S_{C_2}{}^{\omega_o} = S_{R_3}{}^{\omega_o} - \frac{1}{2}$$

$$S_{R_1}{}^{\omega_o} = \frac{-1}{2\omega_o{}^2R_1C_1R_3C_2} \quad S_{R_2}{}^{\omega_o} = \frac{-1}{2\omega_o{}^2R_3C_1R_2C_2}$$

$$S_K{}^Q = \frac{+KQ}{\omega_oR_2C_2} \quad S_{R_1}{}^Q = \frac{-1}{2(1 + R_1/R_2)} + \frac{Q_1}{\omega_oR_1}\left(\frac{1}{C_1} + \frac{1}{C_2}\right)$$

$$S_{R_2}{}^Q = \frac{-1}{2(1 + R_2/R_1)} + \frac{Q}{\omega_oR_2}\left(\frac{1}{C_1} + \frac{1-K}{C_2}\right)$$

$$S_{R_3}{}^Q = \frac{-1}{2} + \frac{Q}{\omega_oR_3C_2} \quad S_{C_1}{}^Q = -\frac{1}{2} + \frac{1}{\alpha\omega_oC_1}\left(\frac{1}{R_1} + \frac{1}{R_2}\right)$$

$$S_{C_2}{}^Q = \frac{-1}{2} + \frac{Q}{\alpha\omega_oC_2}\left(\frac{1}{R_1} + \frac{1}{R_3} + \frac{1-K}{R_2}\right)$$

$$S_K{}^{H_o} = 1 + \frac{Q}{\omega_oR_2C_2}$$

$$S_{R_1}{}^{H_o} = \frac{Q}{\omega_oR_1}\left(\frac{1}{C_1} + \frac{1}{C_2}\right) - 1$$

$$S_{R_2}{}^{H_o} = \frac{Q}{\omega_oR_2}\left(\frac{1}{C_1} + \frac{1-K}{C_2}\right)$$

$$S_{R_3}{}^{H_o} = \frac{Q}{\omega_oR_3C_2}$$

$$S_{C_1}{}^{H_o} = \frac{Q}{\omega_oC_1}\left(\frac{1}{R_1} + \frac{1}{R_2}\right)$$

$$S_{C_2}{}^{H_o} = \frac{Q}{\omega_oC_2}\left(\frac{1}{R_1} + \frac{1}{R_3} + \frac{1-K}{R_2}\right) - 1$$

DESIGN PROCEDURE

The *general* design formulas obtained by solving the network parameter equations for the circuit elements are very complicated. The following design procedure, however, has been found to be useful. It gives a fairly good spread of element values.

Given: Q, $\omega_o = 2\pi f_o$

H_o will be a free parameter,

Choose: $C = C_1 = C_2$, a convenient value

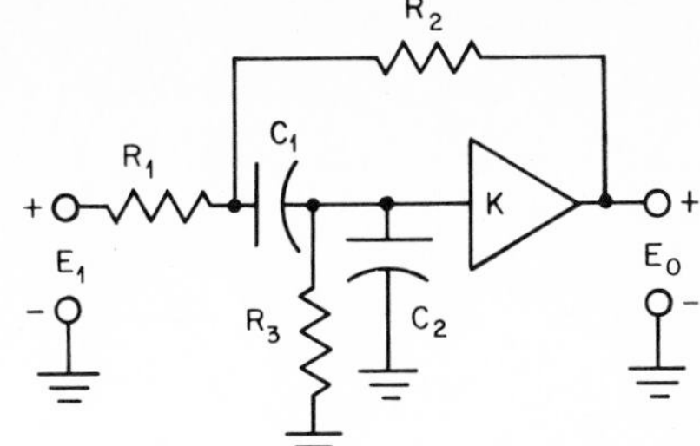

Fig. 8.11 VCVS bandpass network.

Calculate: $K = 5 - \frac{\sqrt{2}}{Q}$

$$R = \frac{\sqrt{2}}{\omega_o C}$$

Then

$$H_o = \frac{5}{\sqrt{2}} Q - 1$$

High-Q circuits will have a large spread of element values and high sensitivities. Q's should be less than 10 for best results.

Four other bandpass circuits are realizable by using the VCVS with $K > 0$. One is obtained by removing C_2 in the circuit of Fig. 8.11 and connecting one terminal to the node formed by $C_1 - R_1 - R_2$ and the other terminal to ground. Two others are generated by interchanging the locations of the resistors and capacitors in the circuit of Fig. 8.11 and the one mentioned above. These two are of less practical interest because they require three capacitors and one of the capacitors is a series capacitor at the input.

Bandpass 2. Still another bandpass realization is illustrated in Fig. 8.12. The voltage transfer function is

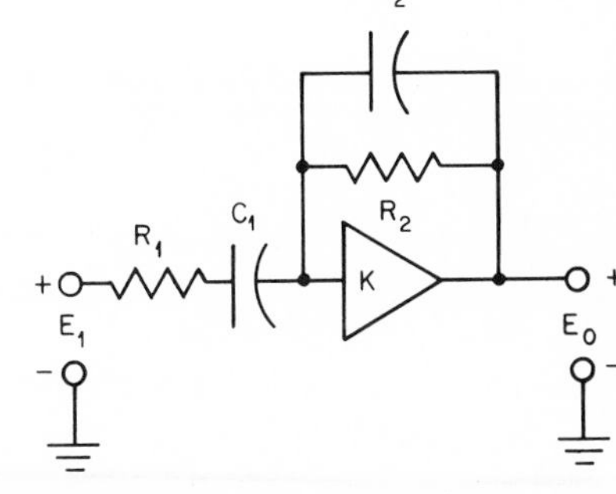

Fig. 8.12 Alternative VCVS bandpass network.

$$\frac{E_o}{E_1}(s) = \frac{s\dfrac{K}{1-K}\dfrac{1}{R_1C_2}}{s^2 + s\left[\dfrac{1}{R_2C_2} + \dfrac{1}{R_1C_1} + \dfrac{1}{R_1C_2(1-K)}\right] + \dfrac{1}{R_1R_2C_1C_2}}$$

$$H_o = \frac{K}{(1-K)(R_1/R_2 + C_2/C_1) + 1}$$

$$\omega_o = \left(\frac{1}{R_1R_2C_1C_2}\right)^{1/2}$$

$$\frac{1}{Q} = \alpha = \sqrt{\frac{R_1C_1}{R_2C_2}} + \sqrt{\frac{R_2C_2}{R_1C_1}} - \frac{1}{1-K}\left(\sqrt{\frac{R_2C_1}{R_1C_2}}\right)$$

$$\phi = \pi + \phi_{BP}$$

$$\tau = \tau_{BP}$$

The center frequency can be trimmed by varying R_1 and R_2. If this is done simultaneously so that their ratio remains constant, Q will not change. Q can be trimmed with K. Note that there is a restriction on the *minimum* value K may have for stability. Because of this restriction, the passband gain H_o will be negative. The sensitivities of the network parameters to element value changes are

$$S_{R_1}^{\omega_o} = S_{R_2}^{\omega_o} = S_{C_1}^{\omega_o} = S_{C_2}^{\omega_o} = -\frac{1}{2}$$

$$S_K^{H_o} = 1 + H_o\left(\frac{R_1}{R_2} + \frac{C_2}{C_1}\right)$$

$$S_{R_2}^{H_o} = -S_{R_1}^{H_o} = H_o\frac{1-K}{K}\frac{R_1}{R_2}$$

$$S_{C_1}^{H_o} = -S_{C_2}^{H_o} = H_o\frac{1-K}{K}\frac{C_2}{C_1}$$

$$S_K^Q = \frac{-K}{(1-K)^2}\frac{Q}{\omega_oR_1C_2}$$

$$S_{R_1}^Q = \frac{1}{2} - \frac{Q}{\omega_oR_1}\left[\frac{1}{C_1} + \frac{1}{C_2(1-K)}\right]$$

$$S_{R_2}^Q = \frac{1}{2} - \frac{Q}{\omega_oR_2C_2}$$

$$S_{C_1}^Q = \frac{-1}{2} + \frac{1}{\alpha\omega_oR_1C_1}$$

$$S_{C_2}^Q = \frac{-1}{2} + \frac{1}{\alpha\omega_oC_2}\left[\frac{1}{R_2} + \frac{1}{R_1(1-K)}\right]$$

DESIGN PROCEDURE

Given: Q, $\omega_o = 2\pi f_o$

H_o is a free parameter,

Choose: $C = C_1 = C_2$, a convenient value

Calculate: $R_1 = R_2 = \dfrac{1}{\omega_o C}$

$$K = \frac{3Q - 1}{2Q - 1}$$

Then

$$|H_o| = 3Q - 1$$

Q should be limited to about 10.

A few general comments about controlled-source realizations are in order. The Q (or α) of a circuit may be adjusted independently of ω_o by adjusting K: it is not independent of H_o, however. Networks may be cascaded without interaction occurring between them. The frequency term ω_o can be adjusted independently of α for the low-pass and high-pass cases, as discussed earlier. The characteristics of the network are sensitive to K. The circuit becomes very Q-sensitive to element value changes for high Q's.

8.3.3 Infinite-gain state-variable circuits An infinite-gain state-variable network configuration is illustrated in Fig. 8.13. This configuration makes use of operational amplifiers in the same way they would

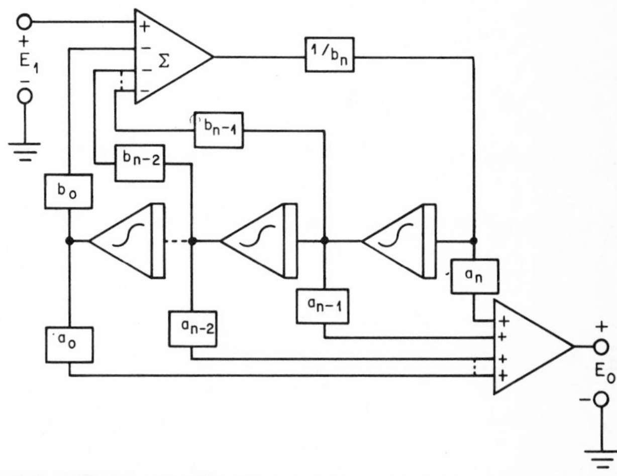

Fig. 8.13 State-variable infinite-gain network configuration.

be used in an analog computer realization of transfer functions (i.e., using integrators and summers). A second-order realization is shown in Fig. 8.14. Here the usual summing amplifier is replaced by a differentially connected operational amplifier to ease the spread in element values.

The voltage transfer function has the form

$$\frac{E_o}{E_1}(s) = \frac{a_o + a_1 s + \cdots + a_{n-1}s^{n-1} + a_n s^n}{b_o + b_1 s + \cdots + b_{n-1}s^{n-1} + b_n s_n}$$

The design procedures used in this section are simplified procedures in that $C_1 = C_2$, $R_1 = R_2$, and $R_5 = R_6$. We set $R_1 = R_2$ and $C_1 = C_2$ in order to scale adequately the output voltages of the operational amplifiers. The condition $R_5 = R_6$ further simplifies design calculations. Note that bandpass, low-pass, and high-pass realizations occur simultaneously. One merely chooses the output at a different point. In addition, one can sum the low-pass and high-pass outputs and form a pair of $j\omega$ axis zeros. The transfer functions are

$$\frac{E_{lp}}{E_1}(s) = \frac{\dfrac{1}{R_1R_2C_1C_2}\dfrac{1 + R_6/R_5}{1 + R_3/R_4}}{s^2 + s\dfrac{1}{R_1C_1}\dfrac{1 + R_6/R_5}{1 + R_4/R_3} + \dfrac{R_6}{R_5}\dfrac{1}{R_1R_2C_1C_2}}$$

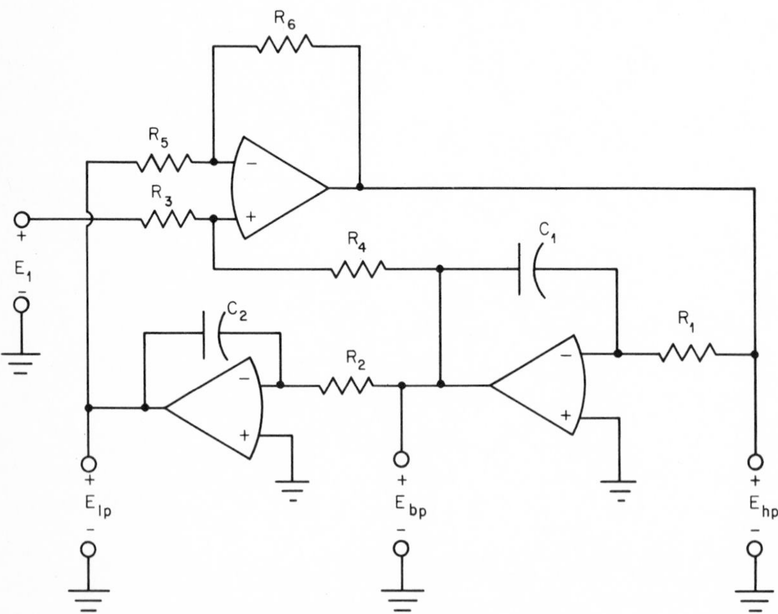

Fig. 8.14 Second-degree state-variable network.

$$\frac{E_{hp}}{E_1}(s) = \frac{s^2 \dfrac{1 + R_6/R_5}{1 + R_3/R_4}}{s^2 + s\dfrac{1}{R_1C_1}\dfrac{1 + R_6/R_5}{1 + R_4/R_3} + \dfrac{R_6}{R_5}\dfrac{1}{R_1R_2C_1C_2}}$$

$$\frac{E_{bp}}{E_1}(s) = \frac{-s\dfrac{1}{R_1C_1}\dfrac{1 + R_6/R_5}{1 + R_3/R_4}}{s^2 + s\dfrac{1}{R_1C_1}\dfrac{1 + R_6/R_5}{1 + R_4/R_3} + \dfrac{R_6}{R_5}\dfrac{1}{R_1R_2C_1C_2}}$$

The state-variable realization in general provides less Q sensitivity to element variation than a single-amplifier realization and for this reason is sometimes used for high-Q bandpass applications ($Q > 50$). Of course it requires three amplifiers, which is a disadvantage. For the low-Q low-pass and high-pass applications, it is a rather expensive circuit to use. Since some filter manufacturers use this circuit as a basic building block, the low-pass and high-pass as well as the bandpass outputs are worth some discussion.

Low Pass. The network parameters for the low-pass function are

$$H_o = \frac{1 + R_5/R_6}{1 + R_3/R_4}$$

$$\omega_o = \left(\frac{R_6}{R_5R_1C_1R_2C_2}\right)^{1/2}$$

$$\alpha = \frac{1 + R_6/R_5}{1 + R_4/R_3}\left(\frac{R_5}{R_6}\frac{R_2C_2}{R_1C_1}\right)^{1/2}$$

$$\phi = \phi_{LP}$$

$$\tau = \tau_{LP}$$

The sensitivities of the network parameters to element value changes are

$$S_{R_5}^{\omega_o} = S_{R_1}^{\omega_o} = S_{R_2}^{\omega_o} = S_{C_1}^{\omega_o} = S_{C_2}^{\omega_o} = -\frac{1}{2} = -S_{R_6}^{\omega_o}$$

$$S_{R_2}^{\alpha} = S_{C_2}^{\alpha} = \frac{1}{2} = -S_{R_1}^{\alpha} = -S_{C_1}^{\alpha}$$

$$S_{R_6}^{\alpha} = -\frac{1}{2} + \frac{R_6/R_5}{R_1C_1\alpha\omega_o(1 + R_4/R_3)} = -S_{R_5}^{\alpha}$$

$$S_{R_3}^{\alpha} = \frac{1}{1 + R_3/R_4} = -S_{R_4}^{\alpha}$$

$$S_{R_3}^{H_o} = -S_{R_4}^{H_o} = \frac{-1}{1 + R_4/R_3}$$

$$S_{R_5}^{H_o} = -S_{R_6}^{H_o} = \frac{1}{H_o}\frac{R_5/R_6}{1 + R_3/R_4}$$

DESIGN PROCEDURE

Given: α, $\omega_o = 2\pi f_o$

H_o is a free parameter.

Choose: $C = C_1 = C_2$, $R_5 = R_6 = R_3$

Calculate: $R_1 = R_2 = \dfrac{1}{\omega_o C}$

$$R_4 = \left(\frac{2}{\alpha} - 1\right) R_3$$

Then

$$H_o = 2 - \alpha$$

High Pass. The network parameters for the high-pass function are

$$H_o = \frac{1 + R_6/R_5}{1 + R_3/R_4}$$

$$\omega_o = \left(\frac{R_4}{R_5R_1R_2C_1C_2}\right)^{1/2}$$

$$\alpha = \frac{1 + R_6/R_5}{1 + R_4/R_3}\left(\frac{R_5}{R_6}\frac{R_2C_2}{R_1C_1}\right)^{1/2}$$

$$\phi = \phi_{HP}$$

$$\tau = \tau_{HP}$$

The sensitivities of the network parameters to element value changes are

$$S_{R_5}{}^{\omega_o} = S_{R_1}{}^{\omega_o} = S_{R_2}{}^{\omega_o} = S_{C_1}{}^{\omega_o} = S_{C_2}{}^{\omega_o} = -\frac{1}{2}$$

$$S_{R_6}{}^{\omega_o} = \frac{1}{2}$$

$$S_{R_1}{}^{\alpha} = S_{C_1}{}^{\alpha} = -\frac{1}{2}$$

$$S_{R_2}{}^{\alpha} = S_{C_2}{}^{\alpha} = \frac{1}{2}$$

$$S_{R_5}{}^{\alpha} = -S_{R_6}{}^{\alpha} = \frac{1}{2} - \frac{R_6/R_5}{R_1C_1\alpha\omega_o(1 + R_4/R_3)}$$

$$S_{R_3}{}^{\alpha} = -S_{R_4}{}^{\alpha} = \frac{1}{1 + R_3/R_4}$$

$$S_{R_3}{}^{H_o} = -S_{R_4}{}^{H_o} = \frac{-1}{1 + R_4/R_3}$$

$$S_{R_5}{}^{H_o} = -S_{R_6}{}^{H_o} = \frac{1}{H_o}\frac{R_6/R_5}{1 + R_3/R_4}$$

DESIGN PROCEDURE

Given: α, $\omega_o = 2\pi f_o$

H_o is a free parameter.

Again a simplified design procedure is described by setting $R_5 = R_6$.

Choose: $C_1 = C_2 = C$

$R_5 = R_6 = R_3$

Calculate: $R_1 = R_2 = \dfrac{1}{\omega_o C}$

$$R_4 = \left(\frac{2}{\alpha} - 1\right) R_3 \qquad \alpha < 2$$

Bandpass. The network parameters for the bandpass case are

$$H_o = \frac{R_4}{R_3}$$

$$\omega_o = \left(\frac{R_6}{R_5 R_1 C_1 R_2 C_2}\right)^{1/2}$$

$$Q = \frac{1}{\alpha} \frac{1 + R_4/R_3}{1 + R_6/R_5} \left(\frac{R_6}{R_5} \frac{R_1 C_1}{R_2 C_2}\right)^{1/2}$$

$$\phi = \pi + \phi_{BP}$$

$$\tau = \tau_{BP}$$

The sensitivities of the network parameters to element values change are

$$S_{R_5}^{\omega_o} = S_{R_1}^{\omega_o} = S_{R_2}^{\omega_o} = S_{C_1}^{\omega_o} = S_{C_2}^{\omega_o} = -\frac{1}{2}$$

$$S_{R_6}^{\omega_o} = \frac{1}{2}$$

$$S_{R_1}^{Q} = S_{C_1}^{Q} = +\frac{1}{2}$$

$$S_{R_2}^{Q} = S_{C_2}^{Q} = \frac{1}{2}$$

$$S_{R_6}^{Q} = S_{R_5}^{Q} = \frac{1}{2} - \frac{R_6/R_5}{R_1 C_1 \alpha_{\omega_o}(1 + R_4/R_3)}$$

$$S_{R_4}^{Q} = -S_{R_3}^{Q} = \frac{1}{1 + R_3/R_4}$$

$$S_{R_3}^{H_o} = -1 = -S_{R_4}^{H_o}$$

DESIGN PROCEDURE

Given: H_o, Q, $\omega_o = 2\pi f_o$

Again the simplified design procedure consists of setting $R_5 = R_6$.

Choose: $C_1 = C_2 = C \qquad R_3 = R_5 = R_6$

$$R_1 = R_2 = \frac{1}{\omega_o C}$$

$$R_4 = R_3(2Q - 1)$$

Note that all these filters can be tuned by varying R_1 and R_2 or C_1 and C_2 simultaneously. The Q can be independently adjusted by R_4; the gain will change, however.

8.3.4 Negative immittance converter circuits A realization for an INIC[1,2] (ideal current-inversion negative immittance converter) using a differential input operational amplifier is shown in Fig. 8.15. The voltage and current relationships are

$$E_1 = E_2$$

$$I_1 = \frac{R_2}{R_1} \qquad I_2 = \frac{1}{K} I_2$$

The sensitivity of K to element value changes is $S_{R_1}{}^K = -S_{R_2}{}^K = 1$.

One reason the INIC realization might be used is its low sensitivity to element value changes as compared with other realizations. However, the INIC realization does not have a low output impedance, and isolating stages must be used if stages are to be cascaded. Since low-pass and high-pass filters have low Q's and, hence, low Q sensitivities for filters up to about six poles, we will discuss only the bandpass realization. It is probably not economical to use the INIC for low-pass and high-pass filters.

Bandpass. The INIC realization for a bandpass filter is shown in Fig. 8.16.

The voltage transfer function is

$$\frac{E_o}{E_1}(s) = \frac{-Ks/R_1C_2}{s^2 + s(1/R_1C_1 + 1/R_2C_2 - K/R_1C_2) + 1/R_1C_1R_2C_2}$$

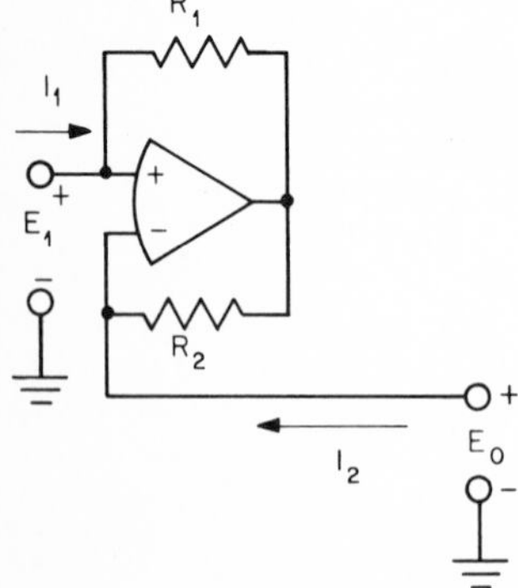

Fig. 8.15 Operational amplifier realization of the INIC.

The network parameters are

$$H_o = \frac{K}{C_2/C_1 + R_1/R_2 - K}$$

$$Q = \frac{1}{\alpha} = \frac{1}{\sqrt{R_1C_1/R_2C_2} + \sqrt{R_2C_2/R_1C_1} - K\sqrt{R_2C_1/R_1C_2}}$$

$$\omega_o = \left(\frac{1}{R_1C_1R_2C_2}\right)^{1/2}$$

$$\phi = \pi + \phi_{BP}$$

$$\tau = \tau_{BP}$$

The sensitivities of the H_o, Q, ω_o network parameters to element value changes are

$$S_K^{H_o} = 1 + H_o$$

$$S_{R_1}^{H_o} = \frac{-R_1/R_2}{C_2/C_1 + R_1/R_2 - K} = -S_{R_2}^{H_o}$$

$$S_{C_1}^{H_o} = \frac{C_2/C_1}{C_2/C_1 + R_1/R_2 - K} = -S_{C_2}^{H_o}$$

$$S_{R_1}^{Q} = \frac{Q}{\omega_o R_1}\left(\frac{1}{C_1} - \frac{K}{C_2}\right) - \frac{1}{2}$$

$$S_{R_2}^{Q} = \frac{Q}{\omega_o R_2 C_2} - \frac{1}{2}$$

$$S_{C_1}^{Q} = \frac{Q}{\omega_o R_1 C_1} - \frac{1}{2}$$

$$S_{C_2}^{Q} = \frac{Q}{\omega_o C_2}\left(\frac{1}{R_2} - \frac{K}{R_1}\right) - \frac{1}{2}$$

$$S_K^{Q} = \frac{QK}{\omega_o R_1 C_2}$$

$$S_{R_1}^{\omega_o} = S_{R_2}^{\omega_o} = S_{C_1}^{\omega_o} = S_{C_2}^{\omega_o} = -\frac{1}{2}$$

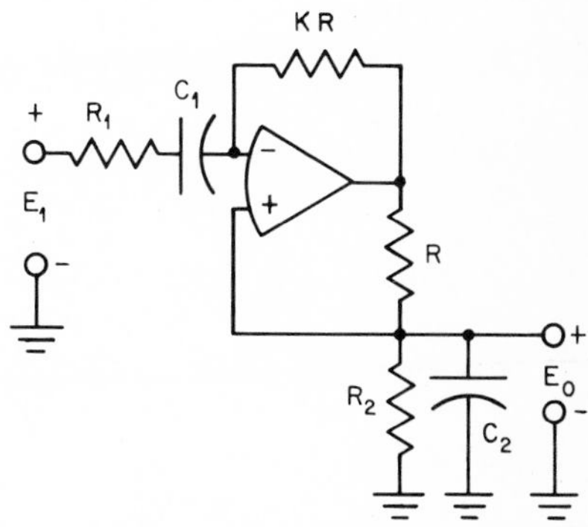

Fig. 8.16 INIC bandpass network.

DESIGN PROCEDURE

Given: Q, $\omega_o = 2\pi f_o$
Choose: $C_1 = C_2 = C$
Let $R_1 = R_2 = R'$
Then

$$K = \frac{2 - 1}{Q}$$

$$R' = \frac{1}{\omega_o C}$$

The value for R in the INIC is relatively arbitrary, but best results are obtained if it is in the 10 to 30 kΩ range.

Note that for this design procedure

$$S_K^Q = 2Q - 1 \qquad Q = \frac{1}{2 - K}$$

$$S_K^{H_o} = 2Q \qquad H_o = 2Q - 1 = \frac{K}{2 - K}$$

$$\omega_o = \frac{1}{R'C}$$

$$S_{R_2}^Q = S_{C_1}^Q = -S_{R_1}^Q = -S_{C_2}^Q = Q - \frac{1}{2}$$

$$S_{C_1}^{H_o} = S_{R_2}^{H_o} = -S_{R_1}^{H_o} = -S_{C_2}^{H_o} = Q$$

As one can see from the schematic diagram, a load at E_o will affect the circuit transfer function. Thus INIC realizations cannot be cascaded without isolating amplifiers between stages.

Note that, in general, if R_1 and R_2 are adjusted by the same percentage, the Q and gain remain constant while the center frequency varies. The same holds true for equal percentage changes in C.

Note also that adjusting K gives a Q adjustment independent of the center frequency; the gain will change, however.

8.4 Tuning Active Filter Stages

This section discusses a technique for tuning the complex-conjugate-pole-pair stage. The single-pole stage is easy to tune and will not be discussed here.

The magnitude response of a low-pass complex pole pair for several

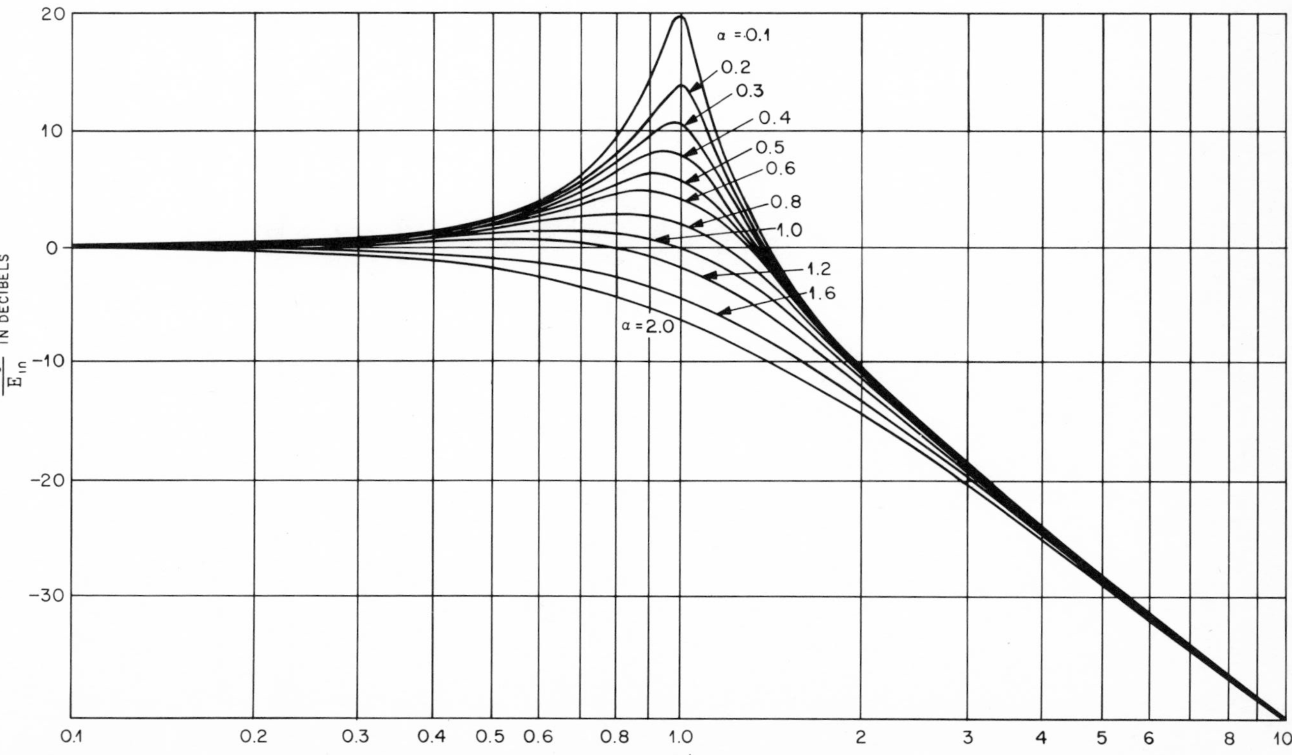

Fig. 8.17 Magnitude response of second-order low-pass filters for several values of $\alpha \leq 2$.

values of α is plotted in Fig. 8.17. Note that at $10\omega_o$ all responses have essentially the same magnitude with respect to the magnitude at direct current. Thus we can tune ω_o independently of α at this frequency. Also at some low frequency, say $0.1\omega_o$, the magnitude of the responses is essentially the same. The curves have a peak at $\omega_\alpha = \omega_o\sqrt{1 - \alpha^2/2}$ for $\alpha < \sqrt{2}$: For $\alpha \geq \sqrt{2}$ there is no peak. The frequency at which the peak occurs will be called the *alpha-peak frequency* and will be designated ω_α. This frequency will be used as a tuning frequency as will be $10\omega_o$ and $0.1\omega_o$. For those stages where $\alpha \geq \sqrt{2}$ the -3-dB frequency will be used instead of ω_α. The particular resistors or capacitors used for tuning or trimming a complex-pole-pair stage were discussed previously for each of the different circuit realizations.

The procedure for tuning the low-pass pole-pair stage is first to measure the response at $0.1\omega_o$ and then to measure the response at $10\omega_o$ and trim the element or elements necessary to adjust ω_o to give the correct response at that frequency. This may affect the response at $0.1\omega_o$, depending on the circuit realization. Now measure the response at ω_α or the -3-dB frequency, whichever is the case, and trim the α of the stage to give the correct response. This may affect the ω_o adjustment. In some realizations α can be set independently of ω_o. The response at $0.1\omega_o$ must be measured again since the α adjustment may have affected it. The gain H_o should be fairly close to the calculated value if there are no parameter effects such as those produced by capacitor leakage and dissipation factor, stray wiring capacitance or amplifier open-loop gain and frequency response limitations.

The tables at the end of this chapter (Sec. 8.7) give the ω_α or the -3-dB frequencies and the magnitude of the peak in decibels with respect to the gain at direct current.

A high-pass stage is tuned in the same manner except *that the tuning frequencies are the reciprocals of those for the low-pass stage.*

Bandpass stage tuning is, conceptually at least, simpler but practically may be more difficult because of interactions among elements. The Q is perhaps a less critical performance parameter than the center frequency. Thus it may be reasonable to adjust the center frequency only and let the Q be what it turns out to be; it will probably be close to the desired value anyway.

Those stages easiest to tune are those where the Q can be adjusted independently from ω_o. Otherwise, one has to achieve the correct values by an iterative process. The Q is adjusted at those frequencies which are -3 dB down from the peak response at ω_o. Those frequencies are

$$f_2 = \frac{f_o}{2Q} + \frac{f_o}{2Q}\sqrt{1 + 4Q^2}$$

$$f_1 = \frac{-f_o}{2Q} + \frac{f_o}{2Q}\sqrt{1 + 4Q^2}$$

where $Q = f_o/\text{bandwidth} = f_o/(f_2 - f_1)$ and $f_1 f_2 = f_o^2$.

For high Q's ($Q > 10$) one can assume arithmetic symmetry of the -3-dB frequencies about the center frequency. Then

$$f_2 = \frac{BW}{2} + f_o$$

$$f_1 = f_o - \frac{BW}{2}$$

where the bandwidth $BW = f_o/Q$.

8.5 How Amplifier Performance Affects Filter Performance

In this section we will examine how certain amplifier performance characteristics affect filter performance. These performance characteristics include dc voltage offset, bias current, voltage and current noise, and open-loop gain and are discussed in more detail in Part 1 and Appendix A.

A dc offset voltage and its drift at the *output* are often important in low-

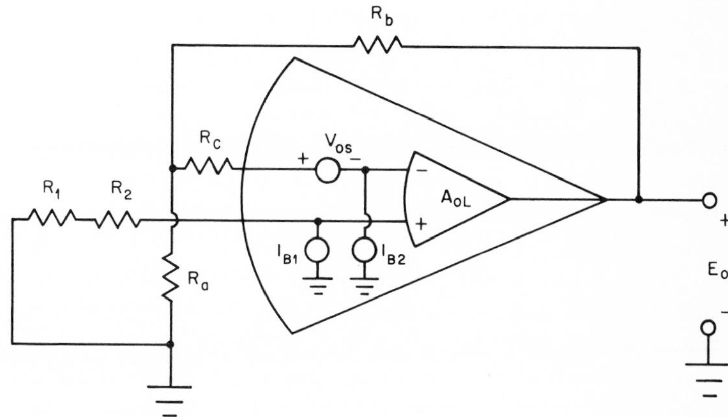

Fig. 8.18 Model for analysis of the effects of offset voltage and bias currents on a filter circuit.

pass filter applications. As an example of this type of analysis, consider the controlled-source low-pass realization shown in Fig. 8.9. At direct current the circuit becomes that of Fig. 8.18. Bias currents and the input offset voltage have been included. An additional resistor R_c has been included and can be used to equalize the current offset drift effects, as will be explained. Analysis reveals

$$E_o = \frac{1}{R_a(1 + 1/A) + R_b/A} \{I_{B2}(R_1 + R_2)(R_a + R_b) - I_{B1}[R_c(R_a + R_b) + R_aR_b] - V_{os}(R_a + R_b)]\}$$

Letting $A \to \infty$

$$E_o = I_{B2}(R_1 + R_2)K - I_{B1}(R_cK + R_b) - V_{os}K$$

where $K = 1 + R_b/\Omega_a$, the ideal closed-loop gain of the controlled-source amplifier of a controlled-source low-pass realization. R_c is used for bias current compensation. R_c may be omitted if impedances are low or if drift and offset are not critically important. Assuming the bias current I_{B1} and I_{B2} are equal, R_c should be

$$R_c = R_1 + R_2 - \frac{R_b}{K}$$

The offset and drift problems associated with other realizations are carried out in the same manner and will not be discussed here.

Output noise of active filter circuits is due to the internal voltage noise and current noise of the operational amplifier. Effects of voltage and current noise can be analyzed by using the noise models of the operational amplifier (see Appendix A). The analysis will not be carried out here since a separate chapter could easily be written on this subject. Rms noise sources are usually assumed, and this is normally the specification given in the data sheets. Low-noise amplifiers sometimes have peak-to-peak noise specified. Current noise may cause a greater noise output than voltage noise if the amplifier noise currents are flowing through large resistances, as is often the case with active filter circuits. FET operational amplifiers have very low bias current and also have low current noise. ; noise.

The effects of open-loop gain characteristics of operational amplifiers on the multiple-feedback bandpass circuit will now be discussed. Open-loop gain effects can be severe in the multiple-feedback circuit and especially for the bandpass realization because of the large amount of *loop gain* required for ideal performance. The open-loop gain of the operational amplifier is neither infinite nor constant for all frequencies. These properties are discussed elsewhere (Appendix A) and will not be covered here. It will be sufficient to say that for our purposes

$$A(s) = \frac{A_O}{1 + s/\omega_o}$$

where ω_o is the -3-dB corner frequency of the operational amplifier.

The exact equation for the infinite-gain multiple-feedback realization is

$$H'(s) = \frac{-Y_1Y_3}{Y_5(Y_1 + Y_2 + Y_3 + Y_4) + Y_3Y_4 + (1/A)[Y_5(Y_1 + Y_2 + Y_3 + Y_4) + Y_3Y_4 + Y_3(Y_1 + Y_2)]}$$

If $A \to \infty$ we have

$$H(s) = \frac{-Y_1Y_3}{Y_5(Y_1 + Y_2 + Y_3 + Y_4) + Y_3Y_4}$$

Then

$$H'(s) = \frac{H(s)}{1 + [1/A(s)][1 - H(s)(1 + Y_2/Y_1)]}$$

Let us define $\beta(s) = 1/[1 - H(s)(1 + Y_2/Y_1)]$ as the *feedback ratio* (output terminal to the $-$ input), so that $A(s)\beta(s)$ is the *loop gain* of the operational amplifier. Now we can rewrite $H'(s)$ as

$$H'(s) = H(s)\left[1 - \frac{1}{1 + A(s)\beta(s)}\right]$$

Thus, the error due to finite loop gain is

$$E(s) = \frac{-H(s)}{1 + A(s)\beta(s)}$$

This equation is completely general for any infinite-gain multiple-feedback realization. Note that the phase of E(s) *is not the phase error of the filter* but is the phase of the error. A plot of magnitude error ($|H(j\omega)| - |H'(j\omega)|$) and phase error $[\phi_H(j\omega) - \phi'_H(j\omega)]$ for a 10-kHz bandpass filter with a gain of 10 and a Q of 20, using an amplifier with a dc gain of 100 dB and a -3-dB corner frequency of 100 Hz (unity-gain bandwidth = 10 MHz), is shown in Fig. 8.19.

Differential input impedance of the operational amplifier also affects filter performance particularly if network element impedances are large. If we include this in the analysis, we insert an admittance Y_6 from the

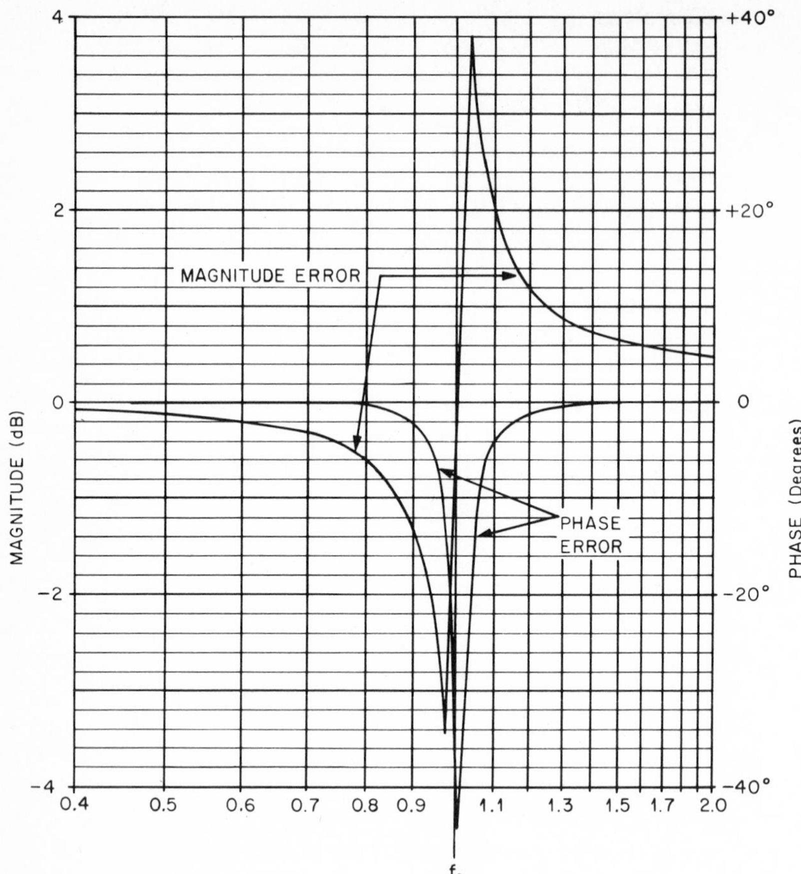

Fig. 8.19 Magnitude and phase error of a bandpass filter due to finite open-loop gain.

minus input of the amplifier to common. In this case

$$\beta(s) = \frac{1}{1 - H(s)[1 + Y_2/Y_1 + (Y_6/Y_1Y_3)(Y_1 + Y_2 + Y_3 + Y_4)]}$$

Both open-loop gain and differential input impedance change with temperature and affect filter performance, especially at low temperatures, since they are both smaller in magnitude. The solution, of course, is to use an amplifier with more gain and higher differential input impedance or reduce circuit element impedance levels.

8.6 Circuit Elements

In this section the salient features of the resistors and capacitors most popularly used as circuit elements will be discussed.

8.6.1 Resistors Three types of resistors most often used are carbon composition, metal film, and wire-wound resistors. Carbon composition resistors have a rather poor temperature coefficient of resistance (200 to 500 ppm/°C) and are used for "room temperature" applications or in filters which may have rather loose performance tolerances with temperature, as in the low-Q stages of two- or three-pole high-pass or low-pass filters. Composition resistors are useful for trimming and padding metal film or wire-wound resistors where the relatively poor temperature coefficient causes only a small percentage change in the overall value. Carbon resistors are relatively inexpensive and are available in a wide range of values.

Bandpass filters and the high-Q stages of a high-pass or low-pass multiple-pole filter require metal film or even wire-wound resistors. Two popular temperature coefficients for metal film resistors are ± 100 ppm/°C (T0) and ± 50 ppm/°C (T2). Metal films can be purchased with a positive- or negative-only temperature coefficient and also with lower temperature coefficients (for example, ± 10 ppm/°C). The metal film resistor is probably the most commonly used resistor for filter applications and is available in a wide range of values. High-Q filters and/or filters which require especially stable parameters with temperature changes may require wire-wound resistors with temperature coefficients as low as a few parts per million per degrees centigrade. High-frequency applications will require noninductive wound resistors.

Integrated-circuit technology offers alternatives to discrete resistors: diffused resistors, thin and thick film.

Base-diffused, emitter-diffused, base pinch, and collector pinch resistors are formed simultaneously with the diffusions for the transistors of the circuit. Temperature coefficients and initial tolerances make this type of resistor marginal for active filter applications unless the filter can be designed so that its parameters depend primarily on resistance ratios.

Thin-film resistors are deposited on ceramic or glass substrates. Materials such as SnO or a SiO-Cr cermet are deposited by silk-screen methods. Others such as Nichrome, tantalum, or cermet may be deposited by evaporation or sputtering. The electrical properties of these resistors are considerably superior to those of diffused silicon resistors. An advantage of thin film over diffused or thick-film resistors is their superior long-term stability.

Thick-film resistors consist of special resistive inks screened and fired on ceramic substrates. Thick-film resistors are trimmable by means of sand blasting or laser techniques. The table following gives typical (untrimmed) parameters for several integrated circuit resistors.

Typical Parameters of Integrated-circuit Resistors

Type	Range, Ω	Temperature coefficient, ppm/°C	Tolerance, %	Matching, %
Base-diffused	100–30 k	500–2000	±10	1
Emitter-diffused	5–100	900–1500	±15	2
Base pinch	5 k–200 k	4000–7000	±50	5
Collector pinch	10 k–500 k	4000–7000	±50	10
Thin film (Ta or Ni-Cr)	30–100 k	0 ± 400	± 2	0.5
Thick film	1–10 M	0 ± 500	±20	10

8.6.2 Capacitors Capacitors present the most severe problem to active-filter designers. Capacitors which have superior characteristics such as polystyrene, Teflon, NPO ceramic, or mica are expensive and large in size. NPO ceramic is available in sizes up to about 0.05 μF for catalog items. Good-quality polystyrene capacitors can be used for the large values (10 μF) in critical applications but then are physically very large. Mica capacitors are available in values up to 0.01 μF but are larger than a Mylar or polycarbonate capacitor of the same value. Physically small ceramic capacitors such as the ceramic disk capacitors and others that have large dielectric constants (from 1,200 to 6,000) have relatively poor characteristics. Capacitance changes with temperature, frequency, voltage, and time amount to several percent. For high-Q applications these changes can make a filter stage unstable or have severe amplitude peaking or attenuation. Such filter stages are usually highly Q-sensitive to element value changes.

The merit of a capacitor dielectric from the point of view of freedom from losses is expressed in terms of the power factor of the capacitor. The *power factor* is the *sine* of the angle by which the current flowing into the capacitor fails to be 90° out of phase with the applied voltage. The *tangent* of this angle is called the *dissipation factor*. The *reciprocal* of the dissipation factor is termed the Q and is the ratio of the capacitor reactance to the equivalent series resistance.

With ordinary dielectrics, phase angle is so small that the power factor, the dissipation factor, and the reciprocal of the capacitor, Q, are, for all practical purposes, equal to each other and to the phase angle expressed in radians. For high-quality capacitors these are practically independent of capacitance, voltage, and frequency. Although the power factor of a capacitor is determined largely by the type of dielectric it is also affected by the environment in which it operates; it tends to increase with temperature and is affected by humidity and by the absorption of moisture.

The effect of a capacitor with its power factor can be taken into account by replacing the capacitor with an ideal capacitor associated with a resistance. This resistance may be represented in series or in parallel. For lower power factors ($R_s \ll 1/\omega C$)R_s is given by

$$\text{Series resistance} = R_s = \frac{\text{power factor}}{2\pi fC}$$

For the parallel resistance we have approximately

$$\text{Parallel resistance} = R_p = \frac{1}{2\pi fC(\text{power factor})}$$

A list of dielectric materials and representative performance features are given in Table 8.1.

Integrated-circuit capacitors are of three types; p-n junctions, MOS structures, and thin-film types. These capacitors have small values, and their values vary greatly with temperature.

The most suitable capacitor for integrated-circuit filters are those utilized in hybrid construction and are NPO ceramic chips or, for low-frequency work, tantalum capacitor chips.

TABLE 8.1 Typical Capacitor Parameters for Different Dielectrics

Dielectric	Power factor	Temperature coefficient of capacitance
Mylar	8×10^{-4}–14×10^{-4}	+250 ppm/°C 0–70°C, larger at extremes
High-quality polystrene	1×10^{-4}–2×10^{-4}	−50 to −100 ppm/°C −60 to +60°C
High-quality mica	1×10^{-4}–7×10^{-4}	0–70 ppm/°C
NPO ceramic	5×10^{-4}–20×10^{-4}	0 ± 30 ppm/°C
Polycarbonate	30×10^{-4}–50×10^{-4}	Non-monotonic Total ±1% 0–70°C, larger at extremes
Teflon	0.5×10^{-4}–1.5×10^{-4}	−250 ppm/°C −60 to 150°C

8.7 Filter Design and Tuning Tables

TABLE 8.2 Butterworth Network Parameters

		Design		Tuning	
Number of poles	Stage	α	ω_o	ω_α or −3 dB* frequency	20 log $G(\omega_\alpha)/G(o)$
2	1	1.414214	1.000000	1.000*	
3	1	a real pole	1.000000	1.000	
	2	1.000000	1.000000	0.707	1.25
4	1	1.847759	1.000000	0.719*	
	2	0.765367	1.000000	0.841	3.01
5	1	a real pole	1.000000	1.000*	
	2	1.618034	1.000000	0.859*	
	3	0.618034	1.000000	0.899	4.62
6	1	1.931852	1.000000	0.676*	
	2	1.414214	1.000000	1.000*	
	3	0.517638	1.000000	0.931	6.02
7	1	a real pole	1.000000	1.000*	
	2	1.801938	1.000000	0.745*	
	3	1.246980	1.000000	0.472	0.22
	4	0.445042	1.000000	0.949	7.25
8	1	1.961571	1.000000	0.661*	
	2	1.662939	1.000000	0.829	
	3	1.111140	1.000000	0.617	0.69
	4	0.390181	1.000000	0.961	8.34
9	1	a real pole	1.000000	1.000*	
	2	1.879385	1.000000	0.703*	
	3	1.532089	1.000000	0.917*	
	4	1.000000	1.000000	0.707	1.25
	5	0.347296	1.000000	0.969	9.32
10	1	1.985377	1.000000	0.655*	
	2	1.782013	1.000000	0.756*	
	3	1.414214	1.000000	1.000*	
	4	0.907981	1.000000	0.767	1.84
	5	0.312869	1.000000	0.975	10.20

* Butterworth filters are frequency-normalized to give −3-dB response at $\omega = 1.0$.

TABLE 8.3 Bessel Network Parameters

Number of poles	Stage	Design: α	Design: ω_o	Tuning: ω_α or -3 dB* frequency	Tuning: $20 \log G(\omega_\alpha)/G(o)$
2	1	1.732051	1.732051	1.362*	
3	1	a real pole	2.322185	2.322*	
	2	1.447080	2.541541	2.483*	
4	1	1.915949	3.023265	2.067*	
	2	1.241406	3.389366	1.624	0.23
5	1	a real pole	3.646738	3.647*	
	2	1.774511	3.777893	2.874*	
	3	1.091134	4.261023	2.711	0.78
6	1	1.959563	4.336026	2.872*	
	2	1.636140	4.566490	3.867*	
	3	0.977217	5.149177	3.722	1.38
7	1	a real pole	4.971785	4.972*	
	2	1.878444	5.066204	3.562*	
	3	1.513268	5.379273	5.004*	
	4	0.887896	6.049527	4.709	1.99
8	1	1.976320	5.654832	3.701*	
	2	1.786963	5.825360	4.389*	
	3	1.406761	6.210417	0.637	0.00
	4	0.815881	6.959311	5.680	2.56
9	1	a real pole	6.297005	6.297*	
	2	1.924161	6.370902	4.330*	
	3	1.696625	6.606651	5.339*	
	4	1.314727	7.056082	2.600	0.08
	5	0.756481	7.876636	6.655	3.09
10	1	1.984470	6.976066	4.540*	
	2	1.860312	7.112217	5.069*	
	3	1.611657	7.405447	6.392*	
	4	1.234887	7.913585	3.857	0.25
	5	0.706560	8.800155	7.623	3.60

* Bessel filters are frequency-normalized to unity delay $\tau(\omega) = 1$ sec at $\omega = 0$.

TABLE 8.4 Chebyshev Network Parameters, Ripple = 0.5 dB, p-p

		Design		Tuning	
Number of poles	Stage	α	ω_o	ω_α or −3 dB* frequency	20 log $G(\omega_\alpha)/G(o)$
2	1	1.157781	1.231342	0.707	0.50
3	1	a real pole	0.626456	0.626*	
	2	0.586101	1.068853	0.973	5.03
4	1	1.418218	0.597002	0.595*	
	2	0.340072	1.031270	1.001	9.50
5	1	a real pole	0.362320	0.362*	
	2	0.849037	0.690483	0.552	2.28
	3	0.220024	1.017735	1.005	13.20
6	1	1.462760	0.396229	0.383*	
	2	0.552371	0.768121	0.707	5.50
	3	0.153543	1.011446	1.005	16.30
7	1	a real pole	0.256170	0.256*	
	2	0.916126	0.503863	0.384	1.78
	3	0.388267	0.822729	0.791	8.38
	4	0.113099	1.008022	1.005	18.94
8	1	1.478033	0.296736	0.283*	
	2	0.620857	0.598874	0.538	4.58
	3	0.288544	0.861007	0.843	10.89
	4	0.086724	1.005948	1.004	21.25
9	1	a real pole	0.198405	0.198*	
	2	0.943041	0.395402	0.295	1.60
	3	0.451865	0.672711	0.637	7.13
	4	0.223313	0.888462	0.223	13.08
	5	0.068590	1.004595	1.003	23.28
10	1	1.485045	0.237232	0.225*	
	2	0.651573	0.487765	0.433	4.21
	3	0.345860	0.729251	0.707	9.35
	4	0.178208	0.908680	0.901	15.02
	5	0.055595	1.003661	1.003	25.10

* These filters are frequency-normalized so that the magnitude response at the pass-band edge passes through the lower boundary of the ripple band at $\omega = 1$.

TABLE 8.5 Chebyshev Network Parameters, Ripple = 1 dB, p-p

		Design		Tuning	
Number of poles	Stage	α	ω_o	ω_α or −3 dB* frequency	$20 \log G(\omega_\alpha)/G(o)$
2	1	1.045456	1.050005	0.707	1.00
3	1	a real pole	0.494171	0.494*	
	2	0.495609	0.997098	0.934	6.37
4	1	1.274618	0.528581	0.229	0.16
	2	0.280974	0.993230	0.973	11.1
5	1	a real pole	0.289493	0.289*	
	2	0.714903	0.655208	0.565	3.51
	3	0.179971	0.994140	0.986	14.93
6	1	1.314287	0.353139	0.130	0.08
	2	0.454955	0.746806	0.707	7.07
	3	0.124942	0.995355	0.991	18.08
7	1	a real pole	0.205414	0.205*	
	2	0.771049	0.480052	0.402	2.96
	3	0.316871	0.808366	0.789	10.09
	4	0.091754	0.996333	0.994	20.76
8	1	1.327947	0.265068	0.091	0.06
	2	0.511120	0.583832	0.544	6.12
	3	0.234407	0.850613	0.839	12.66
	4	0.070222	0.997066	0.312	2.75
9	1	a real pole	0.159330	0.159*	
	2	0.793624	0.377312	0.312	2.75
	3	0.368610	0.662240	0.639	8.82
	4	0.180942	0.880560	0.873	14.88
	5	0.055467	0.997613	0.997	25.12
10	1	1.334229	0.212136	0.070	0.05
	2	0.536341	0.476065	0.440	5.74
	3	0.280859	0.721478	0.707	11.12
	4	0.144161	0.902454	0.898	16.85
	5	0.044918	0.998027	0.998	26.95

* These filters are frequency-normalized so that the magnitude response at the pass-band edge passes through the lower boundary of the ripple band at $\omega = 1$.

TABLE 8.6 Chebyshev Network Parameters, Ripple = 2 dB, p-p

Number of poles	Stage	Design		Tuning	
		α	ω_o	ω_α or -3 dB* frequency	20 log $G(\omega_\alpha)/G(o)$
2	1	0.886015	0.907227	0.707	2.00
3	1	a real pole	0.368911	0.369*	
	2	0.391905	0.941326	0.904	8.31
4	1	1.075906	0.470711	0.305	0.85
	2	0.217681	0.963678	0.952	13.30
5	1	a real pole	0.218308	0.218*	
	2	0.563351	0.627017	0.575	5.34
	3	0.138269	0.975790	0.971	17.21
6	1	1.109145	0.316111	0.196	0.70
	2	0.351585	0.730027	0.707	9.22
	3	0.095588	0.982828	0.981	20.40
7	1	a real pole	0.155340	0.155*	
	2	0.607379	0.460853	0.416	4.75
	3	0.243009	0.797114	0.785	12.35
	4	0.070027	0.987226	0.986	23.10
8	1	1.120631	0.237699	0.145	0.65
	2	0.394841	0.571925	0.549	8.24
	3	0.179098	0.842486	0.836	14.97
	4	0.053512	0.990141	0.989	25.43
9	1	a real pole	0.120630	0.120*	
	2	0.625114	0.362670	0.325	4.53
	3	0.282589	0.654009	0.641	11.06
	4	0.137959	0.874386	0.870	17.23
	5	0.042225	0.992168	0.992	27.49
10	1	1.125921	0.190388	0.115	0.63
	2	0.414283	0.466780	0.446	7.84
	3	0.214523	0.715385	0.707	13.42
	4	0.109773	0.897590	0.895	19.20
	5	0.034169	0.993632	0.993	29.33

* These filters are frequency-normalized so that the magnitude response at the pass-band edge passes through the lower boundary of the ripple band at $\omega = 1$.

TABLE 8.7 Chebyshev Network Parameters, Ripple = 3 dB, p-p

Number of poles	Stage	Design		Tuning	
		α	ω_o	ω_α or −3 dB* frequency	20 log $G(\omega_\alpha)/G(o)$
2	1	0.766464	0.841396	0.707	3.00
3	1	a real pole	0.298620	0.298*	
	2	0.325982	0.916064	0.891	9.85
4	1	0.928942	0.442696	0.334	1.70
	2	0.179248	0.950309	0.943	14.97
5	1	a real pole	0.177530	0.178*	
	2	0.467826	0.614010	0.579	6.84
	3	0.113407	0.967484	0.964	18.92
6	1	0.957543	0.298001	0.219	1.51
	2	0.289173	0.722369	0.707	10.87
	3	0.078247	0.977154	0.976	22.14
7	1	a real pole	0.126485	0.126*	
	2	0.504307	0.451944	0.422	6.23
	3	0.199148	0.791997	0.784	14.06
	4	0.057259	0.983099	0.982	24.85
8	1	0.967442	0.224263	0.164	1.45
	2	0.324695	0.566473	0.551	9.89
	3	0.146518	0.838794	0.834	16.71
	4	0.043725	0.987002	0.987	27.19
9	1	a real pole	0.098275	0.098*	
	2	0.519014	0.355859	0.331	6.00
	3	0.231548	0.650257	0.641	12.77
	4	0.112754	0.871584	0.869	18.97
	5	0.034486	0.989699	0.898	29.25
10	1	0.972004	0.179694	0.131	1.42
	2	0.340668	0.462521	0.449	9.48
	3	0.175474	0.712614	0.707	15.15
	4	0.089664	0.895383	0.894	20.96
	5	0.027897	0.991638	0.991	31.09

* These filters are frequency-normalized so that the magnitude response at the passband edge passes through the lower boundary of the ripple band at $\omega = 1$.

REFERENCES

There has been a massive amount of literature written on the subject of active filters. Extensive bibliographies are given in the two references listed below.

1. L. P. Huelsman, *Theory and Design of Active RC Circuits,* McGraw-Hill Book Company, New York, 1968.
2. L. P. Huelsman, *Active Filters: Lumped, Distributed, Integrated, Digital, and Parametric,* McGraw-Hill Book Company, New York, 1970.

9

ANALOG/DIGITAL, DIGITAL/ANALOG, AND SAMPLING NETWORKS

In this chapter we will discuss the use of operational amplifiers in applications involving both digital and analog signals. In some cases, the digital signals are used to control the acquisition of samples of analog signals. In other applications, the digital signal is being converted to an analog voltage or, conversely, an analog signal is being converted to a series of digital words. In nearly all these applications the desired characteristics of the operational amplifiers are high input impedance, low bias current, low voltage drift, and fast response to transient signals. The degree to which each of these requirements must be met depends, of course, upon the accuracy and speed requirements of the particular application. In the following discussions, these requirements will be examined in more detail.

9.1 Multiplexers[1,12,13]

Multiplexers are used to reduce the number of components and/or the weight required to process more than one analog signal and to facilitate computer control of data acquisition. The number of channels in a multiplexer can vary from two to several hundred, subject to certain practical limitations. The number of samples per second that must be taken of each analog input signal is determined by the highest frequency component of the signal. At least two samples per period must be taken according to the Nyquist sampling theorem.[18] Thus, one limitation on the maximum number of channels in a multiplexer system arises from limited switching rate of the multiplexer channels and the required minimum number of samples per second. Usually, multiplexers are designed with a fixed number of channels on a plug-in card, or in a plug-in module. Two methods are presented in Sec. 9.1.5 for expanding the number of channels through interconnection of two or more multiplexer modules.

The most commonly used switches in modern multiplexer designs are the JFET transistor and the MOSFET transistor. The preference for these switches is due to the excellent dc isolation between the switch driver circuitry and the analog signal path which they provide. Other desirable characteristics of such switches are zero offset voltage, low-leakage current, and a very large OFF-to-ON impedance ratio. Some systems employ relays if fast switching is not important. Only the FET-type transistor switches will be discussed in this section. In addition to discussing the operation and properties of these devices we shall also present a treatment of differential input multiplexers and multitiered multiplexers.

9.1.1 Multiplexer with MOSFET switches The first type of multiplexer to be considered is one using MOSFET switches. Figure 9.1 shows the circuit diagram of a single-ended input multiplexer using N-channel depletion-mode MOSFET transistor switches. The output of each switch is tied to a common node which is the multiplexer output. Each switch driver circuit applies voltage to the gate of a MOSFET switch. This voltage controls the state of the MOSFET: a -15-V level turns it to OFF and a $+15$-V level turns it to ON. With ± 15-V power supplies and an analog input range of ± 10 V, the MOSFET transistor must have a gate-to-source cutoff voltage of -5 V maximum and gate-to-source and gate-to-drain breakdown voltages of ± 25 V minimum.

The output node of the multiplexer must be connected to a high impedance load such as a high input impedance sample-hold module to prevent part of the analog input from being dropped across the drain-to-source resistance of the ON channel. If the multiplexer output must be loaded, a buffer amplifier as shown in Fig. 9.1 should be used to isolate the load from the common node. The amplifier should provide a very accurate gain of unity, which requires that the common-mode rejection and open-loop gain be high. A gain accuracy of 0.01 percent requires that both parameters be in excess of 80 dB. The dc input uncertainty is caused

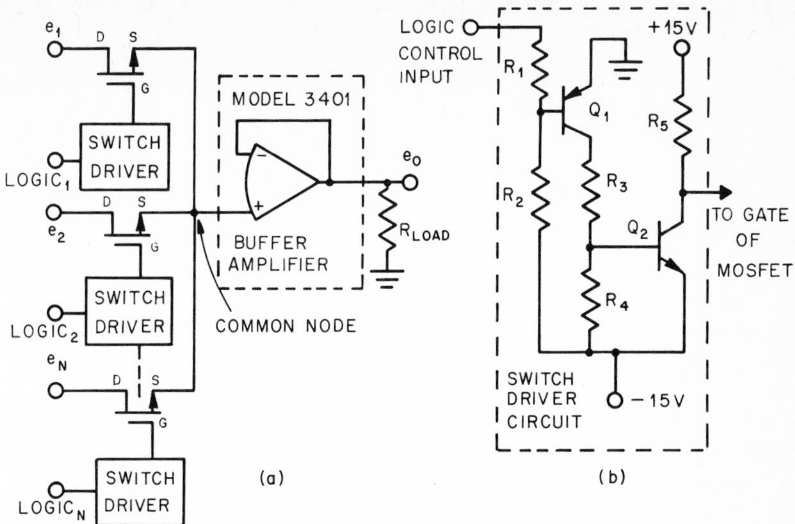

Fig. 9.1 Single-ended multiplexers: (a) using MOSFET transistors; (b) simple switch driver circuit.

by the inherent offset voltage of the input stage of the amplifier and by the bias current flowing from the amplifier through the impedances of the source and the ON channel switch. The time required for the amplifier to settle within a desired percentage (for example, 0.01 percent) of its final output voltage after the application of a step input (settling time) will determine the maximum allowable sampling rate of the multiplexer.

9.1.2 Multiplexer with JFET switches The second type of multiplexer to be considered is one using JFET (junction FET) transistors. With a few changes, the circuit of Fig. 9.1 can be modified to use N-channel JFET transistors. Since the gate-to-source voltage of a JFET must be zero when the device is turned to ON, the gate must somehow be made to follow the analog input. To keep the analog input isolated from the gate of the JFET and the switch driver, the gate is bootstrapped by R_B from the output of a buffer amplifier, as shown in Fig. 9.2. When Q_1 and Q_2 of the gate driver, shown in Fig. 9.1, are turned to ON, the output of the buffer amplifier must supply any output load current, plus $(N - 1)25/R_B(k\Omega)$ mA through the $N - 1$ bootstrap resistors to the -15-V supply (through Q_2) when the input of the ON channel is at $+10$ V. When Q_1 and Q_2 are OFF, the diode (D_1, for instance) is reverse-biased and the FET turns to ON. The gate-to-source cutoff voltage of the JFET must be -5 V maximum (for inputs up to ± 10 V), and the buffer amplifier must have the characteristics previously described if its

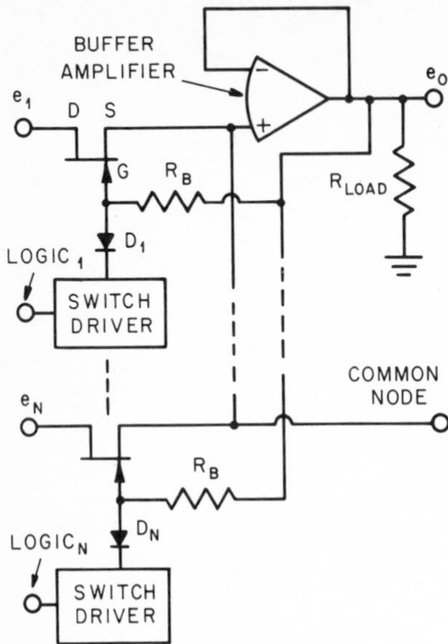

Fig. 9.2 Single-ended multiplexer using JFET transistors.

output is used for the multiplexed signal. The high common-mode rejection and offset requirements for the buffer amplifier can be relaxed if the output can be taken at the common node. The buffer then is a very simple switch-biasing follower whose gain and offsets are not as critically important.

9.1.3 Other design considerations The simple switch driver shown in Fig. 9.1 can be used with either type of FET switch. This circuit has the virtue of simplicity, and it will rapidly turn an FET to OFF (since Q_1 and Q_2 turn to ON rapidly). However, the storage time and the RC time constants associated with Q_1 and Q_2 cause the switch driver to be slower when switching the FET to ON. This break-before-make action may be desirable in some applications, but it leaves the + input of the buffer amplifier open momentarily, thus allowing ac signal pickup such as spikes from the switch driver feeding through the interelectrode capacitances of the FET switches. This problem can be reduced somewhat by using a more complicated switch driver circuit which employs an active pull-up transistor instead of R_5.

The interelectrode capacitances of the FET switches and stray wiring capacitance are responsible for input signal feedthrough from the N − 1 OFF channels to the output (crosstalk). Crosstalk is directly proportional to the frequency and amplitude of the signals applied to the OFF channels and the signal source impedance of the ON channel.

The leakage currents that flow in any input channel of a multiplexer when that channel is ON differ from the currents flowing when that channel is OFF. The leakage current of an ON channel is composed of the gate-to-source and drain-to-source leakages of the N − 1 OFF channels, the leakage from the gate of the ON channel, the leakage (or bias) current from the + input of the buffer amplifier (when it is used), and the leakage current from a load connected to the common node (if this connection is made). The leakage current of each OFF channel is composed of the source-to-drain and the gate-to-drain leakages of that channel.

The input capacitance of a multiplexer is generally small enough ($\approx$30 pF) so that it will not significantly degrade the rise and fall time plus the settling time of a step input signal, provided the input signal source impedance is not too large ($\leq$1 kΩ). It takes about 10 time constants for a step input to settle to 0.01 percent of its final value in a simple RC system. Steps occur when switching from one channel to another channel of a different input voltage.

The frequency response, however, is highly dependent on the input signal source impedance (R_s) and the input capacitance (C_i). The output amplitude will be down by 0.1 percent at approximately $f_{3\,dB}/30$ and by 0.01 percent at approximately $f_{3\,dB}/100$, where $f_{3\,dB} = \frac{1}{2}\pi R_s C_i$. The error due to phase shift is even more severe. These effects are in addition to the gain amplitude error and phase shift of the buffer amplifier.

9.1.4 Differential input multiplexers The multiplexers described in Secs. 9.1.1 and 9.1.2 are all designed for single-ended circuits. In the following paragraphs we consider differential input and output multiplexers. Figure 9.3 shows a simplified diagram of a differential input-output multiplexer which includes a differential DC amplifier instead of a single-input buffer amplifier. There are two basic types of differential multiplexer circuits: the two-wire system and the three-wire system, both shown in Fig. 9.4. In a three-wire system the shield is connected to the common point at the transducer so that each shield is driven by the common-mode voltage (the voltage generated between the common point at the input signal source transducer and the common point of the differential amplifier). Since the common-mode voltage appears at both differential inputs, bootstrapping the shield with the common-mode signal means that no currents due to the common-mode voltage will flow between the differential inputs and the shield. If these currents were allowed to flow, any unbalance in the impedance seen by each differential input line would limit the common-mode rejection of the differential amplifier. In the three-wire system, the shield is switched along with the two input lines. The shield node is then connected to the guard input of the differential amplifier. In the two-wire system, each shield

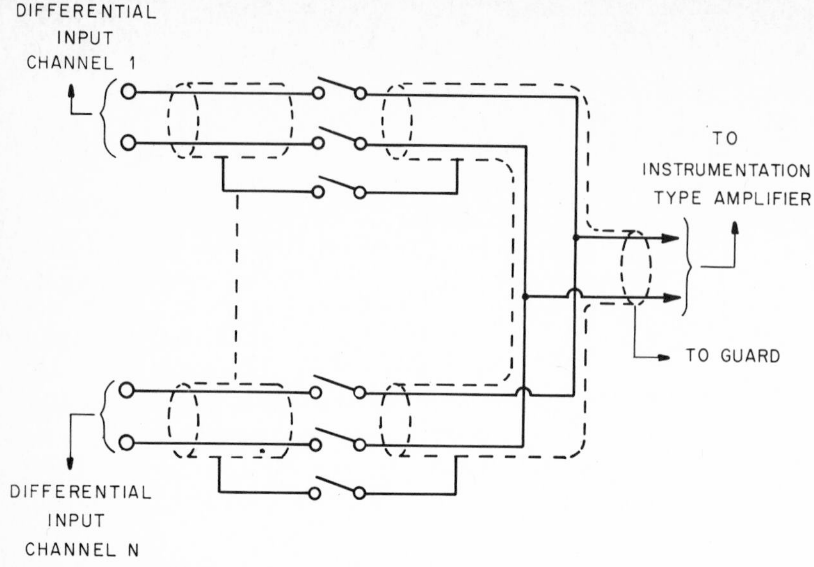

Fig. 9.3 Differential input multiplexer.

is connected to signal common at the signal source, but each shield is not connected at the multiplexer input. If a guard is needed around the multiplexer or the differential amplifier, the common-mode signal can be extracted from the voltages on each differential input, namely, $\frac{1}{2}(e_1 + e_2)$. Often this signal is readily available at a point in the differential amplifier. This method of splitting the shield reduces the complexity of a differential multiplexer by reducing the number of parts required and the number of input pins required on a module.

The CMR is limited by the common-mode input impedance of the differential amplifier and by signal source impedance unbalances. Any mismatch in the R_{ON} resistance of the two multiplexer switches of a differential channel will be interpreted by the instrumentation amplifier as a signal source impedance unbalance. Consequently, the R_{ON} of the FETs in any channel should be matched.

The same considerations and limitations apply to differential multiplexer FET switches as for the single-ended input multiplexer switches described in Secs. 9.1.1 to 9.1.3.

9.1.5 Multitiered multiplexing There are two common ways to combine more than one multiplexer module or plug-in card to expand the number of channels in a system. The method used may depend upon the type of decoding logic that is used to drive the multiplexers. If a 10-channel BCD (binary-coded decimal) to decimal decoder is used, the

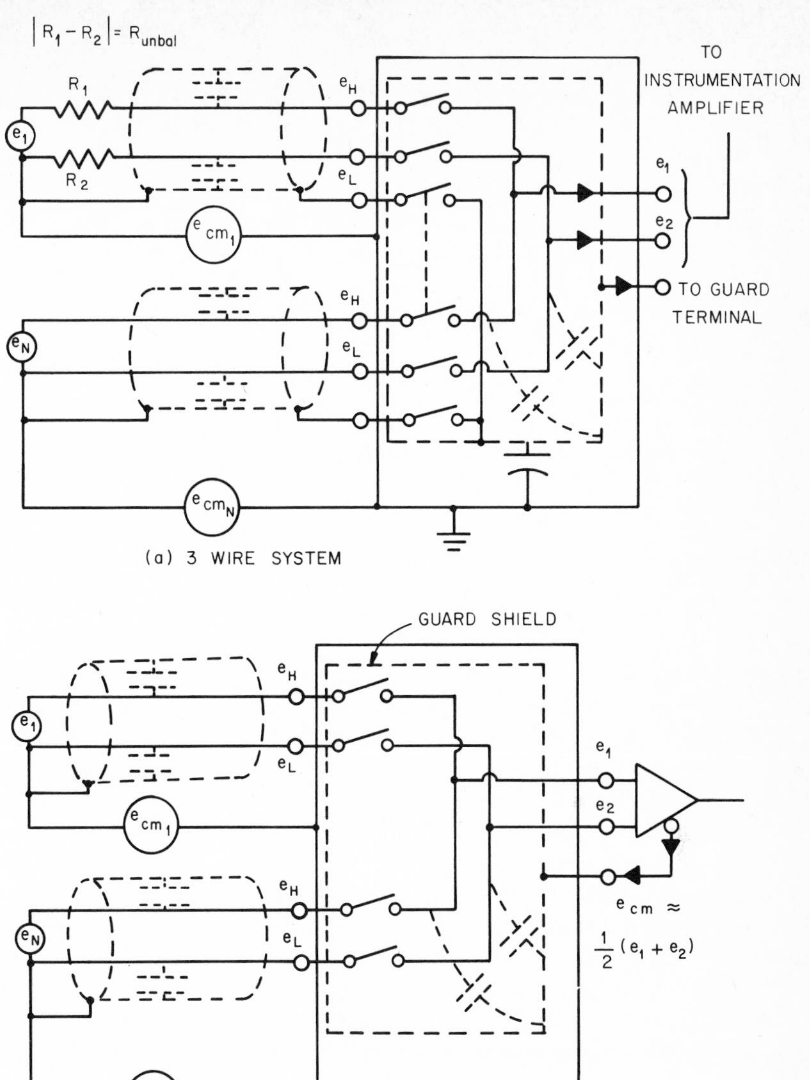

Fig. 9.4 Differential multiplexers: (a) three-wire system; (b) two-wire system with split guard.

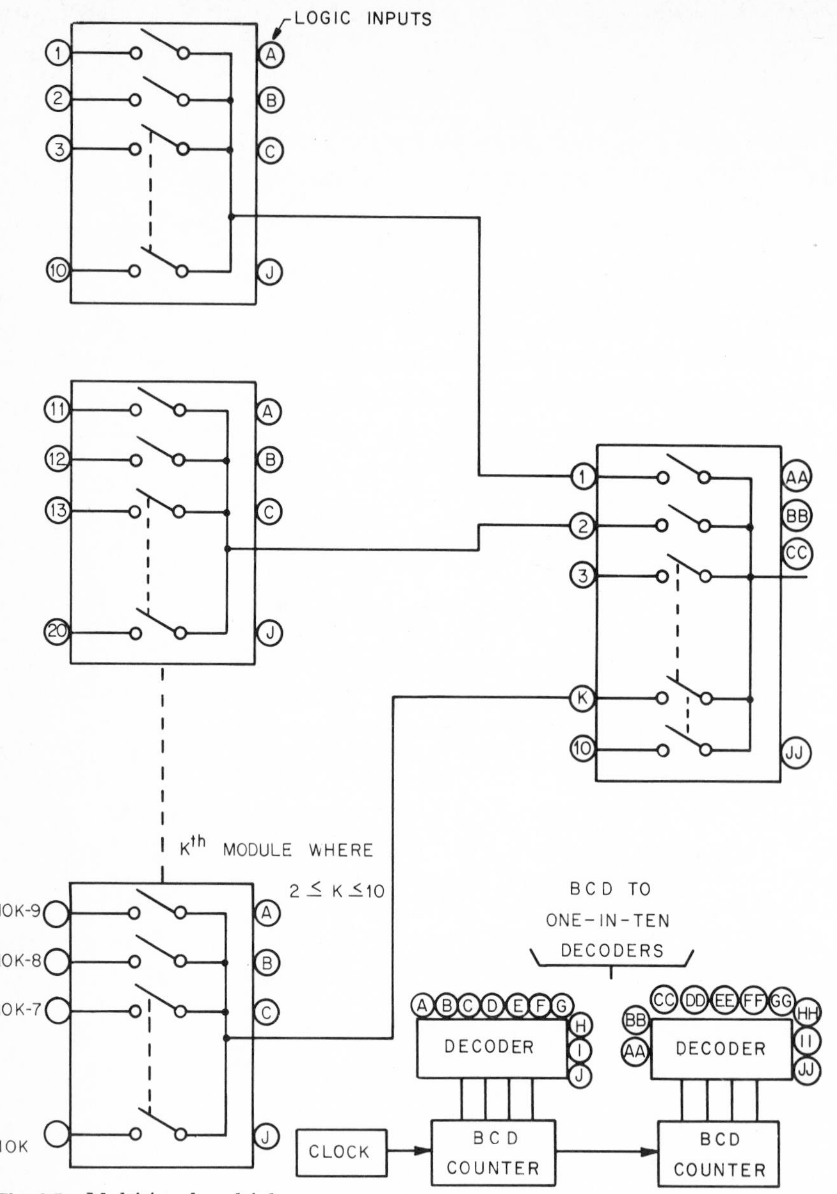

Fig. 9.5 Multitiered multiplexer.

pyramid structure can be used, as shown in Fig. 9.5 for a 10-channel multiplexer. The same multitiered connections also apply to the differential-type multiplexer, although only the single-ended type is shown in the figure. Outputs of the units counter drive the K module logic inputs all in parallel. The 10s counter drives the output accumulator. This technique greatly simplifies the logic required to drive the multiplexers. Up to 100 channels can be built using only 11 modules. The technique can be extended to over 100 channels by adding another BCD counter and decoder for the hundreds position. When less than 100 channels are used, some scheme must be devised to reset the counters to zero when the last channel is reached. This can be accomplished by connecting gates to the four outputs of the 10s and units counters to sense the highest required count. The output of these gates then resets the counters to zero. In some systems binary coding is used. In this case the number of inputs per module would be eight.

9.2 Digital-to-Analog Converters[5,6]

Communicating between the analog and digital worlds requires devices that can translate the language of the two worlds. A digital-to-analog (D/A) converter accepts a digital word as its input and translates, or converts, this word to an analog voltage. Converters can be built to accept the digital word in a variety of codes such as BCD or the binary code. The analog output of a unipolar n-bit binary D/A converter is given by the expression

$$E_o = V_R(a_1 2^{-1} + a_2 2^{-2} + a_3 2^{-3} + \cdots a_n 2^{-n}) \tag{9-1}$$

where V_R is an analog reference voltage and the coefficients a_1 through a_n are equal to 0 if a bit is at OFF and equal to 1 if a bit is at ON. The weight of the most significant bit (MSB) is $V_R/2$ and the weight of the least significant bit (LSB) is $V_R/2^n$. When all bits are turned to ON (all binary inputs at a logical 1), the analog output will be equal to $V_R(1 - 2^{-n})$. When an operational amplifier is used as the output amplifier of a D/A converter, the gain of the amplifier can be adjusted to make the term V_R in Eq. (9-1) equal to 10.240 V. For a 10-bit converter ($2^{10} = 1{,}024$) the LSB comes out an even 10.240 V/1,024 or 10 mV. When all bits are at ON the analog output will be 10.230 V, that is, V_R minus the weight of the LSB. Likewise, for a 12-bit converter, the LSB would be 2.5 mV when $V_R = 10.240$ V. Although making $V_R = 10.240$ V makes the output levels become easy-to-remember combinations of the powers of 2, the analog output is often scaled to be 10.00 V when all bits are at ON.

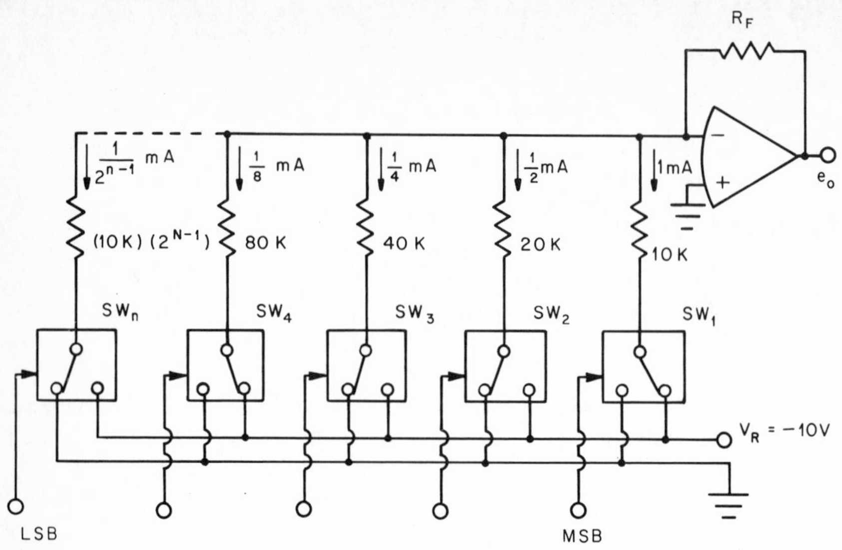

Fig. 9.6 D/A converter using binary-weighted resistors.

9.2.1 D/A converter designs The basic elements of a D/A converter are the resistor network, the current or voltage switches, the reference supply, and the output operational amplifier. Figure 9.6 shows the diagram of a parallel entry binary-coded D/A converter with a binary-weighted resistor network. Binary-weighted currents set by the resistor network and the reference voltage (V_R) are summed by the operational amplifier which is operating as a low output impedance current-to-voltage converter. Choosing the feedback resistor (R_F) equal to 5 kΩ makes the weight of the MSB equal to 5.00 V, the next bit equal to 2.50 V, the next bit 1.25 V, and so on to the LSB. If R_F were 5.12 kΩ, the weight of the bits starting at the MSB would be 5.12 V, 2.56 V, 1.28 V, etc. The voltage switches can be FETs or bipolar transistors connected to form single-pole double-throw switches. The resistor network requires only one resistor per bit, but the resistors have a wide range of values, making temperature coefficient matching difficult.

A disadvantage of the circuit shown in Fig. 9.6 is the wide range of resistance values required. The commonly used R, 2R ladder network of Fig. 9.7 overcomes the wide resistance range disadvantages at the expense of two resistors per bit. Also, the absolute accuracy of the resistors in the R, 2R ladder is not critical, but their ratio is, because the ladder is a precision current splitting device. To understand the operation of this circuit we may consider the weight of the different bits one at a time and then apply superposition. With the MSB turned to ON and all other bits turned to OFF, current I_1 flows out of node 1. Because the resistance

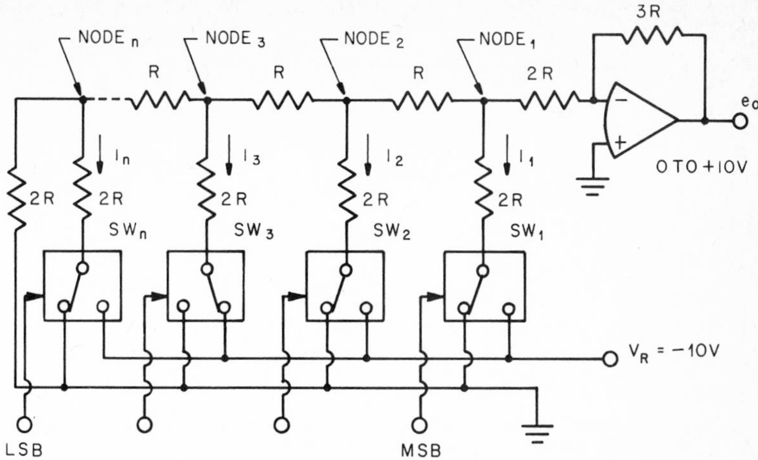

Fig. 9.7 D/A converter using an R, 2R ladder network.

from node 1 to ground is 2R looking to the left and also 2R looking to the right, the current I_1 will split equally to the left and to the right. This holds true at any node since the resistance looking to the left, to the right, or toward the switch is always 2R. Considering node 1 as part of a voltage divider as shown in Fig. 9.8a, the voltage at node 1 will be $V_R/3$. The gain of the operational amplifier from node 1 is $-3R/2R$; therefore, the weight of the MSB is $E_o = (V_R/3)(-3/2) = -V_R/2$. When the second bit is turned to ON and all other bits are turned to OFF, the voltage at node 2 will also be $V_R/3$. The equivalent circuit from node 2 looking to the right (Fig. 9.8b) shows that the voltage at node 1 will be one-half the voltage at node 2, or $V_R/6$. Using the gain of the operational amplifier, the second bit gives $E_o = -V_R/4$. Using this same procedure for the third bit gives $E_o = -V_R/8$, and so on to the LSB which gives $E_o = -V_R/2^n$. Many other resistor network schemes exist, but all have the same purpose: to produce binary-weighted currents or voltages that can be switched to ON or OFF in accordance with a digital input word.

9.2.2 Sources of error Most of the sources of error (such as the finite ON resistance and the voltage and current offsets of the ladder switches) can be compensated at room temperature, so that drift with temperature becomes the major concern. To obtain a drift coefficient for a D/A converter of the order of 10 to 20 ppm/°C requires a very stable reference supply voltage and ladder and feedback resistors with either low temperature coefficients or else well-matched temperature coefficients. The errors caused by the operational amplifier are due mainly to the voltage

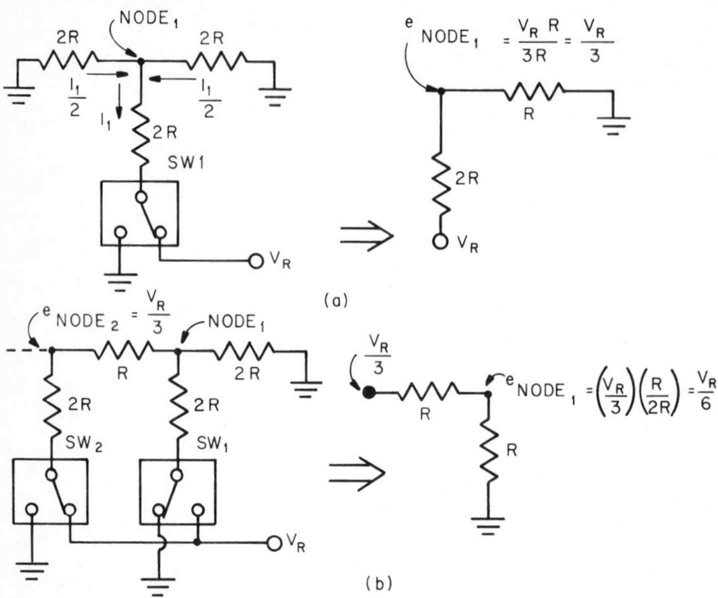

Fig. 9.8 Equivalent circuits: (a) voltage at node, with switch SW_1 at ON; (b) voltage at node, with SW_2 at ON.

drift and the offset current drift. Figure 9.9 shows the equivalent circuit of an amplifier used in a D/A converter. R_{EQ} is the equivalent resistance seen by the amplifier looking back into the ladder network. For example, the resistance looking back into the ladder network of Fig. 9.7 is 3R. I_{B_1} and I_{B_2} are the bias currents of the $-$ and $+$ inputs, respectively. A resistor R_1 equal to the parallel combination of R_F and R_{EQ} is connected from the $+$ input to ground to minimize the offset caused by the bias currents. The output voltage error is produced by the offset current

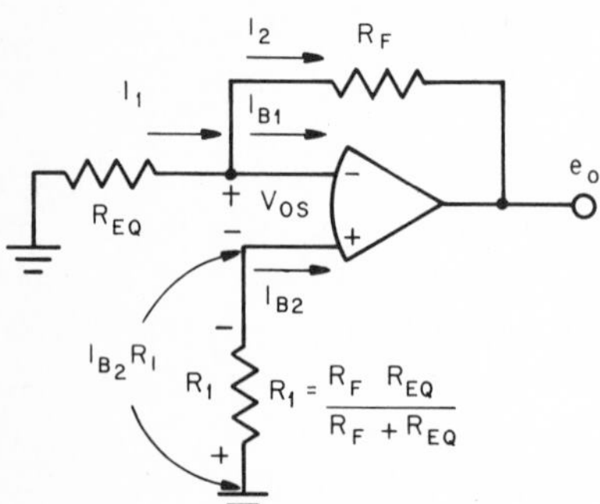

Fig. 9.9 Errors caused by the bias current and offset voltage of the operational amplifier.

$(I_{B_1} - I_{B_2})$ and the voltage offset,

$$e_o = R_F(I_{B_1} - I_{B_2}) + V_{os}(1 + R_F/R_{EQ})$$

These two errors are independent of the digital input word and can be nulled at room temperature, but their effects at other temperatures must be considered. The closed-loop gain error due to the operational amplifier in a D/A converter of 12 bits or less can be neglected when the open-loop gain is 80 dB or greater. Since a change in the digital input causes a step change in input current to the operational amplifier, the step response of the amplifier is important. The slew rate and settling time of the amplifier are of prime importance in high-speed D/A converters. In many designs, large resistance values are used in the resistive input network in order to minimize the errors caused by the ON resistance of the switches. In order to maintain a full-scale output of at least 10 V, the feedback resistor must also be large. Circuit capacitances will, therefore, affect the speed of the D/A converter. Also, some operational amplifiers may tend to oscillate if the feedback resistor is too large, and special compensation will then be needed.

9.2.3 Bipolar operation So far, only unipolar D/A converters have been discussed. Bipolar D/A converters require both positive and negative reference voltages. To represent negative numbers, the sign-magnitude method would seem to be the most logical approach. With this method, the reference voltage polarity is selected in accordance with the sign bit of the digital word. This approach, however, is not the most convenient since the switch that selects the reference supplies must have extremely low impedance. A more commonly used method produces the offset binary code by supplying a constant offset current to the − input of the operational amplifier as shown in Fig. 9.10. This current

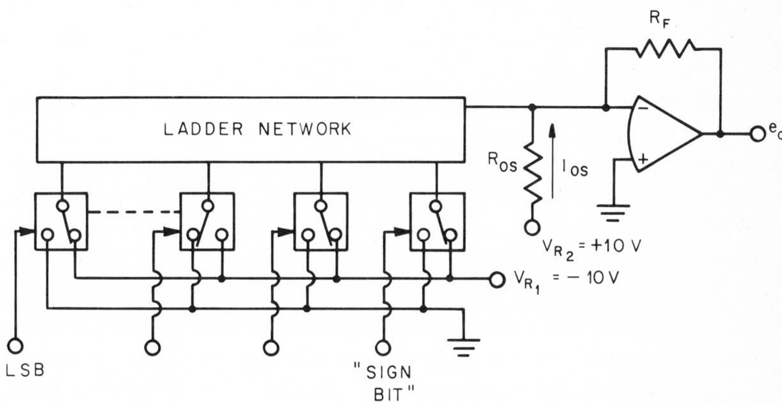

Fig. 9.10 Bipolar D/A converter with offset binary code.

is adjusted to be equal to the current of the MSB. Whether or not the MSB is turned to ON determines the sign of the output; therefore it is called the *sign bit.* The offset binary code is the same as the two's complement code except that the sign bit is reversed. Since the digital inputs to a D/A converter often come from a storage register, the sign bit can be connected to the opposite output of the storage register flipflop, thus making the necessary conversion from the two's complement code to the offset binary code. Using the same approach in Fig. 9.10, a D/A converter can be changed to the one's complement code by decreasing the weight of the offset current and the weight of the sign bit by the weight of the LSB. The one's complement or the two's complement codes can be implemented directly without having to reverse the sign bit, if the opposite reference voltage is connected to the switch of the sign bit only. This eliminates the need for R_{os} but complicates the design of the switch for the sign bit. When these last two methods are used to obtain bipolar operation, the full-scale output will be halved. To keep the original full-scale output, the feedback resistor must be doubled. In a high-speed D/A converter, doubling the feedback resistor can slow down the response of the operational amplifier. Doubling the feedback resistor also makes the resolution of an n-bit (including sign bit) converter half that of an n-bit unipolar converter.

9.2.4 Multiplying D/A converter A multiplying D/A converter (MDAC) has two inputs: one an analog voltage and the other a digital word. The analog output is the product of the two inputs. Any D/A converter is a special case of the MDAC since the reference supply is a fixed analog input. If the reference supply is made to follow an analog input restricted between the range of V_R and 0 V, then a unipolar D/A converter becomes a one-quadrant MDAC. Since an operational amplifier has low output impedance, it can be used to drive the reference supply input. The accuracy of the MDAC will be determined mainly by errors introduced by the switches in the resistor network (assuming no errors are introduced by the amplifier driving the reference voltage line) since the current through the switches will vary from the full-scale value to zero.

Two methods for building a two-quadrant MDAC exist. The first method uses a bipolar D/A converter which employs a unipolar analog input and a bipolar digital input code. The second method uses a unipolar D/A converter which will accept a bipolar analog input and a unipolar digital input code. The method chosen will depend upon which of the two inputs is bipolar. The second method requires switches that will operate with bipolar currents; however, the first method requires an additional unity-gain inverting operational amplifier to produce the opposite reference voltage. When both of these methods are combined a four-quadrant MDAC is the result.

9.3 Analog-to-Digital Converters[3–6]

The analog-to-digital (A/D) converter translates the language of the analog world into the language of the digital world. The analog signal is presented to the input of the A/D converter and after a finite amount of conversion time the digital output is available for use by a digital computer. Several methods of converting analog data to digital data exist. Only two basic techniques will be discussed here: first, A/D converters that use parallel entry D/A converters in their feedback loop and, second, the dual-slope integrator type.

9.3.1 A/D converters that use a D/A converter At least three types of A/D converters use a parallel entry D/A converter in their feedback loop. The three types are the counter ramp (or precision-ramp comparator), the continuous counter ramp (the up-down counter, also called the servo), and the successive approximation. These types are shown in Figs. 9.11, 9.12, and 9.13, respectively. All three have several features in common in addition to the D/A converter. The comparator which is identical in each of the three converters compares the magnitude of the analog input current against the magnitude of the current provided by the D/A converter. The comparator output and the clock input control the operation of the digital logic. The outputs of the digital logic section control the switches of the D/A converter, as well as being the binary output lines of the A/D converter. Whenever any of the outputs of the digital logic are at a logical 1, the corresponding bit in the D/A converter will be turned to ON. The purpose of the digital logic is to turn to ON the proper bits of the D/A converter so that the current i_3 is equal to the current i_1 (see Fig. 9.11). The equation for the currents at the summing junction of the comparator is $i_2 = i_1 - i_3$, where $i_1 = e_1/R_{in}$. When i_2 is greater than zero, the comparator output will be LOW (−0.6 V). This means that the digital word is low in magnitude since i_3 is less than i_1. Likewise, when i_2 is less than zero, the comparator output will be HIGH (15 V) which says that the digital word is too large. The three types of converters differ in the methods used in the digital logic section to find the digital word that will make i_3 equal to i_1. In reality, i_3 may never equal i_1 exactly, since i_3 exists in 2^n discrete steps, each step being equal to the weight of the LSB. For this reason A/D converters have a built-in uncertainty of at least $\pm\frac{1}{2}$ LSB.

The A/D converters shown in Figs. 9.11 to 9.13 accept unipolar inputs; however, they will also operate with bipolar input when an offset current equal to the weight of the MSB is injected into the summing junction of the comparator. The MSB will then be the sign bit. See Sec. 9.2.3 for more information. In the following parts of this section we will consider the three types of A/D converters that use a D/A converter in more detail.

9.3.2 The counter ramp A/D converter The first of the three types of A/D converters using a D/A converter that we will consider is the counter ramp A/D converter shown in Fig. 9.11. The digital logic portion of this converter is a resettable binary counter. The CONVERT command resets the counter to zero. At the beginning of the conversion period, i_1 will be greater than i_3, making i_2 positive and the output of the comparator LOW. The NAND gate will allow the clock pulses to advance

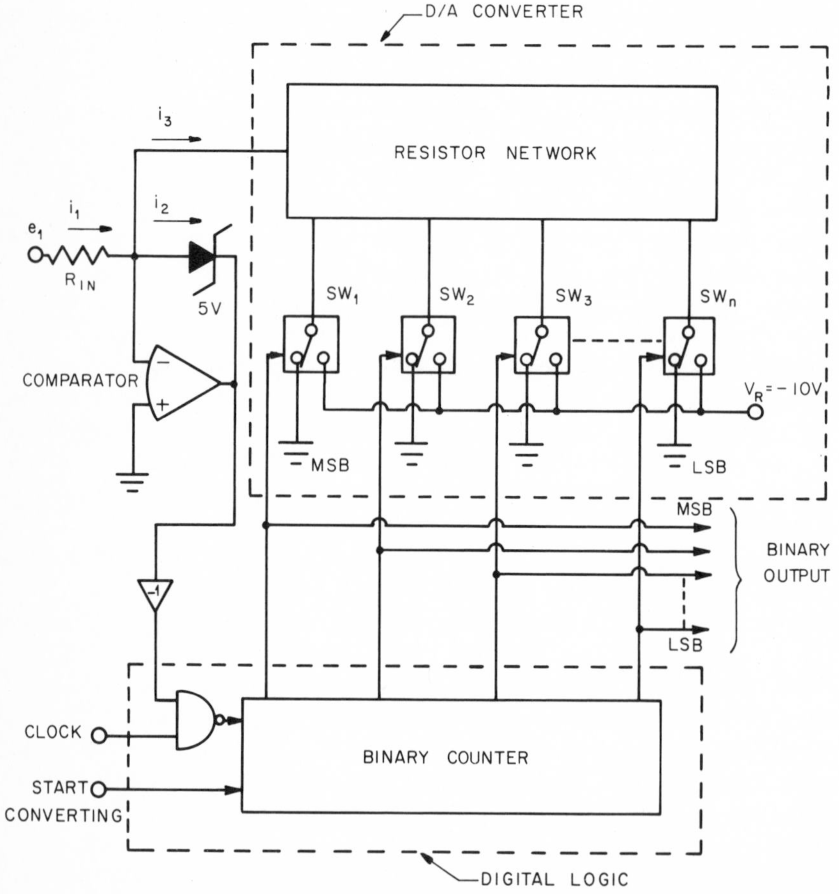

Fig. 9.11 Counter ramp A/D converter.

the counter. The current i_3 will increase in a staircase fashion, until i_3 becomes greater than i_1, the comparator output will go to HIGH, and the NAND gate will inhibit the clock pulses, thus stopping the counter.

The acquisition time of the ramp converter will depend on the magnitude of the analog input and on the clock frequency. Since the D/A converter output changes by the weight of the LSB each time the clock makes a positive transition, the slew rate of the converter will be LSB/$(1/f_c)$, where f_c is the clock frequency expressed in hertz and the LSB is expressed in volts. For example, the LSB of a 10-bit converter with a full-scale input of 10.230 V is 10 mV. When the clock frequency is 1 MHz, the slew rate of the converter is (10 mV)$(1 \times 10^6$ Hz) or 10 mV/μs. For a full-scale input at 10.230 V, the converter would take (10.230 V)/(0.01 V/μs) = 1,023 μs to count from the digital word of 0000000000 to 1111111111.

The logic for a counter ramp converter is straightforward, but the conversion time is long since the binary counter must always start at zero at the beginning of the conversion. The converter described in the next section overcomes some of the disadvantages of the counter ramp converter at the expense of increased complexity in the digital logic circuitry.

The basic accuracy of this type of converter is determined by the voltage offset and bias current of the operational amplifier performing the comparator function.

9.3.3 The continuous counter ramp A/D converter The second of the three types of A/D converters using a D/A converter that we shall consider here is the continuous counter ramp A/D converter shown in Fig. 9.12. The continuous counter ramp or servo A/D converter contains a reversible up-down binary counter. When the output of the comparator is LOW (this says that $i_3 < i_1$), the counter counts up so as to increase i_3. When the comparator output becomes HIGH ($i_3 > i_1$), the counter will reverse and count down. This type of converter is always seeking a null; hence its principle of operation is the same as a servo system. The servo A/D converter can follow a continuously changing input signal provided that the rate of change of the input signal does not exceed the slew rate of the converter. The slew rate of this type of converter is the same as for the counter ramp. For small step changes in the input signal the servo converter is fast since it does not need to be reset to zero before each conversion. In fact, this type of converter has no need for a START CONVERTING command since it converts continuously. A unique feature of the servo A/D converter results because i_3 changes in discrete steps. The digital output will oscillate by the weight of the LSB as soon as the converter acquires the input signal.

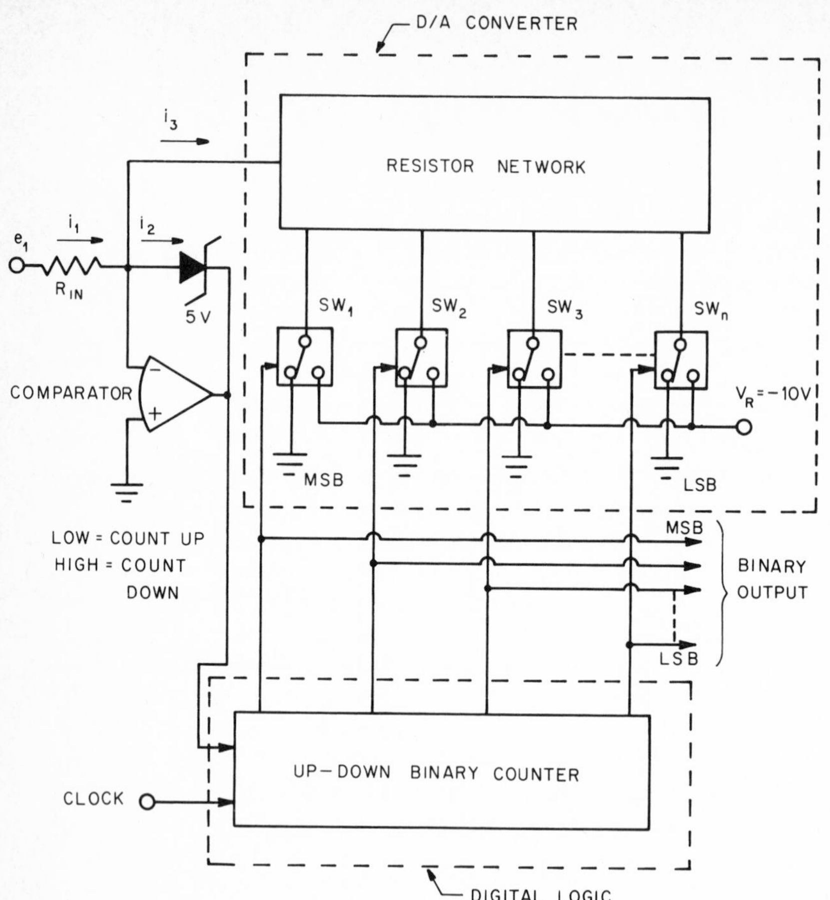

Fig. 9.12 A/D converters with up-down binary counter.

This is analogous to a servo motor "hunting for a null" because of the finite resolution of the wire-wound feedback potentiometer. The increased complexity of the digital logic in a servo A/D converter is not a major disadvantage, because of the availability of integrated-circuit up-down counters. When conversion time is at a premium, neither the servo converter nor the counter ramp method is recommended, especially if the input signal comes from a multiplexer. This is because of the large step changes when switching from one channel to another channel and the consequently long conversion time.

9.3.4 Successive approximation A/D converter The last of the three types of A/D converters using a D/A converter that we shall consider here is the successive approximation A/D converter shown in Fig. 9.13.

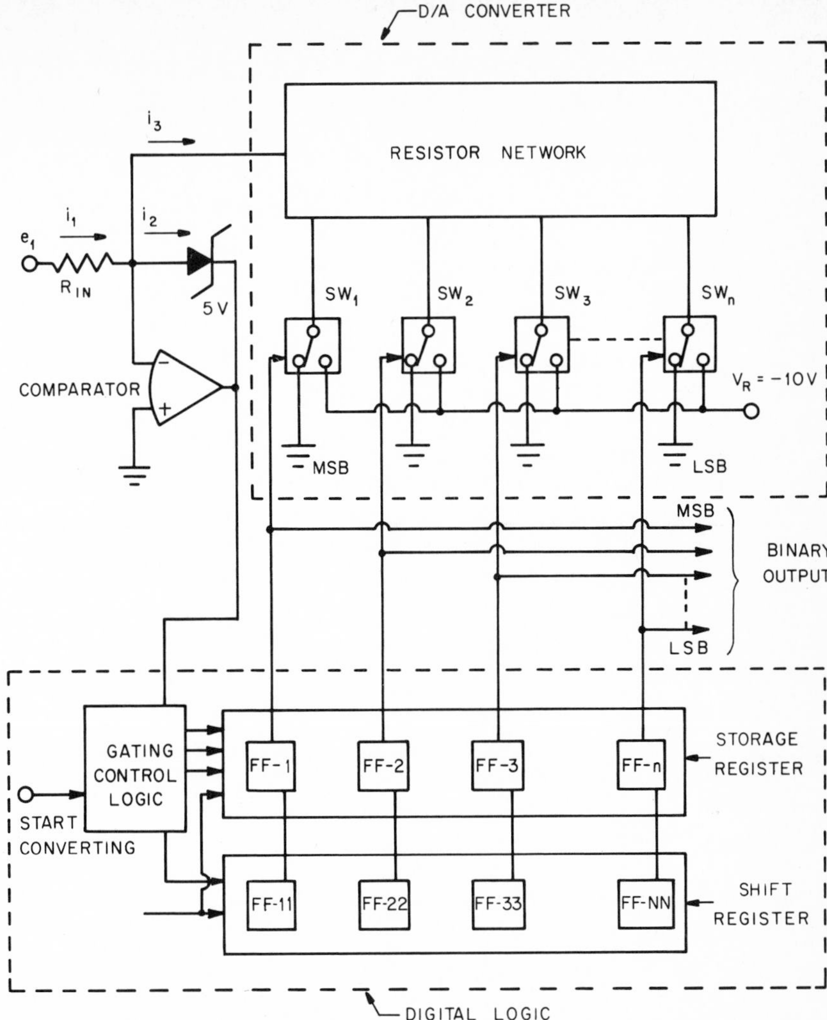

Fig. 9.13 Successive approximation A/D converter.

The n-bit successive approximation A/D converter requires n clock periods to complete a conversion, regardless of the magnitude of the input signal. For general-purpose applications, the successive approximation approach is considerably faster than the previous two methods, but the digital logic is also more complex. The conversion process is not continuous, as for the servo converter, but begins with a CONVERT pulse and ends n clock periods later when the digital output is valid. The input signal must remain constant during the conversion process or the digital output at the end of the conversion period may not be valid. Therefore, it is common practice to precede the A/D converter with a sample-hold circuit.

The converter in Fig. 9.13 requires a CONVERT pulse which resets the flipflops of the shift register and all the flipflops driving the D/A switches to 0 except for flipflop F_1 which is set to 1. A 1 is also shifted into flipflop F_{11} of the shift register. The MSB is now turned to ON, causing the comparator output to go to LOW if i_1 is greater than half-scale or to HIGH if i_1 is less than half-scale. Three events happen at the next negative-going clock: the 1 will be left in flipflop F_1 if the comparator output is LOW or flipflop F_1 will be reset to 0 if the comparator output is HIGH, flipflop F_2 is set to a 1, and a 1 is shifted into flipflop F_{22}. Again, on the next negative-going clock, F_2 will remain a 1 or be reset to 0, depending upon the comparator output, F_3 is set to a 1, and a 1 is shifted into F_{33}. This process continues until the 1 in F_n is either removed or left alone. The conversion is now complete and the digital output is accurate to $\pm\frac{1}{2}$ LSB.

The maximum speed of operation for the three converter designs discussed thus far will depend primarily upon the time required for the comparator to change levels, plus the time required for switching transients to settle out. This is true because the digital logic and the D/A converter will usually be much faster than the comparator. The voltage and current offsets of the comparator will affect the comparison point and thus the accuracy of the A/D converter. (Section 9.6 gives a complete description of comparators.) Errors that are introduced by the D/A converter portion of the A/D converter are caused by drift with time and temperature in the reference voltage, the switch offsets, and the resistor network.

9.3.5 Dual-slope integrator A/D converter The three A/D converters discussed in Secs. 9.3.2 to 9.3.4 are all characterized by requiring the use of a D/A converter. In the following paragraphs we shall discuss a different type of A/D converter, the dual-slope integrator A/D converter. Figure 9.14a shows the block diagram of a dual-slope integrator A/D converter. The essential parts of the converter are the integrator, the zero-crossing comparator, the reference voltage, and the digital logic. This converter is restricted to negative input voltages.

When the converter receives the RESET pulse, the flipflops of the binary counter are reset to 0. A 0 input to the switch drive circuit opens S_2 and closes S_1, connecting the input to the integrator. As soon as the integrator output crosses zero the comparator output will go to LOW, allowing clock pulses through the NAND gate to the counter. During time interval T_1 the counter counts up to 01111.....1, and on the next clock pulse the counter will switch to 10000.....0. The 1 in the last

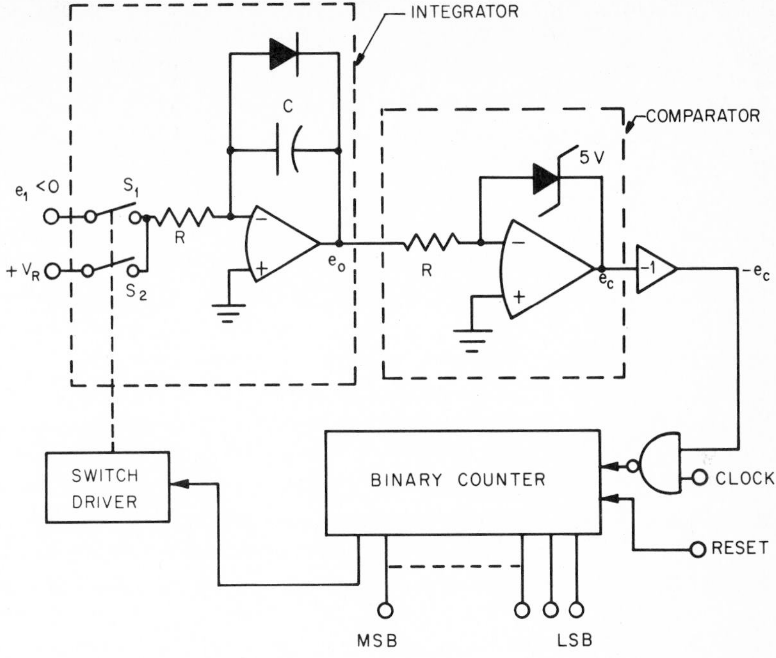

(a) BLOCK DIAGRAM

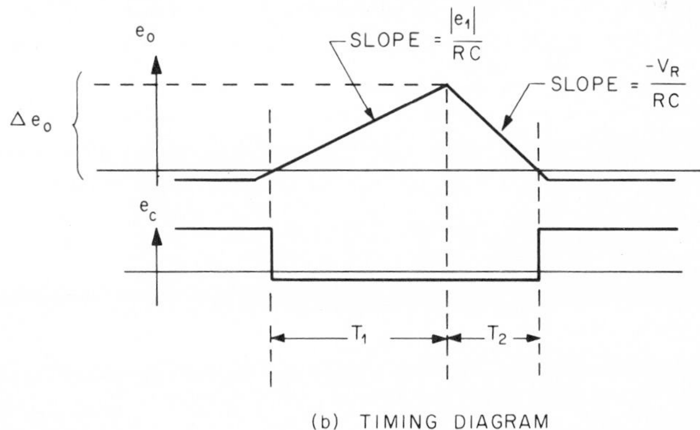

(b) TIMING DIAGRAM

Fig. 9.14 Dual-slope integrator A/D converter. (a) Block diagram; (b) timing diagram.

flipflop causes the switch drive to open S_1 and close S_2. This connects the reference voltage which causes the integrator to integrate down. When its output crosses zero the comparator output will go to HIGH, inhibiting clock pulses to the counter. This completes the conversion and the digital output is now valid. The applicable equations are

$$\Delta e_o = -\frac{1}{RC}\int_0^{T_1} e_1\,dt = -\frac{e_1 T_1}{RC} \qquad e_1 < 0$$

$$\Delta e_o = \frac{1}{RC}\int_0^{T_2} V_R\,dt = \frac{V_R T_2}{RC} - \frac{e_1 T_1}{RC} = \frac{V_R T_2}{RC}$$

or

$$T_2 = \frac{|e_1| T_1}{V_R}$$

Thus, the counter actually measures the interval T_2 which is proportional to e_1. The voltage e_1 is assumed constant during the measurement interval. If e_1 varies, the digital output represents its average value over the interval T_1. This type of converter has the advantage that the tolerances of R and C do not affect its accuracy. Long-term aging of clock frequency also does not affect the accuracy. The voltage and current offsets of the comparator will cause the comparator to switch at a voltage different from 0 V, but this error is automatically compensated, since its input signal crosses zero twice. The voltage and current offsets of the integrator, unfortunately, are not compensated and will limit the accuracy. Drift in the reference voltage will also degrade the achievable accuracy.

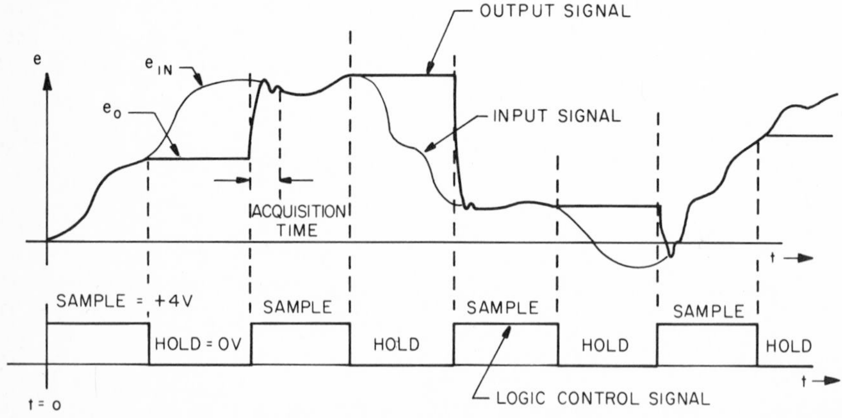

Fig. 9.15 Input-output waveforms of a sample-hold module.

9.4 Sample-hold Modules[2,15]

A sample-hold module is a device that tracks an input signal and then holds the instantaneous input value upon command by a logic control signal, as illustrated in Fig. 9.15. Sample-hold modules are often used with measuring devices that cannot tolerate a time-varying input signal, such as some types of analog-to-digital converters. Other applications are analog delay, phase sensing, period measurement, and measurement of short-term parameter stability.

9.4.1 Sample-hold fundamentals The simplest sample-hold circuit is the switch and capacitor of Fig. 9.16. Two important specifications can be easily illustrated by using the basic circuit of Fig. 9.16. These specifications are aperture time and acquisition time.

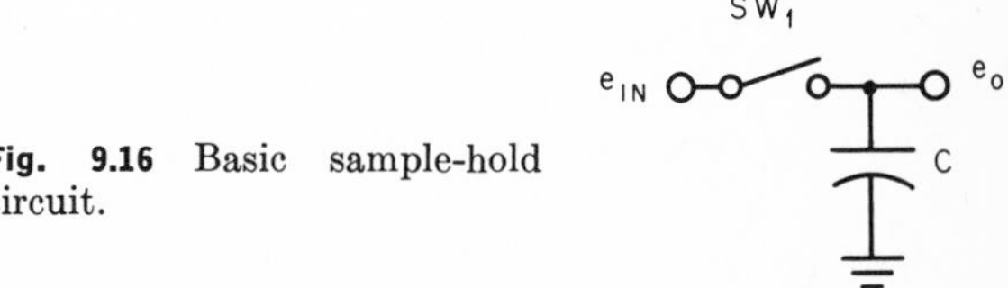

Fig. 9.16 Basic sample-hold circuit.

Aperture time is the delay (reaction time) between the time the control logic tells switch SW_1 to open and the time this actually happens. When very long (milliseconds) aperture time can be tolerated, SW_1 can be a relay. With FET or bipolar transistor switches aperture times of less than 100 ns are feasible. Figure 9.17 illustrates the holding error caused by aperture time.

In time-varying systems the input signal to a sample-hold changes while the sample-hold is holding a value, and so the time required for the sample-hold to acquire the new value of input signal (to within a stated

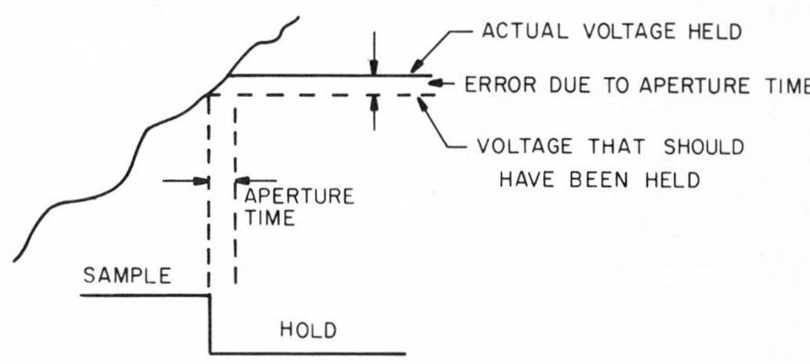

Fig. 9.17 Aperture-time error.

accuracy) when the sample-hold is switched from HOLD to SAMPLE, is important. This is the acquisition time. The worst case occurs when the output of the sample-hold must change full scale, for example, from −10 to +10 V or from +10 to −10 V. If a circuit like the one in Fig. 9.16 is used, acquisition time will depend upon the driving source and the available charging current $C\, de_o/dt < I_{max}$, the maximum current available from the signal source. If e_i has a source impedance (R_s), then e_o will change exponentially with the time constant $\tau = R_sC$. For e_o to settle to within 0.01 percent of the input will require approximately $9R_sC$ (seconds). If e_i is the output of an operational amplifier, the acquisition time will be determined by the output current capability, the slew rate, and the settling time of the operational amplifier. Since a sample-hold is a combination of switching circuits and analog circuits, switching spikes will occur because of the interelectrode capacitance of switches and stray capacitance. In some systems, switching spikes can be disastrous, especially in servo or display systems where the load is inductive. When a sample-hold module switches into HOLD, a small amount of charge is transferred from the holding capacitor, because of the interelectrode capacitance of the switch. The voltage change associated with this charge offset is known as the sample-to-hold offset error. During the HOLD mode a small portion of the input signal feeds through the capacitance of the switch to the output. This feedthrough increases with increasing input frequency, but the effect can be decreased by making C larger. When a sample-hold is in the HOLD mode, leakage currents will cause the output voltage to drift at a rate of

$$\frac{\Delta E}{\Delta t} = \frac{I_{leakage}}{C}\left(\frac{\text{volts}}{\text{second}}\right)$$

These leakage currents can be the bias current of an operational amplifier, the OFF leakage current of a switch, or the internal leakage of the holding capacitor. If the operational amplifier has an FET input stage and the switch is an FET device, the leakage current (and therefore the drift) will double for every +10°C rise in temperature.

As the temperature deviates from 25°C, the input offset of such an amplifier will drift, causing an output offset during the SAMPLE mode. The gain accuracy in SAMPLE mode may also change with temperature.

9.4.2 Sample-hold circuit There are two types of sample-hold circuits, namely, inverting and noninverting. The inverting sample-hold circuit of Fig. 9.18 shows a simple sample-hold circuit that responds to step inputs with the time constant $\tau = RC$. When e_i is a step input (this occurs, effectively, when switch SW_1 is closed), C is charged to $e_o = -e_i$.

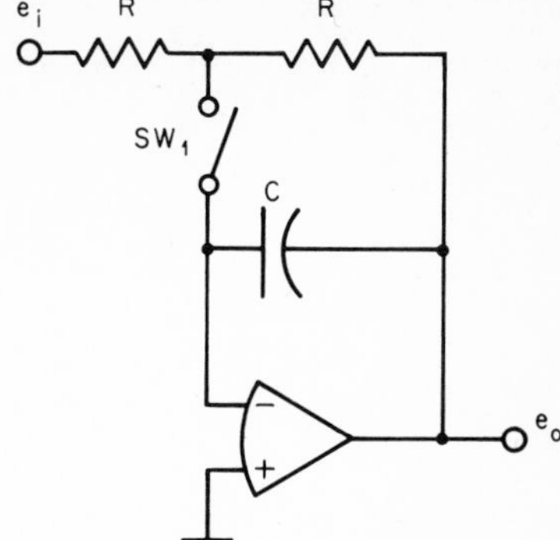

Fig. 9.18 Sample-hold circuit.

During the time switch SW_1 is open, the time-varying input signal has changed to another value, e_i'. Therefore when switch SW_1 is closed, e_o will proceed to the new value of input according to the equation

$$e_o = -e_i + (e_i - e_i')(1 - e^{-t/RC})$$

Switch SW_1 can be a relay, FET, or diode bridge switch. The acquisition time of the circuit in Fig. 9.18 can be speeded up considerably by using a switch which has current gain as shown in Fig. 9.19. Analysis of this

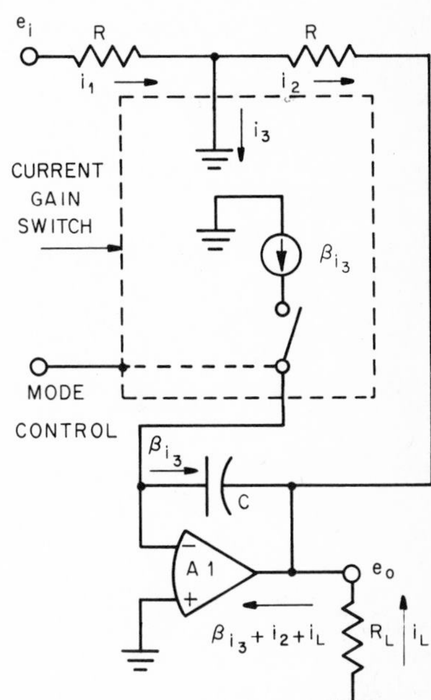

Fig. 9.19 Sample-hold with current-gain switch.

circuit shows that its performance is governed by the following equations:

$$i_1 - i_3 = i_2$$

$$i_1 = \frac{e_i}{R}$$

$$i_2 = \frac{-e_o}{R}$$

$$e_o = \frac{-1}{C}\int \beta i_3\,dt$$

and so

$$I_3(s) = \frac{-E_o(s)Cs}{\beta}$$

For a step function input,

$$E_i(s) = \frac{E}{s}$$

$$e_o(t) = -E(1 - e^{-t/RC\beta})$$

Since the amplifier and the current switch can deliver only a finite current, following a step input, e_o will slew at a rate $de_o(t)/dt = I/C$, where I is the maximum current that the switch will deliver. (The amplifier must also be able to supply this current, plus the current i_2 and any load current to R_L.) The slew continues until the current through C drops below the maximum current that the current-gain switch will deliver. Then e_o will settle exponentially, with time constant RC/β. The current-gain switch will have nonzero voltage offset and input bias current. These effects can be compensated by summing a small current into the input of the current switch. The inverting sample-hold has the advantage that it has very low output impedance and therefore can drive loads without affecting the decay in the HOLD mode. The disadvantage is that the input impedance is fairly low, being equal to R.

As an example of a noninverting sample-hold device, consider the circuit shown in Fig. 9.20. This circuit has very high input impedance and has an acquisition time determined by the time constant $R_{ON}C$,

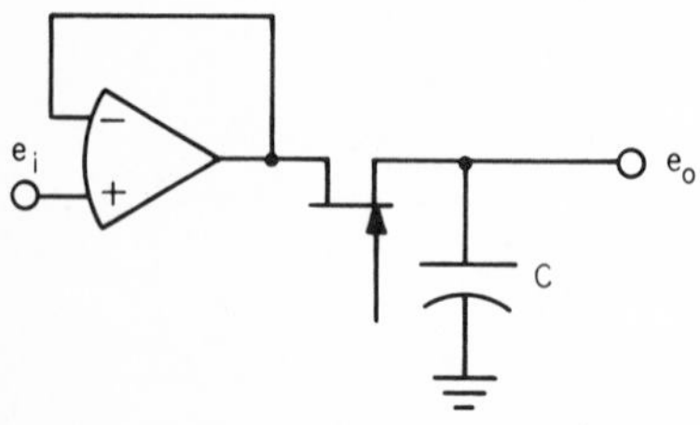

Fig. 9.20 Single-operational-amplifier noninverting sample-hold.

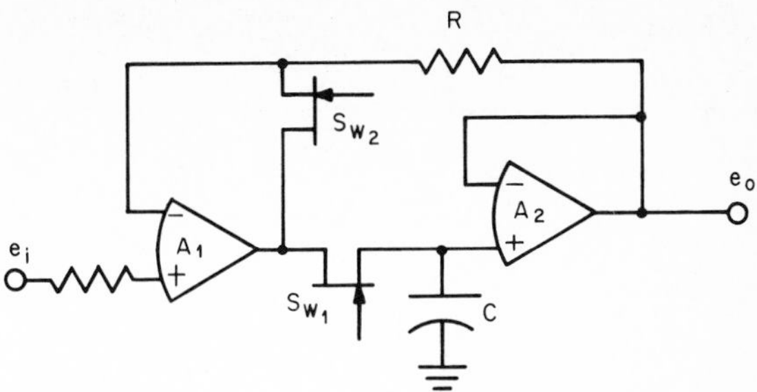

Fig. 9.21 Noninverting two-operational-amplifier sample-hold.

provided $C\ de_o/dt < I_{max}$, where R_{ON} is the ON resistance of the FET switch and where I_{max} is the maximum current the amplifier will deliver or the I_{DSS} current of the FET (whichever is smaller). This circuit has the disadvantage that the output cannot be loaded if small decay in HOLD is required. An extension of this circuit will provide output buffering and will eliminate the $R_{ON}C$ time-constant limitation. Such a circuit is shown in Fig. 9.21. Placing switch SW_1 inside the feedback loop enables A_1 to deliver its maximum current through SW_1 until C is fully charged. In the HOLD mode, switch SW_1 is opened and SW_2 is closed. Switch SW_2 provides feedback for A_1. This noninverting sample-hold has the advantage that the input impedance is very high. The gain accuracy of this circuit is determined by the open-loop gain linearity and CMR of A_1. For open-loop gain and CMR in excess of 80 dB, the closed-loop gain error may be less than 0.01 percent.

9.5 Peak Detectors[16,17]

A peak detector is a special kind of sample-hold circuit. The input signal is tracked until the input reaches a maximum value and then the peak detector automatically holds the peak value. For a noninverting unity-gain peak detector designed to detect positive-value peaks, the output and input waveforms are expressed graphically, as shown in Fig. 9.22.

Peak detectors can save considerable expense, especially if the only alternative is to combine an analog-to-digital converter and a digital computer to find peak values. Typical applications of peak detectors include transient waveform analysis and repetitive waveform analysis. Two specific applications include the output waveform analysis of gas chromatographs and mass spectrometers.

9.5.1 Design considerations The simplest peak detector circuit is the diode and capacitor circuit shown in Fig. 9.23. In the PEAK DETECT

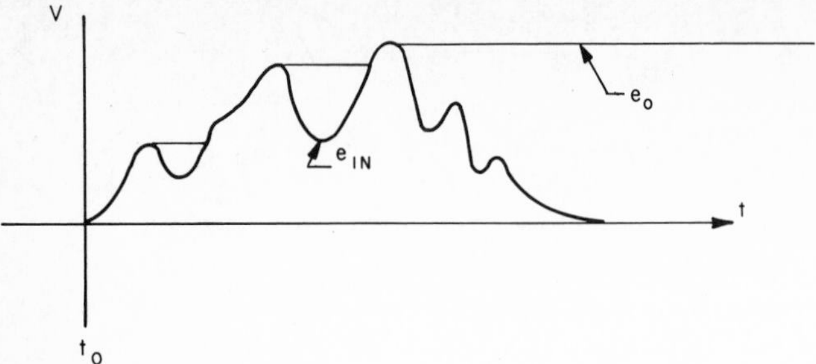

Fig. 9.22 Input and output waveforms of a peak detector.

mode, switch SW_1 is closed and SW_2 is open. Diode D_1 allows current to flow in one direction only to charge the holding capacitor, C_1. When e_i becomes less than e_o (ignoring the drop across D_1), D_1 becomes reverse-biased, and C_1 holds the peak value. To reset the circuit (RESET mode), switch SW_1 is opened and SW_2 is closed. A third mode of operation, the HOLD mode, is desirable in some applications. Both SW_1 and SW_2 are opened so that C_1 holds the "desired" peak value while ignoring other larger value inputs. Because of the leakage current I_d of D_1 the voltage stored on C_1 will decay at a rate given by $\Delta e_o/\Delta t = I_{d\,leakage}/C_1$(volts/second), when switch SW_1 is closed. Any shunt impedance loading C_1 causes an exponential decay of time constant $\tau = R_{load}C_1$. Another output, often useful, tells the status of the peak detector. The status output is a bilevel digital signal which changes state at the precise time that D_1 stops conducting.

The circuit shown in Fig. 9.23 has several drawbacks. For precision measurements, the nonlinear drop across D_1 is difficult to compensate, especially if the accuracy of the detector is to be independent of temperature variations and the frequency and amplitude response of the input

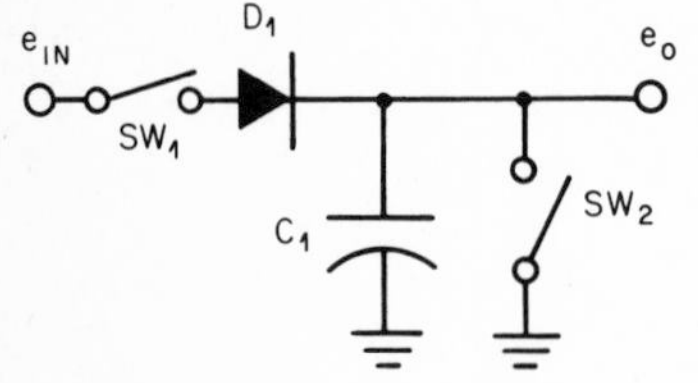

Fig. 9.23 Basic peak detector circuit.

signal. When the circuit of Fig. 9.23 is enclosed inside the feedback loop of an operational amplifier, the disadvantages of the basic circuit are largely eliminated. However, certain precautions must be observed when using operational amplifiers. A peak detector circuit must be either critically damped or overdamped, because any overshoot in the step response will be held as the peak value. Eliminating overshoot can be difficult in circuits where two or more operational amplifiers are enclosed inside the same feedback loop. Also, the amplifiers must be stable when driving a capacitive load. Precautions must be taken to prevent the operational amplifiers from overloading after a peak is detected, since the feedback loop is broken when D_1 becomes reverse-biased.

9.5.2 Noninverting peak detector circuits Some of the simplest implementations of the basic peak detector circuits used in Sec. 9.5.1 are the peak detector circuits illustrated in Figs. 9.24 and 9.25. These are connected as unity-gain noninverting followers. First let us consider the single-amplifier circuit shown in Fig. 9.24. Since D_1 is inside the feedback loop, its forward voltage drop is divided by the open-loop gain of the operational amplifier and can be expressed as an equivalent offset in series with the input of an ideal peak detector. A_1 serves two other useful purposes. The input signal source needs to supply only the input bias current of A_1. The output rise time is not determined by the time constant of the ON resistance of D_1 times C_1 but is dependent only on the output current capability I_{max} of A_1. The output slew rate is given by $\Delta e_o/\Delta t = I_{max}/C_1$(volts/second), provided this is not larger than the specified maximum slew rate of A_1. D_2 is necessary to prevent A_1 from overloading at negative saturation voltage when E_i is less than E_o. D_2, however, must withstand the short-circuit current of A_1. A_1 should be an FET input amplifier to minimize the decay rate after detecting a peak since the input bias current of the inverting input will discharge C_1. Also, the input stage of the amplifier will not conduct when E_i is less than E_o. If the output is to be loaded, an output buffer amplifier is required to prevent the load from discharging C_1.

The two-amplifier circuit of Fig. 9.25 overcomes the problems of the single-amplifier circuit. A_1 never locks up, and the output buffer amplifier A_2 is part of the circuit. The operation of this circuit is practically the same as the first circuit. In the two-amplifier circuit, however, A_2 operates as a unity-gain follower inside the overall feedback loop. When e_i becomes less than e_o, D_2 conducts, supplying feedback for A_1. This prevents A_1 from overloading. Capacitor C_f is required to stabilize the loop and prevent overshoot for a step input signal. In both circuits A_1 must be stable while driving the capacitive load C_1, and A_1 must have

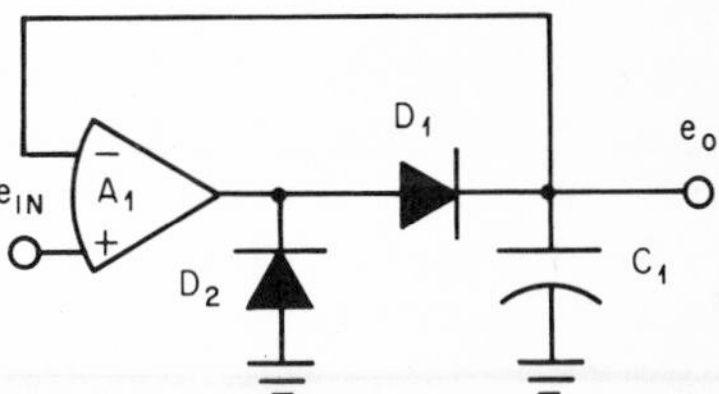

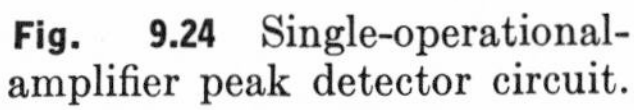
Fig. 9.24 Single-operational-amplifier peak detector circuit.

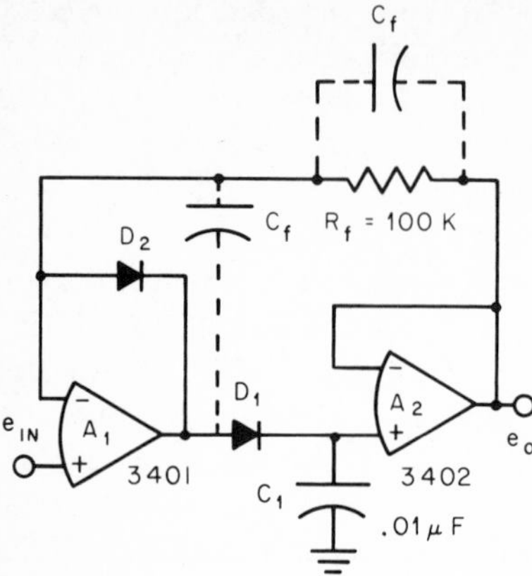

Fig. 9.25 Improved circuit with two operational amplifiers.

good CMR. In the improved circuit, A_1 need not be an FET input amplifier, but A_2 should have an FET input if long holding times with small decay are required.

9.5.3 Inverting peak detector circuits In Sec. 9.5.2 we considered peak detector circuits which were noninverting. In this section we consider ones which are inverting. Two such circuits are shown in Figs. 9.26 and 9.27. Although the peak detecting circuit of Fig. 9.26 takes three operational amplifiers, closed-loop stability is easier to obtain than in the noninverting circuits since A_2 is connected as an integrator and does not

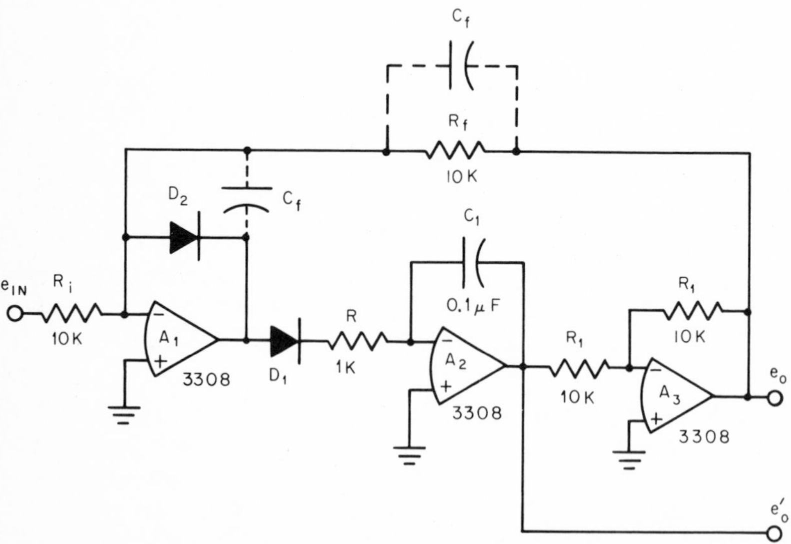

Fig. 9.26 Three-operational-amplifier peak detector.

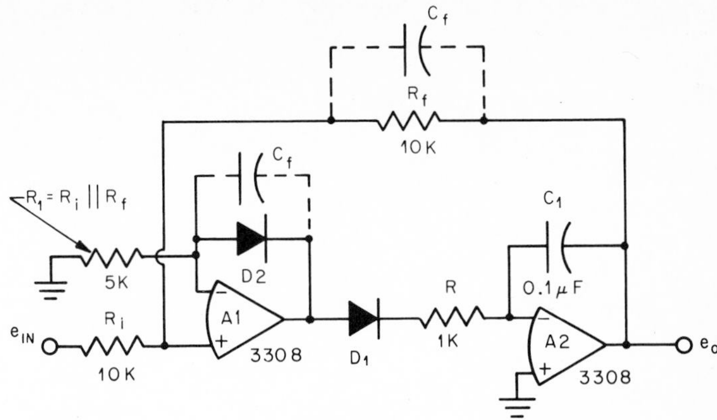

Fig. 9.27 Detector circuit of Fig. 9.26 reduced to two operational amplifiers.

have to drive a capacitive load to ground. Diode D_1 ensures that C_1 can be charged in one polarity only. When tracking the input signal, the output signal is the negative of the input, that is, $e_o = -e_i(R_f/R_i)$. Usually one makes $R_f = R_i$ for unity-gain operation. The input impedance to this circuit is R_i, which requires that the input signal source impedance be small. D_2 provides feedback for A_1 after a peak is detected. A_3 is connected as a unity-gain inverter to provide the proper phase relationship from input to output, and C_f stabilizes the loop. The voltage $e_o' = +e_i(R_f/R_i)$ is available at the output of A_1.

The three-operational-amplifier circuit of Fig. 9.26 can be reduced to the two-amplifier circuit shown in Fig. 9.27 without sacrificing any flexibility. The required inversion from input to output is maintained by making the + input of A_1 the input summing junction. Since the summing junctions of each of the operational amplifiers in Figs. 9.26 and 9.27 are at virtual ground when in PEAK DETECT mode, the CMR of the amplifiers is not important. It should be noted that both of the inverting peak detector circuits described above provide output isolation between the holding capacitor and an output load.

Any of the positive peak detector circuits shown can be converted to detect negative input peaks by simply reversing the direction of the diodes.

9.5.4 RESET and HOLD mode circuits Manual switches like the ones shown in Fig. 9.23 can be added to any of the peak detector circuits. In the circuits of Figs. 9.26 and 9.27, switch SW_2 is placed in parallel with C_1. For the RESET mode SW_1 is opened and switch SW_2 is closed; for the HOLD mode both SW_1 and SW_2 are opened. For electronic operation switches SW_1 and SW_2 can be JFET or MOSFET transistor

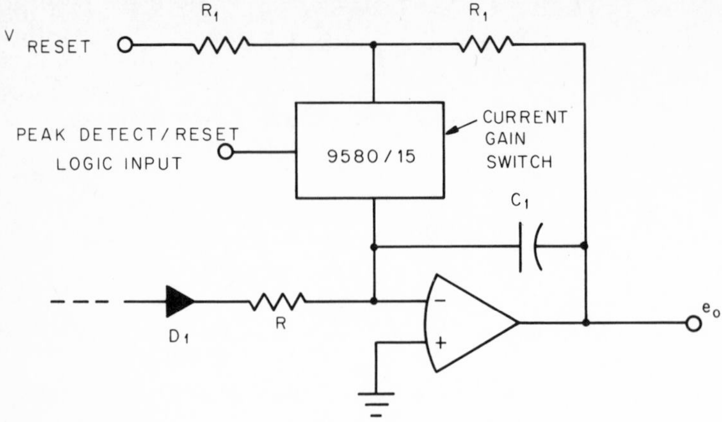

Fig. 9.28 Fast reset circuit for inverting-type peak detectors.

switches. If fast resetting is needed, C_1 and the ON resistance of SW_2 must be as small as practical. Another method of implementing the HOLD mode is to place a switch in series with the input signal to "gate" it to OFF. When only the PEAK DETECT and RESET modes are needed, switch SW_1 can be eliminated if SW_2 has very low impedance such as that of relay contacts. Figure 9.28 shows a very fast method for resetting the inverting-type peak detectors with a current-gain sample-hold switch. (See Sec. 9.4.)

This circuit has the advantage that the switch in series with D_1 is not necessary since the sample-hold switch has very low output impedance. To reset C_1 to 0 V, V_{RESET} must be zero. Capacitor C_1 can be reset to +10 V by making V_{RESET} equal to −10 V so that input peaks can be detected over the full −10 to +10 V range.

9.5.5 Peak-to-peak detector Figure 9.29a shows a method for measuring the peak-to-peak value of a signal that swings both positive and negative in amplitude. If only positive peak detectors are available, the circuit can be built using one extra amplifier as shown in Fig. 9.29b.

9.6 Comparators[15–17]

Comparators are used as analog/digital (hybrid) building blocks, since the digital output signal is simply the answer to the question: Is the analog input signal greater than or less than the analog reference signal? The input and reference signals can come from voltage or current sources or from a combination of the two types of sources. When operational amplifiers are used as comparators, there are usually one or more summing resistors connected to the inverting input (summing junction) of the operational amplifier. One can think of the circuit as comparing currents

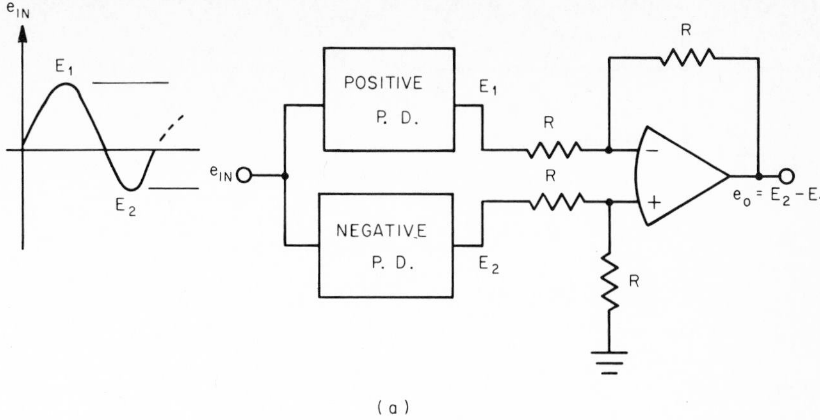

(a)

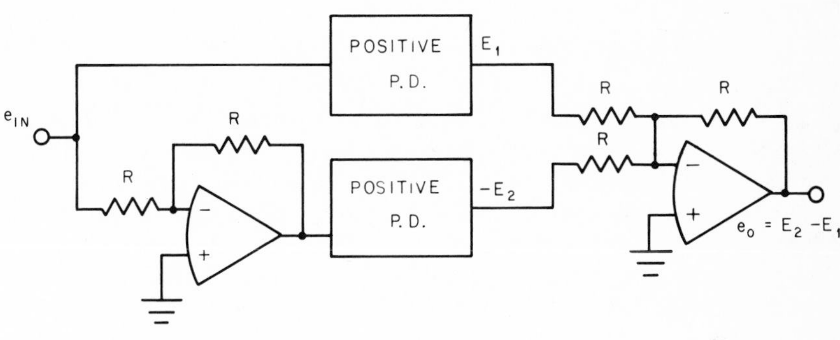

(b)

Fig. 9.29 Peak-to-peak detector circuits: (a) using one positive and one negative peak detector; (b) using two positive peak detectors.

or voltages. In this section we shall consider several types of such comparators.

9.6.1 Zero-crossing detector The simplest comparator is the zero-crossing detector which answers the question: Is the input signal greater than or less than zero? A typical circuit for such a detector is shown in Fig. 9.30. The limit circuit shown in the figure produces one output level when i_3 is positive and a different output level when i_3 is negative. Since the limit circuit changes state when i_3 changes sign, the comparison point occurs when $i_3 = 0$ as shown below (assume the summing junction potential is zero):

$$i_1 = I_2 + i_3 \tag{9-2}$$

$$\frac{e_1}{R} = I_2 + i_3 \tag{9-3}$$

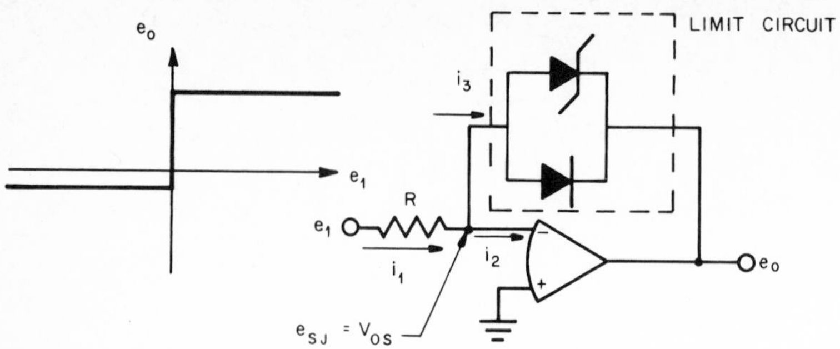

Fig. 9.30 Zero-crossing comparator.

When $i_3 = 0$,

$$\text{Current comparison:} \qquad \frac{e_1}{R} = i_1 = I_2 \tag{9-4}$$

or

$$\text{Voltage comparison:} \qquad e_1 = I_2 R \tag{9-5}$$

Equations (9-4) and (9-5) show that the comparison point occurs when the input current balances the input bias current I_2. To eliminate the error due to I_2, a resistor equal to R can be connected from the + input of the operational amplifier to ground. Similarly a comparison-point error will occur if the voltage offset of the operational amplifier is not zero, as shown below:

$$i_1 = I_2 + i_3 \tag{9-6}$$

$$\frac{e_1 - V_{os}}{R} = I_2 + i_3 \tag{9-7}$$

When $i_3 = 0$,

$$e_1 = V_{os} + I_2 R \tag{9-8}$$

The voltage offset can be adjusted to zero at a specific temperature such as +25°C but should be considered at other temperatures because of the unavoidable voltage drift. The error caused by the bias current is I_2R. As discussed above, this error can be eliminated by connecting a resistance of value R between the + input and ground, provided the bias currents of the + and − inputs are equal. If they are not equal, the difference is called the differential bias current (current offset), and Eq. (9-8) can still be used except I_2 is now the differential bias current.

The input impedance of the zero-crossing comparator is R. The circuit has the disadvantage that noise on the input signal will cause i_3 to be "noisy" and therefore e_o will "chatter" when i_3 is changing sign. A solution to this problem will be discussed in Sec. 9.6.4. Many different limit circuits can be used with the comparators discussed in this section. For a more complete discussion of limiters, see Sec. 7.2.

9.6.2 Level detector To make a comparison at some level other than 0 V, the circuit of Fig. 9.30 can be modified in one of the ways shown in Fig. 9.31. In Fig. 9.31a two summing resistors are required, but the reference voltage can conveniently be the positive or negative power supply, if the supplies are well regulated. The comparison point is scaled by the ratio R_1/R_2. Voltage offset causes the error shown below. When $i_3 = 0$,

$$e_1 = -\frac{R_1}{R_2}V_{REF} + \overbrace{V_{os}\left(1 + \frac{R_1}{R_2}\right)}^{\text{error term}}$$

Even though R_3 is used to provide bias current compensation, the differential bias current (offset current) will cause an error in the comparison point equal to $I_{os}R_1$, where I_{os} is the offset current.

In Fig. 9.31b only one summing resistor is required, but V_{REF} must be equal to the desired comparison level. The voltage on the inverting input has two major error components: the voltage offset of the amplifier and another error voltage resulting from the finite common-mode rejection of the operational amplifier. Resistor R is needed at the noninverting input if the bias current I_2 causes a significant error. The circuit of Fig. 9.32 is another which may be used to make a comparison at some voltage or current level other than zero. It operates in the following manner. Assume that the limit circuit contains two 6-V zener diodes. When the input signal is negative and is approaching zero, the output voltage e_o will be positive since the currents i_1 and i_3 are negative. The voltage at the two inputs of the operational amplifier will be $+6[R_3/(R_3 + R_4)]$ V; therefore, i_3 will not go to zero until the input voltage is also equal to $+6[R_3/(R_3 + R_4)]$ V. The limit circuit then begins changing to the opposite state (-6 V) and the voltage on the inputs will become $-6[R_3/(R_3 + R_4)]$ V. This action is regenerative because of the positive feedback to the + input (inverting), thus increasing the switching speed. The switching can be speeded up even more by placing a small capacitor (10 to 100 pF) in parallel with R_4. Any noise on the input signal at the time this switching occurs will not trigger the circuit to its original state unless the noise on the input exceeds $-2\{6[R_3/(R_3 + R_4)]\}$ V. Now that the input signal is positive no comparator action will occur until the input crosses zero again and becomes equal to $-6[R_3/(R_3 + R_4)]$ V. This transfer function is shown in Fig. 9.32b.

The summing junction voltage is

$$e_{SJ} = \frac{e_oR_3}{R_3 + R_4}$$

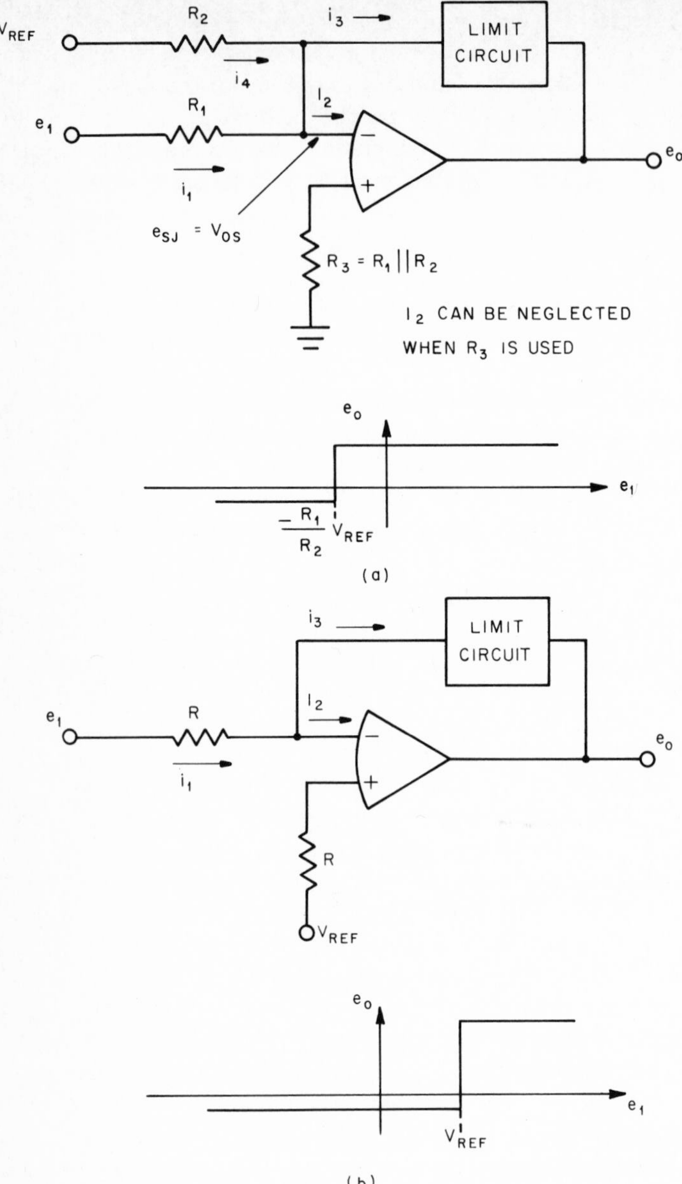

Fig. 9.31 Two-level detector circuits: (a) summing type; (b) differential type.

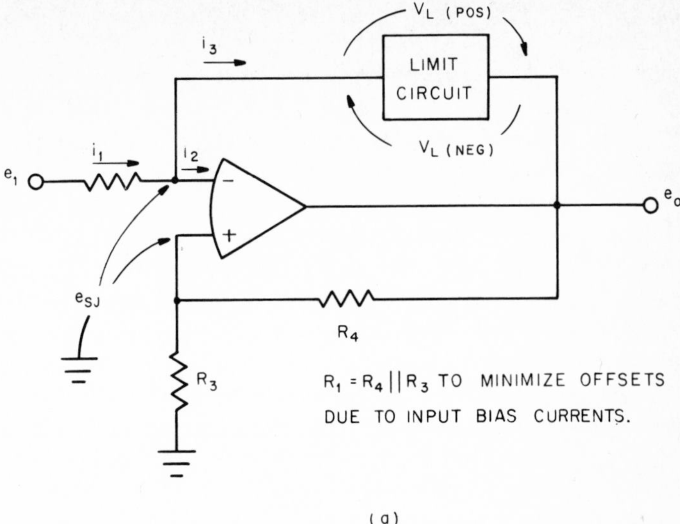

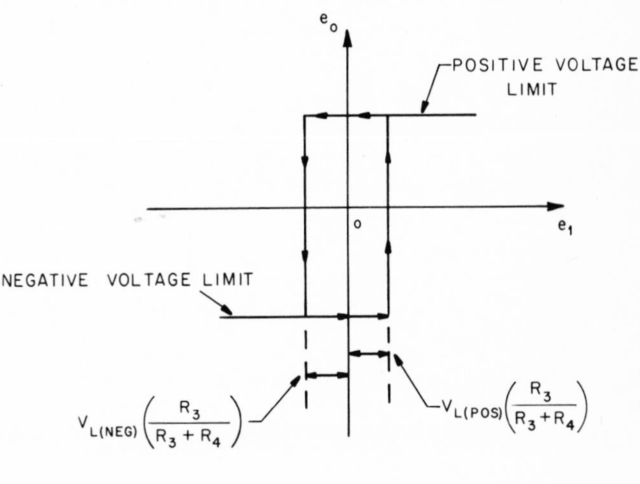

Fig. 9.32 Zero-crossing detector with hysteresis. (a) Circuit diagram; (b) transfer curve.

but

$$e_o = e_{SJ} \pm V_L$$

Thus

$$e_{SJ} = \pm V_L \frac{R_3}{R_4} \qquad \text{(hysteresis)}$$

and

$$e_o = \pm V_L\left(1 + \frac{R_3}{R_4}\right)$$

If hysteresis is added to the circuits of Fig. 9.31a and b, the hysteresis can still be calculated by using the above equations, except that the hysteresis will be centered about $-(R_1/R_2)V_{REF}$ in Fig. 9.31a and about V_{REF} in Fig. 9.31b. The disadvantage of using hysteresis is that the comparison points do not occur at the zero reference or level. Therefore, a tradeoff must be made between the degree of noise immunity required and the error at the comparison points. Prefiltering the input signal to reduce the input noise may be helpful. In order to have symmetrical comparison points about zero, the output voltage limits must be equal in magnitude. In some applications the positive and negative limits will not be the same magnitude, resulting in asymmetry about the nominal comparison point.

9.6.3 Window comparator The final type of comparator that we shall consider in this section is the window comparator. Figure 9.33 shows the circuit diagram and the transfer function for such a comparator. The center of the window is set by the negative of the input V_2, and the window width is twice the input ΔV. Thus the window can be shifted while maintaining constant window width by varying only one voltage (V_2). This feature is useful for probability studies (see Fig. 9.34) where the window width is swept at a constant rate across the analog input range. Similarly, the window width can be varied by a single voltage, without affecting the center of the window.

The window comparator operates in the following way. When $e_1 + V_2 < 0$, D_2 of A_1 is conducting and D_1 is reverse-biased. Therefore the output of A_1 does not contribute to the output of A_2 since the voltage at the junction of R/2 and R/4 is zero. The limiter circuit of A_2 changes sign when the current i_d changes sign, or

$$\frac{e_1}{R} + \frac{V_2}{R} + \frac{\Delta V}{R} = i_d = 0 \tag{9-9}$$

so that

$$e_1 = -V_2 - \Delta V \tag{9-10}$$

Equation (9-10) gives the lower comparison point of the window. When $e_1 + V_2 > 0$, D_1 will be conducting so that the output of A_1 will be $-\frac{1}{2}(e_1 + V_2)$. Another comparison point will be given by

$$\frac{e_1}{R} + \frac{V_2}{R} + \frac{\Delta V}{R} - \frac{\frac{1}{2}(e_1 + V_2)}{R/4} = i_d = 0 \tag{9-11}$$

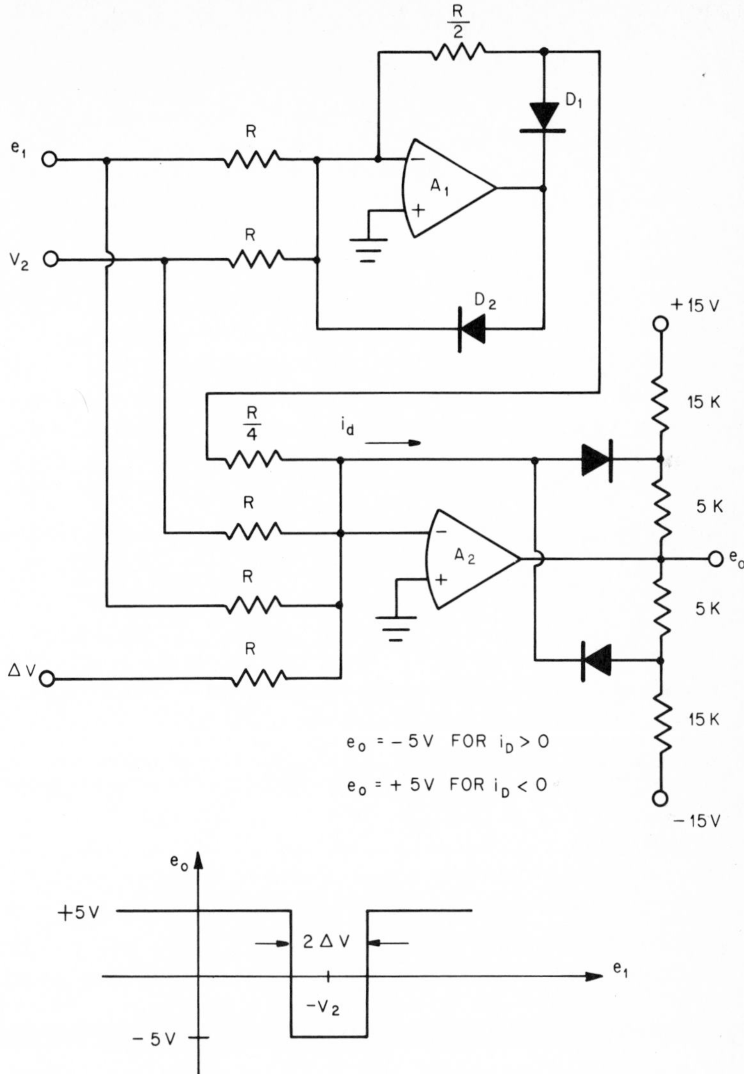

Fig. 9.33 Circuit diagram for a window comparator.

so that

$$e_1 = -V_2 + \Delta V \tag{9-12}$$

which is the upper comparison point.

By adding the appropriate logic gates at the outputs of A_1 and A_2 a window comparator can be given three logic outputs called the GO, HIGH, and LOW outputs. Whenever the input signal is inside the

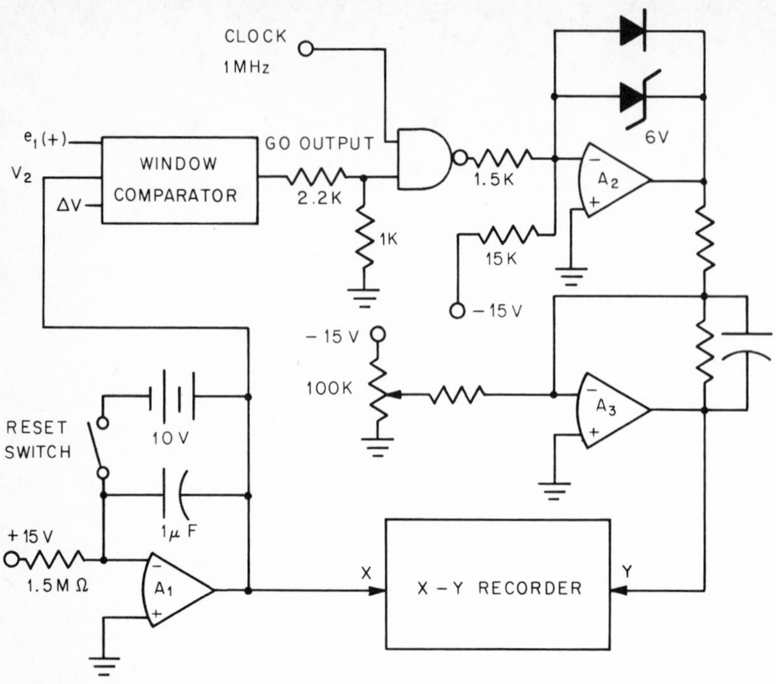

Fig. 9.34 Probability density analyzer.

window, the GO output will be a logical 1 and the other two outputs will be at a logical 0. If the input signal drops to a value below the window, the LOW output will switch to a logical 1 and the GO output will drop to a logical 0. When the input signal exceeds the window, the HIGH output will be at a logical 1 and the other outputs at a logical 0.

The window comparator is shown in Fig. 9.34 as part of a probability density analyzer. The GO output of the window comparator goes to HIGH each time the input signal is inside the window. When the clock and the GO output are both HIGH, the output of the first NAND goes to LOW. A_2 is a comparator that is used to generate precision voltage levels for the RC averaging filter. The 100 kΩ potentiometer is adjusted to null the offset, since the LOW output of A_2 is not 0 V. A_1 is connected as an integrator that sweeps the window center from -10 to $+10$ V at a rate of 1V/s. The window width (2 ΔV) and the rate the window is swept will be a function of the input being analyzed.[1]

9.6.4 Amplitude classifier Often it is required to sort items into many different bins. An example of this is grading apples into different groups according to size. Figure 9.35 shows how a system can be implemented

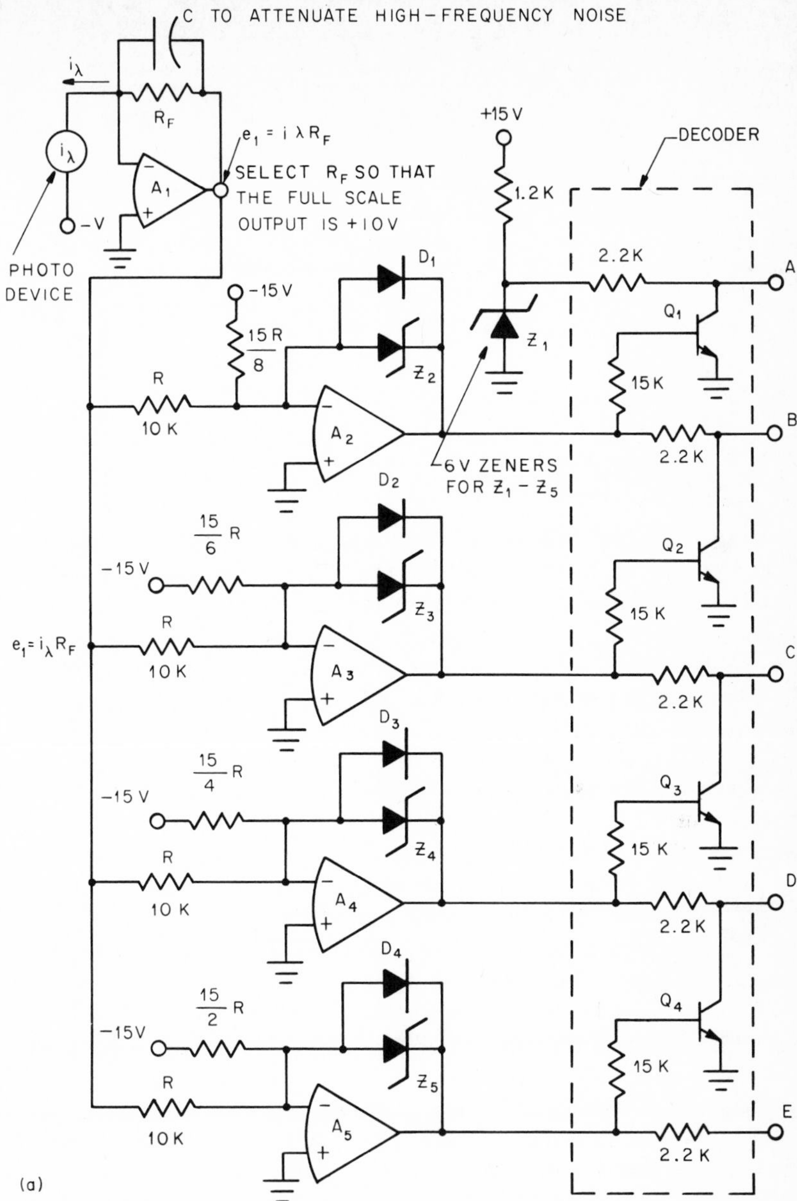

Fig. 9.35 Five-level amplitude classifier. (a) Circuit diagram.

	OP AMP OUTPUTS					DECODER OUTPUTS				
INPUTS e_1	Z_1	A_2	A_3	A_4	A_5	A	B	C	D	E
$8 < e_1 \leq 10$	1	0	0	0	0	1	0	0	0	0
$6 < e_1 \leq 8$	1	1	0	0	0	0	1	0	0	0
$4 < e_1 \leq 6$	1	1	1	0	0	0	0	1	0	0
$2 < e_1 \leq 4$	1	1	1	1	0	0	0	0	1	0
$0 \leq e_1 \leq 2$	1	1	1	1	1	0	0	0	0	1

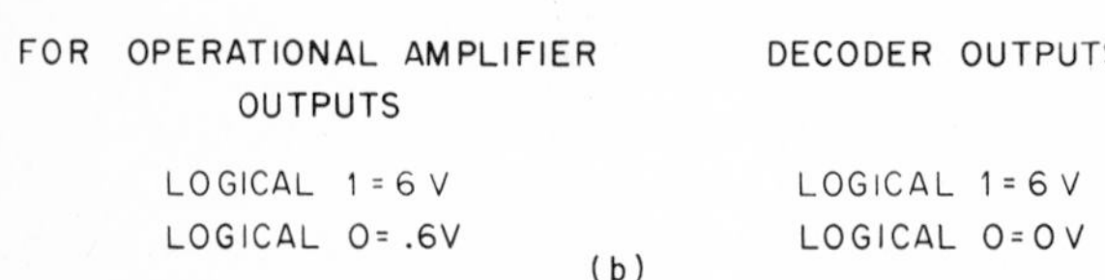

Fig. 9.35 (b) Table of output codes.

by using operational amplifiers as comparators. Amplifier A_1 operates as a current-to-voltage converter for the light-sensitive photodiode array which is used to detect the size of the apples. Amplifiers A_2 to A_5 operate as biased comparators with a simple clamp circuit. The decoder ensures that only one logic output is at HIGH at a time. This decoder can drive TTL (transistor-transistor logic) or DTL (diode-transistor logic) directly or an n-p-n switching transistor which can control larger currents such as the coil current of a relay. Hysteresis can be added to each of the comparators for noise immunity if desired.

REFERENCES

1. Analog-computer Multiplexer, *Columbia Univ. Eng. Center, Final Rep. F/123*, August, 1955.
2. R. N. Close and G. O. Thogersen, Using an Electronic Analog Memory, *Electron. Design*, December, 1955.
3. T. Truitt, A High-speed Analog-Digital Converter, *PCC Rep.* 134, Electronic Associates, Inc., Princeton, N.J., 1958.
4. A. K. Susskind, *Notes on Analog-Digital Conversion Techniques*, John Wiley & Sons, Inc., New York, 1958.
5. *Analog/Digital Conversion Handbook*, Digital Equipment Corp., Maynard, Mass., 1962.
6. Hermann Schmid, A Practical Guide to A/D Conversion, *Electron. Design*, Dec. 5 and 19, 1968.
7. P. F. M. Gloess, Binary Coding by Successive Approximation, U.S. Patent 2,569,927, Oct. 2, 1957.
8. G. J. Herring, Electronic Digitizing Techniques, *J. Br. IRE*, July, 1960, pp. 513–517.
9. R. W. Gilbert, Pulse Time Encoding Apparatus, U.S. Patent 3,074,057, Jan. 15, 1963.

10. F. W. Cheney, Analog-to-Digital Conversion with Threshold Circuits, *IRE Trans. Electron. Computers,* March, 1961, pp. 100–101.
11. D. Savitt, A High-speed Analog to Digital Converter, *IRE Trans. Electron. Computers,* March, 1959.
12. W. N. Moody, A Comparison of Low-level Commutators, *Proc. 1963 Natl. Telemetering Conf.,* Albuquerque, N.M., Sec. 2-4, pp. 1–12, May, 1963.
13. R. E. Mahan, Low-level Multiplexing, *Instrum. Control Syst.,* October, 1969.
14. C. M. Edwards, Precision Electronic Switching with Feedback Amplifiers, *Proc. IRE,* vol. 44, p. 1613, 1956.
15. G. A. Korn and T. M. Korn, *Electronic Analog and Hybrid Computers,* McGraw-Hill Book Company, New York, 1964.
16. *Applications Manual for Operational Amplifiers,* Philbrick/Nexus Research, Dedham, Mass., 1965.
17. *Handbook and Catalog of Operational Amplifiers,* Burr-Brown Research Corporation, Tucson, Ariz., 1969.
18. H. S. Black, *Modulation Theory,* chap. 4, D. Van Nostrand Company, Inc., Princeton, N.J., 1953.

10
WAVEFORM GENERATORS

In this chapter we continue our discussion of the application of operational amplifiers by treating the subject of waveform generation. The chapter is broken down into several sections according to circuits which may be used to generate specific waveforms. Thus, Sec. 10.1 treats square-wave generators, Sec. 10.2 treats square-wave and triangle-wave generators, etc. Generating nonlinear waveforms such as triangle waves, ramps, sawtooths, square waves, etc., generally requires an integrator, a comparator, and a latching (or memory) logic circuit. A generalized block diagram is shown in Fig. 10.1. Even though three blocks are shown in the figure, all three functions may often be performed using only one or two operational amplifiers. Additional operational amplifiers, however, often improve the flexibility and generality of the circuit.

10.1 Square-wave Generators[3–5]

In this section we shall discuss circuits which may be used as square-wave generators. Three examples of such circuits are considered. These may

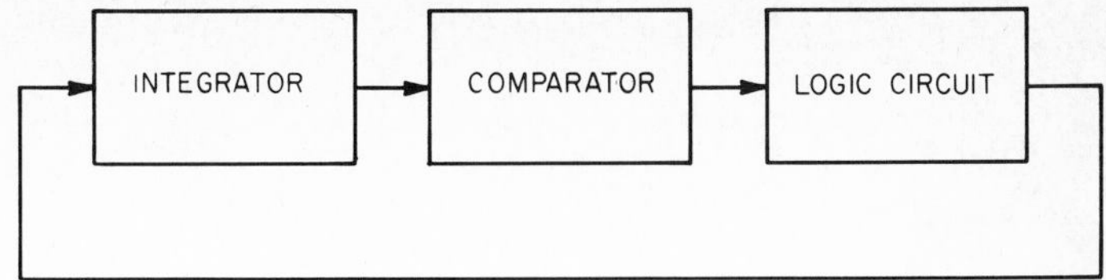

Fig. 10.1 Block diagram of a nonlinear waveform generator.

be applied to different special requirements as indicated in the discussion of the circuits.

10.1.1 Square-wave generator using one operational amplifier The circuit shown in Fig. 10.2 forms a simple and inexpensive square-wave generator. The output voltage of this circuit is limited by the back-to-back zener diodes to either $+V_Z$ or $-V_Z$ volts as shown in Fig. 10.3. The elements R_F and C provide the integrating, or timing, function. The operational amplifier serves the function of comparison. The required regenerative action comes from feeding the noninverting input of the operational amplifier with a fraction of the output voltage. Most operational amplifiers are input-protected, and the impedance across the amplifier inputs will be very low under the conditions imposed by this circuit. The R_1 resistors maintain a high input impedance across the amplifier's input under all conditions.

To see how this circuit operates, assume that, in Fig. 10.2, $R_1 \gg R_3$ and R_4, and that $\beta = R_4/(R_3 + R_4)$. Suppose that the previous output voltage was negative and that the voltage across the capacitor C has just reached $-\beta V_Z$. When the voltage e_1 becomes more negative than $-\beta V_Z$, the operational amplifier will flip from saturation in the negative direction to saturation in the positive direction. The operational amplifier will stay saturated in the positive direction because the voltage βe_o is now

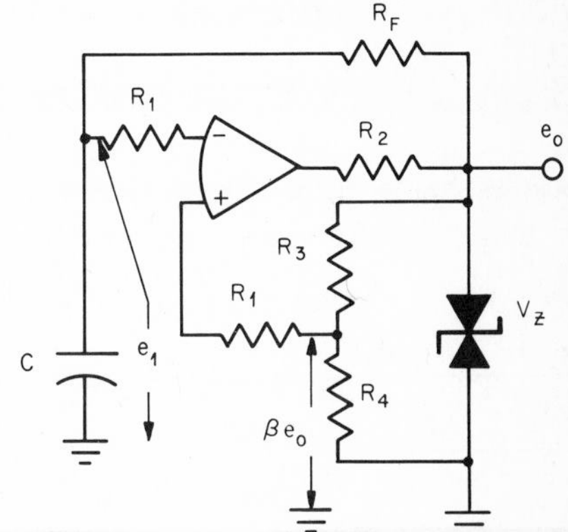

Fig. 10.2 Simple square-wave generator.

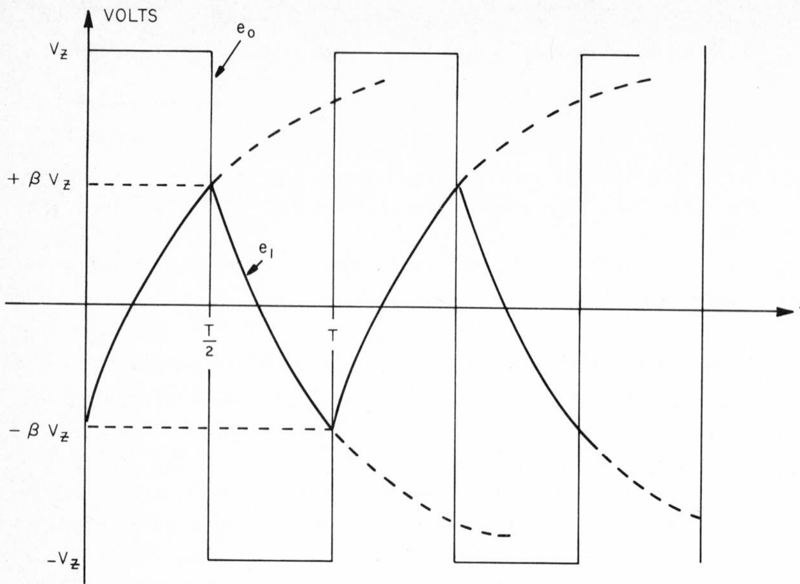

Fig. 10.3 Square-wave generator waveforms.

positive. Reversal will again occur when e_1 becomes equal to $+\beta V_Z$. In a typical cycle, if $e_1 = -\beta V_Z$ at $t = 0$, then, for the following half-cycle,

$$e_1(t) = (1 + \beta)\left(1 - e - \frac{t}{R_F C}\right) V_Z - \beta V_Z$$

Defining T as the period, we require that

$$e_1\left(\frac{T}{2}\right) = V_Z - [(1 + \beta)e^{-T/2R_F C}]V_Z = \beta V_Z$$

Solving for T, we obtain

$$T = 2R_F C \ln \frac{1 + \beta}{1 - \beta}$$

If β is chosen to be 0.473, then $T = 2R_F C$ and $f = 1/2R_F C$.

The following comments may be made concerning this circuit:

1. It is excellent for fixed-frequency applications in the audio frequency range.
2. The frequency may be trimmed by varying R_F.
3. Frequency stability depends primarily upon the capacitor and zener diode stability; even inexpensive operational amplifiers will contribute very little frequency drift.

4. Waveform symmetry in both amplitude and time depends upon the symmetry of the back-to-back zener diode voltages.

In general, the choice of amplifier is not at all critical for frequencies in the range from 10 Hz to 10 kHz. The amplifier must have only enough output current capability to drive the zeners, the divider network, and any external load and to charge the capacitor C. At low frequencies, however, the amplifier bias current, input noise level, and input impedance become significant and should be considered. At high frequencies, the delay time of the amplifier in coming out of saturation becomes significant, particularly if the delay time is unequal for positive and negative saturation. In addition, the amplifier slew rate becomes important. This circuit can be used for frequencies in the 10 to 100 kHz range only if the operational amplifier is chosen very carefully.

10.1.2 High-performance square-wave generator The performance of the circuit shown in Fig. 10.2 can be improved by replacing the resistor R_F with a transistor current source circuit. The voltage e_1 will now be a triangle wave, and the waveform symmetry is now adjustable. The triangle wave itself may also be used as an output.

In the circuit shown in Fig. 10.4 the back-to-back zener diodes and the resistor R_2 may be omitted if the amplifier output, when saturated, is constant and the positive and negative limits are equal. For example, the Burr-Brown 3401 or 3402 are fast FET input amplifiers with excellent saturation characteristics which may be used in such a modified circuit.

10.1.3 Low-cost version The saturation characteristics of inexpensive operational amplifiers are sometimes ill defined. However, if an operational amplifier that has reasonably symmetrical and stable saturation characteristics is used and if some overshoot and ringing is allowable, the back-to-back zeners shown in Fig. 10.2 may be omitted. Although this eliminates the cost of the zeners, the operational amplifier cost may go up. Thus the circuits of Fig. 10.2 or 10.4 are generally preferred. However, in some situations where performance requirements are not stringent, the circuit of Fig. 10.5 will serve adequately.

10.2 Square- and Triangle-wave Generators[1–5]

In this section we shall discuss circuits which may be used to generate a square wave or a triangle wave or both. The circuits which are presented include the following: a low-cost single operational-amplifier circuit, a general-purpose function generator which will generate other waveforms

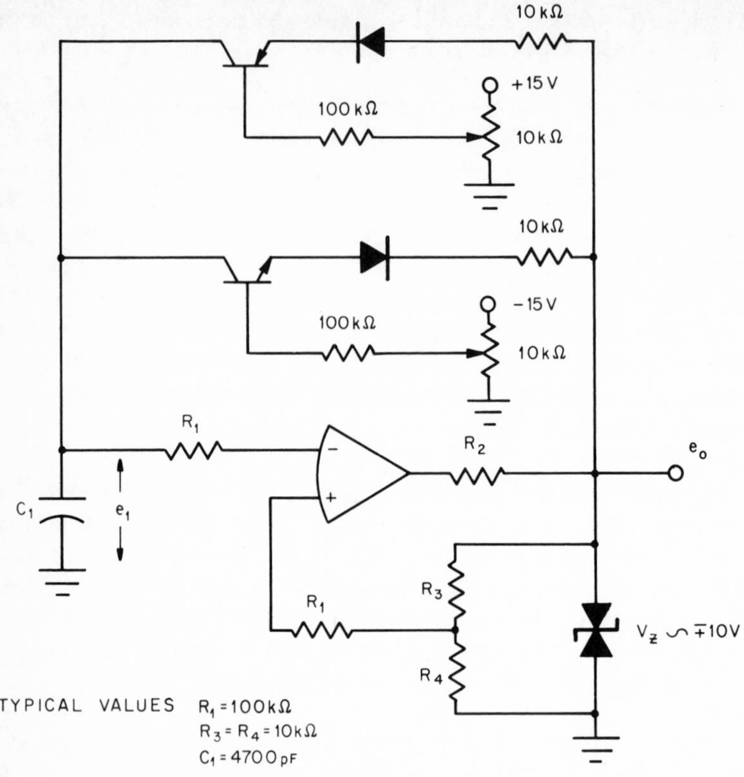

Fig. 10.4 Improved square-wave generator.

as well as square and triangle waves, and a diode-bridge triangle-waveform generator.

These circuits may be used to generate square waves, triangle waves, or both. Circuits using one, two, and three operational amplifiers are discussed. In general, adding operational amplifiers increases the com-

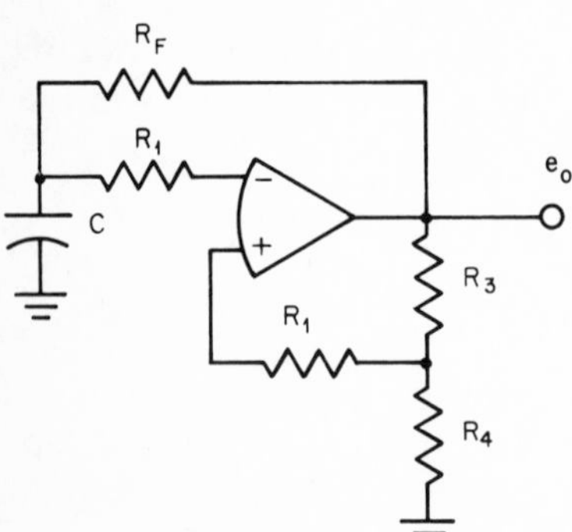

Fig. 10.5 Simple square-wave generator.

ponent cost somewhat but provides a better range of operation, higher performance, and less dependence upon "tweaking" of the circuit.

10.2.1 Square- and triangle-wave generator using one operational amplifier Only one operational amplifier and three transistors are required in this circuit. This is an excellent, low-cost circuit for laboratory-type use, but the temperature stability is little better than the classical two-transistor astable multivibrator.

The circuit is shown in Fig. 10.6. The function of the transistors is as follows: Transistor Q_1 acts as a comparator and inverter; Q_2 inverts the signal and provides a symmetrical signal output; and Q_3 (an emitter follower) is used to provide a symmetrical output impedance. The operation of the circuit may be understood by noting that the square-wave output is fed back to the comparator input (the base of Q_1), thus providing positive feedback and hysteresis. This action is analogous to a Schmitt trigger circuit. The operational amplifier is connected as an integrator and will have a triangle-wave output. Potentiometer P_1 adjusts the frequency of oscillation; P_2 adjusts the zero-crossing point of the comparator (which in turn varies the amplitude symmetry of the triangle wave), and P_3 adjusts the time symmetry of the triangle wave. The output amplitude of the square wave is adjusted by P_4. P_5 provides extra control over frequency and triangle-wave amplitude.

10.2.2 General-purpose function generator In the above paragraphs we discussed a circuit which had the capability of generating square and

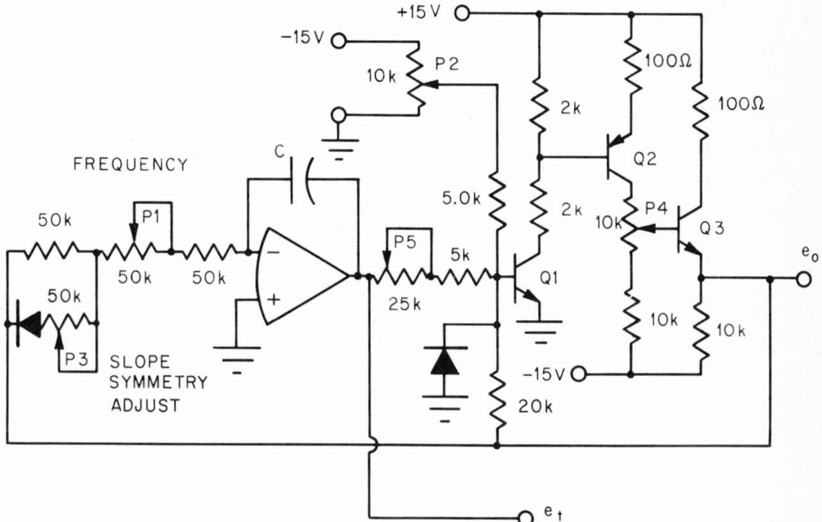

Fig. 10.6 Square- and triangle-wave generator.

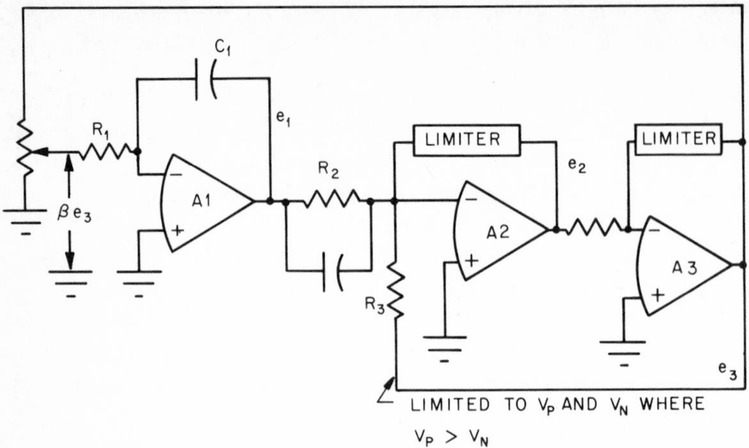

Fig. 10.7 Square- and triangle-wave generator.

triangle waves. Sometimes, however, more complex waveforms are needed. Sawtooths, triangle waves, pulse trains, ramps, and almost any other waveform of interest can be generated by circuits which use only a few operational amplifiers. Also, many of these circuits have several outputs available. An example of such a circuit is shown in Fig. 10.7. It can be used to generate simultaneously a triangle and square wave or to generate pulse trains. The circuit contains three operational amplifiers. Amplifier A_1 is an integrator that derives its input from the output of the comparator circuit A_2 and A_3. The integrator input will be βV_P or $-\beta V_N$. An important feature of this circuit is the positive feedback around the A_2, A_3 combination. The regenerative action provides rapidly uniform switching even though the oscillation frequency may be changing.

To see how this circuit operates, assume $e_1 = -(R_2/R_3)V_P$ at $t = 0$ and the circuit is just switching to a positive-going ramp out of e_1. Then,

$$e_1(t) = -\frac{\beta V_N}{R_1C_1}t - \frac{R_2}{R_3}V_P \qquad 0 < t < T_1$$

where T_1 is the time at which the comparator switches. Evaluating the above equation at $t = T_1$ we observe that

$$-\frac{R_2}{R_3}V_N = -\frac{\beta V_N}{R_1C_1}T_1 - \frac{R_2}{R_3}V_P$$

$$T_1 = \frac{R_2}{R_3}(V_N - V_P)\frac{R_1C_1}{\beta V_N}$$

At $t = T_1$, when the comparator switches, the integrator output becomes a negative-going ramp. The next switching occurs at $e_1 = -(R_2/R_3)V_P$. If we call this time T_2 we may derive

$$-\frac{R_2}{R_3}V_P = \frac{-\beta V_P}{R_1C_1}T_2 - \frac{R_2}{R_3}V_N$$

$$T_2 = \frac{R_2}{R_3}(V_P - V_N)\frac{R_1C_1}{\beta V_P}$$

This completes a full cycle, and so the period T is $T_1 + T_2$. Substituting from the above relations for T_1 and T_2, we obtain

$$T = \frac{R_2}{R_3}\frac{R_1C_1}{\beta}\left(\frac{V_N - V_P}{V_N} + \frac{V_P - V_N}{V_P}\right)$$

$$T = \frac{R_2}{R_3}\frac{R_1C_1}{\beta}\left(2 - \frac{V_P}{V_N} - \frac{V_N}{V_P}\right)$$

Many special circuits may be derived from this general relationship. Two examples follow:

1. *Square- and Triangle-wave Generator.* The general circuit described above and shown in Fig. 10.7 may be used to realize a square- and triangle-wave generator. To see this, let $R_2 = R_3$ and $V_P = -V_N$. Then,

$$T = \frac{4R_1C_1}{\beta}$$

$$f_o = \frac{\beta}{4R_1C_1}$$

where f_o is the frequency of oscillation. As an application of these results, consider the circuit shown in Fig. 10.8. In addition to the basic circuit configuration shown in Fig. 10.7 this circuit includes a sine function shaping circuit (Burr-Brown 4118/25 Sin-Cos function generator) which uses the triangle wave generated by the basic circuit as an input. The result is a very good ultra-low-frequency sine wave. In addition, if voltage-controlled frequency is desired, the potentiometer is simply replaced by a multiplier. Then β may be made a function of an input control voltage.

2. *Sawtooth Generator.* The general circuit shown in Fig. 10.7 may also be used to realize a sawtooth wave generator of excellent linearity. The detailed circuit for achieving this is shown in Fig. 10.9. A diode and resistor are added so that the retrace may be very rapid. This circuit is also an excellent pulse generator. To see how this circuit operates, note that with the output of amplifier A_3 limited to ± 10 V, the ramp time (positive going) will be $T_1 = 2R_1C_1$. Allowing 0.7 V for the drop across diode CR_1, the reset time T_2 will be $T_2 = 2.15R_aC_1$. If $R_a \ll R_1$, then

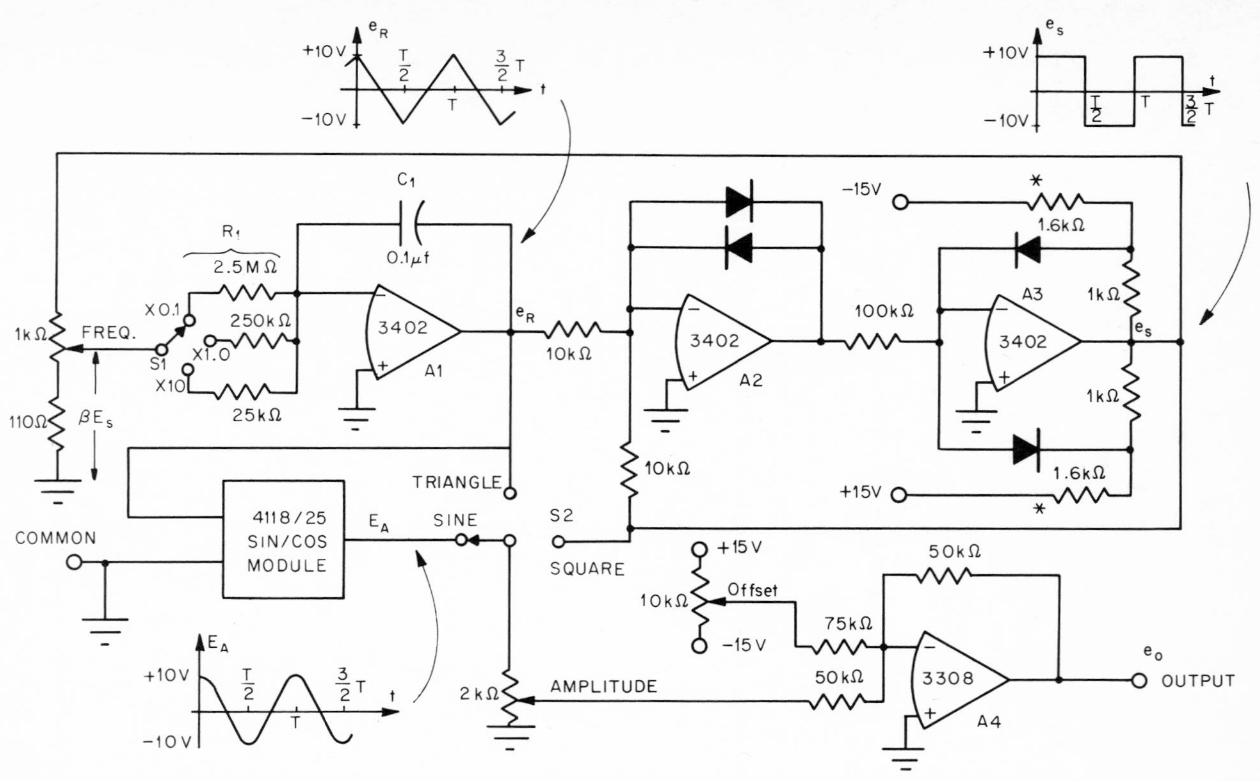

Fig. 10.8 Low-frequency function generator.

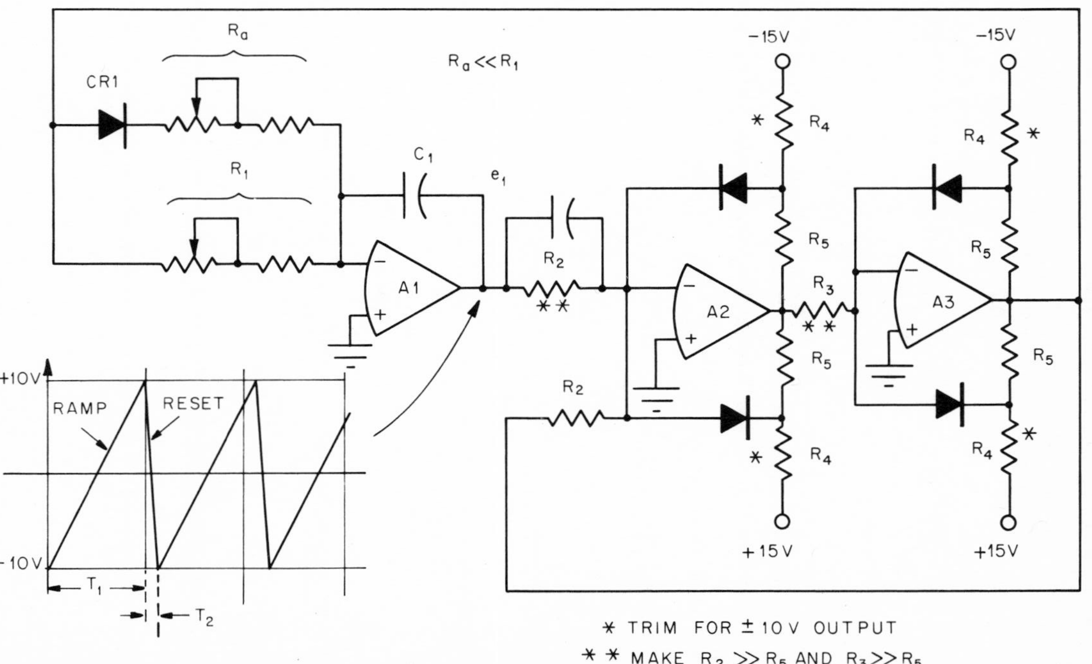

Fig. 10.9 Ramp generator.

$T_2 \ll T_1$. The only limit on reset time T_2 is the output current rating of A_1 and A_3, and so very fast resetting is possible.

10.2.3 Diode-bridge triangle-wave generator A final circuit for generating a triangle wave is shown in Fig. 10.10. It is referred to as a diode-bridge triangle-wave generator. This circuit does not have quite the high-frequency capability of the three-operational-amplifier triangle generator, but it is somewhat more economical and is an excellent circuit for many applications. For best results, the diodes in the bridges should all be of the same type. Potentiometer P_1 adjusts the amplitude of the triangle wave. This also affects the frequency.

In analyzing this circuit it should be noted that the diode bridges act as current gates; when e_1 is positive, D_2 and D_4 are blocking and current flows through both D_3's into the summing junctions. The voltage waveforms at e_1 and e_o, assuming all diode drops to be 0.6 V, are shown in Fig. 10.11. Typical circuit values for the components shown in Fig. 10.10

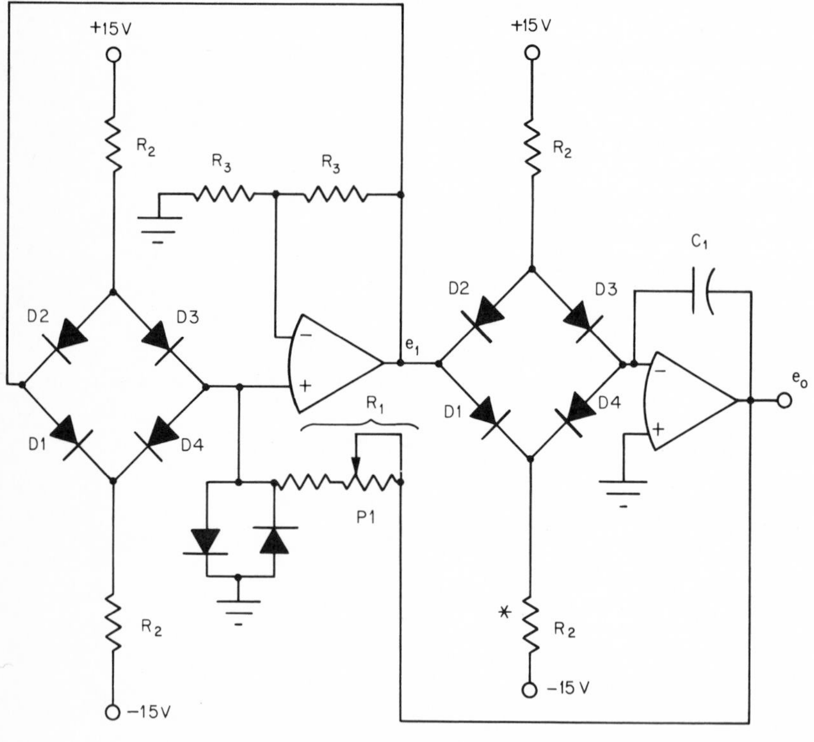

Fig. 10.10 Triangle-wave generator using diode bridges.

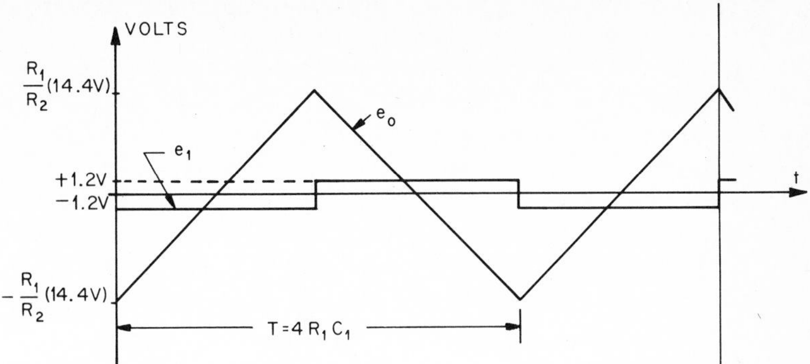

Fig. 10.11 Triangle generator waveforms.

would be $R_2 = 10\ k\Omega$, $R_1 = 5\ k\Omega$, $R_3 = 100\ k\Omega$, $C_1 = 0.1\ \mu F$. These values will provide a triangle wave of approximately +7.2 to −7.2 V swing at a frequency of 500 Hz.

10.3 Sine-wave Generators[3–5]

One of the most important waveforms that an engineer may be called upon to generate is the sinusoidal waveform. In this section we shall treat a variety of techniques which may be used to perform this task. Specifically we shall investigate the use of Wien-bridge oscillators, quadrature oscillators, and phase-shift oscillators. For the Wien-bridge and the quadrature oscillator case several different circuits are considered. These may be applied to different special requirements as indicated in the discussion of the circuits.

10.3.1 Wien-bridge oscillator—general description A Wien bridge may be combined with an operational amplifier to form an excellent sine-wave generator. Some sort of automatic gain control is generally used to stabilize the magnitude of the output sinusoid. A general schematic of a Wien-bridge oscillator is shown in Fig. 10.12. To see how this circuit operates let us assume that the output e_o is a sinusoid; then the feedback ratio of the bridge is given by

$$\frac{Z_2}{Z_1 + Z_2} = \frac{R_2}{R_1 + R_2(1 + C_2/C_1) + j(\omega R_1 R_2 C_2 - 1/\omega C_1)}$$

where $Z_1 = R_1 + 1/j\omega C_1$ and $Z_2 = R_2/(1 + j\omega R_2 C_2)$. The operational amplifier will maintain 0 V between its input terminals; thus,

$$\beta \hat{E}_o = \frac{Z_2}{Z_1 + Z_2} \hat{E}_o$$

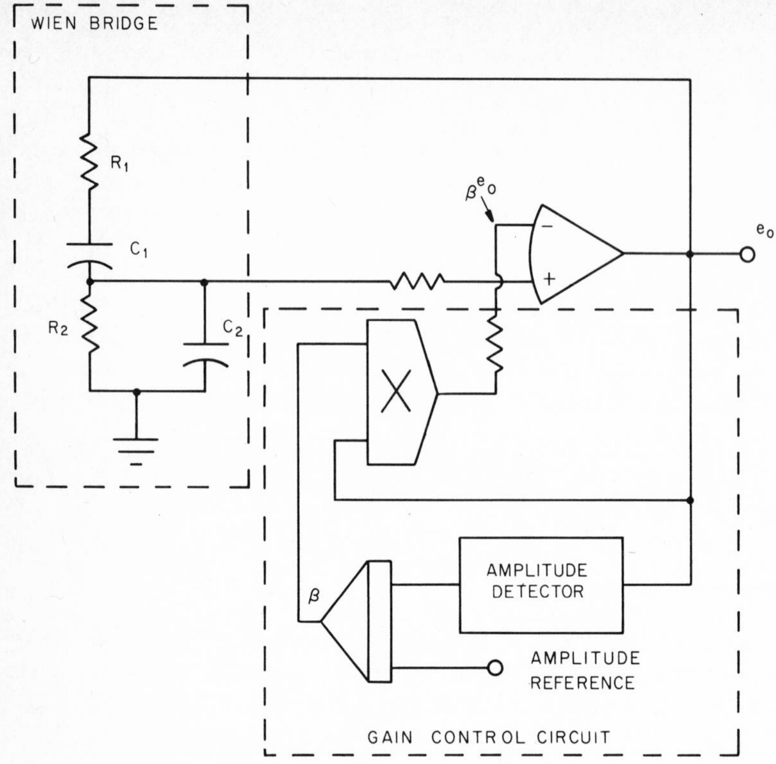

Fig. 10.12 Wien-bridge oscillator.

where $\hat{E}_o$ is a phasor representing the voltage $e_o(t)$. The condition for oscillation is

$$\omega_o R_1 R_2 C_2 - \frac{1}{\omega_o C_1} = 0$$

$$\omega_o = \frac{1}{\sqrt{R_1 R_2 C_1 C_2}}$$

If we make $R_1 = R_2$ and $C_1 = C_2$, then

$$\omega_o = \frac{1}{R_1 C_1} \qquad \text{and} \qquad \beta = \frac{1}{3}$$

If $\beta = \frac{1}{3}$ and the condition of $R_1 = R_2$ and $C_1 = C_2$ is met, then the output will be a sinusoid of frequency $1/2\pi RC$.

It should be noted that, so long as β is $\frac{1}{3}$, the circuit will oscillate at any amplitude. Also, if β is less than $\frac{1}{3}$, the oscillation will diverge and if β is more than $\frac{1}{3}$ the oscillation will converge. Thus it is common practice to provide some sort of automatic amplitude control. This is usually done by varying the negative feedback gain (β) to stabilize the oscillator. Incandescent lamps, thermistors, FETs, diode bridges, or general-purpose multipliers can all be used for such gain control purposes.

10.3.2 Precise Wien-bridge oscillator[4,5] As a typical implementation of the general Wien-bridge oscillator diagram shown in Fig. 10.12, consider the circuit shown in Fig. 10.13. The actual Wien bridge is formed by R_1, C_1, R_2, and C_2. The oscillatory output of amplifier A_1 is amplified by A_2, and the output level is sensed by the absolute-value circuit of A_3 and A_4. The amplifier A_4 acts as an error integrator and will stabilize only when the absolute value of the input equals the reference amplitude. A diode bridge is used for varying the negative feedback of A_1. An FET can be used for gain control rather than the diode bridge if desired.

The integrator gain is set by capacitor C_3. The choice of C_3 is a tradeoff between response time and distortion. Small values of C_3 will allow the circuit to reach its stable value very rapidly. Also, response to any disturbance is rapid. On the other hand, making C_3 large will minimize distortion. The frequency of oscillation, as discussed previously, will be

$$f_o = \frac{1}{2\pi R_1 C_1}$$

where $R_1 = R_2$ and $C_1 = C_2$. Frequencies in the range of 10 Hz to 10 kHz are practical for this circuit. Distortion of less than 0.1 percent and excellent frequency stability are readily achieved. The circuit will operate at frequencies above 10 kHz, but the type of operational amplifier must be carefully chosen and stray capacitances should be considered.

Although, in the circuit shown in Fig. 10.13, five operational amplifiers are used, similar circuits are available in miniature encapsulated packages. In such packages, integrated-circuit operational amplifiers are usually used to minimize the size.

10.3.3 Low-cost Wien-bridge oscillator The Wien-bridge oscillator circuit presented in the preceding paragraphs has the disadvantage of requiring five operational amplifiers. In Fig. 10.14 a circuit diagram for a Wien-bridge oscillator which requires only one operational amplifier is given. The primary virtue of this circuit is that very few components are required. Distortion will be greater than with the previously discussed Wien bridge. But, depending upon care of adjustment, distortion will be in the range of 1 to 5 percent. This circuit has high output impedance, and any loading at e_o will shift the operating point of the diodes, which will in turn change the amplitude. Thus this circuit *must* be used with either a fixed load at e_o or a buffer must be added. As with

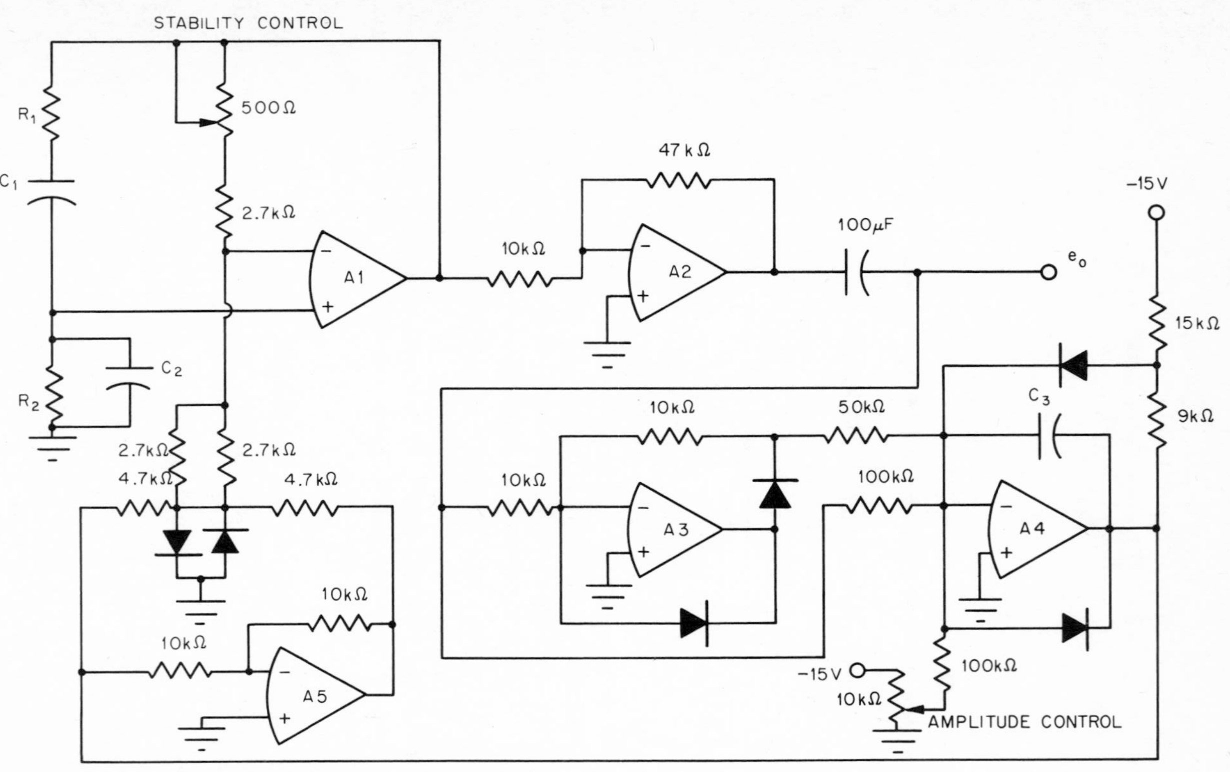

Fig. 10.13 Wien-bridge oscillator, diode gain control.

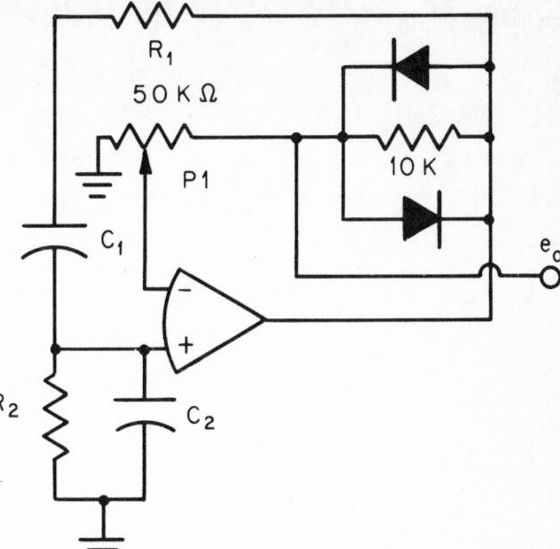

Fig. 10.14 Low-cost Wien-bridge oscillator.

the previous Wien-bridge circuit, R_1 is set equal to R_2 and C_1 is set equal to C_2. Then

$$f_o = \frac{1}{2\pi R_1 C_1}$$

Potentiometer P_1 is adjusted until the oscillations just start to diverge. At that condition, the inverting input to the operational amplifier will be about $\frac{1}{3}e_o$. As the oscillations grow, the diodes start to conduct and the impedance across the diodes lowers. This raises the amount of negative (or degenerative) feedback. Adjustment of P_1 will vary the output amplitude at which amplitude stability occurs. Unlike the circuit shown in Fig. 10.13, the amplitude, amplitude stability, and distortion of this circuit all interact somewhat. The control over amplitude is indirect since P_1 must be set so that distortion is minimized. Distortion is lower as amplitude is made greater. Also, using matched diodes will minimize distortion. Frequency stability depends primarily on the quality of the Wien-bridge components, and so good frequency stability is easily obtained with this simple circuit.

Much lower distortion may be obtained by using amplitude limiting circuits which are thermally limited. The limiting elements of such circuits may be thermistors or incandescent lamps.

10.3.4 Quadrature oscillators The sine-wave generator circuits presented so far in this section have been based on Wien-bridge techniques. In the following paragraphs we present a quite different technique for generating sinusoidal waves, namely, the use of a quadrature oscillator. The quadrature oscillator has two important advantages over the Wien-bridge oscillator:

1. A cosine and sine term are simultaneously available as outputs.

2. Stabilizing the oscillation without introducing excessive distortion is relatively easy.

Quadrature oscillator circuits are similar to Wien-bridge circuits in that they are most attractive for fixed-frequency applications in the range of 10 Hz to 10 kHz, rather than variable-frequency applications. Frequencies outside this range are obtainable, but the components, particularly the operational amplifiers, must be chosen with some care. The general form of the circuit is shown in Fig. 10.15.

The basic principle of the quadrature oscillator is to implement a loop that solves the differential equation

$$\ddot{X} + \omega_o^2 X = 0$$

The steady-state solution (ignoring phase angle) is

$$X = A \sin \omega_o t$$

As with the Wien-bridge oscillator, some means of amplitude stabilization is generally required. Two general methods are used:

1. Design for a slightly divergent oscillation; then use nonlinear amplitude limiting to keep the output bounded.
2. Sense the output amplitude and compare it with a reference. Use the resultant error signal for automatic gain control purposes.

Method 1 is often satisfactory for fixed-frequency oscillators, particularly if distortion is not critical and if the desired frequency is somewhere in the range of 1 Hz to 10 kHz. Method 2 provides better performance but at the cost of increased circuit complexity. Circuits which include the two different approaches outlined above are described in the following paragraphs.

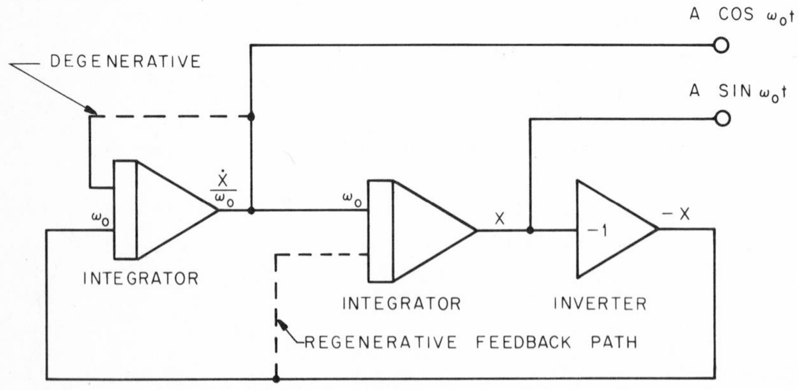

Fig. 10.15 Quadrature oscillator, general form.

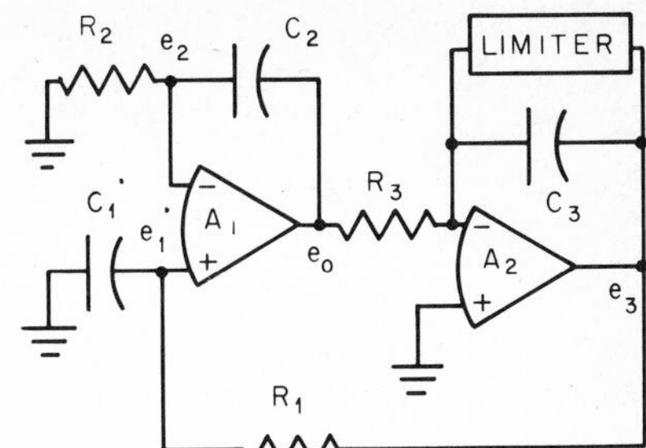

Fig. 10.16 Quadrature oscillator with amplitude clipping.

1. ***Quadrature Oscillator with Nonlinear Amplitude Limiting.*** A quadrature oscillator circuit which uses amplitude limiting is shown in Fig. 10.16. Two operational amplifiers are used as integrators in this circuit. Some form of amplitude limiting, such as that discussed previously in Chapter 7, is applied to amplifier A_2. The behavior of the circuit is best understood if the nonlinear limiting on A_2 is first not considered. After the linear behavior is described, the effect of the nonlinearity can be considered. In considering the linear behavior, let us assume that there is an initial voltage of V_1 on capacitor C_1 and all other initial conditions are zero; then the Laplace transforms of voltages e_1 and e_2 are given by

$$E_1(s) = \frac{1}{R_1C_1s + 1} E_3(s) + \frac{V_1}{s}$$

$$E_2(s) = \frac{R_2C_2s}{R_2C_2s + 1} E_o(s) \qquad \text{and} \qquad E_3(s) = \frac{1}{R_3C_3s} E_o(s)$$

Assuming ideal operational amplifiers, E_1 and E_2 will be equal. Defining $\tau_2 = R_2C_2$, and $\tau_3 = R_3C_3$, the output E_o will be

$$E_o(s) = \frac{(s + 1/\tau_1)(s + 1/\tau_2)V_1}{s^3 + (1/\tau_1)s^2 + (1/\tau_1\tau_3)s + 1/\tau_1\tau_2\tau_3}$$

If $\tau_1 = \tau_2$, then

$$E_o(s) = \frac{(s + 1/\tau_1)^2V_1}{s^3 + (1/\tau_1)s^2 + (1/\tau_1\tau_3)s + 1/\tau_1{}^2\tau_3}$$

$$= \frac{(s + 1/\tau_1)V_1}{s^2 + 1/\tau_1\tau_3}$$

The solution as a function of time is found by taking the inverse Laplace transformation. Thus, we obtain

$$e_o(t) = V_1\sqrt{\tau_3 + 1}\,\sin\left(\frac{1}{\sqrt{\tau_1\tau_3}}t + \psi\right)$$

where

$$\psi = \tan^{-1} \frac{\tau_1}{\sqrt{\tau_1 \tau_3}}$$

Now if $\tau_1 = \tau_3$, then

$$e_o(t) = V_1 \sqrt{\tau_1 + 1} \sin \left(\frac{1}{\tau_1} t + 45° \right)$$

In this case the frequency of oscillation is

$$f_o = \frac{1}{2\pi RC}$$

where $R = R_1 = R_2 = R_3$ and $C = C_1 = C_2 = C_3$.

In a practical circuit, slight mismatching of components will cause the circuit to slowly converge or diverge. If R_1C_1 is deliberately made slightly greater than R_2C_2, the oscillator output amplitude will diverge. But if limiters clip the output of A_2, the output amplitude will stabilize. The output distortion will be roughly proportional to the degree of mismatch between R_1C_1 and R_2C_2. The distortion will generally be lower at the output for e_o rather than the output for e_3. A practical version of this circuit is shown in Fig. 10.17.

2. *Quadrature Oscillator with Amplitude Control.* The second type of quadrature oscillator to be discussed in this section is the one in which the output amplitude is sensed and used to control the loop damping. A schematic of such a circuit is shown in Fig. 10.18. To understand the operation of this circuit, let us assume that all initial conditions are zero and that e_R and e_3 are slowly varying relative to e_1 and e_2. The circuit equation found from using Laplace transforms is

$$\left(R_1{}^2C_1{}^2s^2 + \frac{E_3}{10} R_1C_1s + 1 \right) E_1(s) = 0$$

Setting $\omega_o = \sqrt{400 - E_3{}^2/20R_1C_1}$ and $\alpha = E_3/20R_1C_1$, the solution in the time domain is

$$e_1(t) = \frac{1}{\omega_o} e^{-\alpha t} \sin \omega_o t$$

Since

$$\alpha = \frac{e_3}{20R_1C_1}$$

the oscillation will tend to diverge if $e_R > |e_1|_{DC}$ and converge if $e_R < |e_1|_{DC}$. The oscillation will stabilize at an amplitude where

$$e_R = \text{avg}\ \{|e_1|\}$$

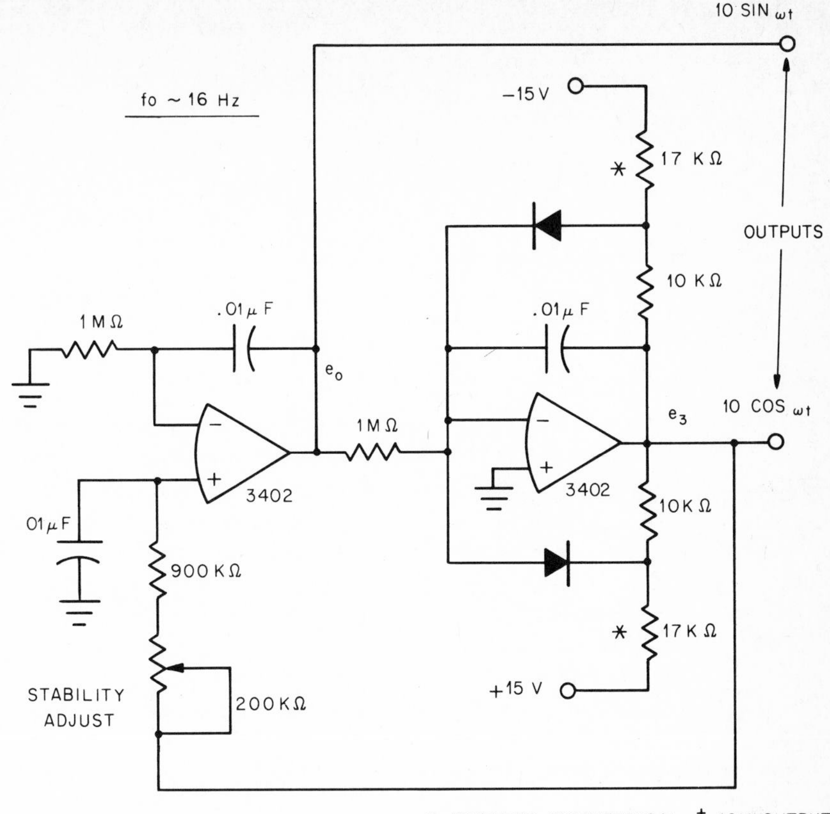

Fig. 10.17 Circuit for quadrature oscillator.

In other words, e_3 is an error voltage and will be driven to near zero. The time constant R_2C_2 must be much greater than the time constant R_1C_1. Note that the frequency of oscillation approaches $1/2\pi R_1C_1$ as e_3 approaches zero.

$$\omega_o = \frac{400 - e_3^2}{20R_1C_1} \sim \frac{1}{R_1C_1} \qquad \text{as } e_3 \to 0$$

Stabilizing the amplitude control loop is sometimes a problem with this circuit. Capacitor C_2 must be large enough to provide adequate low-pass filtering for the rectified current proportional to e_1, but if C_2 is too large the control loop may go into a slow limit-cycle oscillation. Adjusting the control loop gain by varying P_1 will generally stabilize the loop, but then the amplitude stability may suffer. In a practical circuit, each decade of frequency will generally require different values of C_2, and P_1

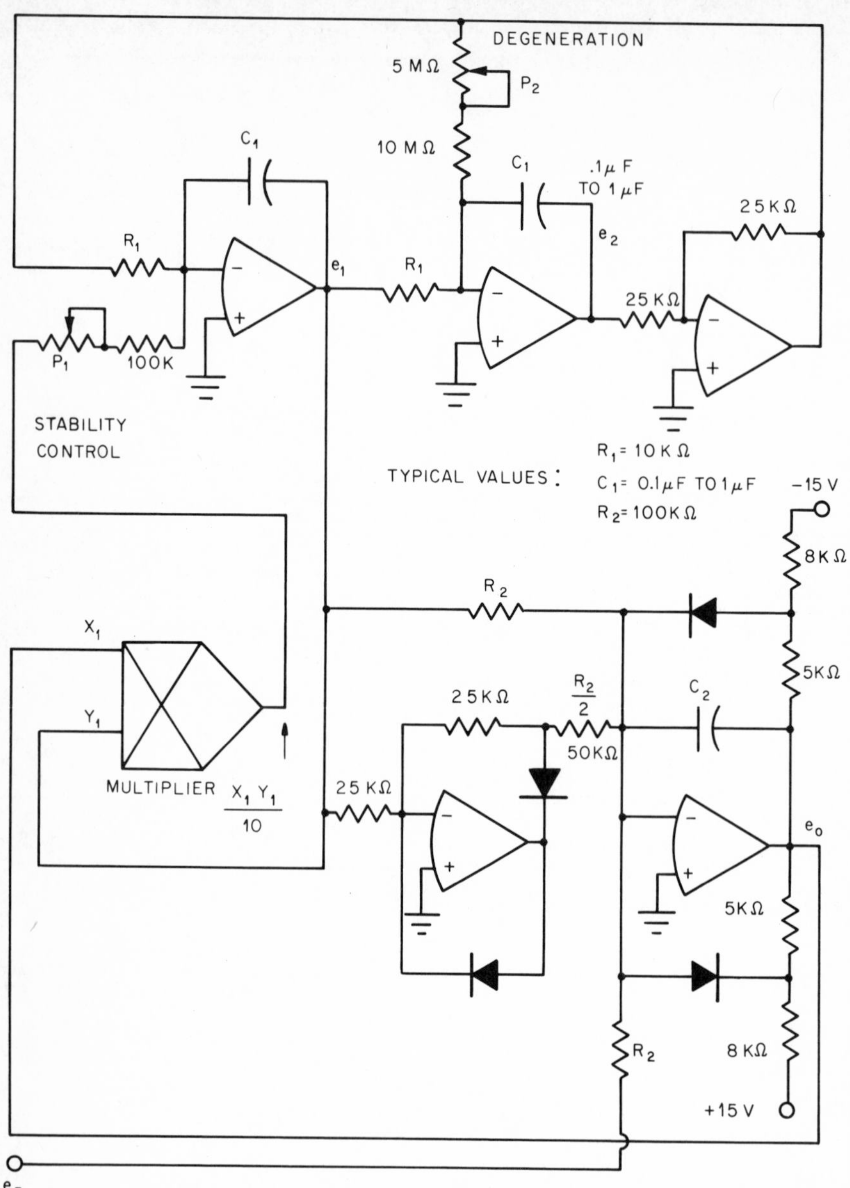

Fig. 10.18 Quadrature oscillator with gain control.

will have to be readjusted. P_2 may need adjusting also. Another amplifier can be added to allow for more complex amplitude-loop phase compensation.

10.3.5 Phase-shift oscillators In the following paragraphs we present a discussion of the generation of sinusoidal waveforms by a phase-shift oscillator. Although similar in concept to the previously discussed Wien-bridge and quadrature oscillator circuits, the phase-shift oscillator offers several advantages:

1. Only one amplifier is required to generate the sine term. In addition, the cosine term is easily obtained by adding a single additional amplifier, if desired.
2. Either differential input *or* single-ended input operational amplifiers may be used.

The primary disadvantage of the circuit is that three matched capacitors are required. Changing the frequency of oscillation is not easy with this circuit, but it is often satisfactory for generating a fixed-frequency sine wave. The clipping will cause some distortion, and so this circuit will generally have lower performance than the circuits discussed in the preceding paragraphs.

To see how this type of oscillator operates, consider the circuit shown in Fig. 10.19. The loop equation is

$$(R_F R^2 C^3 s^3 + 3R^2C^2s^2 + 4RCs + 1)E_o(s) = 0$$

This equation is satisfied if

$$R_F = -\frac{3R^2C^2s^2 + 4RCs + 1}{R^2C^3s^3}$$

Since R_F is real and constant, s must be a fixed value $j\omega_o$. The real portion of the right member of the above equation must equal R_F and

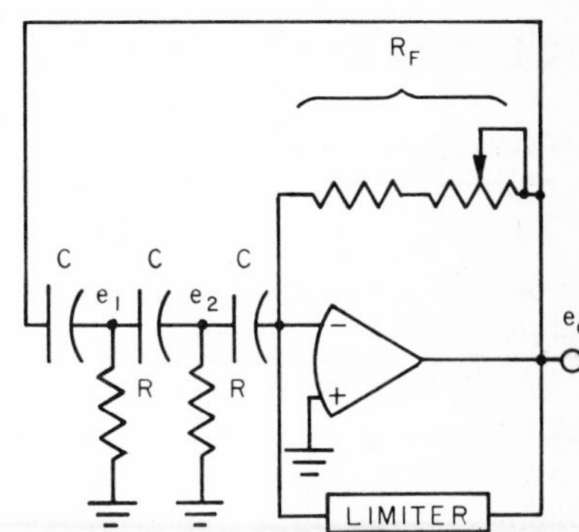

Fig. 10.19 Phase-shift oscillator, general form.

the imaginary portion must be zero. Substituting $s = j\omega_o$ we obtain

$$R_F = \frac{4RC\omega_o + j(3R^2C^2\omega_o^2 - 1)}{R^2C^3\omega_o^3}$$

Thus

$$R_F = \frac{4}{RC^2\omega_o^2} \qquad \text{and} \qquad \omega_o^2 = \frac{1}{3R^2C^2}$$

If R_F is made equal to 12R, this circuit will oscillate at frequency $\omega_o/2\pi$, where $\omega_o = 1/\sqrt{3}\,RC$.

To obtain a stable oscillation, R_F should be made slightly greater than 12R. A limiter circuit is then used to contain the divergence. It should be noted, however, that the distortion increases as R_F is made greater than 12R. But the circuit stabilizes more rapidly and is more stable in amplitude as R_F is made larger. Both the limiter circuit and R_F are generally made variable when using this circuit.

The cosine function may be obtained by adding one more amplifier to the circuit shown in Fig. 10.19 to see this. The voltage $\hat{E}_2$ is

$$\hat{E}_2 = -\frac{1}{R_FC(j\omega_o)}\hat{E}_o$$

where E_o is the phasor $A < 0°$, and $e_o(t) = A \sin \omega_o t$. Thus we may write

$$e_2(t) = A\frac{3}{12}\cos \omega_o t$$

To prevent undesired loading of the phase-shift network, the input resistor to the cosine amplifier is made equal to R and the R to common is removed. A version of this circuit with oscillation frequency of 920 Hz is shown in Fig. 10.20.

10.4 Pulse Circuits—Monostable Multivibrators[4–6]

In the preceding sections of this chapter we have discussed operational amplifier circuits for generating square, triangle, and sine waves. In this section we present some operational amplifier monostable multivibrator circuits for generating pulses. A monostable multivibrator circuit (sometimes referred to as "single shot") can also be designed with active elements other than operational amplifiers. The primary advantage of the operational amplifier circuits over conventional transistor or vacuum-tube circuits is the wide range of pulse-width adjustment and the improved stability with temperature.

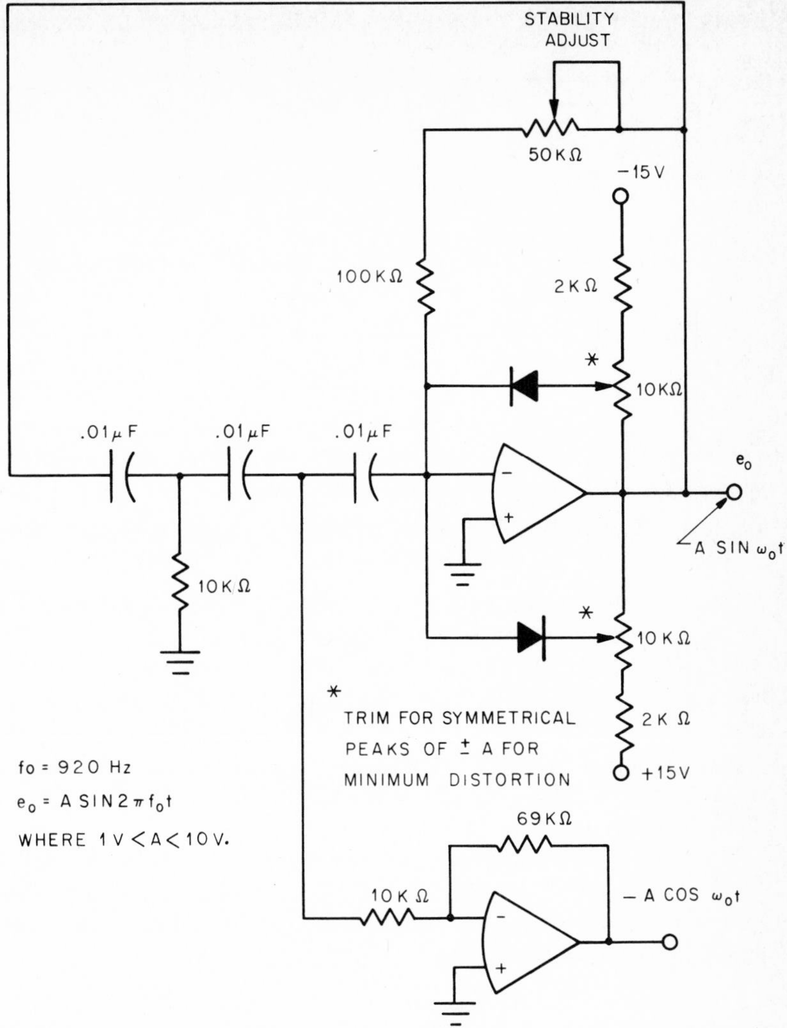

Fig. 10.20 Phase-shift oscillator circuit diagram.

10.4.1 Monostable multivibrator using one amplifier A relatively simple monostable multivibrator may be constructed using only one operational amplifier. The circuit is shown in Fig. 10.21a. Although it is a very simple circuit, it will be precise if the temperature stability of the back-to-back zener diode is good. If a back-to-back zener diode is not available, a diode bridge circuit with one conventional zener may be substituted for the back-to-back zener, as shown in Fig. 10.21b. To see how the circuit shown in Fig. 10.21a operates, note that in the

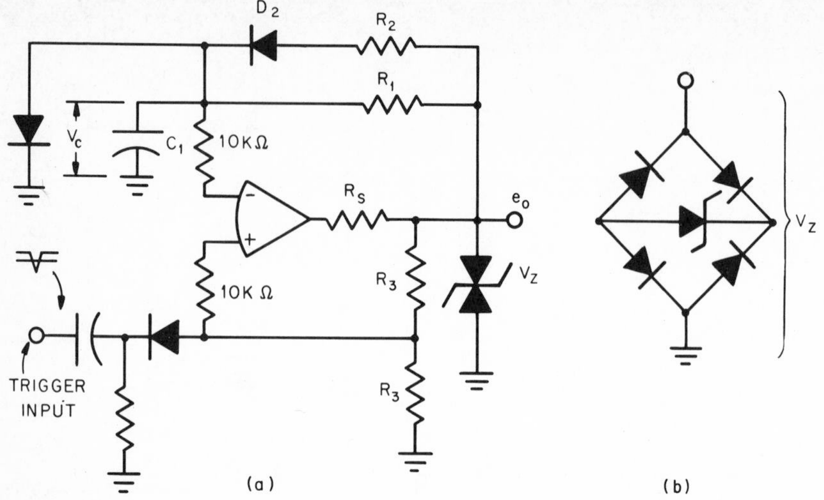

Fig. 10.21 Monostable multivibrator. (a) Circuit diagram; (b) bridge limiter.

stable state the output is at $+V_Z$ and the capacitor voltage V_C is clamped at about +0.6 V. A negative trigger of greater than $-V_Z/2$ will cause the output to flip negative to $-V_Z$. The capacitor then starts charging through R_1 towards $-V_Z$. But when V_C is more negative than $-V_Z/2$, the output will flip back to $+V_Z$. This completes the single pulse. To reset for the next pulse, C_1 is charged through R_2 and D_2. By making $R_2 \ll R_1$, the reset time can be much shorter than the output pulse width.

For applications where the pulse duty rate is very low and reset time is not critical, the R_2, D_2 portion of the circuits shown in Fig. 10.21a may be omitted. In addition, if an amplifier with good saturation characteristics (such as the Burr-Brown 3401 or 3402) is used, the zener diode and R_s may be omitted. The voltage $V_C(t)$ is given by

$$V_C(t) = (V_Z + 0.6)e^{-t/R_1C_1} - V_Z$$

The pulse width is

$$T = R_1C_1 \ln \frac{2(V_Z + 0.6)}{V_Z} \sim R_1C_1 \ln 2$$

$$T \sim 0.7\, R_1C_1$$

This circuit provides good performance for time constants of about 10 ms or longer. The switching speed of the amplifier from saturation to saturation becomes critical for shorter time constants.

10.4.2 Precise wide-range monostable multivibrator A more complex monostable multivibrator circuit is shown in Fig. 10.22. Although three

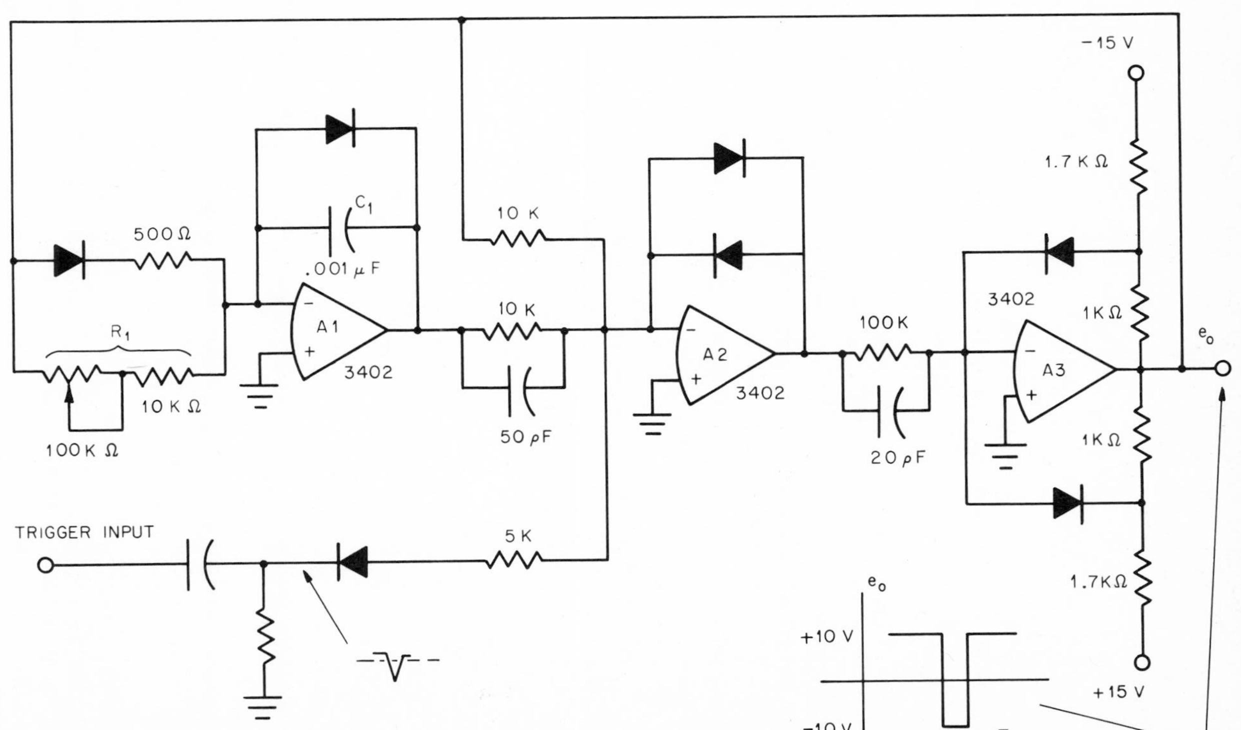

Fig. 10.22 High-performance monostable multivibrator.

amplifiers are required, this circuit provides adjustable trigger sensitivity, very wide range of pulse width adjustment, and very flat well-controlled output pulses. Virtually any operational amplifier can be used in this circuit. But for very high-speed narrow pulses, amplifiers with fast slew rate, high output-current rating, and good settling-time characteristics should be used. In concept, the circuit is very similar to the triangle-wave generator circuit of Fig. 10.10. The primary difference is that the integrator is "locked up" until a trigger pulse starts the cycle of operation.

To see how this circuit operates, note that in the quiescent state the output e_o is at $+10$ V and the input to amplifier A_1 is held at about -0.6 V by the diode across the feedback. Amplifier A_2 is also at about -0.6 V. Now if a negative-going spike of at least -5 V is applied through the 5 kΩ resistor to A_2, A_2 will switch to positive and A_3 will flip over to a -10-V output. This switching action is regenerative. The trigger input should be much narrower than the desired output pulse width but must be long enough for the amplifiers to switch states. When the output of A_3 switches to negative, the integrator A_1 will integrate from -0.6 to $+10$ V at a rate determined by R_1C_1. When the output of A_1 reaches $+10$ V, the comparator A_2 will flip to negative and the output will jump back to $+10$ V. This completes the "single-shot" pulse. The integrator is driven back to -0.6 V very rapidly through the 500-Ω input resistor, and the circuit is then ready for another trigger input. The pulse width is given by

$$T \sim 1.06 R_1 C_1$$

With the component values shown, pulses of 10 to 100 μsec are readily achieved.

REFERENCES

1. G. A. Korn and T. M. Korn, Modern Servomechanism Testers, *Electron. Eng.*, September, 1950.
2. R. M. Howe and C. Leite, Low-frequency Oscillator, *Rev. Sci. Instrum.*, vol. 24, 901, 1953.
3. G. A. Korn and T. M. Korn, *Electronic Analog and Hybrid Computers*, McGraw-Hill Book Company, New York, 1964.
4. *Applications Manual for Operational Amplifiers*, Philbrick/Nexus Research, Dedham, Mass., 1965.
5. *Handbook and Catalog of Operational Amplifiers*, Burr-Brown Research Corporation, Tucson, Ariz., 1969.
6. J. M. Pettitt, *Electronic Switching, Timing, and Pulse Circuits*, McGraw-Hill Book Company, New York, 1959.

11

MODULATION AND DEMODULATION

General Theory

Since, as discussed in the preceding chapter, operational amplifiers are used for generating sine waves, pulse trains, and many other functions, it is natural to consider operational amplifiers for signal *modulation* and *demodulation*. These functions can generally be implemented with combinations of integrators, multipliers, comparators, and precision gates. In this chapter we will present operational amplifier circuits which may be used to achieve three types of modulation, namely, amplitude, frequency, and pulse width modulation. In addition, a discussion of circuits which may be used for demodulation will be given.

11.1 Amplitude Modulation

11.1.1 Using multipliers for amplitude modulation[6] The most direct means of amplitude modulation is to use a multiplier as described in Chapter 7. The general form of a circuit for accomplishing this is shown

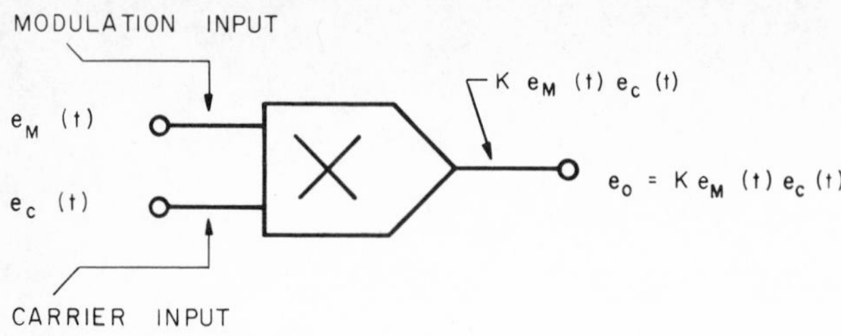

Fig. 11.1 Amplitude modulation.

in Fig. 11.1. Some special cases of this type of modulation occur fairly often. For example, the carrier is often a sinusoid. If e_M varies positively and negatively with respect to zero, the carrier is suppressed. If the modulation input is also a sinusoid, then

$$\begin{aligned} e_O &= Ke_M(t)e_C(t) = k(A \sin \omega_M t)(B \sin \omega_C t) \\ &= \tfrac{1}{2}kAB[\cos(\omega_C - \omega_M) - \cos(\omega_C + \omega_M)] \end{aligned}$$

From the above equation we see that there are two sidebands present at the output and that the carrier is suppressed. We conclude that using multipliers is a simple and direct means of amplitude modulating.

11.1.2 Pulse amplitude modulation Although multipliers provide the most general means of amplitude modulation, amplitude-modulating a pulse train is somewhat easier, in that the carrier has only two states: +V and zero. Thus pulse amplitude modulation (PAM) may be done by using gating circuits. In the following paragraphs we present several circuits for performing PAM.

1. ***PAM Using Transistor Gating.*** A transistor gate circuit for performing PAM is shown in Fig. 11.2a. In this circuit, the pulse train input e_C switches transistor Q_1 between the ON and OFF states. The modulation voltage e_M is always negative and varies from 0 to −10 V. When e_C is +10 V, Q_1 is biased OFF and the output e_O is $-e_M - V_B$. When e_C switches down to approximately 0 V, Q_1 is saturated ON by the 33 kΩ to −15 V bias source. Then e_O will be equal to $-V_B - 2V_{CE(sat.)}$. The voltage $V_{CE(sat.)}$ may be made very small by choosing a switching transistor with low $V_{CE(sat.)}$ and by making the $R_1/2$ resistors fairly large in value. Typically $V_{CE(sat.)}$ will be from 20 to 200 mV using switching transistors.

Typical signal waveforms for the circuit are shown in Fig. 11.2b. An FET may be used in place of the bipolar transistor if desired. Offset in the ON state will generally be lower with FETs, but the capacitive feedthrough of switching transients may be worse. Also, offset may be made lower by inverting the transistor (reverse the collector and emitter). However, the dynamic range is somewhat different then.

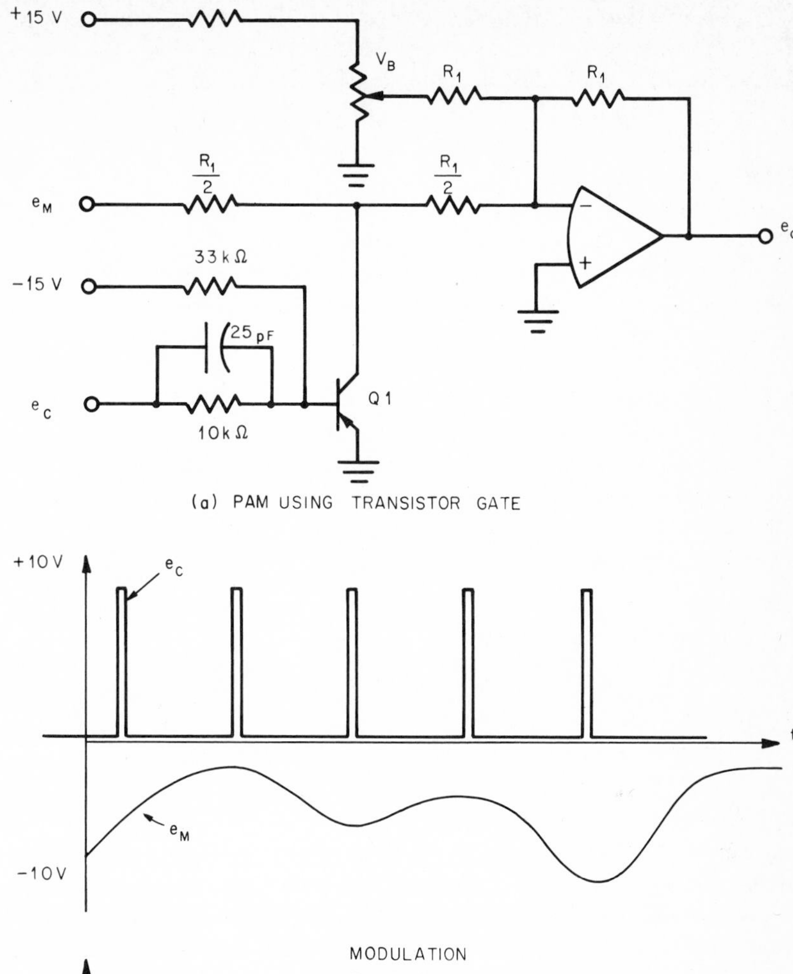

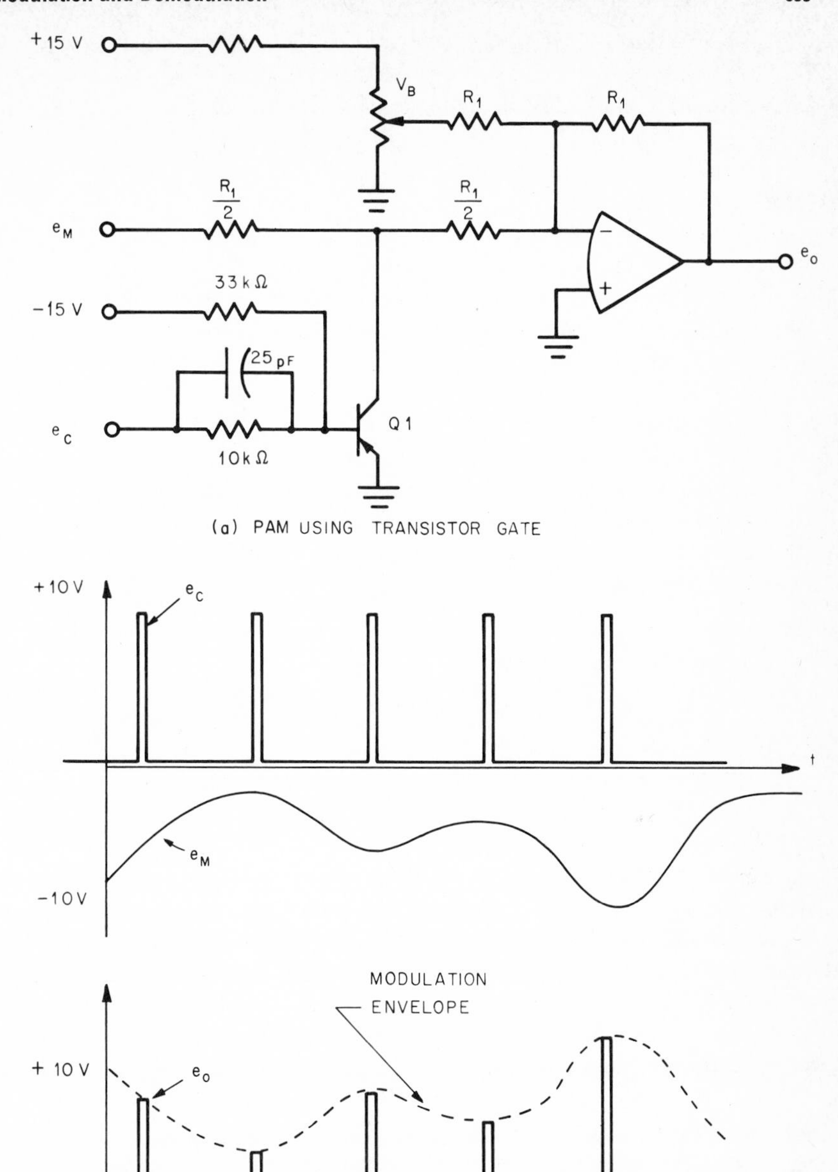

Fig. 11.2 Pulse amplitude modulation. (a) PAM using a transistor gate; (b PAM waveforms.

2. ***PAM Using a Precision Limiter.***[1,3] A second method of performing PAM is through the use of diodes. Although diode bridges can be used for gating, the accuracy is dependent upon the diode characteristics. However, by using the previously discussed precision rectifier circuit (Sec. 7.2.3), a very precise pulse amplitude modulator of wide dynamic range may be designed.

Such a circuit is shown in Fig. 11.3a. In this circuit assume that e_C is a pulse train (Fig. 11.3b) that is switching from a low voltage of zero to a positive high voltage E_H that is approximately $+10$ V. The modulation voltage e_M is symmetrical about zero and varies between ± 4 V. Amplifier A_1 has two possible outputs: $e_1 = (-e_C - e_M - V_B)$ if $(-e_C - e_M - V_B) < 0$ or $e_1 = 0$ if $(-e_C - e_M - V_B) > 0$. Amplifier A_2 also has two possible outputs: $e_o = -e_C - (-e_C - e_M - V_B) = e_M + V_B$ if $e_1 < 0$, or $e_O = -e_C$ if $e_1 = 0$. If E_H is more positive than $|e_M + V_B|$, then e_1 will be negative and the output will be $(e_M + V_B)$. If $(e_M + V_B) < 0$, then e_1 will be zero when e_C is low and the output will then be zero. Typical waveforms are shown in Fig. 11.3b. The bias voltage V_B is set at -5 V for the signal levels shown.

11.2 Frequency Modulation[6]

In this section we shall discuss the use of operational amplifier circuits for performing frequency modulation. This may be accomplished by the use of a voltage-controlled oscillator or by a voltage-to-frequency converter.

11.2.1 Voltage-controlled oscillator

Let us first consider frequency modulation through the use of voltage-controlled oscillators. A voltage-controlled oscillator (VCO) has a sinusoidal output with a frequency that is proportional to a dc control voltage. The amplitude may, or may not, be variable. Also, the sine wave may be very undistorted, or a high level of distortion may be present and acceptable. Key specifications are the *linearity* of frequency change with control voltage input and the *dynamic range* of the frequency deviation.

A simple approach to designing VCOs is to control the frequency of a square wave and then filter the square wave to obtain the fundamental sine-wave output. If the voltage-to-frequency circuits are teamed up with some filter circuits (see Chapter 8), a VCO can be obtained.

1. ***High-performance VCO.*** In telemetry systems, a VCO with both an in-phase *and* quadrature output is sometimes needed. Also, in some test or instrumentation applications it is desirable to start the oscillation at some known phase angle. The circuit shown in Fig. 11.4 has all these features and is also a very precise circuit. Unlike the other amplitude-controlled oscillators discussed earlier, this circuit does *not* depend upon a

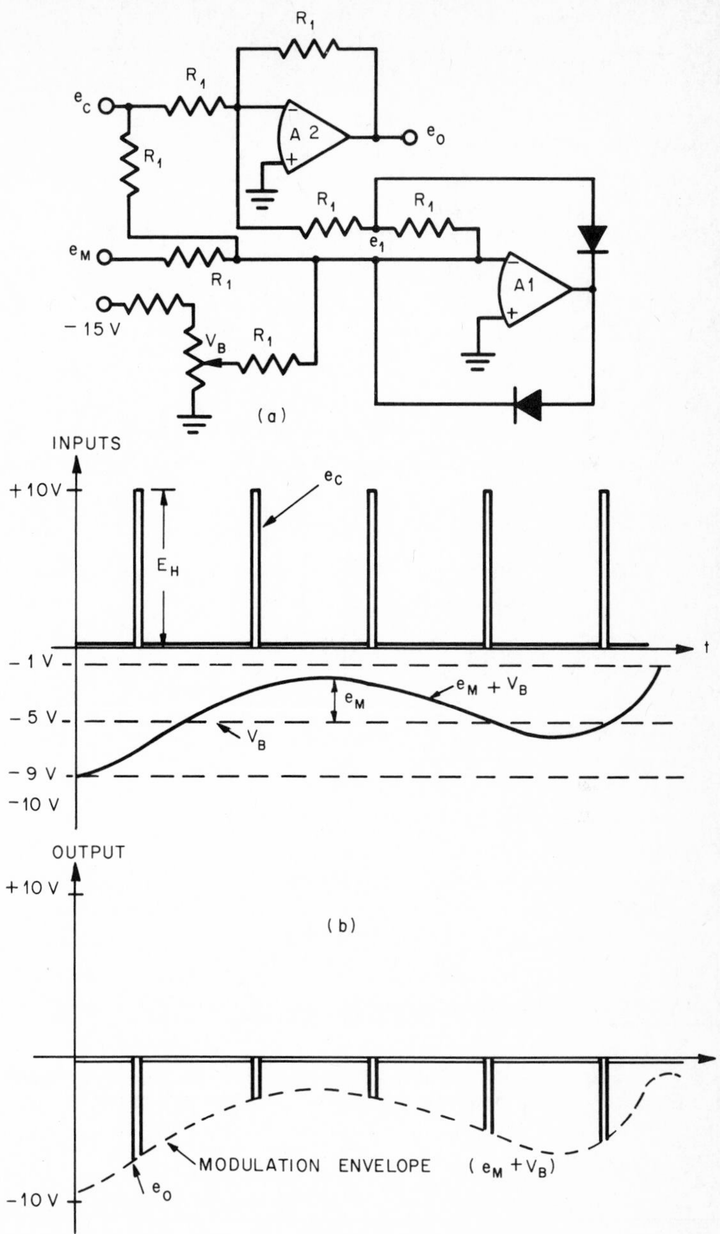

Fig. 11.3 Precision pulse amplitude modulator. (a) Precise pulse amplitude modulator; (b) PAM waveforms.

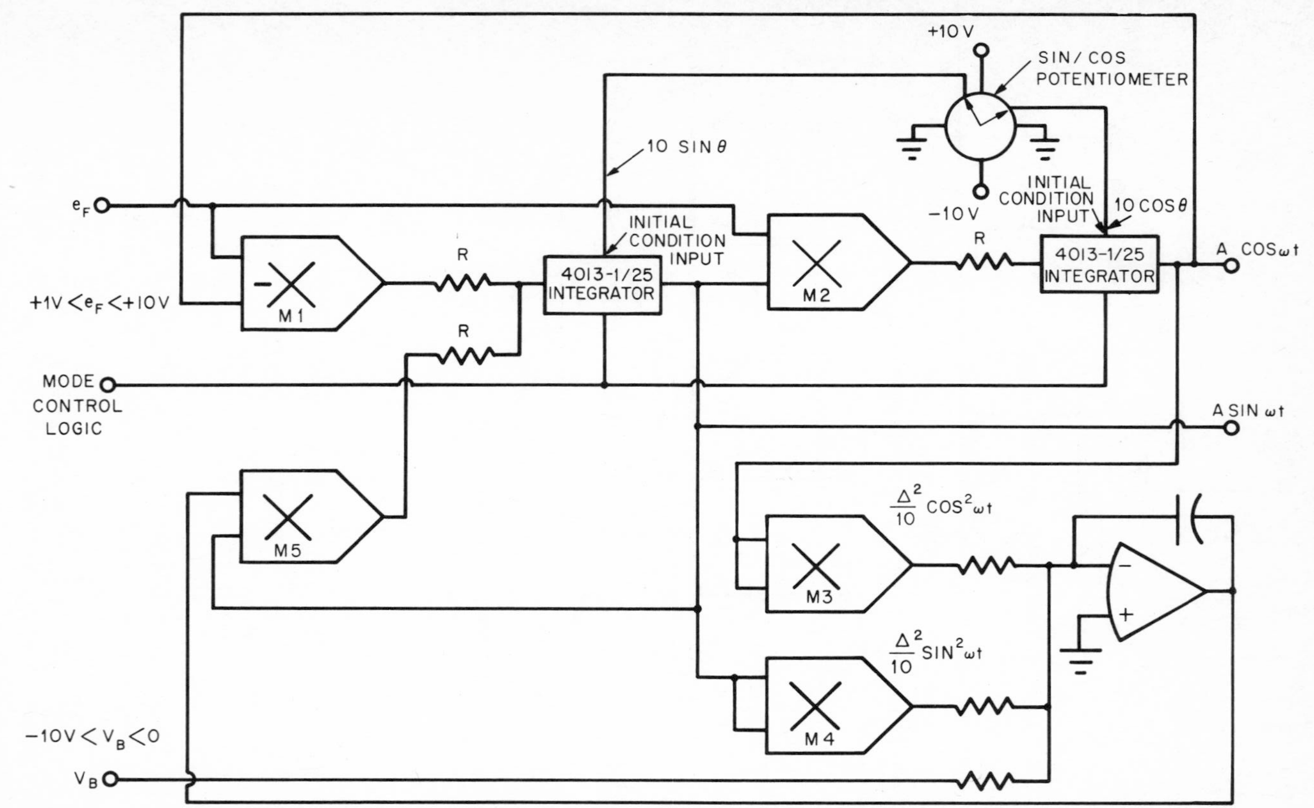

Fig. 11.4 High-performance VCO.

time-averaging technique to obtain the amplitude. The output signals instantly assume the proper amplitudes, independent of the frequency setting. This circuit uses switched integrators of the type described in Sec. 6.3. The initial conditions for the integrators come from the sine-cosine potentiometer. Once the integrators are switched to ON, the circuit functions as a quadrature oscillator. (Several conventional quadrature oscillators were described in Sec. 10.3.4.) Multipliers M_1 and M_2 vary the gains of the two integrators, and thus the frequency varies. Multipliers M_3 and M_4 in effect demodulate the oscillator's amplitude. The A sin ωt and A cos ωt outputs are both squared, and their sum is compared with the reference voltage V_B. This difference controls the magnitude and polarity of the feedback through multiplier M_5 for the A sin ωt output. Multiplier M_5 controls the loop amplitude stability, and both regeneration and degeneration can be applied by it.

2. *Wide-range VCO.* Almost all conventional VCO circuits operate on one of two principles:

1. A square wave is generated and its frequency is controlled by the dc input voltage. The square wave is then filtered to obtain a sine wave.
2. A sine-wave oscillator is developed; then the frequency of oscillation is varied by varying the oscillator loop gain.

Still another approach to designing VCOs is to generate a *triangle wave* that may be controlled in frequency. The triangle wave is then put through a shaping network that exhibits a sinusoidal gain. One advantage to this approach is that the sine gain-shaping circuit can easily be designed to operate from direct current up into the high audio range.

A typical circuit using this technique is shown in Fig. 11.5. As shown, the circuit will provide about a 100:1 dynamic range at any switch setting, with 1 percent linearity. The frequency of oscillation will be

$$f = \frac{e_1}{40R_1C_1} \qquad \text{Hz}$$

The variation in frequency with e_1 is illustrated in Fig. 11.6.

11.2.2 Voltage-to-frequency converters The second basic method of achieving frequency modulation is through the use of a voltage-to-frequency converter. The term voltage-to-frequency converter (VFC) implies that the frequency of some periodic signal is made proportional to an analog control voltage. The output may be any periodic waveform, such as a square wave, a pulse train, a triangle wave, or a sine wave. Pulse-train or square-wave outputs are generally desired if the output is destined to drive a counter of some sort. A VCO is, of course, also a voltage-to-frequency converter. But allowing a pulse or square-wave

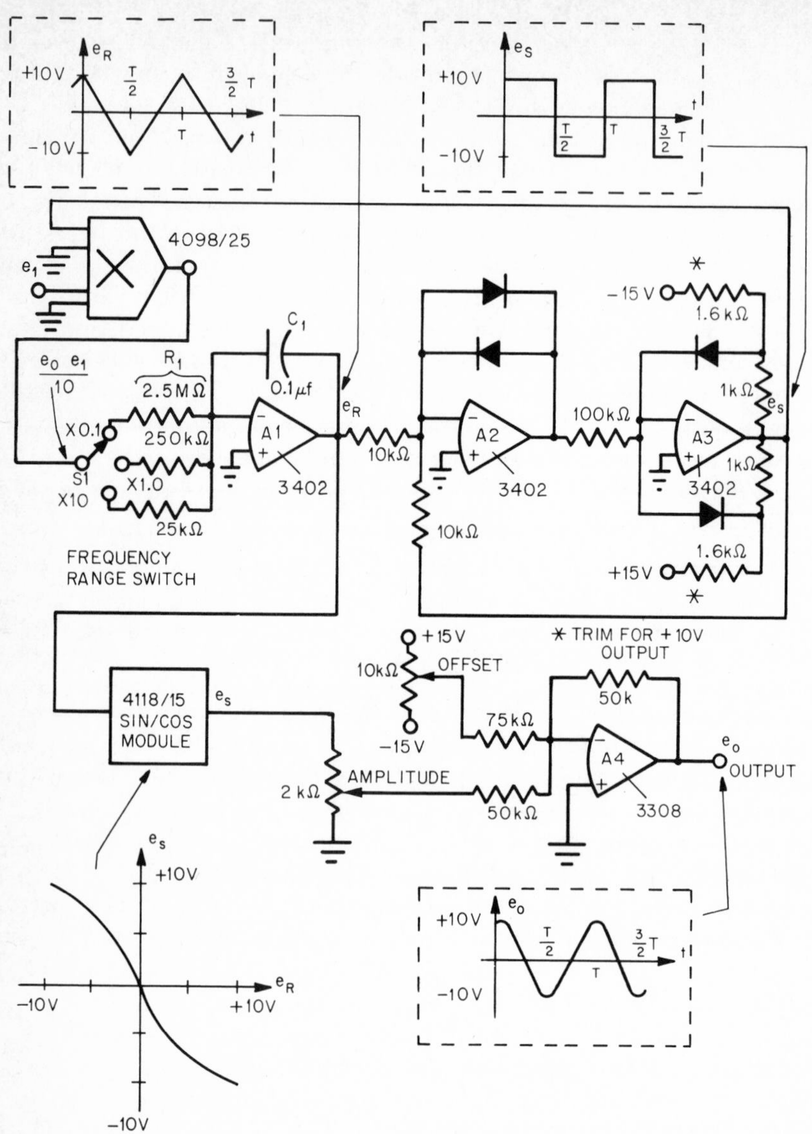

Fig. 11.5 Wide-range VCO circuit.

output will generally simplify the design, particularly if a wide dynamic range of frequency is desired. In the following paragraphs a few typical voltage-to-frequency converter circuits will be discussed. The primary differences lie in the linearity and dynamic range of these circuits.

1. ***Square-wave Output VFC.*** Two circuits which perform the function of a square-wave output VFC are shown in Figs. 11.7 and 11.8.

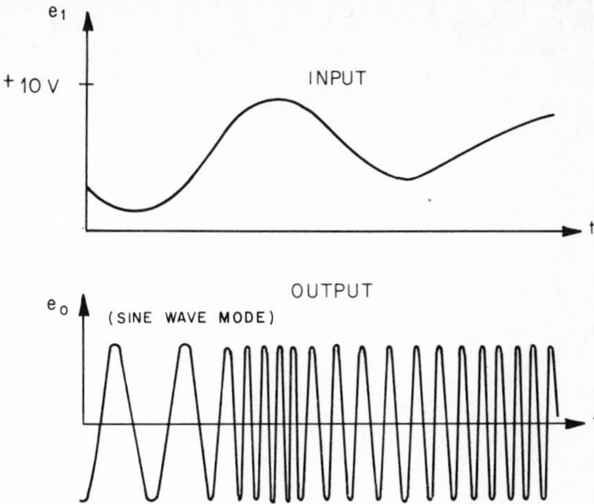

Fig. 11.6 Response of wide-range VCO circuit.

These two circuits are very similar to the triangle- and square-wave generators discussed previously in Sec. 10.2.2. In both circuits, the magnitude of the voltage into the integrator is controlled by a dc input voltage. In the circuit of Fig. 11.7, a multiplier is used to modulate this amplitude and, consequently, the frequency. A wide-bandwidth multiplier with good step response should be used for best results. In the circuit shown in Fig. 11.8 a diode bridge limiter is used to modulate the square-wave input to the integrator, thus varying the frequency. Since the input to the integrator is known to be a square wave, the diode bridge may be used to modulate the amplitude by alternately gating $+e_1$ and $-e_1$ to the input of the integrator.

2. *Pulse-train Output VFC.* A circuit which performs the function of a pulse-train output VFC is shown in Fig. 11.9. Only two operational amplifiers are required for this voltage-to-frequency converter. The operational amplifiers should have good saturation characteristics, high input impedance, and good slew rate capability for best results. An inexpensive wideband FET input operational amplifier, such as the Burr-Brown 3402, is a representative choice. Low-cost bipolar IC (internal connection) operational amplifiers may be used, but the component values will differ and range of operation may be more limited.

To understand the operation of this circuit, first consider the amplitude adjust potentiometer. If this potentiometer is adjusted for an output of -8 V, the output of A_2 will be negative if e_2 is more positive than -8 V. Since the collector of Q_1 is at 0 V, the base is being driven by the output of A_2 to about -11 to -12 V; then Q_1 is indeed OFF as assumed. The input voltage is positive, and so the integrator A_1 will integrate in the

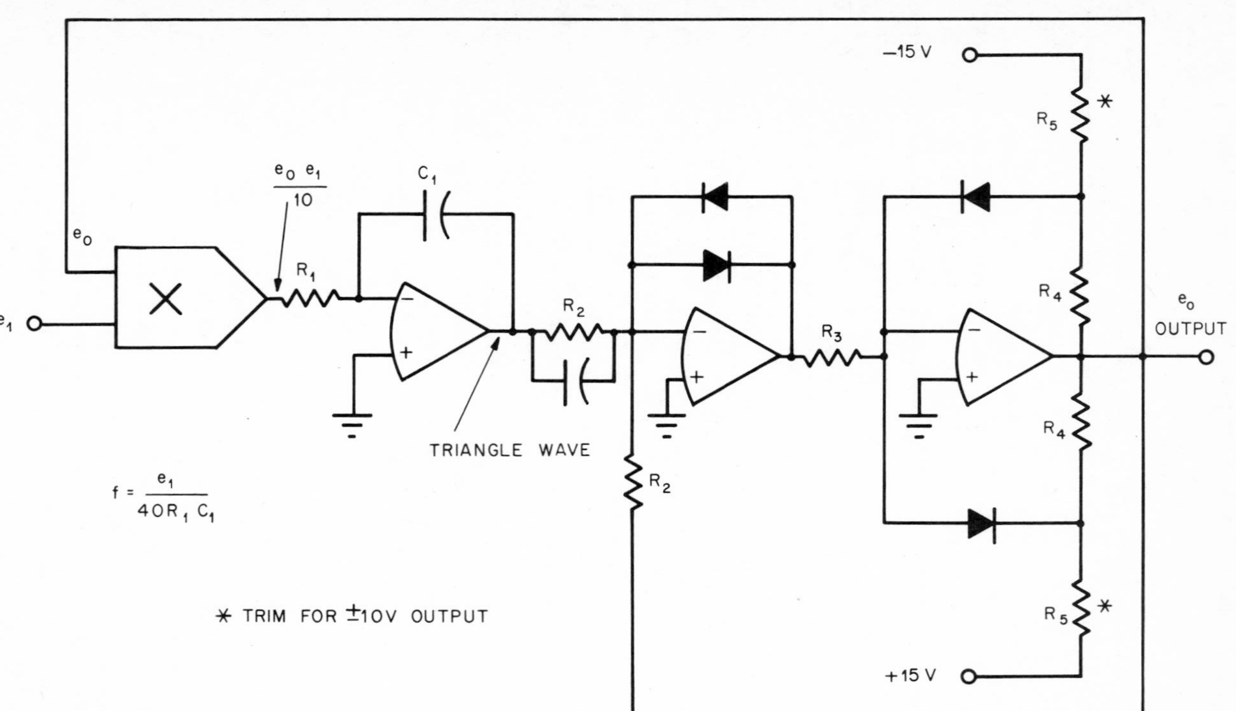

Fig. 11.7 Voltage-to-frequency converter using a multiplier.

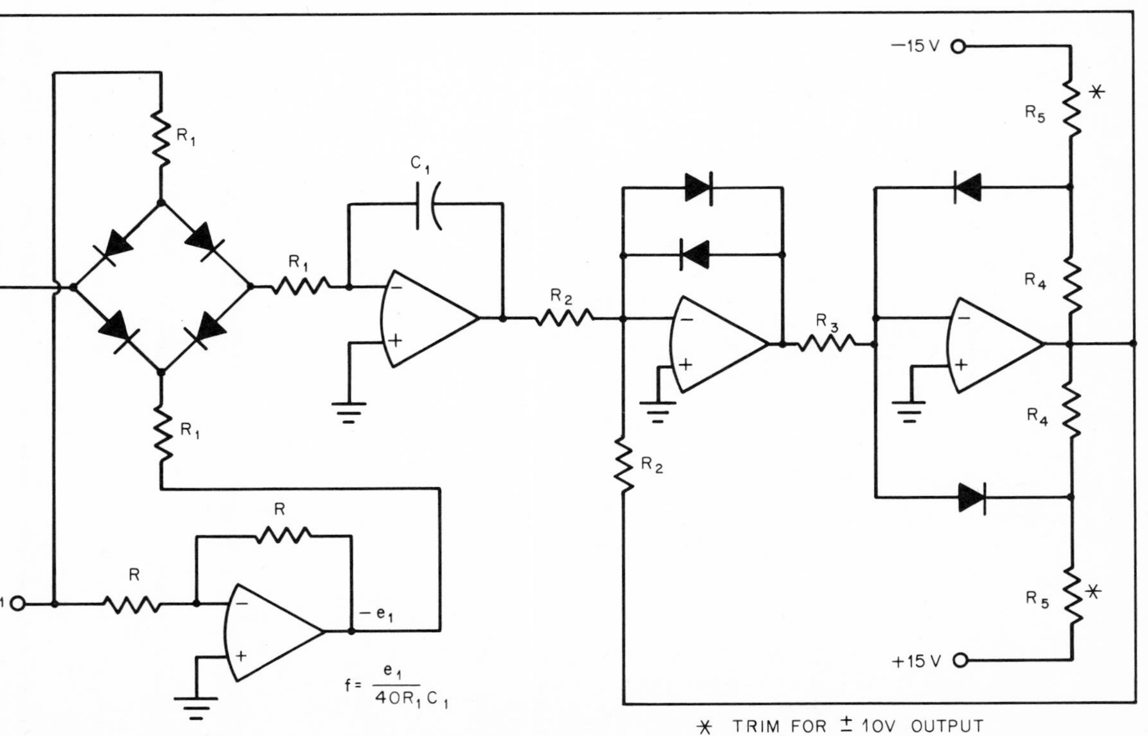

Fig. 11.8 Voltage-to-frequency converter using a diode bridge.

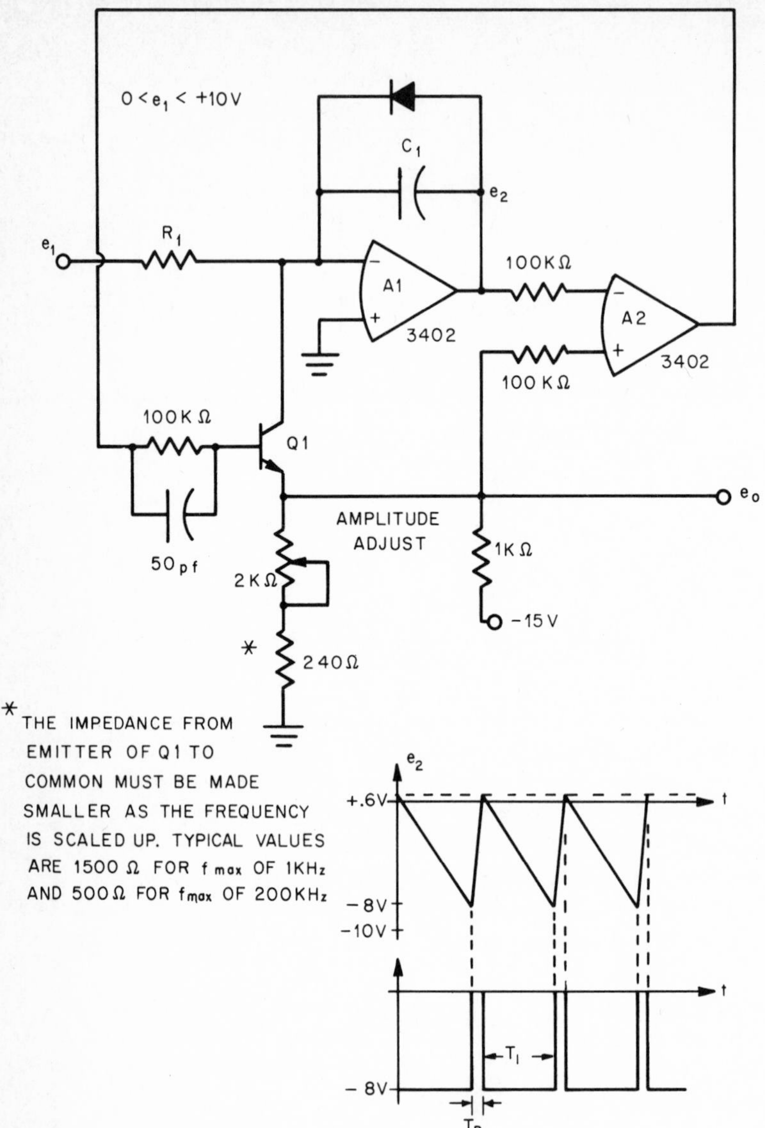

Fig. 11.9 Voltage-to-frequency converter, pulse-train output.

negative direction. When the A_1 output reaches -8 V, the output of A_2 will switch to a positive output. This will gate Q_1 to ON, and since the collector of Q_1 is at approximately 0 V the output e_o will also be very nearly 0 V. More exactly, it will be at $-V_{CE(sat.)}$, or about 0.1 V. With Q_1 at ON, it will act as a current source to the summing junction of the integrator and A_1 will integrate rapidly in the positive direction. When the integrator output exceeds zero, the comparator A_2 will switch to the negative saturation condition. The diode across the integrator prevents overshoot and reduces delay time. A_2 going negative will switch Q_1 back into the OFF condition. The cycle is then completed and will start over again.

The delay time as A_2 switches from one saturated state to the other limits the practical frequency range for this circuit. This delay time will be considerably different for various types of operational amplifiers. The pulse width T_P may be as short as 20 to 100 μs using a fast FET input amplifier such as the Burr-Brown 3402.

The conversion factor of this circuit is found from

$$\frac{e_1T_1}{R_1C_1} = 8.6 \qquad \text{and} \qquad f = \frac{1}{T_I + T_P}$$

$$f = \frac{1}{8.6R_1C_1/e_1} + T_P$$

If $T_P \ll T_I$, then

$$f = \frac{e_1}{8.6R_1C_1}$$

Linearity of this circuit is fairly good at low frequencies, where T_P is much smaller than T_I, but is poor at high frequencies. But if T_P is constant, the nonlinearity is predictable. Since T_P is determined by the slewing time of the integrator and by the switching delay time of the comparator, it will be essentially constant.

The circuit is sensitive to changes in external loading at the output e_o. Thus, if this voltage-to-frequency converter is to drive a load of less than 100 kΩ, an emitter follower or some other form of buffering should be added to the output.

3. *High-performance VFC.* A very wide-range, linear voltage-to-frequency converter can be designed using high-performance function modules. This circuit makes use of a high-speed integrator with current-amplified resetting capability (see Sec. 6.3 for a description of this type of integrator). Although somewhat more expensive than the circuits previously discussed, this circuit offers excellent performance. It is shown in Fig. 11.10a.

To understand the operation of this circuit, note that the integrator A_1 is controlled by the switched current amplifier. When the voltage at pin 3 is high (approximately +4 V), the switch is at OFF and A_1 integrates at a rate determined by R_1, C_1, and E_1. When the voltage at pin 3 is low (approximately +0.6 V), the integrator will very rapidly

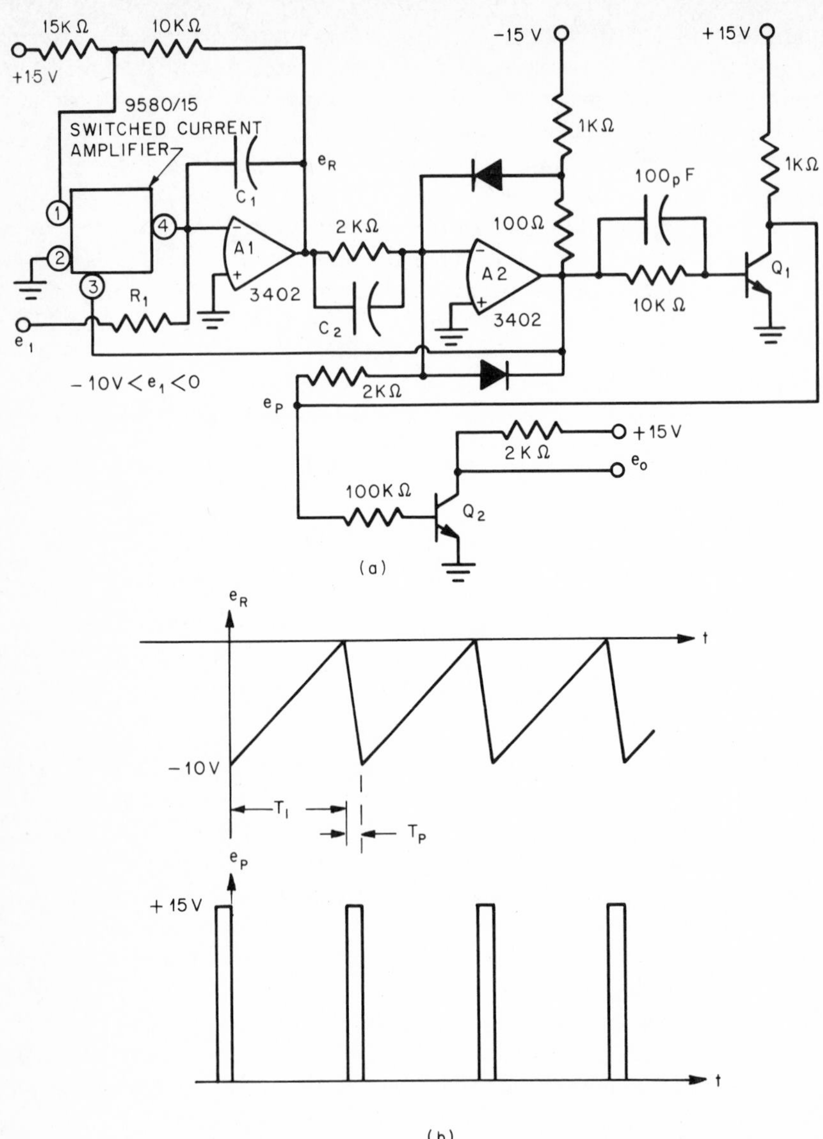

Fig. 11.10 High-performance VFC: (a) voltage-to-frequency circuit; (b) typical waveforms.

reset to −10 V. Amplifier A_2 and transistor Q_1 act as a high-speed comparator. Q_1 and Q_2 may be almost any silicon switching transistor. Q_2 is an optional buffer output stage. Typical waveforms are shown in Fig. 11.10b. The integrator input resistor R_1 or the capacitor C_1 can be varied to provide the desired scale factor of conversion. The equations of operation are

$$10 = \frac{e_1 T_1}{R_1 C_1} \qquad \text{and} \qquad f = \frac{1}{T_I + T_P}$$

$$f = \frac{1}{10R_1C_1/e_1 + T_P}$$

The reset time T_P will be about 3 to 4 μs using the component values shown. If $T_P \approx 3\ \mu$s and $C_1 = 0.01\ \mu$F, then

$$f = \frac{e_1}{R_1 \times 10^{-7} + 3e_1 \times 10^{-6}}$$

If, for example, a scale factor of 1 kHz/V were desired, then R_1 should be about 10 kΩ. Then

$$f \approx e_1 \times 10^3 \text{ Hz}$$

and $0 < f \leq 10$ kHz for $-10\text{ V} < e_1 \leq 0$. This circuit also makes an excellent voltage-controlled ramp generator. The output voltage is then e_R.

Using the 3402 operational amplifier, the circuit shown in Fig. 11.10a can be operated up to 100 kHz by making C_1 equal to 1,000 pF. However, if frequencies under 10 kHz are of interest, an inexpensive amplifier, such as the Burr-Brown 3308/12C, may be substituted for the 3402s. With reasonable care, a dynamic range of 1,000:1 with 1 percent linearity is feasible with this circuit.

11.3 Pulse Width Modulation[1,3]

In this section we shall discuss circuits which may be used for pulse width modulation. In this type of modulation system, a dc or slowly varying voltage may be used to control the width of pulses. The pulse repetition rate is usually fixed, and the carrier input is often in the form of a square wave or pulse train. Pulse widths may be modulated by many means, but two common methods that offer good linearity over a wide range of operating frequencies are discussed in the following paragraphs.

11.3.1 Voltage-to-pulse-width modulator with square-wave carrier input

If the carrier input can be converted to the form of a square or triangle wave, this method is very simple to apply. The sine-wave carrier input is amplified and clipped and then converted to a triangle wave by an integrator. The modulation input biases the triangle and thus modulates the pulse width about the 50 percent duty cycle condition. A circuit for performing such a function is shown in Fig. 11.11a. Typical wave-

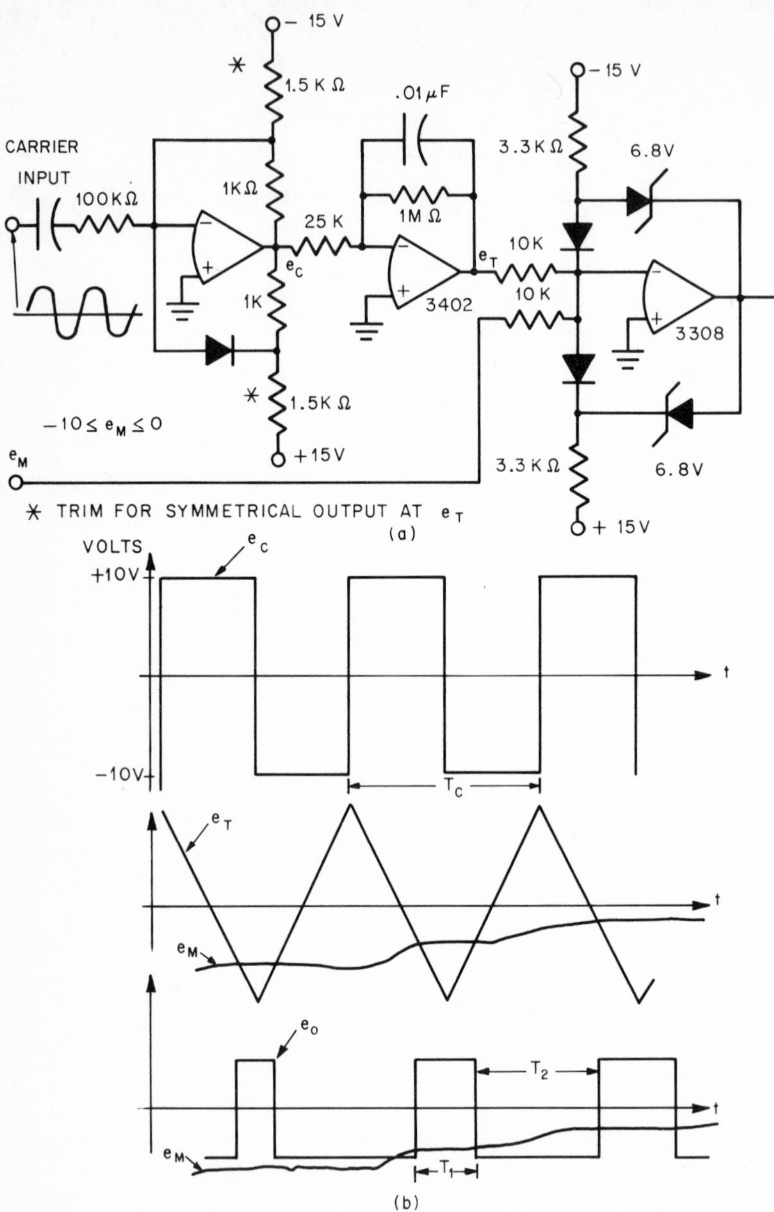

Fig. 11.11 Pulse width modulator. (a) Circuit diagram; (b) waveforms.

forms are shown in Fig. 11.11b. The pulse width T_I is given by

$$T_I = \frac{10 + e_M}{20} T_C$$

where $-10 \leq e_M \leq 0$.

11.3.2 Voltage-to-pulse-width converter A switched integrator can also be used to obtain a very linear and stable synchronized pulse width modulator. A pulse train provides the clocking signal, and the output is a pulse train synchronized to the clock pulse input. Pulse width of the output is a linear function of the input voltage. A circuit for providing such modulation is shown in Fig. 11.12. The values of V_R, C_1, and R_1 in the circuit must be chosen according to the pulse repetition rate and the desired dynamic range. For example, assume that the clock frequency is 1 kHz and the input voltage varies from 0.1 to 10 V. If V_R is +10 V, then

$$0.01R_1C_1 < T_P < R_1C_1$$

T_P must be less than T_C to avoid ambiguity, and so R_1C_1 must be less than T_C. If R_1C_1 is chosen to be $0.9T_C$, then R_1C_1 is 0.9 ms. If C_1 is chosen to be 0.01 μF, then R_1 would need to be 90 kΩ. We now have $T_P = 0.09e_1$ ms. In ratio form, $T_P/T_C = 0.09e_1$. See Sec. 6.3 for a description of integrators using switched current amplifiers.

11.4 Demodulation

The previous sections of this chapter have discussed the use of operational amplifier circuits to perform various types of modulation. In this section we discuss the inverse process, namely, demodulation. Several treatments are given covering amplitude, frequency, and pulse width demodulation.

11.4.1 Amplitude demodulation To accomplish amplitude demodulation, demodulators, or discriminators, are needed to recover the low-frequency signal information that has modulated a high-frequency carrier signal. The carrier is often in the form of a pulse train or sine wave. Many commonly available transducers, such as synchros, have an output that is of the suppressed-carrier amplitude-modulated type. Synchro demodulators must be phase-sensitive, they should have a positive output when the input is in phase with the carrier reference signal and a negative output when the input is 180° out of phase. Some typical demodulator circuits will now be considered.

1. ***Phase-sensitive Demodulation of Suppressed Carrier Signals.*** Two circuits for performing phase-sensitive demodulation of suppressed

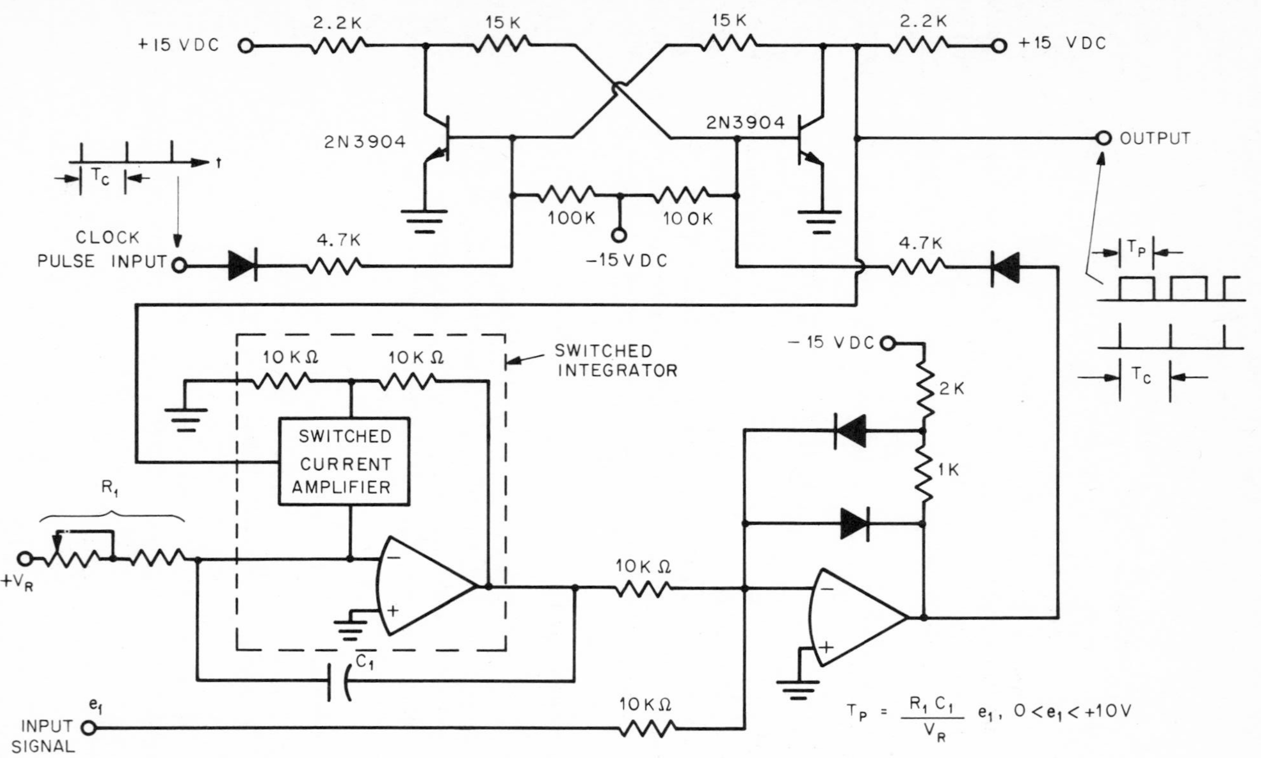

Fig. 11.12 Simplified diagram of a voltage-to-pulse-width converter.

carrier signals will be considered. The input is of the form

$$e_1(t) = e_S(t) \sin \omega_C t$$

$e_S(t)$ is the low-frequency signal that is to be recovered, and ω_C is the carrier frequency. Now if $e_1(t)$ is multiplied by a square wave of amplitude A that is in phase with the carrier reference signal, the low-frequency portion of the output will be proportional to $e_S(t)$. If the square wave is $e_2(t)$, then, using Fourier series,

$$e_O(t) = e_1(t)e_2(t) = Ae_S(t)\left[\frac{2}{\pi} - \frac{4}{\pi}\left(\frac{1}{3}\cos 2\omega_C t + \frac{1}{3.5}\cos 4\omega_C t + \frac{1}{5.7}\cos 6\omega_C t + \cdots\right)\right]$$

Now if $e_O(t)$ is low-pass filtered, the output will be

$$\text{Output} = \frac{2A}{\pi} e_S(t)$$

The other terms constitute ripple. Notice that the lowest frequency of ripple is at twice the carrier frequency. A block diagram of the desired circuit for performing such a demodulation is shown in Fig. 11.13. Two circuits for implementing such a phase-sensitive demodulator are shown in Figs. 11.14 and 11.15.

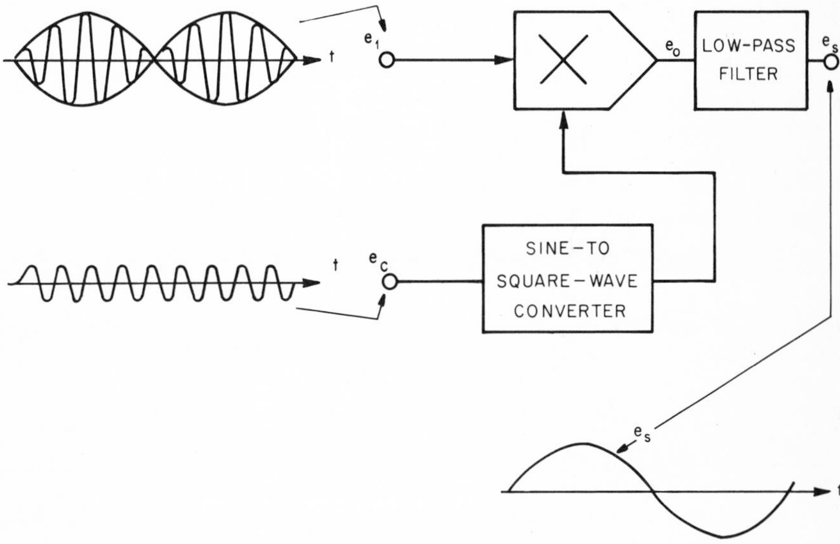

Fig. 11.13 Block diagram of a full-wave, phase-sensitive demodulator.

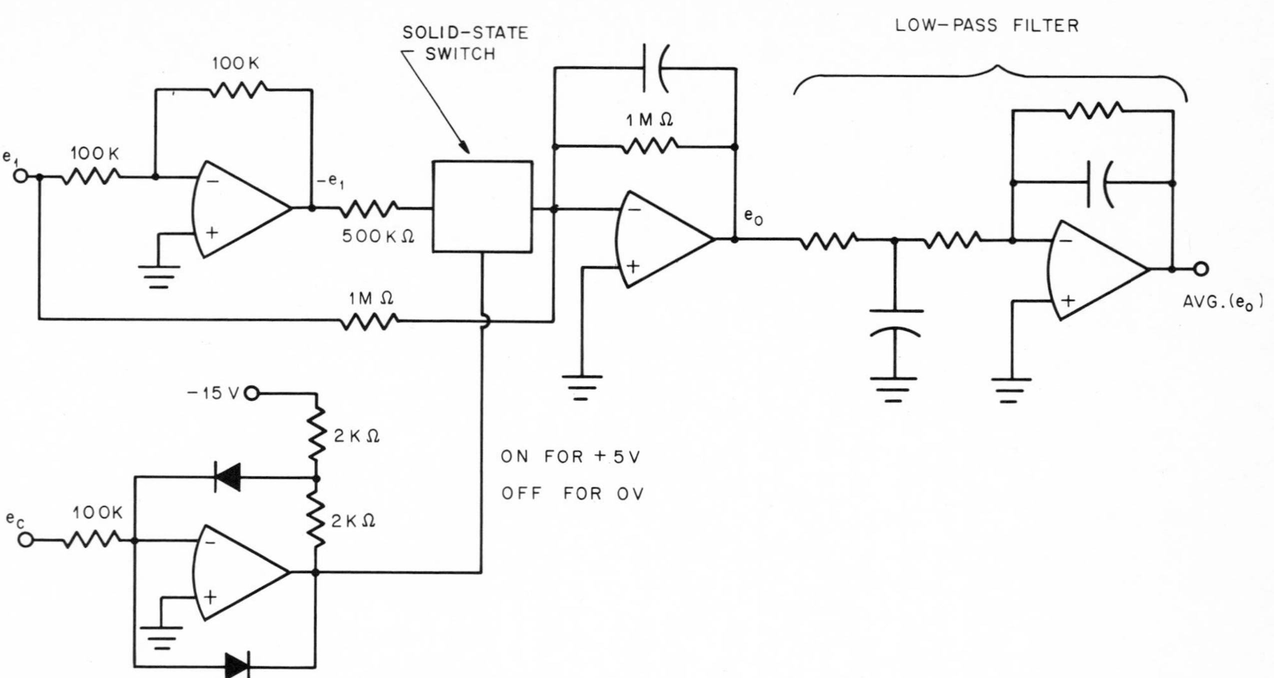

Fig. 11.14 Phase-sensitive demodulator using switching technique.

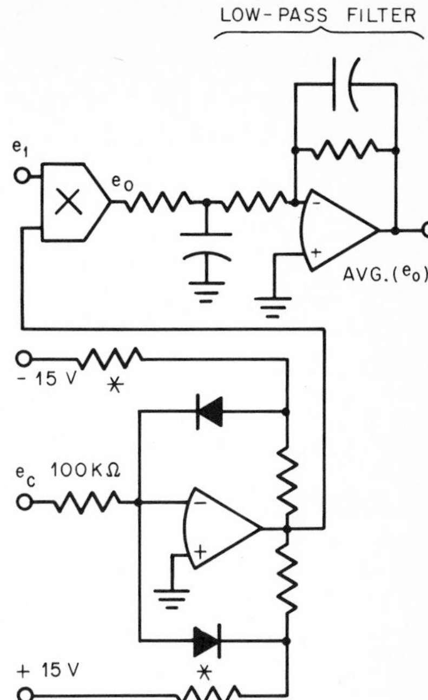

Fig. 11.15 Phase-sensitive demodulator using a multiplier.

2. ***Pulse Sample Demodulator.*** Another means of amplitude demodulation is to sample-and-hold the desired amplitude information. For example, if the peak values of an ac suppressed carrier signal are sampled and held, the output will be a sequence of steps. The fundamental of the output will be the desired low-frequency signal, although delayed in time by one-half cycle of the carrier because of the sampling process. A half-wave 400-Hz pulse sample demodulation circuit is shown in Fig. 11.16a. The waveforms are shown in Fig. 11.16b. It should be noted that the width of the sampling pulse is not critical. With a 400-Hz carrier, a pulse width of about 70 μs centered on the sine wave will sample the voltage within $\pm 5°$ of the peak.

The technique described above can also be used for full-wave demodulation if the input is first full-wave, phase-sensitive detected. Full-wave sampling will reduce the time delay to one-quarter cycle, but at the expense of greater circuit complexity.

11.4.2 Demodulation of FM signals In the above paragraphs we have considered circuits for amplitude demodulation. Now let us consider the frequency demodulation problem. There are three basic techniques, namely, time-averaging demodulation, demodulation by measuring the

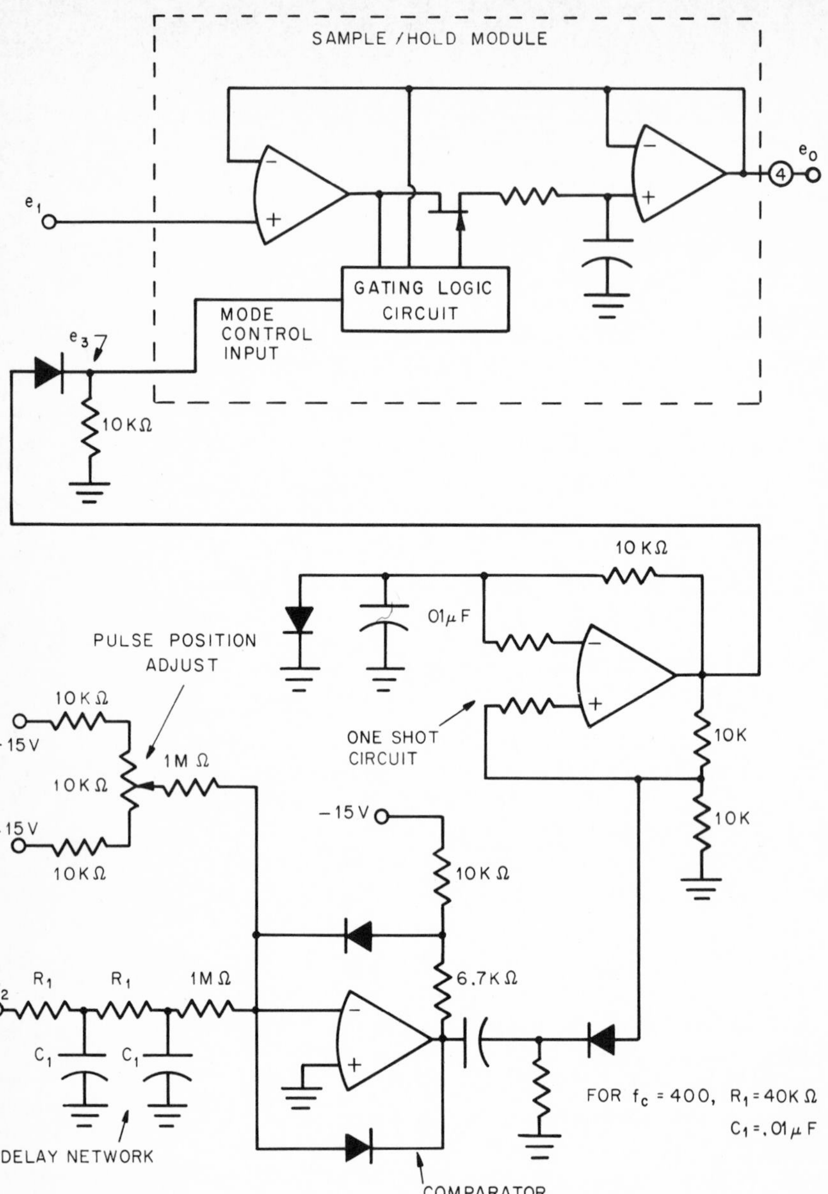

Fig. 11.16 Pulse sample demodulator, 400-Hz carrier. (a) Circuit diagrams.

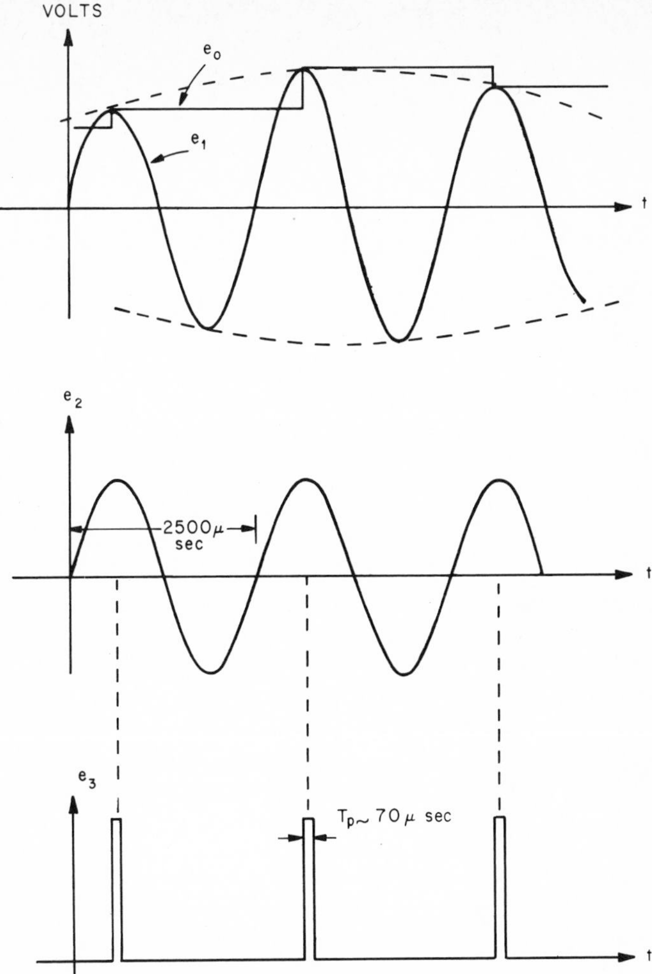

Fig. 11-16 (b) Waveforms.

period, and demodulation using phase-lock techniques. Details of these three methods are given in the following paragraphs.

1. ***Time-averaging FM Demodulation.*** If the modulation frequency is much lower than the carrier frequency, a simple time-averaging technique may be used to measure frequency. The signal input waveform is converted into a train of uniform pulses. The pulse train is then low-pass-filtered to obtain the dc term. The dc output is proportional to the frequency of the input. To adequately filter out the ripple, the filter time constant must be very large. Thus the response rate of this type of demodulator is rather slow. A typical operational amplifier circuit for time-averaging frequency demodulation is shown in Fig. 11.17a.

2. *FM Demodulation by Measuring the Period.* If the modulation frequency is rapidly varying relative to the carrier, it may be preferable to measure the period of one cycle and to continuously compute 1/T. This will provide a cycle-to-cycle computation of frequency rather than averaging a large number of cycles over a long period of time.

The essence of this approach is first to convert the input signal into a train of narrow pulses and then to measure the time between pulses by means of a gated integrator of some sort. The integrator input is a constant reference, and so the voltage out of the integrator is proportional to time. A general block diagram of the process is shown in Fig. 11.18. The design of a circuit to implement the process is straightforward since most of the blocks have been discussed previously. The primary limitation of this type of FM demodulation is its dynamic range. However, a frequency range of 10:1 may be easily designed and a range of 100:1 is possible. Two sources of error are most significant:

1. The pulses of the pulse train e_2 must have a finite width. Each pulse must first stop the integrator and then transfer the peak value V_P into the sample-and-hold circuit; then the integrator must be reset. Although functions may be accomplished very rapidly, they do pose a limit on the resolution of measuring the period.
2. Most dividers commercially available today are limited in dynamic range. Typically, the error increases as the denominator is made smaller. Thus, as the range of V_P is made greater, the divider will generally contribute more error.

Even with these limitations, FM demodulation by measuring the period is a valuable technique of converting frequency information to dc voltage form. The method does not depend on time averaging over a number of cycles, and so the circuit responds very rapidly to changes in frequency.

3. *FM Demodulation Using Phase-lock Techniques.* The methods of FM demodulation discussed previously all depend upon having a noise-free input. Wave shape is not critical, but any noise that interferes with detection of zero crossings would cause error. Where noise is a problem, a phase-lock technique may be used to good advantage. The block diagram of such a technique is shown in Fig. 11.19.

To analyze the circuit operation, assume an input signal e_1 and a VCO signal $e_O(t)$ having the form

$$e_1(t) = V_S \sin(\omega_S t + \theta_1)$$
$$e_O(t) = V_O \cos(\omega_S t + \theta_O)$$

When $\theta_1 - \theta_O$ is zero, the output is 90° out of phase with the input.

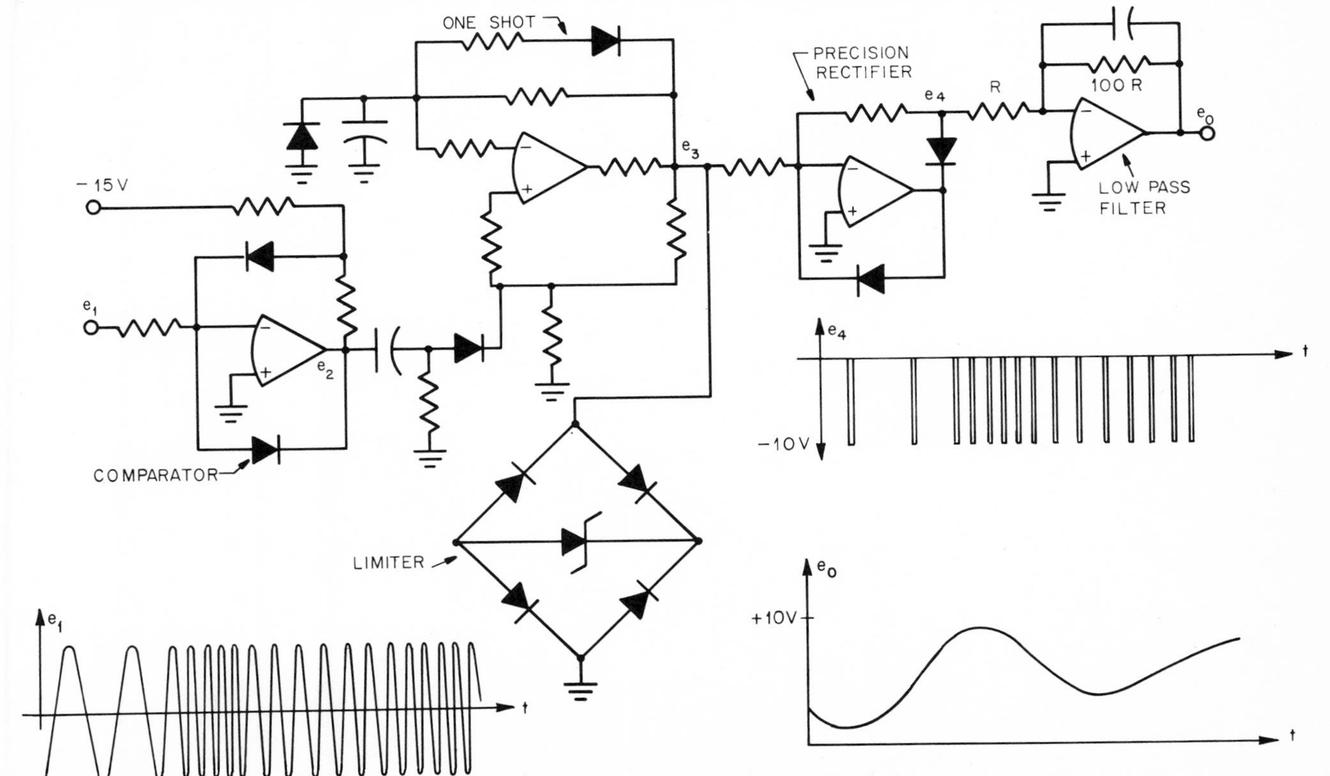

Fig. 11.17 Frequency demodulation. (a) Circuit diagram; (b) waveforms.

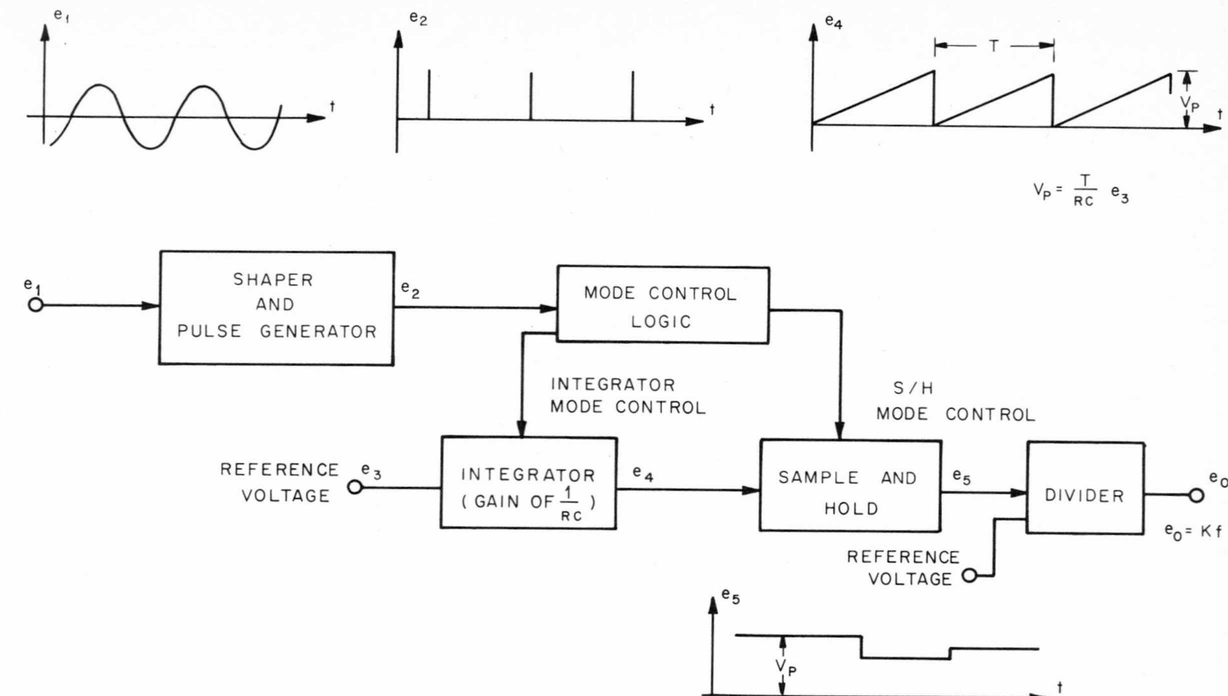

Fig. 11.18 Block diagram of a frequency-to-dc converter.

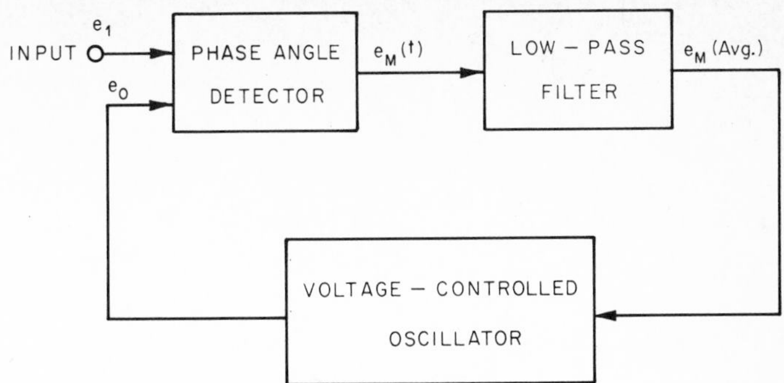

Fig. 11.19 Block diagram of a phase-locked loop.

A multiplier can be used as a phase detector. The output of the multiplier e_M is $e_O e_1/10$, and so

$$e_M(t) = \frac{V_S V_O}{10} [\sin (\omega_S t + \theta_1) \cos (\omega_S t + \theta_O)]$$

$$= \frac{V_S V_O}{10} (\sin \omega_S t \cos \omega_S t + \cos \theta_1 \cos \theta_O$$

$$+ \cos^2 \omega_S t \sin \theta_1 \cos \theta_O - \sin^2 \omega_S t \cos \theta_1 \sin \theta_O$$

$$- \cos \omega_S t \sin \omega_S t \sin \theta_1 \sin \theta_O)$$

Now if e_M is low-pass-filtered to remove all double-frequency terms, this reduces to

$$e_{M\text{ avg}} = \frac{V_S V_O}{20} (\sin \theta_1 \cos \theta_O - \cos \theta_1 \sin \theta_O)$$

$$= \frac{V_S V_O}{20} \sin (\theta_1 - \theta_O)$$

If this low-pass-filtered signal is used to control the frequency of the VCO, the frequency of e_O will be varied until the phase angle is minimized. Thus the two signals will phase-lock. The output of the VCO will be in quadrature with the input and will track the input signal in frequency. The averaged signal from the low-pass filter controls the VCO and is proportional to the input signal phase so long as the loop is "locked" ($\theta_1 - \theta_O \ll 90°$).

The design of an operating phase-lock loop is not difficult, but optimizing the loop for a given application and predicting the noise rejection, acquisition rate, probability of locking, etc., are very difficult. These topics are discussed in other books.[5]

A typical phase-lock loop using operational amplifiers and analog function modules is shown in Fig. 11.20. For the values shown, frequency

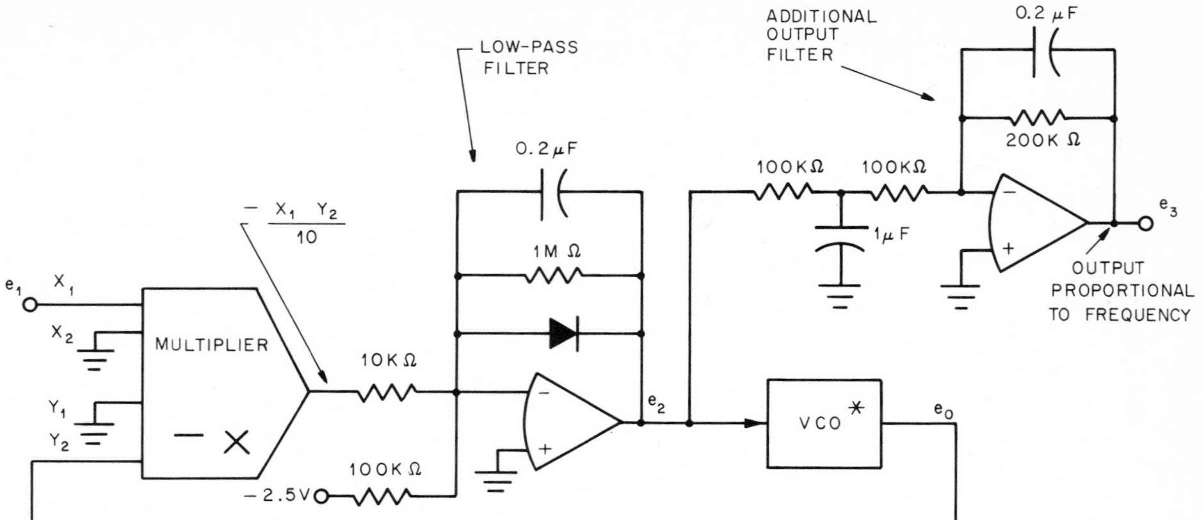

* ONE OF THE CIRCUITS FROM SECTION 11.2A OF THIS CHAPTER MAY BE USED FOR THE VCO, OR A COMMERCIALLY AVAILABLE VCO COULD BE USED. FOR THE CIRCUIT VALUES SHOWN, THE VCO MUST HAVE AN OUTPUT FREQUENCY OF 1K Hz PLUS 1K Hz/0.5V OVER A 20:1 RANGE.

Fig. 11.20 FM demodulator using a phase-locked loop.

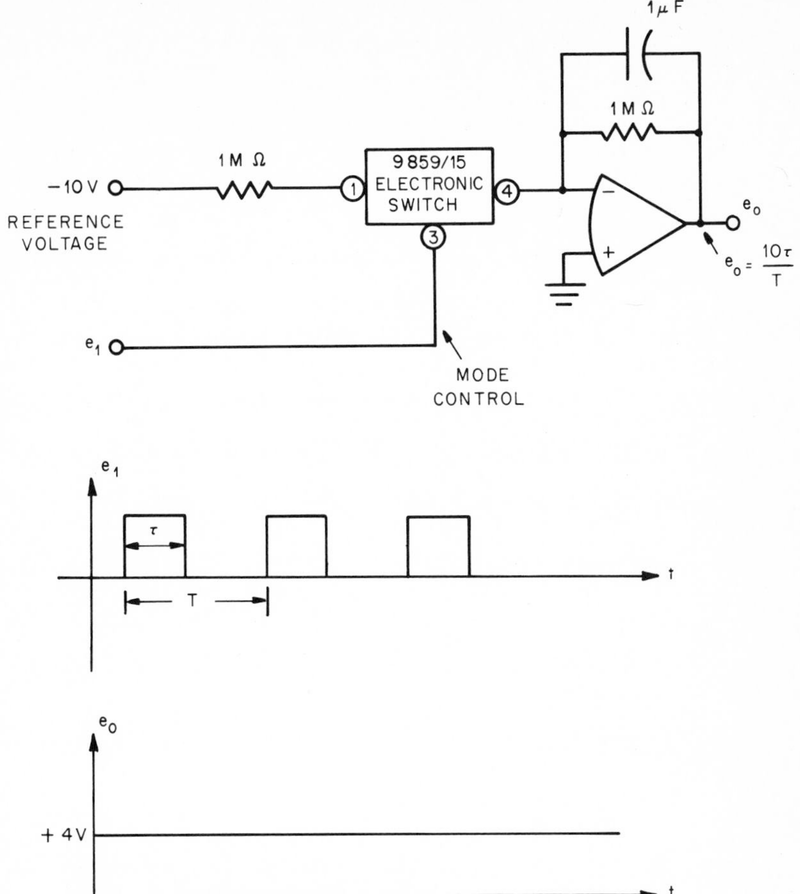

Fig. 11.21 Pulse width demodulation.

may vary over a 20:1 range, and signal amplitudes may also vary over a 20:1 range without losing lock.

11.4.3 Pulse width demodulation A pulse-width-modulated pulse train is easily converted to a dc voltage. Simple low-pass filtering will provide a voltage proportional to the pulse width. If the amplitude of the pulse-width-modulated waveform is not accurate and stable, the circuit in Fig. 11.21 may be used. The pulse height is then proportional to the reference voltage. The variation of pulse width of e_1 causes a corresponding variation in the ON and OFF time of the switch. The output level is independent of fluctuations in the height of the e_1 pulses.

REFERENCES

1. G. A. Korn and T. M. Korn, *Electronic Analog and Hybrid Computers*, McGraw-Hill Book Company, New York, 1964.
2. G. A. Korn, Exact Design Equations for Operational Amplifiers with Four-terminal Computing Networks, *IRE Trans. Electron. Computers*, February, 1962.
3. *Applications Manual for Operational Amplifiers*, Philbrick/Nexus Research, Dedham, Mass., 1965.
4. J. M. Pettitt, *Electronic Switching, Timing, and Pulse Circuits*, McGraw-Hill Book Company, New York, 1959.
5. Floyd M. Gardner, *Phaselock Techniques*, John Wiley & Sons, Inc., New York, 1965.
6. T. Cate, Designing with Packaged Analog Multipliers, *EEE*, May, 1969.
7. T. Cate, Designing with Nonlinear Function Modules, *EEE*, September, 1969.

APPENDIX A

FUNDAMENTAL CIRCUIT THEORY

In this appendix, the basic properties of operational amplifiers are presented. Each of the performance characteristics discussed here is a result of amplifier design factors discussed in detail in Chapters 1 to 5. Test methods for each performance characteristic are given in Appendix B.

A.1 Basic Concepts

The operational amplifier is simply a high-gain, direct-coupled amplifier. It is usually designed to amplify signals extending over a wide frequency range and is normally used with external feedback networks. Many operational amplifiers have a single input terminal, but the greater number have a differential input. Nearly all have a single output terminal. Thus most operational amplifiers may be represented by the symbol of Fig. A.1. Single-ended amplifiers may be treated as the special case where + input is grounded.

There are certain "ideal properties" of operational amplifiers toward which their design is directed. These properties are never realized in practice, of course, but the assumption of such idealness allows rapid preliminary analysis of feedback circuits involving these amplifiers.

The idealized amplifier properties which are usually assumed are

$$\begin{aligned}
&\text{Gain} = \infty \quad (A \rightarrow \infty) \\
&e_o = 0 \text{ when } e_1 = e_2 \\
&\text{Input impedance} = \infty \quad (Z_i \rightarrow \infty) \\
&\text{Output impedance} = 0 \ (Z_o \rightarrow 0) \\
&\text{Bandwidth} = \infty \ (\text{response delay} = 0)
\end{aligned}$$

When feedback is applied, the characteristics of the amplifier are determined largely by the feedback network. This is illustrated in the following analysis of the two most common operational amplifier feedback circuits: the inverting and noninverting circuits.

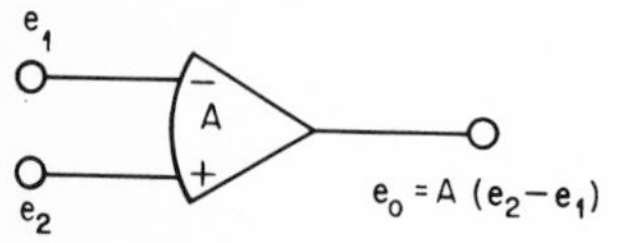

Fig. A.1 Symbol of an operational amplifier.

A.2 Fundamental Inverting Circuit

The circuit shown in Fig. A.2 is representative of the general class of inverting circuits. The common feature of these circuits is that the noninverting input is connected to signal common. In analyzing such circuits, using the ideal amplifier properties, it is noted that no current flows into the amplifier and gain A is assumed to be arbitrarily large. This leads to the circuit equations

$$i_1 = \frac{e_1 - e_s}{Z_1} = \frac{e_s - e_o}{Z_F} = i_F$$

$$e_o = -Ae_s \qquad A \rightarrow \infty$$

Solving for e_o/e_1 and eliminating terms which approach zero give the overall (closed-loop) gain or transfer function,

$$\frac{e_o}{e_1} = -\frac{Z_F}{Z_1} \tag{A-1}$$

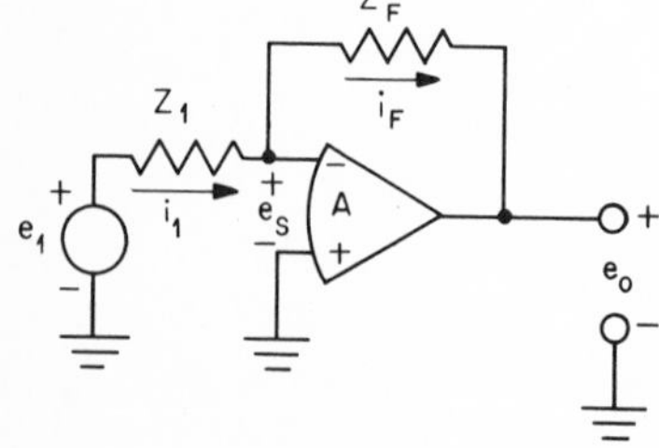

Fig. A.2 Inverting feedback amplifier.

Note that the closed-loop gain includes a sign inversion and has a magnitude determined solely by the ratio of the feedback elements. Note also that the summing point voltage e_s approaches zero as the amplifier gain A becomes arbitrarily large,

$$e_s = \frac{-e_o}{A} \rightarrow 0 \qquad \text{as } A \rightarrow \infty$$

This circuit condition is described by referring to the summing point as a "virtual ground." With the summing point at ground potential, the current through Z_1 is

$$i_1 = \frac{e_1}{Z_1}$$

Thus, i_1 is independent of the value of Z_F. However, this input signal current does flow through Z_F, since no current flows into the operational amplifier inputs ($Z_i = \infty$). Since one end of Z_F is at ground potential (the summing point), the other end must be a voltage of $-i_1Z_F = e_o$, the amplifier output voltage. Input impedance of the circuit is simply Z_1.

The simple inverting amplifier may be modified by adding additional signal sources and impedances as shown in Fig. A.3. The summing point remains at ground potential, and the various input currents are independent of on one another:

$$i_1 = \frac{e_1}{Z_1} \qquad i_2 = \frac{e_2}{Z_2} \qquad i_3 = \frac{e_3}{Z_3}$$

The sum of the currents flows in the feedback element, Z_F, generating the output voltage:

$$e_o = -i_FZ_F = -\left(e_1\frac{Z_F}{Z_1} + e_2\frac{Z_F}{Z_2} + e_3\frac{Z_F}{Z_3}\right) \qquad \text{(A-2)}$$

Thus the circuit of Fig. A.2 functions as a summing amplifier where each

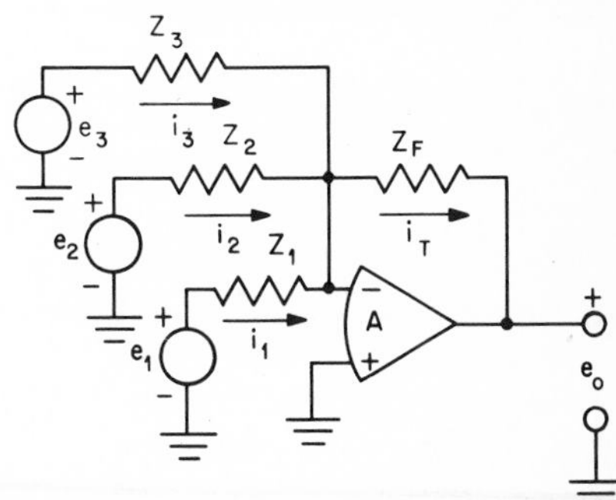

Fig. A.3 Summing amplifier.

input voltage term in the summation may be operated on by a different scale factor or linear operator.

In the inverting circuits of Figs. A.2 and A.3 the input and feedback elements need not be single components (such as resistors or capacitors) but may be more complex elements composed of a number of linear or nonlinear elements. The Z_i's then represent the short-circuit transfer functions of these elements. Regardless of the complexity of the input and feedback networks, the same principles will be found to hold:

1. The summing point is a virtual ground.
2. No current flows into the amplifier inputs; current flowing into the summing point from the input networks must flow through the feedback network.

A.3 Noninverting Circuits

The inverting feedback circuits discussed in the preceding section may be realized with either single-ended or differential input amplifiers. However, those to be discussed here require operational amplifiers having a noninverting input for signals and an inverting input for feedback voltages. Usually such operational amplifiers are differential input types, although there are amplifiers which operate only in the noninverting mode.

The general noninverting circuit is shown in Fig. A.4. The signal is applied to the noninverting input and a portion of the output signal is "fed back" to the inverting input. This feedback network then determines the overall closed-loop transfer function. When the loop is closed the following equations apply:

$$e_1 = i_1 Z_1 = e_o \frac{Z_1}{Z_1 + Z_F} \qquad \text{assuming that amplifier input current is zero}$$

$$e_o = A(e_2 - e_1)$$

Combining the above equations and allowing the amplifier gain A to

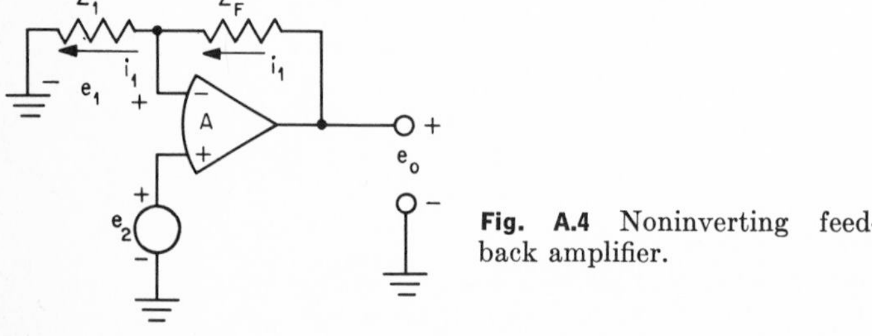

Fig. A.4 Noninverting feedback amplifier.

become arbitrarily large yields

$$\frac{e_o}{e_1} = \frac{Z_1 + Z_F}{Z_1} \qquad \text{and} \qquad e_2 = e_1 \tag{A-3}$$

Thus, the closed-loop gain is always greater than, or equal to, unity and is determined by Z_1 and Z_F. If Z_F is equal to zero (and $Z_1 = \infty$, open circuit), the gain is exactly 1.0 and the amplifier acts as a *voltage follower;* the output voltage follows the input voltage exactly. The advantage of such a voltage follower, and of noninverting circuits in general, is the impedance buffering property, i.e.,

$$Z_i \rightarrow \infty$$
$$Z_o \rightarrow 0$$

Such amplifier circuits are widely used to provide isolation of signal source and load, thus preventing undesired interactions or "loading" effects.

Note that the summing point conditions must be generalized slightly in extending the analysis to noninverting circuits:

1. When the operational amplifier is operating linearly with feedback, the potentials at the two inputs are equal.
2. No current flows into either input of the amplifier.

As in the inverting circuits, the feedback network need not be a simple voltage divider but may be a combination of linear and nonlinear elements selected to yield a desired transfer function.

Extension of the analysis technique to more complex circuits is relatively easy. Although idealized amplifiers were used in deriving the basic feedback circuit transfer functions, the results are usually quite good as a first approximation. However, since real amplifiers can only approximate the ideal over limited ranges of voltage, current, and frequency, it is necessary to analyze the effects of each amplifier parameter. This is the purpose of the following sections.

A.4 Open-loop Gain

The open-loop gain is the magnitude of the amplification factor, A Although the dc and low-frequency gain of such amplifiers may be extremely large (typically 10^5), it is nevertheless finite and therefore contributes a small error term to the closed-loop transfer functions previously derived. If a finite gain is assumed, but the other idealized amplifier conditions are assumed to hold, the following expressions are obtained for closed-loop gain:

Inverting circuit: $$\frac{e_o}{e_i} = \frac{-Z_F/Z_1}{1 + 1/A\beta} = \frac{1 - 1/\beta}{1 + 1/A\beta} \tag{A-4}$$

Noninverting circuit: $$\frac{e_o}{e_i} = \frac{1 + Z_F/Z_1}{1 + 1/A\beta} = \frac{1/\beta}{1 + 1/A\beta} \tag{A-5}$$

where

$$\beta = \frac{1}{1 + Z_F/Z_1} \tag{A-5a}$$

The term $A\beta$ is usually referred to as the loop gain since it may be thought of as the gain around the "loop" formed by the amplifier and its feedback network. Note that for both circuits the gain may be expressed as

$$\frac{e_o}{e_i} = \frac{\text{ideal gain}}{1 + 1/A\beta} = A_{CL} \tag{A-6}$$

Thus the accuracy of the closed-loop gain is limited by the amount of loop gain available. Usually, $A\beta$ will be much larger than unity in order to obtain stable closed-loop gain. The expression for closed-loop gain then is closely approximated by the expression

$$\frac{e_o}{e_i} = (\text{ideal gain})\left(1 - \frac{1}{A\beta}\right) \tag{A-7}$$

Gain error is given by the $1/A\beta$ term. Since β is usually fixed by the desired circuit function, the gain error is made acceptably low by choice of an amplifier with the required value of open-loop gain A. Although it is possible to adjust the elements of the feedback network to compensate for this gain error term, the temperature sensitivity of the open-loop gain places a practical limitation on the ultimate accuracy achievable by this technique.

A.5 Frequency Response and Stability

At direct current and low frequencies it is usually sufficient to regard the open-loop gain of the operational amplifier as a number, A_o, which is sometimes expressed in decibels:

$$\text{Open-loop gain in dB} = 20 \log A_o$$

However, for higher frequencies it is necessary to consider the frequency-sensitive character of the open-loop gain. The open-loop gain can be approximated by a rational function with one or more poles and (possibly) zeros. For example,

$$A(s) = \frac{A_o(1 + \tau_a s)}{(1 + \tau_1 s)(1 + \tau_2 s)(1 + \tau_3 s)}$$

A typical plot of the magnitude of $A(j\omega)$ versus frequency (Bode plot) appears in Fig. A.5.

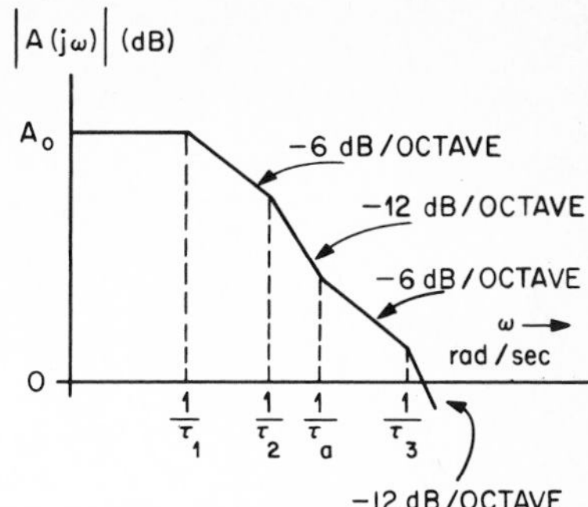

Fig. A.5 Typical open-loop gain response of an operational amplifier.

The closed-loop stability of such an amplifier is established by application of Bode's criterion. According to Bode's criterion, the rate of closure between the open-loop frequency response, A(jω), and the reciprocal of the feedback gain $1/\beta(j\omega)$, must be less than 12 dB per octave for a stable closed-loop system. To guarantee sufficient phase margin in the most common feedback circuits, the rate of open-loop gain rolloff usually is made approximately 6 dB per octave by internal compensation of the amplifier. The curves of Fig. A.6 illustrate stable and unstable situations. Further stability considerations are discussed in Chapter 5.

When the open-loop response of the operational amplifier has a −6 dB per octave rolloff rate, it can be represented by a transfer function having

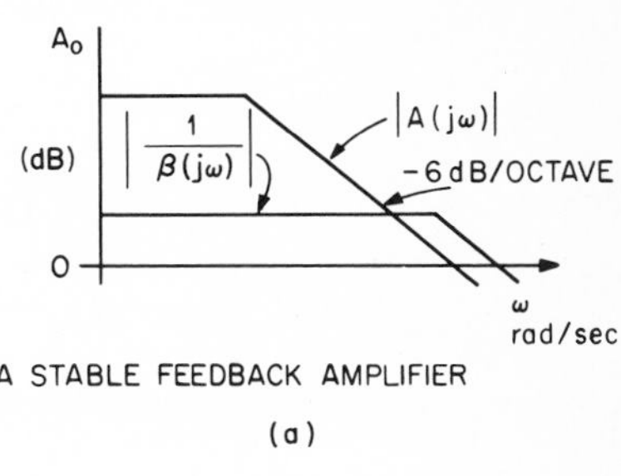

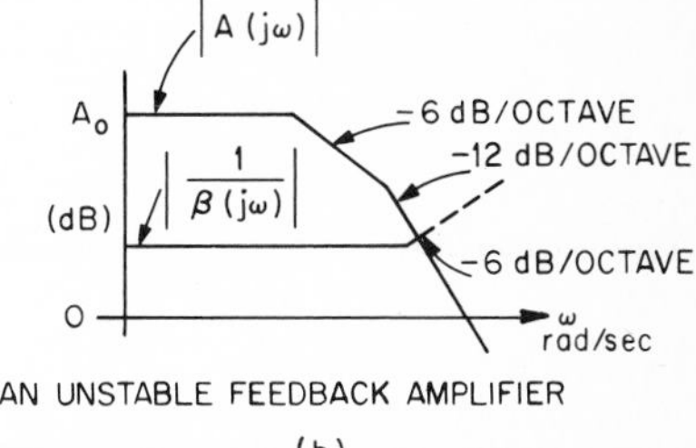

Fig. A.6 Typical Bode plots for stable and unstable systems. (a) A stable feedback amplifier; (b) an unstable feedback amplifier.

one pole on the real axis,

$$A(s) = \frac{A_o}{1 + \tau_o s} \tag{A-8}$$

where

$$A_o = \text{open-loop gain at } s = 0 \qquad \text{(dc gain)}$$

This expression may be substituted into the closed-loop gain expression [Eq. (A-6)] to yield

$$A_{CL} = \frac{\text{ideal gain}}{1 + (1 + \tau_o s)/A_o\beta} \approx \frac{\text{ideal gain}}{1 + (\tau_o/A_o\beta)s}$$

Inverting circuit: $$A_{CL} = \frac{1 - 1/\beta}{1 + \tau_o/A_o\beta s} \tag{A-9}$$

Noninverting circuit: $$A_{CL} = \frac{1/\beta}{1 + (\tau_o/A_o\beta)s} \tag{A-10}$$

The loop gain factor $A\beta$ determines the accuracy of the closed-loop gain, as discussed earlier. At low frequencies both $A(j\omega)$ and $\beta(j\omega)$ may be regarded as real numbers and the product $A_o\beta$ is generally very large. However, for higher frequencies, $A(j\omega)$ begins to decrease in magnitude and exhibit considerable phase shift. The loop gain thus decreases as a function of frequency and the closed-loop gain becomes less accurate. This behavior is illustrated in Fig. A.7 for simple resistive feedback where

$$\beta = \frac{R_1}{R_F + R_1}$$

Several points illustrated by the plots of Fig. A.7 are worthy of mention:

1. The corner frequency (−3-dB point) of the closed-loop gain is given by the intersection of the $A(j\omega)$ and $1/\beta$ curves.
2. The open-loop unity-gain bandwidth, ω_c, is related to the closed-loop −3-dB bandwidth, ω_o, by the equation

$$\omega_c = \frac{\omega_o}{\beta} = \text{constant}$$

Thus the closed-loop bandwidth is proportional to β. The constant ω_c is sometimes called the gain-bandwidth product.

3. At low frequencies (where loop gain is high) the closed-loop gain is determined by the feedback network. At high frequencies (where loop gain is small) the closed-loop gain curve approaches the open-loop gain curve asymptotically.
4. At low frequencies the error term $1/A\beta$ is a real number and is equal to the gain error. If $A\beta$ is 100 (40 dB), the closed-loop gain error is 1 percent.

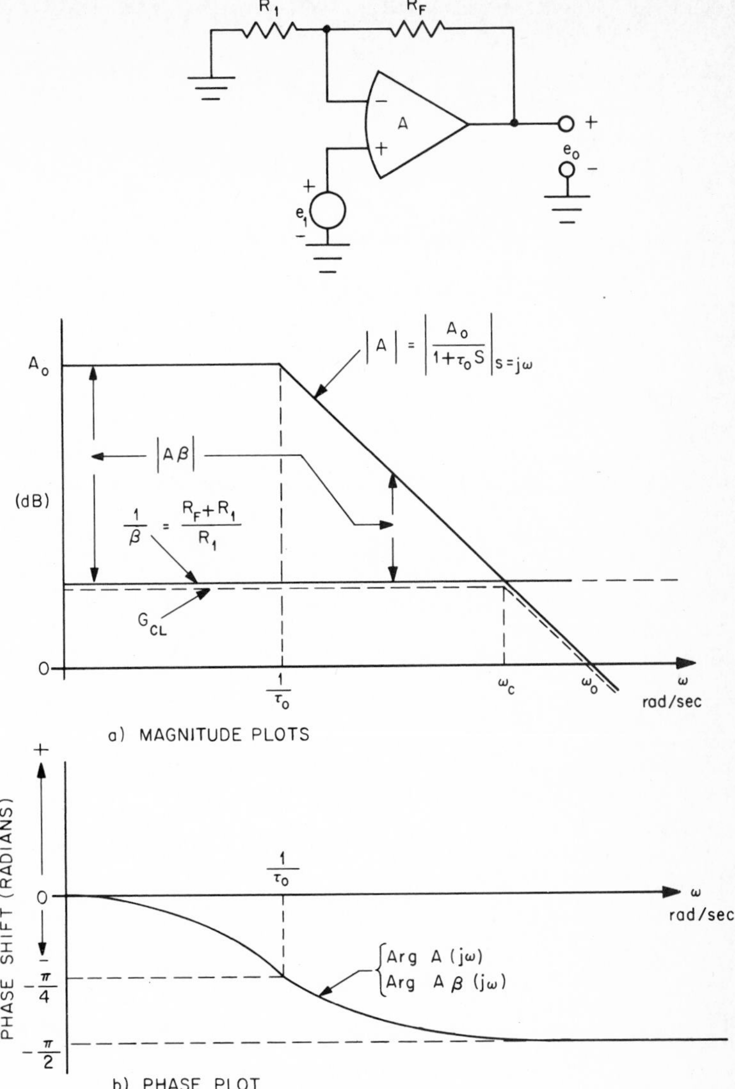

Fig. A.7 Illustration of open-loop, closed-loop, and loop gain. (a) Magnitude plots; (b) phase plot.

5. At high frequencies the error term $1/A\beta$ is about $(1/A\beta)\underline{/+90°}$. The magnitude of the closed-loop gain is then

$$G_{CL} = \frac{\text{ideal gain}}{1 + |1/A\beta|^2}$$

Thus for $A\beta = 10\underline{/-90°}$ the closed-loop gain magnitude error is approximately 1 percent. Phase-shift error, however, is larger. In mathematical terms

$$\frac{|C_{CL}| - \text{ideal gain}}{\text{Ideal gain}} \approx 0.01$$

$$\frac{|G_{CL} - \text{ideal gain}|}{\text{Ideal gain}} \approx 0.10$$

A.6 Common-mode Signal Considerations

For differential input amplifiers, the voltage at both inputs can be raised above ground potential. The common-mode voltage e_{cm} is defined as the average of the two input voltages (Fig. A.8a).

$$e_{cm} = \frac{e_1 + e_2}{2}$$

An operational amplifier responds, ideally, only to the difference voltage $(e_2 - e_1)$ and produces no output voltage for a common-mode input voltage. However, in practical amplifiers an input common-mode voltage e_{cm} generates an output voltage e_{ocm}, as discussed in Secs. 1.3, 1.4, and 4.1. Thus a common-mode gain may be defined

$$A_{cm} = \frac{e_{ocm}}{e_{cm}}$$

This output voltage may be referred to the input and may be represented by the circuit model of Fig. A.8b.

It is customary to define a common-mode rejection ratio (CMRR) as the ratio of open-loop (differential) gain to common-mode gain.

$$\text{CMRR} = \frac{A}{A_{cm}} \tag{A-11}$$

or, in decibels,

$$\text{CMR(dB)} = 20 \log \text{CMRR}$$

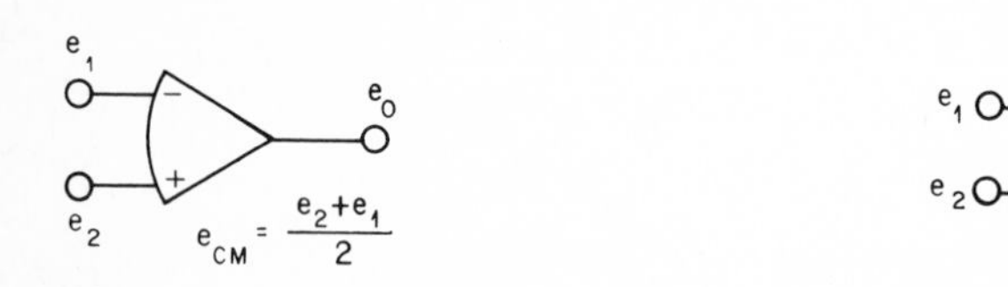

Fig. A.8 Common-mode rejection concepts. (a) Definition of common-mode voltage; (b) circuit model of common-mode error.

These definitions tend to obscure the fact that both open-loop gain and common-mode gain may be nonlinear functions of signal level. Usually the specified value of CMRR represents an average over the signal range. Further discussion appears in Appendix B.

The significance of common-mode rejection ratio in feedback circuits will be illustrated by a simple example, the noninverting unity-gain circuit (voltage follower) of Fig. A.9. Note that the equivalent common-mode input error voltage is included. Equations describing the circuit operation are

$$e_{cm} = \frac{e_2 + e_o}{2} \approx e_2$$

$$e_o = A\left(e_2 - e_o - e_2\frac{1}{\text{CMRR}}\right)$$

$$e_o(1 + A) = e_2\left(A - \frac{A}{\text{CMRR}}\right)$$

$$\frac{e_o}{e_2} = \frac{1 - 1/\text{CMRR}}{1 + 1/A} \tag{A-12}$$

Thus the finite common-mode rejection ratio contributes an additional closed-loop gain error term, in addition to that due to finite open-loop gain. In many cases the finite common-mode rejection ratio is the major contributor to gain error and nonlinearity, particularly in the voltage follower configuration.

A.7 Input Offset Voltage

The ideal operational amplifier develops zero output voltage when both inputs are at zero potential. Any output voltage which appears under the above condition can be replaced by an equivalent dc input voltage, V_{OS}, referred to as the input offset voltage of the amplifier. The effect of this dc offset voltage may be analyzed in a feedback circuit such as that of Fig. A.10. All signal voltages sources are replaced by short circuits. Since it is a dc analysis, all capacitors are replaced by open

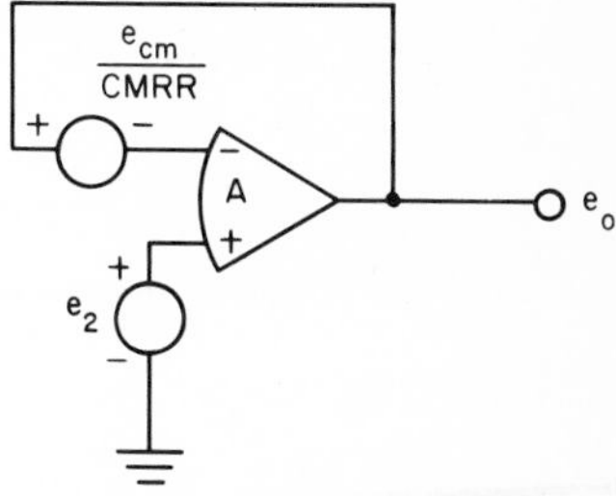

Fig. A.9 Common-mode rejection in a voltage follower.

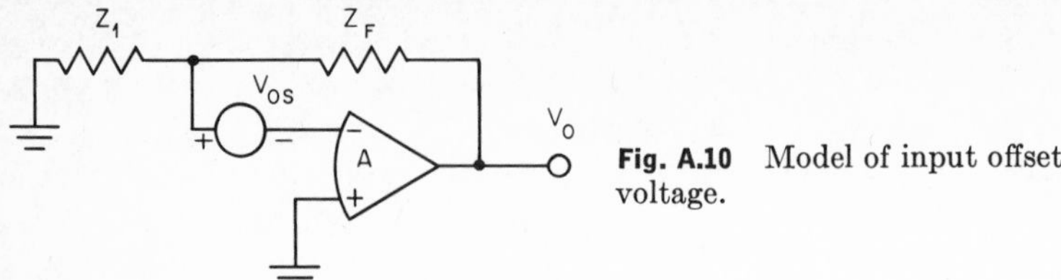

Fig. A.10 Model of input offset voltage.

circuits and all inductors by short circuits. $Z_F(0)$ and $Z_1(0)$ then represent the effective resistive components of the feedback network.

The resulting output offset voltage is

$$V_o = \frac{Z_F(0) + Z_1(0)}{Z_1(0)} V_{os} = \frac{V_{os}}{\beta(0)} \tag{A-13}$$

where $\beta(0)$ is the feedback ratio at direct current.

The expression applies whether the amplifier is used inverting or non-inverting with the signal source. The offset voltage and its sensitivities to various parameters (temperature, power supply voltage, time, etc.) represent some of the most important sources of error in operational amplifier circuits. Its origin and thermal sensitivities are discussed in Chapter 2. Definitions of these sensitivities ("drift") are given in Appendix B.

A.8 Input Bias Current

Another characteristic of practical operational amplifiers is the need for a "bias" current to flow in each input lead. This current usually represents the base or gate current, or a portion of it, required by the amplifying elements of the input stage, as discussed in Sec. 2.3. These dc bias currents may be represented by the current generators shown in Fig. A.11. Analysis of this feedback circuit yields an expression for dc output voltage error due to bias currents:

$$V_o = I_{B1}R_F - I_{B2}R_2\left(1 + \frac{R_F}{R_1}\right) \tag{A-14}$$

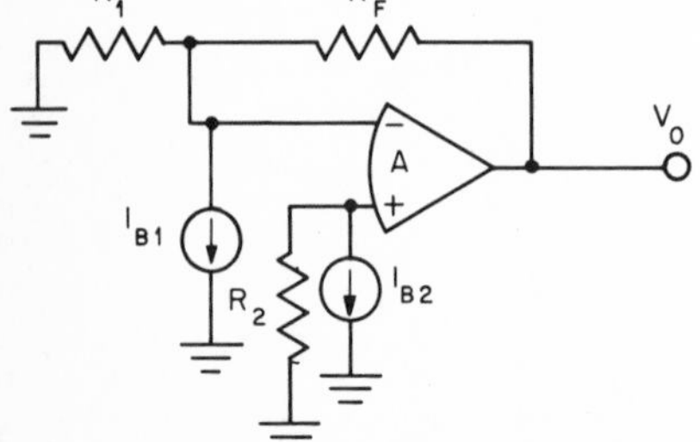

Fig. A.11 Model for input bias currents.

In most operational amplifiers the bias currents to I_{B1} and I_{B2} will be similar in magnitude and will tend to track one another as temperature varies. Thus it is possible to minimize the effects of bias currents by equating the coefficients of I_{B1} and I_{B2} in Eq. (A-14). This yields the equations

$$R_2 = \frac{R_F R_1}{R_1 + R_F} \tag{A-15}$$

$$V_o = (I_{B1} - I_{B2})R_F = I_{OS}R_F \tag{A-16}$$

The difference current, $I_{OS} = (I_{B1} - I_{B2})$, is termed the input offset current of the amplifier and is usually much smaller than the input bias currents. It is not always possible, however, to choose R_1, R_2, and R_F to satisfy the condition of Eq. (A-15). The relationship of Eq. (A-14) is more general and applies for all values of these resistors.

A.9 Input Noise, Voltage, and Currents

The inherent noise of the operational amplifier, generated internally by resistors and active elements as outlined in Secs. 2.4 and 4.3, may be represented by equivalent voltage and current noise generators at the amplifier inputs. This model is shown in Fig. A.12 where the amplifier is operating in a typical feedback circuit. The total output noise is given by

$$e_{no} = \left(1 + \frac{Z_F}{Z_1}\right)\sqrt{(Z_2\overline{i_{ni2}{}^2} + \overline{e_n{}^2}) + \overline{i_{ni1}{}^2}Z_F} \tag{A-17}$$

The noise is random and has a variety of spectral characteristics depending on the active elements used in the amplifier, the internal circuit design, and the bandwidth of interest. The bandwidth of the closed-loop amplifier determines the amount of noise transmitted to the output. The greater the closed-loop bandwidth, the greater will be the noise. Closed-loop bandwidth may be limited by the frequency characteristics

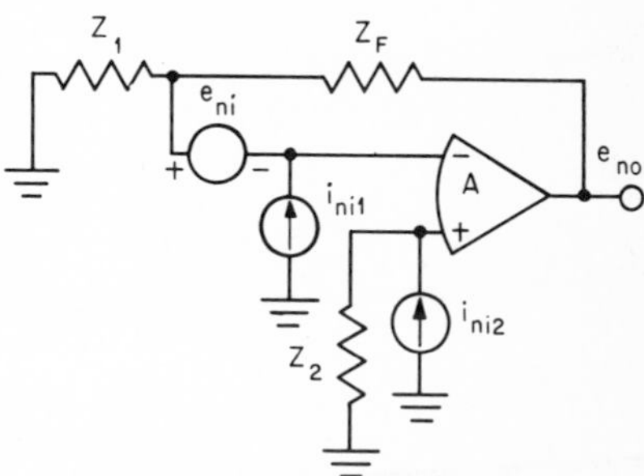

Fig. A.12 Model for input noise.

of the feedback network or may be limited by the gain-bandwidth product of the operational amplifier. The effects of current noise can be reduced by decreasing the impedance levels. This, of course, is not always possible since some of the impedance elements may be determined by other considerations.

A.10 Output Impedance

Since practical amplifiers will have nonzero output impedances, the effects of such output impedances must be taken into account. In order to evaluate these effects, the circuit model and feedback network of Fig. A.13 are analyzed. The resulting equations are

$$e_o' = A(e_s - \beta e_c)$$

where

$$\beta = \frac{Z_1}{Z_1 + Z_F}$$

$$e_o = e_o' \frac{1}{1 + Z_o/Z_L + Z_o/(Z_1 + Z_F)}$$

or

$$e_o = A'(e_s - \beta e_o)$$

where

$$A' = A \frac{1}{1 + Z_o/Z_L + Z_o/(Z_1 + Z_F)} \tag{A-18}$$

Thus the effective open-loop gain is reduced by the factor

$$\frac{1}{1 + Z_o/Z_L + Z_o/(Z_1 + Z_F)}$$

as a result of the nonzero output impedance and its interaction with the load and feedback networks.

Effective closed-loop output impedance of the amplifier (Z_{CL}) of Fig. A.13 is analyzed by assuming an incremental change in load current, Δi_L.

$$\Delta e_o = \Delta i_L Z_{CL}$$

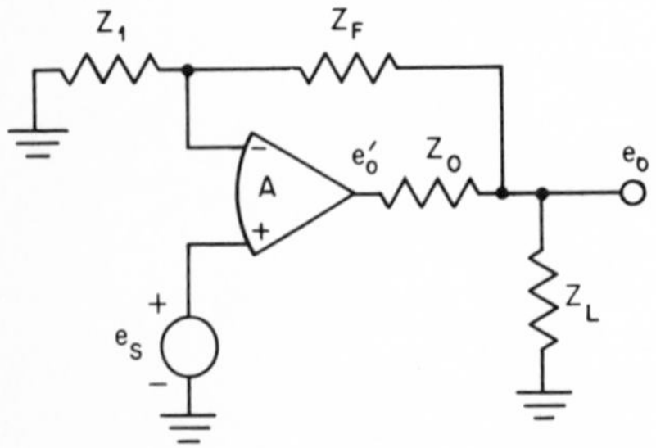

Fig. A.13 Model for open-loop output impedance.

Also

$$A\beta(\Delta e_o) = \Delta e_o' = \Delta e_o + \Delta i_L Z_o$$

Thus

$$\Delta e_o = \Delta i_L \frac{Z_o}{A\beta - 1} = \Delta i_L Z_{CL}$$

and

$$Z_{CL} = \frac{Z_o}{A\beta - 1} \approx \frac{Z_o}{A\beta} \tag{A-19}$$

if

$$A\beta \gg 1$$

In words, the equation states that the effective closed-loop output impedance is less than the open-loop output impedance by a factor equal to the reciprocal of loop gain.

A.11 Input Impedance

In the idealized model of the open-loop operational amplifier, the input impedance is assumed infinite. In practical operational amplifiers, there are two components of input impedance which must be considered. The impedance between the two input terminals is the *differential input impedance* Z_{id}. Impedance from either input to common is designated *common-mode input impedance* Z_{icm}.

The effects of these finite impedances may be evaluated for both inverting and noninverting operation, using the circuit of Fig. A.14. The expression for output signal becomes

$$e_o = \frac{e_2(1 + Z_F/Z_1 + Z_F/Z_{icm})}{1 + 1/A\beta'} - \frac{e_1(Z_F/Z_1)}{1 - 1/A\beta'} \tag{A-20}$$

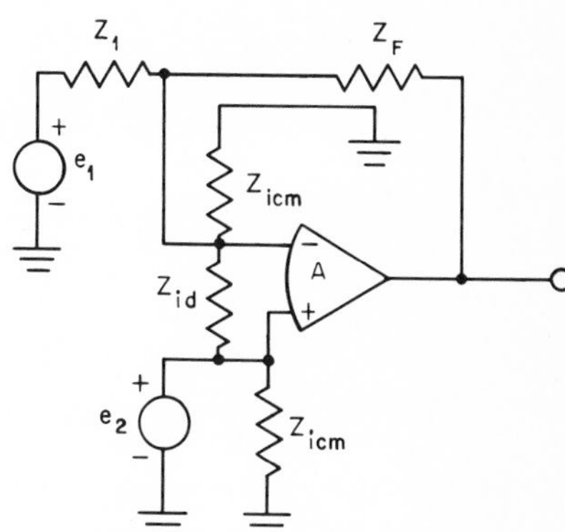

Fig. A.14 Model for open-loop input impedance.

where

$$\beta' = \frac{1}{1 + Z_F/Z_1 + Z_F/Z_{id} + Z_F/Z_{icm}} \tag{A-21}$$

Note that the feedback factor is modified by consideration of these finite input impedances. Also, the expression for noninverting gain is slightly modified [compare Eqs. (A-20) and (A-21) with (A-4), (A-5), and (A-5a)].

The signal current flowing into the noninverting input is

$$i_2 = \frac{e_2(1 + Z_F/Z_1 + Z_F/Z_{icm})}{A(1 + 1/A\beta')Z_{id}} + \frac{e_2}{Z_{icm}}$$

(for $e_1 = 0$). Examination of this expression with a few simplifying assumptions allows a better understanding of its meaning. If

$$|A\beta'| \gg 1.0 \qquad Z_F \ll Z_{icm}$$
$$\beta \approx \beta'$$

then

$$i_2 \approx \frac{e_2}{A\beta Z_{id}} + \frac{e_2}{Z_{icm}}$$

These two terms represent input impedances in parallel. Thus the noninverting input impedance is expressed as

$$Z_i = \frac{1}{1/A\beta Z_{id} + 1/Z_{icm}} \tag{A-22}$$

The input current for inverting operation is

$$i_1 = \frac{e_1}{Z_1}\left(1 - \frac{Z_F/Z_1}{(1 + 1/A\beta')A}\right)$$

(for $e_2 = 0$). For $|A\beta'| \gg 1.0$, this reduces to

$$i_1 \approx \frac{e_1}{Z_1}$$

and the inverting input impedance is given approximately by

$$Z_i \approx Z_1 \tag{A-23}$$

For most feedback circuits, the effects of finite input impedances on closed-loop gain are negligible at low frequency. However, at higher frequencies the input capacitance can be very significant—sometimes dominant—in determining closed-loop frequency response.

A.12 Other Parameters

A good many other parameters are generally specified for operational amplifiers. Many of these are specified as sensitivities of the parameters already discussed to temperature, power supply voltage, and time. Others simply specify design limitations such as rated output voltage and current, rated common-mode voltage, power supply drain current, etc. Slew rate and full power response frequency are also commonly specified. These are defined and discussed elsewhere (Appendix B and Chapter 5).

APPENDIX B

DEFINITION AND MEASUREMENT OF PERFORMANCE CHARACTERISTICS

Defined in this appendix are the parameters commonly used to characterize operational amplifier performance. In each case practical test circuits for parameter measurement are presented and described in conjunction with the common measurement conditions. The conditions under which a given measurement is made will sometimes vary between manufacturers, and such differences are examined. In the following discussion the operational amplifier parameters are presented in the four categories of open-loop differential characteristics, output signal response, input error signals, and common-mode characteristics.

B.1 Open-loop Differential Characteristics

By itself an operational amplifier is an open-loop device, and its characteristics measured under this condition determine much of its performance in feedback applications. As outlined in Appendix A, many closed-loop characteristics can be predicted from the basic amplifier open-loop parameters and the feedback loop gain. Open-loop parameters then serve as reference points from which the associated characteristics in almost any closed-loop application can be computed. Defined and described in this section are the open-loop voltage gain, output resistance, input resistance, input capacitance, and unity-gain bandwidth.

B.1.1 Open-loop voltage gain A This is the ratio of output signal voltage to differential input signal voltage.

The dc open-loop gain level A_O is commonly specified; however, it is measured with an ac signal in order to discriminate the signal from the amplifier dc offset voltage. As long as the measurement frequency is much lower than the first amplifier pole frequency f_{p1}, the measured gain will equal the dc level. Since this gain is typically very high, careful shielding must be used to avoid hum and noise which can overshadow the microvolt-level differential input signal. In some cases the open-loop voltage gain is measured under a specific load. Although this automatically includes the effect of this particular load in the gain measurement, it complicates the computation of gain under different load conditions. The open-loop gain measured without a load provides a convenient reference from which the effect open-loop gain under a given load can be found by considering its loading effect upon the open-loop output resistance. When the open-loop gain known is that for a specific load, it must first be translated into the unloaded value before the effective gain under a different load can be found.

Although the ac testing avoids measurement error from the dc input offset voltage, the effect of this voltage must still be removed if the measurement is to be made in the open-loop state since high gain amplification of this voltage results in output saturation. In the test circuit of Fig. B.1a the input offset voltage is counterbalanced by a dc input voltage supplied with a potentiometer through a voltage divider. Alternatively the effect of the dc input offset voltage can be removed by applying heavy feedback at very low frequencies, as in Fig. B.1b. In this case the closed-loop dc gain is zero and the offset is not amplified while the feedback is chosen so that it does not constrain the amplifier gain at the test frequency. However, the charging of the large capacitor C to the input offset voltage level through the high feedback resistance R slows this test. In each above circuit an attenuator is used to reduce the input signal voltage to an appropriate level for output swings of at least 20 percent of the output voltage rating. Also using these circuits the gain magnitude frequency response is drawn from a series of measurements at increasing frequencies. At higher frequencies the output signal level must be decreased to avoid distortion, and the amplified noise begins to interfere with the measurement. This problem is removed by applying feedback to limit the amplifier gain presented to the noise. As long as the closed-loop gain established by the feedback is around 50 or more times the

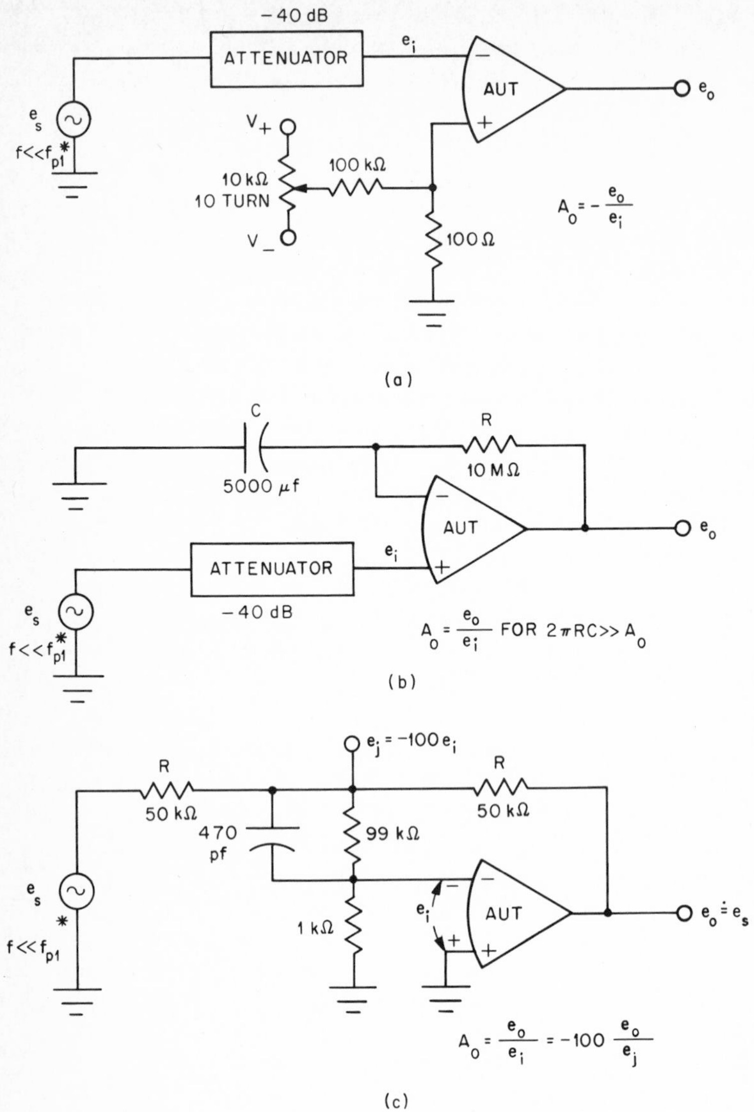

Fig. B.1 Open-loop test circuit.

open-loop gain at the measurement frequency, the response measurement is not affected. Such feedback is readily applied to the circuit of Fig. B.1b by connecting a gain-limiting resistor in series with the capacitor C.

Open-loop voltage gain can also be measured under closed-loop conditions with a specified output level by measuring the associated small differential input signal voltage of the amplifier. For such a measurement the effect of the input offset voltage is again constrained by feedback, but no capacitance charging delay results. As indicated in Fig. B.1c, the output level for this measurement is fixed by the unity-gain feedback to a level very nearly equal to the input signal voltage. In order to obtain a more accurate measurement of the small differential input voltage E_i the voltage divider shown is inserted between the summing junction and the amplifier input. As a result, feedback forces the summing junction to be an amplified replica of E_i which is far easier to measure. As long as the summing junction signal is very small in comparison with the output signal, the divider introduces only a small error.

Frequently the first amplifier response pole occurs below 50 Hz, especially for amplifiers with a continuous -6 dB per octave response slope. Gain measure with an ac signal frequency well below this pole frequency is complicated by the low-frequency limitations of common meters and test oscillators. To permit gain measurement at 1 Hz or lower, the test circuit of Fig. B.2 can be used without the need for a test oscillator or ac meter. As shown, the amplifier under test is connected in the feedback configuration of Fig. B.1c, which develops an amplified replica of the differential input voltage at the junction of the summing and feedback resistors. In this case the test signal is supplied by the 1-Hz square-wave generator formed with A_1, and the output of the amplifier being tested would be a square wave of rated output level. The associated summing junction signal is separated from its dc offset voltage by the 10-μf coupling capacitors and the FET switches. By switching one end of a coupling capacitor to ground on one half cycle, the signal swing at that point is referenced to zero on the other half cycle. As a result, the inputs to the differential instrumentation amplifier are two ground-referenced square waves as shown with amplitudes which equal that of the summing junction signal. These square waves are of opposite phase and polarity since they result from opposite phase switching. By amplifying the difference of these two signals, the instrumentation amplifier produces a dc output related to the amplitude of the summing junction signal and, thereby, to the dc gain as expressed. This dc output is produced in only one cycle since the switches rapidly charge the coupling capacitors. To avoid response sag on this signal the instrumentation amplifier should be a high input impedance FET type.

B.1.2 Output resistance R_O This is the effective output source resistance when operated open loop.

Using the open-loop parameter test circuits of Figs. B.1 and B.2, the output resistance of an operational amplifier is measured by observing the low-frequency gain decrease produced by the load. The gain decrease results from the output voltage division across the output resistance and

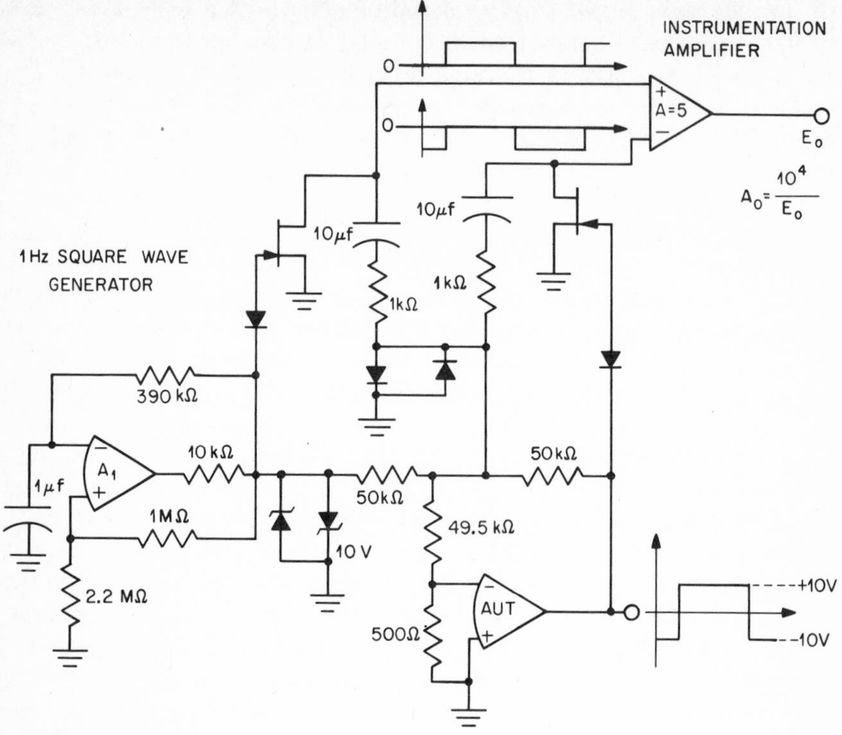

Fig. B.2 A 1-Hz open-loop test circuit.

the load resistance, and the loaded gain is

$$A'_O = \frac{R_L}{R_O + R_L} A_O$$

Then the output resistance will be

$$R_O = \left(\frac{A_O}{A'_O} - 1\right) R_L$$

Note that for the closed-loop measurement circuit of Figs. B.1c and B.2 the feedback resistance also represents a load to the output. From the open-loop output resistance found in this way, the analogous resistance under closed-loop conditions can be approximated by the open-loop value divided by the loop gain as indicated in Appendix A.

B.1.3 Differential input resistance R_I This is the effective resistance between the two inputs when operated open loop.

This resistance characteristic is also measured by observing the open-loop gain loss from a voltage divider formed with the resistance. With the open-loop test circuit of Fig. B.1a the gain loss resulting from insertion of the two source resistances R_G in Fig. B.3 is measured. Either of the other open-loop gain measurement circuits can also be used. The capacitors paralleling the source resistances help to reduce high-frequency noise. When the switches are opened, the gain will decrease because of input loading, as indicated by a drop in e_o to a new value of

$$e_o' = \frac{R_I}{2R_G + R_I} e_o$$

as long as the test frequency is much less than $1/2\pi R_G C_G$. From this measurement the input resistance can be found as

$$R_I = \frac{e_o'}{e_o - e_o'} 2R_G$$

Two source resistances R_G are inserted to balance out the effect of stray output signal coupling to the inputs by the added resistance. With equal resistances added to the two inputs the signals coupled to each input will be nearly equal and thereby eliminated by the amplifier common-mode rejection. However, this measurement remains extremely sensitive to such stray feedback which makes shielding, short input leads, and low values of R_G necessary. For the high input resistance provided by an FET or varactor input stage this measurement is not feasible, but it is seldom necessary to predict accurately such a high input resistance. From the results of Appendix A, the open-loop input resistance found in this test can be related to that of the closed-loop case. The closed-loop resistance is approximated by the open-loop value multiplied by the loop gain. However, the closed-loop differential input resistance is limited to the level of the common-mode input resistance which shunts the inputs.

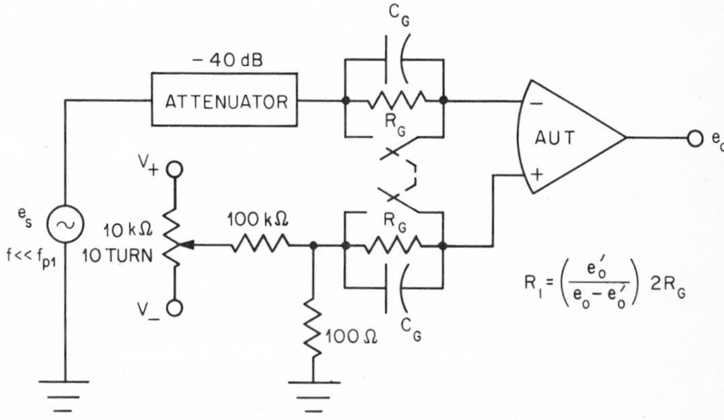

Fig. B.3 Test circuit for differential input resistance and capacitance.

B.1.4 Differential input capacitance C_I This is the effective capacitance between the two inputs when operated open loop.

This capacitance can be measured in the same manner as was the differential input resistance above. However, the stray coupling problem is even more severe at the higher frequencies used here, and a well-shielded test fixture is imperative. The frequency at which the test is to be performed should be chosen considering the frequency-independent nature of the value of the input capacitance. In general, the differential input capacitance is largely Miller-effect capacitance at low frequencies. As described in Sec. 1.2, this capacitance drops to a much lower value at higher frequencies when the first-stage gain falls, reducing the Miller effect. The high capacitance at low frequencies may not represent a significant shunt to the input resistance or to normal source resistance levels. In this general case the higher frequency input capacitance is of more interest.

B.1.5 Unity-gain bandwidth f_c This is the frequency range from direct current to that frequency at which the open-loop gain crosses unity.

Because of slewing rate limiting only small-signal response is achieved at this frequency, and the output test signal should be observed to ensure that the amplifier is in linear operation. The small output signal in this test would be significantly affected by the highly amplified noise common to the open-loop test circuits. For this reason the unity-gain bandwidth is measured in a lower-gain closed-loop circuit such as Fig. B.4. As long as the closed-loop gain limit imposed by the feedback is far greater than unity the response near the unity-gain point is not significantly altered by the feedback.

B.2 Output Signal Response

Many factors limit the output signal performance of operational amplifiers. To characterize this performance, the parameters frequently specified include the rated output, slewing rate, full power response, settling time, and overload recovery time. Each of these is defined in this section with associated test circuits.

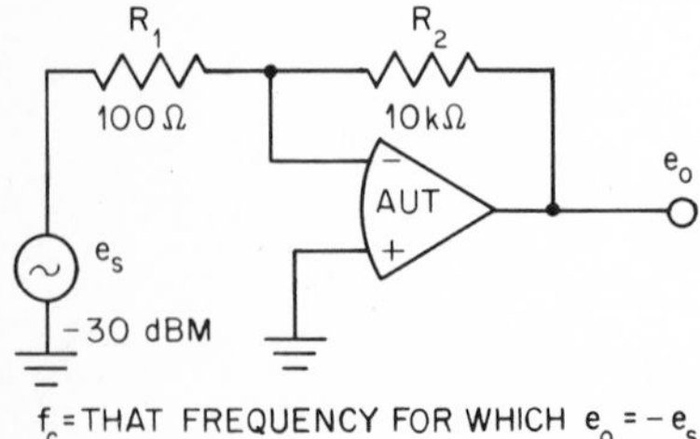

Fig. B.4 Unity-gain bandwidth test circuit.

B.2.1 Rated output Rated output is the peak values of output voltage and current which can be simultaneously supplied.

This parameter is simply measured using a low-frequency output voltage which swings to the positive and negative limits. An output load is connected which, in conjunction with the feedback resistor, would raise the peak output current to its rated level at the rated peak voltage. The resulting output swing limits define the amplifier output capabilities. Since output current limiting can limit the output voltage swing under load, the two output signal ratings are interdependent and specified together.

B.2.2 Slewing rate S_r This is the maximum rate of change of output voltage when supplying the rated output.

In general, slewing rate is measured in the unity-gain voltage follower circuit of Fig. B.5a as this is most often the worst-case condition. With this circuit the amplifier common-mode swing limitations which would affect slewing rate are also included in the test. Alternatively the measurement can be performed in an inverting circuit such as Fig. B.5b. The inverting test is most commonly used for single-ended input amplifiers. Once again the feedback resistor R_2 acts as part of the output load. With either circuit the amplifier is driven by a high-frequency square wave of sufficient magnitude to drive the output beyond its rated level. The output is overdriven in both directions as represented in Fig. B.6 in order to remove the rounded peaks from the measurement interval as these portions are not slewing-rate-limited. From this response the slewing rate is found as the slope of the transition between the rated output extremes. Frequently the positive and negative swings will have different slewing rates, and both must be examined. In such a case the lower slewing rate is commonly specified.

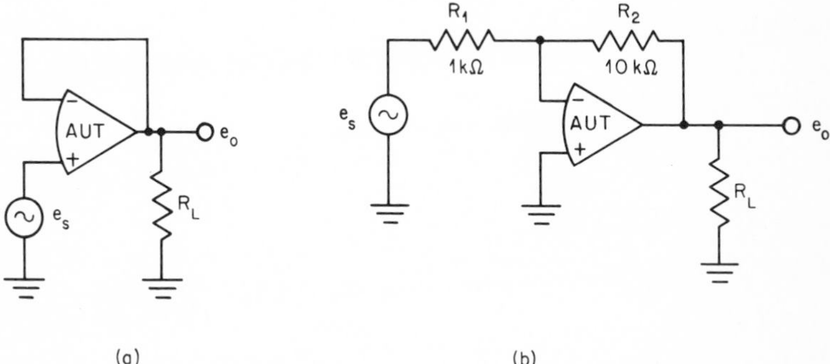

Fig. B.5 Test circuits for slewing rate and full power response measurement.

B.2.3 Full power response f_p This is the maximum frequency at which rated output can be supplied without significant distortion.

This response limit is often a result of the rate limiting which determines slewing rate above. In this case the rate limiting affects sinusoidal signals by limiting their slopes. By equating the maximum slope of a sine wave to the slewing rate, full power response is related to S_r by

$$f_p = \frac{S_r}{2\pi E_{or}}$$

where E_{or} is the rated output voltage. The full power response can be independently measured using a sinusoidal signal with the circuits of Fig. B.5, which were described above, for the slewing rate test. For measurement the signal frequency is increased until the maximum frequency is found for which the rated output can be maintained without significant distortion. In general, the distortion is eliminated by feedback until the limiting frequency is reached, and so a rough visual evaluation is satisfactorily accurate to define f_p. Distortion levels of a few percent can typically be detected in this test, because of the obvious effect of the limiting. When measuring the f_p of fast amplifiers in the noninverting gain circuit of Fig. B.5b, it may become necessary to increase the input signal amplitude as the open-loop gain drops below the desired closed-loop gain level. Alternatively, unity-gain feedback can be used to avoid output signal decrease from the open-loop gain drop.

B.2.4 Settling time t_s Settling time is the time following application of a step input required for output voltage settling to within a specified percentage of its final value.

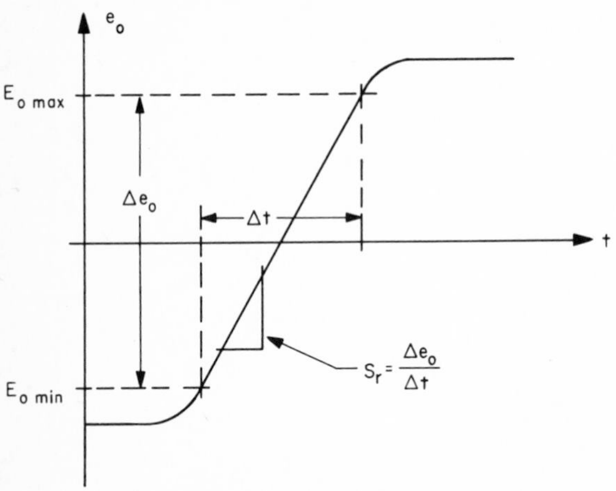

Fig. B.6 Slewing rate test signal.

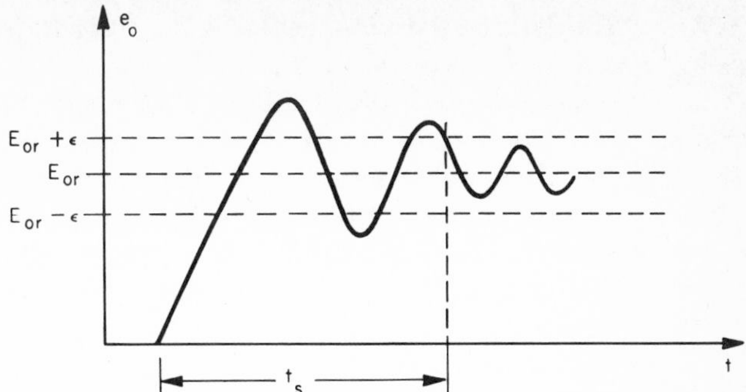

Fig. B.7 Output signal settling time.

As defined, the settling time is a measure of the time for which an operational amplifier does not provide an accurate output following a rapid signal change. In general, t_s is specified for rated output voltage and current transitions in a unity-gain circuit and for an error band of 0.1 or 0.01 percent of the rated voltage level. Such a measurement is represented on the signal waveform of Fig. B.7. Note that the rise time controlled by slewing rate is included in the settling time, as this is part of the time for which the output is in error. In practice the specified error band is too small to observe with the output signal directly, and it is necessary to separate the error from the signal for measurement. This is achieved in unity-gain measurement circuits by subtracting the output signal from the input signal.

When a unity-gain inverting test circuit is used as in Fig. B.8, this subtraction is readily achieved with the added R_2 resistors which sum the input and output signals. Only the error signal attenuated by a factor of 2 appears at the junction of these resistors if they are matched and if

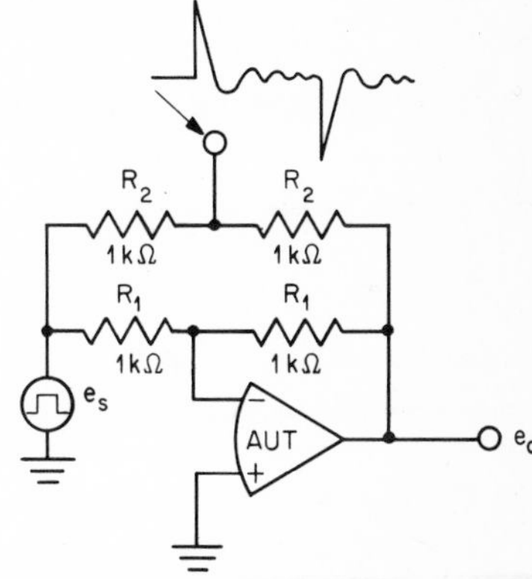

Fig. B.8 Settling time measurement circuit.

the gain is exactly unity. This error signal can be observed on an oscilloscope, but the oscilloscope must have an overload recovery time which is small compared with t_s in order to accommodate the large initial error signal. Care must be taken to avoid stray lead capacitances which can greatly alter the results of this test. The resistors R_1 and R_2 are made small to further reduce the effects of such capacitances, and these resistors serve as the load resistance. When the settling time of the noninverting configuration is to be measured, a high-speed differential amplifier is used to subtract the output signal from that at the input. Such differential amplifiers are included in some high-speed oscilloscopes, but fast overload recovery must also be assured.

B.2.5 Overload recovery time This is the time required for the output to return to linear operation from saturation following removal of an overdrive signal.

The recovery time depends somewhat upon the degree of overload and the feedback impedance. A common test condition is that achieved with the circuit of Fig. B.5b using a square-wave input signal of a level which is twice that needed to reach output saturation. This is a 100 percent overdrive condition. Measurement is made by observing the time delay between the input and output signal.

B.3 Input Error Signals

Input error signals which limit the signal sensitivity of operational amplifiers are produced by dc biasing and noise as described in Chapters 2 and 4. The dc biasing errors are represented by the equivalent input offset voltage, input bias currents, input offset current, and their thermal drifts. As described in the same chapters, noise is represented by the equivalent input noise voltage and input noise currents. By defining these error signals as equivalent input signals, a representation results which is valid for any feedback configuration.

B.3.1 Input offset voltage V_{OS} This is the differential dc input voltage required to provide zero output voltage with no input signal or source resistance.

As will be described later, the input offset voltage varies with temperature and power supply voltage. The offset is most commonly specified at room temperature for rated power supply voltage. Frequently a maximum offset over the operating temperature range is also specified. To facilitate the measurement of the input offset voltage a high-gain test circuit is used to amplify the offset as indicated in Fig. B.9.

B.3.2 Input offset voltage drift This drift is the rate of change of input offset voltage with temperature, power supply voltage, or time.

Each of these drift characteristics can be measured by using the input offset voltage test circuit of Fig. B.9 to measure V_{OS} at several temperatures, power supply levels, or times. The drifts are commonly measured as average drifts over a specified range to simplify measurement and specification. Testing is then required at only a few points of the specified range to provide the average drift figure. However, care must be taken to avoid averaging two large but opposing drifts over different portions of the specified range. The average of such a U-shaped drift curve would be deceptively small. For this reason a measurement is commonly made at an intermediate point of the range. From this intermediate measurement and two end point tests the two separate drifts are computed, and their magnitudes are averaged to define the average drift. The input offset voltage drift as a function of temperature is defined in this way by

$$\left(\frac{dV_{OS}}{dT}\right)_{Av} = \frac{|V_{OS}(T_1) - V_{OS}(25°C)| + |V_{OS}(T_2) - V_{OS}(25°)|}{|T_1 - T_2|}$$

where the specified temperature range is from T_1 to T_2 and the intermediate point is 25°C. Similar expressions can be used to evaluate drift versus power supply voltage and time.

B.3.3 Input bias current I_B This current is the dc biasing current required at either input to provide zero output voltage with no input signal or offset voltage.

As described in Chapter 2, the input bias current is the base current of bipolar input transistor or the gate leakage current of an input FET. This current is commonly specified at room temperature and over the operating temperature range. Generally the specified limits are maxima applying to either input bias current; however, the limits are sometimes applied to the less definite average of the two currents. The input bias currents are measured by forcing them to flow in large resistors, as in the

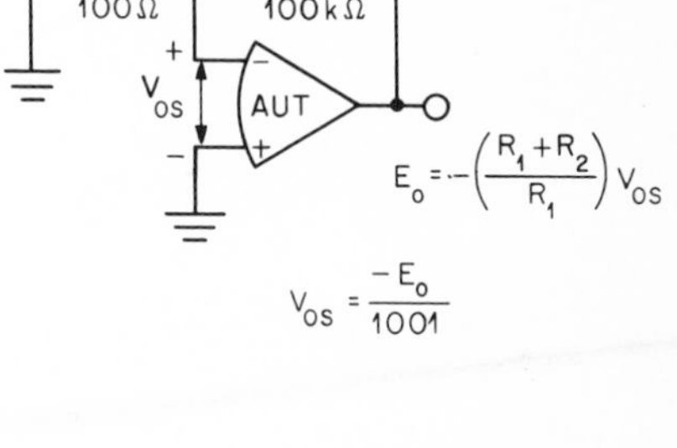

Fig. B.9 Input offset voltage test circuit.

test circuit of Fig. B.10, which are bypassed to reduce noise. Because of the large resistors the output voltage created by the input bias currents overshadows that due to V_{OS} with bipolar transistor input amplifiers. The output voltage is essentially the product of one of the resistors and the associated current. For FET input, chopper-stabilized, and varactor amplifiers the higher resistance levels are used to measure the very low gate currents, and it is typically necessary first to null the input offset voltage.

B.3.4 Input bias current drift This is the rate of change of input bias current with high temperature, power supply voltage, or time.

The average drifts are generally measured by using the test circuit of Fig. B.10 with the average defined in the same way as was the average input offset voltage drift above.

B.3.5 Input offset current I_{OS} This is the difference between the two input bias currents.

This difference current is measured as indicated in Fig. B.10. Since the measurement takes the difference between two currents which are of the same order of magnitude, it is necessary to match the two resistors to within about 0.1 percent.

B.3.6 Input offset current drift This drift is the rate of change of input offset current with temperature, power supply voltage, or time.

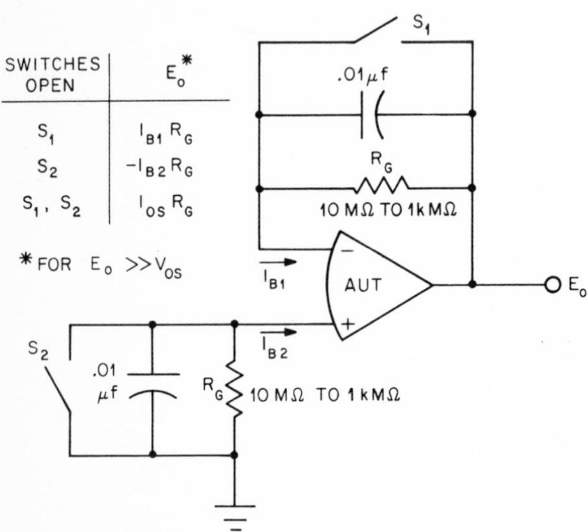

Fig. B.10 Measurement circuit for input bias currents and input offset current.

The average drifts are generally measured by using the test circuit of Fig. B.10, with the average drift defined in the same way as was the average input offset voltage drift.

B.3.7 Input noise voltage e_n This is the equivalent differential input noise voltage which would reproduce the noise at the output if all amplifier noise sources were set to zero when the source resistances are zero.

The small input noise voltage is most conveniently measured in a high-gain circuit which amplifies the noise, such as that of Fig. B.11. An output filter is used to limit the measurement to that noise within the specified bandwidth. Note that the closed-loop response of the amplifier must encompass this bandwidth for accurate amplification of the noise in question. Although a truly rectangular filter passband including only the specified frequency range is not possible, a known filter response can be related to an effective rectangular passband as discussed in Sec. 4.3. For a single-pole low-pass filter with a pole at f_p an effective noise bandwidth is defined by Eq. (4-36) as

$$f_e = \frac{\pi}{2} f_p$$

For measurement the circuit requires an exceptionally well-shielded environment, particularly if the measurement passband includes the ac

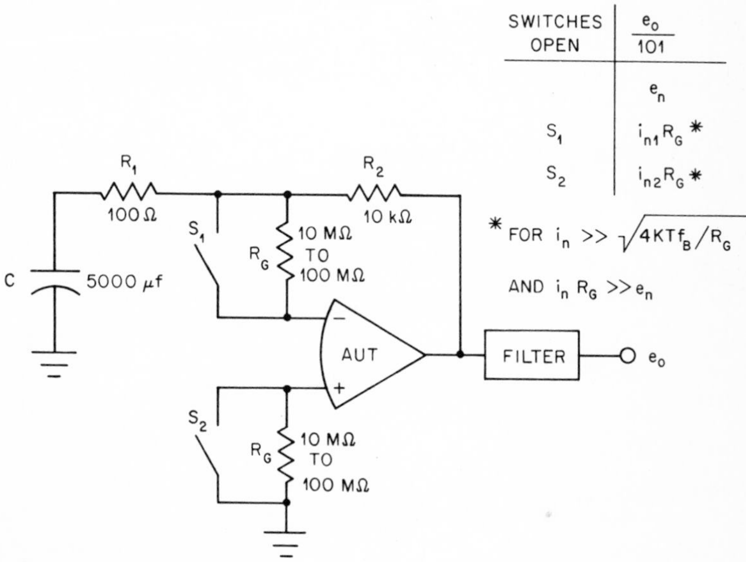

Fig. B.11 Measurement circuit for equivalent input noise voltage and input noise currents.

power frequency. An enclosed test fixture should be used with a low-noise power supply such as a battery. Care should be taken to avoid noise coupling from adjacent electrical equipment or wiring.

B.3.8 Input noise current i_n This current is the equivalent input noise current at either input which would reproduce the output noise if the amplifier noise sources were set to zero when the associated source resistance is large compared with e_n/i_n.

By inserting large source resistances in series with the inputs, as shown in Fig. B.11; an output noise can be developed which is essentially the product of either noise current and a source resistor R_S. However, care must be taken to ensure that the output noise due to i_n is greater than the thermal noise of R_S as well as greater than e_n. As discussed in Sec. 4.3 and displayed in Fig. 4.3, the thermal noise of R_G can mask that created by i_n over certain source resistance ranges. Although the noise currents are generally distinguishable for bipolar transistor input amplifiers, the smaller noise currents of chopper-stabilized, varactor, and FET input operational amplifiers can be masked by the thermal noise of R_G. The latter noise from Sec. 1.4 is

$$i_n = \sqrt{\frac{4KTf_B}{R_G}}$$

where f_B is the measurement bandwidth. Because of this thermal noise, the above current noise test is not generally recommended or needed for FET input or varactor amplifiers. The same shielding outlined for the input noise voltage measurement should be observed here.

B.4 Common-mode Characteristics

The input ground isolation and noise rejection provided by the common-mode signal capabilities of an operational amplifier are defined by four basic parameters. These are the common-mode input resistance, input capacitance, rejection ratio, and voltage range.

B.4.1 Common-mode rejection ratio CMRR This is the ratio of the differential voltage gain to the common-mode voltage gain.

As expressed, the CMRR is a figure of merit comparing the gain received by differential signals with that received by common-mode signals. The common-mode gain is often a nonlinear function of the common-mode voltage level, especially for FET input amplifiers. For this reason the full common-mode voltage swing must be used in measuring CMRR to result in a figure which applies over the rated common-mode voltage range. This is achieved by using the difference amplifier circuit of Fig.

B.12. For well-matched or balanced resistors as indicated, the signal at the two inputs is essentially a common-mode signal. However, the common-mode unbalance of the amplifier produces an output error voltage and an associated differential input voltage $e_i = e_o/A_d$. Then the common-mode rejection ratio can be written

$$\mathrm{CMRR} = \frac{A_d}{A_{cm}} = \frac{e_o/e_i}{e_o/e_{cm}} = \frac{e_{cm}}{e_i}$$

This can be rewritten considering

$$e_o = \frac{R_1 + R_2}{R_1} e_i$$

and

$$e_{cm} \doteq e_s \qquad \text{for } R_2 \gg R_1$$

The common-mode rejection ratio is then expressed simply in terms of the input and output signals by combining the last three relationships to get

$$\mathrm{CMRR} = \frac{R_1 + R_2}{R_1}\frac{e_s}{e_o}$$

B.4.2 Common-mode input resistance R_{Icm} This resistance is the effective resistance between either input and common.

This test is analogous to that employed for the measurement of differential input resistance. By inserting a bypassed resistance in series with the input in Fig. B.13, a voltage divider is formed with R_{Icm} which attenuates the signal as expressed by

$$e_o' = \frac{R_{Icm}}{R_{Icm} + R_G} e_o$$

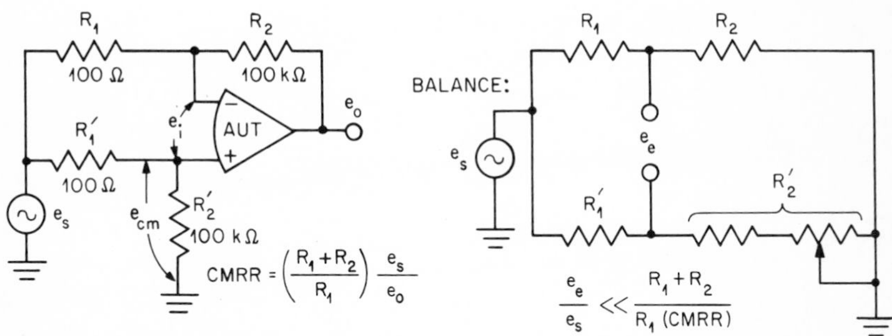

Fig. B.12 Common-mode rejection measurement circuit.

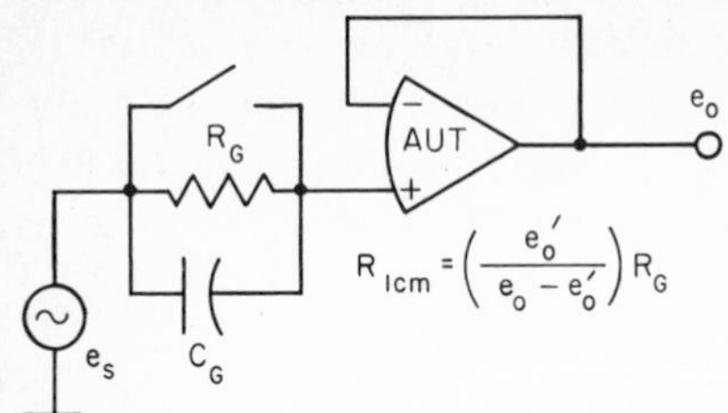

Fig. B.13 Common-mode input resistance test circuit.

The common-mode input resistance is found by noting the output signal change and using this change in the resulting expression for R_{Icm}:

$$R_{Icm} = \frac{e_o'}{e_o - e_o'} R_G$$

For FET input and varactor amplifiers this measurement is severely disturbed by noise and stray capacitance signal coupling, because of the very large value of R_G required. However, it is seldom necessary to know accurately this large resistance, and it is generally sufficient to consider the levels predicted from component characteristics.

B.4.3 Common-mode input capacitance C_{Icm} This capacitance is the effective capacitance between either input and common.

The capacitive portion of common-mode input impedance is measured in a manner analogous to that of Fig. B.13 for the input resistance. Greater care is needed in shielding to avoid feedback signal coupling through stray capacitances.

B.4.4 Rated common-mode input voltage This is the peak value of common-mode input voltage which can be applied for linear operation.

The common-mode voltage range is typically limited by saturation of the input stage. The saturation point is readily detected in the CMRR test circuit of Fig. B.12 by observing the input voltage level which results in a sudden, large increase in output voltage.

APPENDIX C
SENSITIVITY OF ACTIVE FILTERS

C.1 Sensitivity Fundamentals

Sensitivity is a measure of the change in some performance characteristic of a network resulting from a change in value of one or more of the elements of the network. Thus sensitivity functions are useful for evaluating changes in network performance due to element value tolerances or to element value changes with time and temperature. In this appendix the sensitivities of the magnitude of the voltage transfer function $|H(j\omega)| = G(\omega)$, of the phase $\phi(\omega)$, and of the group delay $\tau(\omega)$ with respect to changes in H_o, α, and ω_o are derived. Then in each section describing specific circuits, the sensitivities of these parameters with respect to circuit element changes are presented. The sensitivities of the magnitude of the filter voltage transfer function $G(\omega) = |H(j\omega)|$ and the phase $\phi(\omega)$ are of interest because these quantities are used to specify and to evaluate filter performance. Furthermore, these are the quantities measured by an ac voltmeter and a phasemeter. Another network function is the group delay $\tau(\omega)$, which is important in applications requiring a linear phase characteristic (constant group delay).

C.1.1 Definition of sensitivity The symbol S is used to denote sensitivity. In addition, a superscript character is used to indicate the performance characteristic that is changing, and a subscript character is used to indicate the specific element or parameter that is causing the change. The sensitivity of a network function $N(\omega)$ with respect to a parameter x is defined by

$$S_x^N = \frac{dN(\omega)/N(\omega)}{dx/x} = \frac{x}{N(\omega)}\frac{dN(\omega)}{dx}$$

It must be emphasized that sensitivity functions are theoretically valid only for infinitesimal changes. However, as a practical matter, the sensitivity function is sufficiently accurate for changes in element values or network parameters of 5 percent, or even 10 percent in some cases.

Note that changes in N and x have been normalized so that the sensitivity function actually specifies percentage changes from the nominal values of N and x.

A modification of this sensitivity definition is useful for filter applications. In such applications the magnitude of the voltage transfer function is usually expressed in decibels,

$$g(\omega) = 20 \log G(\omega)$$

It is perhaps more useful then to have a sensitivity function that answers the question: How many decibels does the magnitude of the filter response change for a given *normalized* element value change? In mathematical terms we want

$$S_x^g = \frac{dg}{dx/x}$$

where S_x^g is the *unnormalized* change of the magnitude response *in decibels* for a *normalized* change in the network parameter or element x. We can find S_x^g in terms of S_x^G.

$$S_x^g = \frac{dg(\omega)}{dx/x} = \frac{d[20 \log G(\omega)]}{dx/x}$$

$$= 20 \log e \frac{d[\log_e G(\omega)]}{dx/x}$$

$$= 8.685880 \frac{dG(\omega)/G(\omega)}{dx/x}$$

$$S_x^g = 8.7 S_x^G$$

The sensitivity of the network functions $G(\omega)$, $\phi(\omega)$, and $\tau(\omega)$, discussed in the preceding section, to changes in the network parameters H_o, ω_o, and α for single real pole and complex pole pair filter sections for low-pass, high-pass, and bandpass will now be presented.

C.1.2 Low-pass sensitivity functions

Single Pole

$$S_{\omega_o}{}^{G} = 1 - \frac{G^2}{H_o{}^2}$$

$$S_{H_o}{}^{G} = 1$$

$$S_{\omega_o}{}^{\phi} = -\frac{\sin 2\phi}{2\phi}$$

$$S_{\omega_o}{}^{\tau} = 2\sin^2\phi - 1 = -\cos 2\phi$$

Complex Pole Pair

$$S_{\omega_o}{}^{G} = 2 - \frac{G^2}{H_o{}^2}\left[2 + \frac{\omega^2}{\omega_o{}^2}(\alpha^2 - 2)\right]$$

$$S_{\alpha}{}^{G} = -\alpha^2 \frac{\omega^2}{\omega_o{}^2}\frac{G^2}{H_o{}^2}$$

$$S_{H_o}{}^{G} = 1$$

$$S_{\omega_o}{}^{\phi} = \frac{\sin 2\phi}{2\phi} + \frac{2\omega_o}{\alpha\omega}\left(\frac{\sin^2\phi}{\phi}\right)$$

$$S_{\alpha}{}^{\phi} = \frac{\sin 2\phi}{2\phi}$$

$$S_{\omega_o}{}^{\tau} = \left(\frac{2\sin 2\phi}{\alpha} - \frac{\omega_o}{\omega}\cos 2\phi\right)\frac{\phi}{\omega_o\tau}S_{\omega_o}{}^{\phi} - \frac{2\sin^2\phi}{\alpha\omega_o\tau}$$

$$S_{\alpha}{}^{\tau} = \left(\frac{2\sin 2\phi}{\omega_o} - \frac{\alpha}{\omega}\cos 2\phi\right)\frac{\phi}{\alpha\tau}S_{\alpha}{}^{\phi} - \frac{2\sin^2\phi}{\alpha\omega_o\tau}$$

C.1.3 High-pass sensitivity functions

Single Pole

$$S_{\omega_o}{}^{G} = -\frac{\omega_o{}^2}{\omega^2}\frac{G^2}{H_o{}^2}$$

$$S_{H_o}{}^{G} = 1$$

$$S_{\omega_o}{}^{\phi} = \frac{+\sin 2\phi}{2\phi}$$

$$S_{\omega_o}{}^{\tau} = +2\cos^2\phi - 1 = \cos 2\phi$$

Complex Pole Pairs

$$S_{\omega_o}{}^{G} = -\frac{\omega_o{}^2}{\omega^2}\frac{G^2}{H_o{}^2}\left(2\frac{\omega_o{}^2}{\omega^2} + \alpha^2 - 2\right)$$

$$S_{\alpha}{}^{G} = -\alpha^2\frac{\omega_o{}^2}{\omega^2}\frac{G^2}{H_o{}^2}$$

$$S_{\omega_o}{}^{\phi} = \frac{2\omega_o}{\alpha\omega\phi}\sin^2\phi + \frac{\sin 2\phi}{2\phi}$$

$$S_{\alpha}{}^{\phi} = \frac{\sin 2\phi}{2\phi}$$

$$S_{\omega_o}{}^{\tau} = \left(\frac{2\sin 2\phi}{\alpha\tau} - \frac{\omega_o}{\omega\tau}\cos 2\phi\right)\frac{\phi}{\omega_o}S_{\omega_o}{}^{\phi} - \frac{2\sin^2\phi}{\alpha\omega_o\tau}$$

$$S_{\alpha}{}^{\tau} = \left(\frac{2\sin 2\phi}{\omega_o\tau} - \frac{\alpha}{\omega\tau}\cos 2\phi\right)\frac{\phi}{\alpha}S_{\alpha}{}^{\phi} - \frac{2\sin^2\phi}{\alpha\omega_o\tau}$$

$$S_{H_o}{}^{G} = 1$$

C.1.4 Bandpass sensitivity functions

Complex Pole Pairs

$$S_{\omega_o}{}^{G} = 1 - \frac{G^2}{H_o{}^2\alpha^2}\left(2\frac{\omega_o{}^2}{\omega^2} + \alpha^2 - 2\right)$$

$$S_{\alpha}{}^{G} = -S_Q{}^{G} = 1 - \frac{G^2}{H_o{}^2}$$

$$S_{\omega_o}{}^{\phi} = \frac{2\omega_o}{\alpha\omega\phi}\sin^2\phi - \frac{\sin 2\phi}{2\phi}$$

$$S_{\alpha}{}^{\phi} = \frac{-\sin 2\phi}{2\phi}$$

$$S_{\omega_o}{}^{\tau} = \left(\frac{\omega_o}{\omega}\cos 2\phi - \frac{2}{\alpha}\sin 2\phi\right)\frac{\phi}{\omega_o\tau}S_{\omega_o}{}^{\phi} - \frac{2\cos^2\phi}{\alpha\omega_o\tau}$$

$$S_{\alpha}{}^{\tau} = \left(\frac{\alpha}{\omega}\cos 2\phi - \frac{2}{\omega_o}\sin 2\phi\right)\frac{\phi}{\alpha\tau}S_{\alpha}{}^{\phi} - \frac{2\cos^2\phi}{\alpha\omega_o\tau}$$

In the bandpass case the sensitivity functions $S_x{}^{\omega_o}$, $S_x{}^{Q}$, and $S_x{}^{H_o}$ are probably of more interest than $S_x{}^{G}$. $S_x{}^{G}$ would be useful for evaluating the sensitivity of stagger-tuned bandpass filters. $S_x{}^{\phi}$ is important for applications when a phase match must be maintained between two filters.

C.1.5 Some sensitivity identities In following sections that describe actual filter realizations, the sensitivity of the network parameters $\alpha = 1/Q$, ω_o, and H_o to element value changes will be given. Using these sensitivity equations along with those sensitivity equations given above, and through the use of several identities relating sensitivity functions in general, we will be able to find the sensitivity of the magnitude, phase, and group delay to individual element value changes as well as develop sensitivity equations for filters involving several pole pair sections.

Some useful identities involving sensitivity functions follow:

1. $S_x^{kx^n} = n$

2. $S_x^y(\omega_1, \omega_2, \ldots, \omega_n) = \sum_{i=1}^{n} S_{\omega_i}^y S_x^{\omega_i} \qquad \omega_i = \omega_i(x)$

3. $S_x^{\prod_{i=1}^{n} y_i(x)} = \sum_{i=1}^{n} S_x^{y_i}$

4. $S_x^{\mu(x)/v(x)} = S_x^{\mu} - S_x^{v}$

5. $S_x^{ky^n(x)} = nS_x^{y(x)}$

6. If $\sum_{i=1}^{n} y_i(x) = \Sigma_i$,

$$S_x^{\Sigma_i} = \frac{1}{\sum_{i=1}^{n} y_i} \sum_{i=1}^{n} (y_i S_x^{y_i})$$

C.2 Application of Sensitivity Functions

As an example of the use of sensitivity functions for one stage, consider a complex pole pair stage with resistors R_1 and R_2 and capacitors C_1 and C_2. It is desired to find out how much the magnitude, phase, and group delay change when *all* these parameters change. Let $\delta_{x_i} = dx_i/x_i$ be the normalized change in element x_i. The new magnitude response will be

$$G_{new} = G_{old}(1 + S_{R_1}^G \delta_{R_1} + S_{R_2}^G \delta_{R_2} + S_{C_1}^G \delta_{C_1} + S_{C_2}^G \delta_{C_2})$$

where we have obtained $S_{R_1}^G$, $S_{R_2}^G$, etc., from

$$S_{R_1}^G = S_{H_o}^G S_{R_1}^{H_o} + S_{\omega_o}^{G_o} S_{R_1}^{\omega_o} + S_{\alpha}^G S_{R_1}^{\alpha}$$

The normalized incremental change in G is then

$$\delta_G = \frac{G_{new} - G_{old}}{G_{old}} = (S_{R_1}^G \delta_{R_1} + S_{R_2}^G \delta_{R_2} + S_{C_1}^G \delta_{C_1} + S_{C_2}^G \delta_{C_2})$$

Thus, in general, we can write for a single stage

$$\delta_G = \sum_{i=1}^{n} S_{x_i}^G \delta_{x_i}$$

where n is the number of elements being considered. Phase and group delay are analyzed in the same manner and yield

$$\delta_\phi = \sum_{i=1}^{n} S_{x_i}^{\phi} \delta_{x_i}$$

$$\delta_\tau = \sum_{i=1}^{n} S_{x_i}^{\tau} \delta_{x_i}$$

Now let us calculate the incremental change in the magnitude response, phase, and group delay of several cascaded stages for changes in element values. Let $\delta_{G_i} = dG_i/G_i$ be the normalized change in G_i. The new gain of M cascaded stages is

$$G_{new} = \prod_{i=1}^{M} G_i(1 + \delta_{G_i}) \qquad G_{old} = \prod_{i=1}^{M} G_i$$

The normalized change in overall gain is then

$$\delta_G = \prod_{i=1}^{M} (1 + \delta_{G_i}) - 1$$

Substituting from above the δ_{G_i} for each stage, the overall normalized change in magnitude for M cascaded stages is

$$\delta_G = \prod_{j=1}^{M} \left(1 + \sum_{i=1}^{n_j} S_{x_i}{}^{G_j} \delta_{x_i}\right) - 1$$

where n_j is the number of elements being considered in each jth stage. For example, consider a four-pole, two-stage filter. We want to calculate the effect of the capacitor tolerance on the magnitude response. First we calculate

$$S_{C_1}{}^{G_1} = S_{H_{o_1}}{}^{G_1} S_{C_1}{}^{H_{o1}} + S_{\alpha_1}{}^{G_1} S_{C_1}{}^{\alpha_1} + S_{\omega_{o_1}}{}^{G_1} S_{C_1}{}^{\omega_{o1}}$$

and similarly $S_{C_2}{}^{G_1}$. The S, $S_{C_i}{}^{H_{oi}}$, $S_{C_i}{}^{\alpha_i}$, and $S_{C_i}{}^{\omega_{oi}}$ are numbers obtained after the particular circuit realization has been designed. The $S_{H_{o_i}}{}^{G_i}$, $S_{\alpha_i}{}^{G_i}$, and $S_{\omega_{o_i}}{}^{G_i}$ were derived earlier in this section for each particular single-pole or complex conjugate pole pair low-pass-, high-pass-, and bandpass-type filter stages. $S_{C_3}{}^{G_2}$ and $S_{C_4}{}^{G_2}$ are also calculated as above. Thus, for the overall filter

$$\delta_G = (1 + S_{C_1}{}^{G_1}\delta_{C_1} + S_{C_2}{}^{G_1}\delta_{C_2})(1 + S_{C_3}{}^{G_2}\delta_{C_3} + S_{C_4}{}^{G_2}\delta_{C_4}) - 1$$

If the tolerances on the capacitor are ± 10 percent, then $\delta_{C_i} = \pm 0.1$. Worst-case choices for the sign δ will depend on the sign of $S_{C_i}{}^{G_i}$ for a particular ω.

Remember δ_G is a function of ω. Once an ω is chosen, δ_G will be determined. Suppose at $\omega = \omega_A$, $S_{C_1}{}^{G_1} = S_{C_2}{}^{G_1} = S_{C_3}{}^{G_2} = S_{C_4}{}^{G_2} = 0.2$ and $\delta_{C_i} = +0.1$. Then

$$\delta_G = \frac{dG}{G} = (1 + 0.02 + 0.02)(1 + 0.02 + 0.02) - 1$$

$$\delta_G = 0.082 \text{ or } 0.71 \text{ dB}$$

For an individual stage the normalized change in phase for normalized changes in element values is

$$\delta_{\phi_i} = \sum_{i=1}^{n_i} S_{x_i}^{\phi}\delta_{x_i}$$

where n_i is the number of elements being considered. For M cascaded stages the new phase is

$$\phi_{new} = \phi(1 + \delta_{\phi_1}) + \phi_2(1 + \delta_{\phi_2}) + \cdots \phi_M(1 + \delta_{\phi_M})$$

$$\phi_{old} = \sum_{i=1}^{M} \phi_{\phi_i}$$

The normalized change in total phase is then

$$\delta_\phi = \frac{\phi_{new} - \phi_{old}}{\phi_{old}} = \frac{\sum_{i=1}^{M} \phi_i\delta_{\phi_i}}{\sum_{i=1}^{M} \phi_i}$$

and substituting the δ_{ϕ_i} for each stage

$$\delta_\phi = \frac{\sum_{i=1}^{M} \phi_i \sum_{j=1}^{n_i} S_{x_j}^{\phi_i}\delta_{x_j}}{\sum_{i=1}^{M} \phi_i}$$

For our four-pole example we first calculate

$$S_{C_1}^{\phi_1} = S_\alpha^{\phi_1}S_{C_1}^{\alpha_1} + S_{\omega_o}^{\phi_1}S_{C_1}^{\omega_{o1}}$$

and similarly $S_{X_2}^{\phi_1}$. Again the $S_{C_i}^{\alpha}$ and $S_{C_i}^{\omega_o}$ depend on the circuit realization, and the $S_{\alpha_i}^{\phi_i}$ and $S_{\omega_{oi}}^{\phi_i}$ are given earlier in this section. $S_{C_3}^{\phi_2}$ and $S_{C_4}^{\phi_2}$ are also calculated as above. Thus, for the overall filter

$$\delta_\phi = \frac{\phi_1(S_{C_1}^{\phi_1}\delta_{C_1} + S_{C_2}^{\phi}\delta_{C_2}) + \phi_2(S_{C_3}^{\phi}\delta_{C_3} + S_{C_4}^{\phi}\delta_{C_4})}{\phi_1 + \phi_2}$$

As a numerical example, suppose

$$S_{C_1}^{\phi_1} = S_{C_2}^{\phi_1} = S_{C_3}^{\phi_2} = S_{C_4}^{\phi_2} = 0.03 \qquad \text{at } \omega = \omega_A$$

so that $\delta_{C_1} = \delta_{C_2} = \delta_{C_3} = \delta_{C_4} = 0.1$, and that $\phi_1(\omega_A) = 0.628$ radian and $\phi_2(\omega_A) = 1$ radian. Then

$$\delta_\phi = \frac{0.628(0.003 + 0.003 + 0.003) + 1(0.003 + 0.003)}{1.628}$$

$$\delta_\phi = 0.006$$

The discussion for phase also applies to the group delay τ since, like phase, the group delays of individual stages add.

For an individual stage then

$$\delta_{\tau_i} = \sum_{i=1}^{n_i} S_{x_i}{}^{\tau} \delta_{x_i}$$

and for M cascaded stages

$$\delta_{\tau} = \frac{\sum_{i=1}^{M} \tau_i \sum_{j=i}^{n_i} S_{s_j}{}^{\tau_i} \delta_{x_j}}{\sum_{i=1}^{M} \tau_i}$$

INDEX